# ADVANCED ELECTRONIC CIRCUIT DESIGN



David J. Comer

Donald T. Comer

**Brigham Young University**

**John Wiley & Sons, Inc.**

| | |
|---|---|
| EXECUTIVE EDITOR | Bill Zobrist |
| PROGRAM ASSISTANT | Jovan Yglecias |
| MARKETING MANAGER | Katherine Hepburn |
| SENIOR PRODUCTION EDITOR | Christine Cervoni |
| SENIOR DESIGNER | Dawn Stanley |
| ILLUSTRATION EDITOR | Anna Melhorn |

This book was set by the author and printed and bound by Hamilton Printing Co. The cover was printed by Phoenix Color Corp.

This book is printed on acid free paper.




ISBN 0-471-22828-1

Printed in the United States of America

10 9 8 7 6 5 4 3 2 1

# PREFACE

This book is written as a follow-up to the textbook *Fundamentals of Electronic Circuit Design.* The first book was intended to teach fundamental principles of design based on the MOSFET and the BJT. Both discrete circuit and integrated circuit design were considered relative to amplifier design and simple logic and switching circuit design.

This book moves to the next higher level of design, that is, the design of more complex circuits based on the MOSFET and the BJT. Sinusoidal oscillators, high-power amplifiers, multipliers and other sophisticated circuits are considered here. A discussion of these circuits require more advanced topics than those needed for the simpler amplifier circuits of the fundamental text. Topics such as amplitude stabilization in oscillators, thermal design for power stages, pulse-width modulation schemes, nonlinear operation in multiplying circuits, and other practical concepts are treated in this text. Of course, the fundamental aspects of electronic circuit design are prerequisite to a study of this text.

**Organization:** The first chapter provides an extensive review of basic MOSFET and BJT operation. This allows a reader to fill in some holes in background before proceeding to the more advanced topics. Much of this material comes from *Fundamentals of Electronic Circuit Design.* Chapter 2 introduces the concept of thermal dissipation of a device and simple heat sink design. The traditional power stages, class-A, class-B, and class-C are discussed in this chapter. Chapter 3 continues with power stages at a more modern level. The lower-power IC stages, the class-G stage, and the class-D or pulse-width modulated output stage is covered in this material. Wideband amplifier design is considered in Chapter 4 using either the op amp for lower frequency work or the BJT and MOSFET for higher bandwidth stages. Chapter 5 proceeds to narrowband design and the considerations that are involved in tuned circuit design. The important sinusoidal oscillator is discussed in Chapter 6 including the crystal-controlled oscillator. The concept of modulation and communication principles are introduced in Chapter 7 wherein amplitude modulation, phase modulation, and frequency modulation are defined. On-off keying, phase-shift keying, and frequency-shift keying are included to stress the importance of digital communications at the present time. Chapters 8 and 9 discuss actual circuits used in amplitude modulation and angle modulation. Such topics allow the introduction of multiplier circuits used in modulators or mixers, phase-locked loops, and automatic gain control circuits. Each of these circuits are important to other areas as well as communications.

Chapters 10 and 11 introduce the field of analog-digital converters, an important component of the mixed-signal circuits area. Both interface circuits and converters are treated in these chapters. Chapter 12 is a condensed treatment of filter design. Chapter 13 discusses some directions in which the electronics field appears to be heading and speculates on the importance of nontraditional materials such as gallium arsenide and gallium nitride.

**Pedagogical Features:** Several pedagogical features of the first text are also incorporated in this book.

**Important Concepts.** Each section opens with a summary of the key points to be included in that section. This helps the student focus on these significant topics as the material is studied.

**Practical Considerations.** Scattered liberally throughout the text are Practical Considerations sections that relate the developed theory to the actual practice of circuit design. A more thorough understanding of practical implications of the theory can be achieved from these sections.

**Practice Problems.** Practice problems with accompanying answers are included to allow the reader to test comprehension before moving to succeeding material.

**Summary.** Each chapter ends with a summary of key concepts developed in the chapter.

**End-of-Chapter Problems.** The end-of-chapter problems are segmented by section headings to help the instructor identify appropriate problems to assign and to help students identify concepts needed for solution. The numbers of the more difficult problems are followed by an asterisk. Design problem numbers are prefaced by the letter D.

**Worked Examples.** Many examples that apply the previously developed theory are provided to solidify concepts for the student.

It is hoped that this text will serve to broaden the students' understanding of the entire electronics field. A basic capability in the areas of power stages, communication circuits, mixed-signal circuits, oscillators, and filters will prepare the graduating engineer to interact effectively with the experienced circuit designer.

**Acknowledgments:** It has been a privilege to work with the enthusiastic and very able staff at John Wiley & Sons, including Bill Zobrist, Executive Editor, Christine Cervoni, Senior Production Editor, and Anna Melhorn, Illustration Editor.

David J. Comer

Donald T. Comer

# BRIEF CONTENTS

# CONTENTS

# Chapter 1

# INTRODUCTION AND REVIEW

## 1.1 ADVANCED ELECTRONIC CIRCUITS

The first course in electronic circuits is devoted primarily to single-stage amplifiers and circuits with relatively few devices. An advanced course in electronic circuits will typically discuss more complex circuits that either include more elements or require more advanced concepts to analyze.

Another difference between the first course and an advanced course is that in the first course amplifiers are used as building blocks for more complex circuits. These stages are not produced and sold as individual circuits. Many of the circuits discussed in the advanced course are production circuits or become production circuits with slight modifications. Power amplifiers, sinusoidal signal generators or oscillators, and multipliers are examples of circuits that may be sold in discrete or integrated circuit implementations.

A broader view of overall circuit operation is important in the advanced course, though it is still important to understand the basic operation of devices. The next section reviews the two most important devices in electronic circuit design: the metal-oxide-semiconductor field-effect transistor (MOSFET) and the bipolar junction transistor (BJT). This material would typically be covered in the earlier course and is included here as a review.

## 1.2 TRANSISTORS

Both the MOSFET and the BJT are very important in the electronics field. The MOSFET is most important in the computer area, especially in personal computers and workstations, whereas the major usage of the BJT in digital circuits is in supercomputers or special-purpose computing.

Although use of the MOSFET in high-frequency communications is increasing, the BJT dominates this area from the low GHz range upward. The BJTs used in these applications are often Si-Ge heterojunction devices or gallium arsenide, indium phosphide, or gallium nitride devices. Fortunately, the same model or equivalent circuit can often be used to analyze the BJT implemented in these various materials.

In this text, both MOSFET and BJT will be discussed in connection with various circuits. A simulation program such as Spice is necessary for useful circuit design. However, the concepts that determine circuit performance are often obscured by the use of a simulation program. Most of the circuits introduced in this text use equivalent circuits for the devices and discuss circuit operation in terms of these models. An understanding of these models is necessary to understand circuit operation. Whereas simulation will be used to fine-tune a circuit, design is almost always done using the equivalent circuits as the basis. The next two sections will provide a brief review of the operation and models of the MOSFET and the BJT.

## 1.3 THE MOSFET

**Important Concepts**

**1.** The MOSFET has three regions of possible operation: the cutoff region, the triode region, and the active region. In most analog applications, the device operates in the active region.

**2.** The triode region is sometimes used to create a resistance with a limited voltage range.

**3.** The basic equation of operation in the triode region approaches that of active region operation at the point of pinchoff.

**4.** Amplifiers typically utilize the active region of the MOSFET.

**5.** Parasitic capacitances exist between each pair of terminals and between each terminal and substrate.

**6.** Some parasitic capacitances are nonlinear with bias voltage. Simulation programs are generally used to analyze the high-frequency response of MOSFET amplifiers.

### 1.3.1 The Triode Region

Although the MOSFET is typically operated in the active region in analog and mixed signal circuits, it is occasionally operated in the triode region when acting as a resistance. The equation expressing drain current in the triode region is

$$I = \frac{\mu W}{L} C_{ox} \left[ (V_{GS} - V_T) V_{DS} - \frac{V_{DS}^2}{2} \right] \tag{1.1}$$

This equation expresses the V-I characteristics of the nMOSFET in the triode region. It is based on the existence of a finite charge density throughout the channel. This implies that the gate-to-channel voltage must exceed $V_T$ over the channel length. Since the minimum gate-to-channel voltage will exist from gate to drain, this equation is valid up to values of $V_{DS}$ that satisfy

$$V_{GS} - V_{DS} = V_T \tag{1.2}$$

The maximum value of $V_{DS}$ in the triode region is called the pinchoff value and is given by

$$V_{DSP} = V_{GS} - V_T \quad \text{at pinchoff} \tag{1.3}$$

For values of $V_{DS}$ less than $V_{DSP}$ and $V_{GS}$ greater than $V_T$, the nMOSFET is in the triode region, that is,

$$V_{GS} > V_T \tag{1.4}$$

and

$$V_{DS} \leq V_{GS} - V_T = V_{DSP} \tag{1.5}$$

This region is also referred to as the linear region, but the variation of $I_D$ with $V_{DS}$ is linear only for small values of $V_{DS}$. From Eq. (1.1), when $V_{DS}$ is small enough that the term $(V_{GS} - V_T)V_{DS}$ is larger than $V_{DS}^2/2$, the equation approximates a linear variation. As $V_{DS}$ increases, the term $V_{DS}^2/2$ becomes larger and decreases the

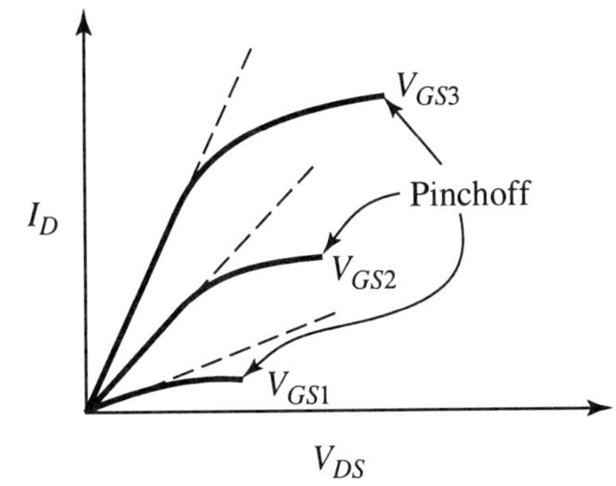


Figure 1.1: Triode region operation for the nMOS device.

value that $I_D$ would have if a linear variation continued. Figure 1.1 shows the V-I characteristics of an nMOS device in the triode region. The dashed lines are included to demonstrate the departure of the curves from a linear variation as the pinchoff voltage is approached. As $V_{GS}$ is increased to generate another curve, the value of $V_{DSP}$ also increases.

The triode region is used in applications requiring a voltage-controlled resistance. The resistance is a function of the applied gate-to-source voltage as seen in Fig. 1.1. Equation (1.1) also shows this linear variation for small values of $V_{DS}$. The triode region is used in some amplifier stages to provide a resistive load for another MOSFET stage operating in the active region.

**Practice Problems**

**1.1** If $V_{GS} - V_T = 1.8$ V in Eq. (1.1), calculate the value of $V_{DS}$ for which the second-order term is 10% of the linear term. *Ans:* 0.36 V.

**1.2** Calculate the percent difference between the actual current given by Eq. (1.1) and the current found by neglecting the second-order term at $V_{DS} = V_{DSP} = 1.8$ V. *Ans:* 100%.

### 1.3.2 The Active Region

Amplification of a signal requires that the amplified signal be linearly proportional to the input signal. This situation can be approximated when the MOSFET operates in the active region. For a constant value of applied $V_{GS}$, if $V_{DS}$ increases above the pinchoff value, the gate-to-drain voltage decreases to the point that the drain end of the channel has a zero electron density. The resistance of an incremental length of the channel at the drain end at pinchoff is very high compared to the resistance of an equal length at the source end of the channel. As $V_{DS}$ is further increased above $V_{DSP}$, most of this increased voltage drops across the pinched-off drain end of the channel. This condition is shown in Fig. 1.2.

Conditions along the channel toward the source change little as $V_{DS}$ increases above pinchoff, thus, the drain current hardly changes with $V_{DS}$. One approximation often used for the MOSFET is that the drain current remains constant at the pinchoff value for any value of $V_{DS}$ above $V_{DSP}$. This region is called the *saturation* region by digital designers, but in the analog area, the preferred term is the *active* region. Figure 1.3 shows the V-I characteristics for an nMOS device in both the triode and the active regions.

The dashed lines show a slight increase in drain current as $V_{DS}$ increases above $V_{DSP}$. These lines represent a better approximation to the actual characteristics of the MOSFET than do the solid lines, especially in short-channel or submicron channel length devices.

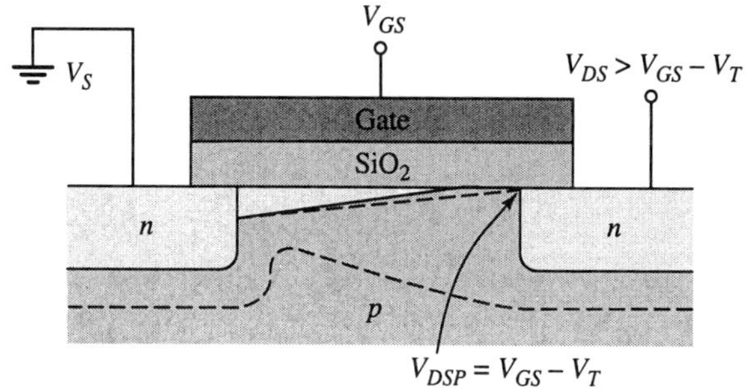


Figure 1.2: Induced electron density at or above pinchoff.

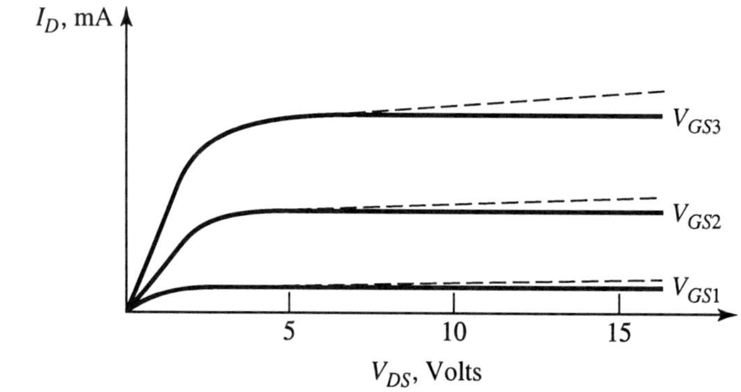


Figure 1.3: V-I characteristics in the active region.

The channel resistance increases dramatically as the electron density at the drain end approaches zero. This situation is referred to as *drain pinchoff* or simply as *pinchoff* and occurs when the voltage from gate to drain equals the threshold voltage as expressed by Eq. (1.3)

$$V_{DSP} = V_{GS} - V_T \quad \text{at pinchoff}$$

The equation for drain current for an nMOSFET in the region where $V_{DS}$ is less than the pinchoff value is given by Eq. (1.1). When pinchoff is reached as shown in Fig. 1.1, Eq. (1.1) can be used with the drain-to-source voltage of Eq. (1.3) to find the current at pinchoff. This gives

$$I_{DP} = \frac{\mu W}{2L} C_{ox} [V_{GS} - V_T]^2 \tag{1.6}$$

Note that the pinchoff current is a function of the ratio of channel width to channel length or $W/L$. This gives the designer some control over the V-I characteristics of the MOSFET.

Equation (1.6) is useful in a great deal of design work, even though it is not a precise equation. It is particularly useful in manual analysis of MOS devices with longer channel lengths, for example, over 1 $\mu$ in length.

Because the actual curves in the active region have a slight upward slope, Eq. (1.6) is sometimes modified to reflect this finite slope by means of a multiplicative factor. The resulting equation for active region current is

$$I_D = I_{DP}[1 + \lambda(V_{DS} - V_{DSP})] = \frac{\mu W}{2L} C_{ox} [V_{GS} - V_T]^2 [1 + \lambda(V_{DS} - V_{DSP})] \tag{1.7}$$

where $\lambda$ is the *channel length modulation factor*. This equation is valid only for values of $V_{DS}$ above $V_{DSP}$. The appearance of $\lambda$ in the equation for drain current leads to a slight upward slope of current with $V_{DS}$. The upward slope is due to the change in channel length as the voltage $V_{DS}$ is increased. When the drain voltage increases above the value that causes pinchoff, the pinched-off end of the channel extends slightly further from the drain toward the source end of the channel. The channel length shortens accordingly, and more current flows through this shorter channel for a given value of $V_{GS}$. This is called the channel length modulation effect and is indicated in Fig. 1.2 by the solid line representing electron distribution in the channel.

### Incremental Parameters

Taking the reciprocal of the upward slope in the V-I characteristics leads to the incremental resistance from drain to source, with $V_{GS}$ held constant. This parameter is designated $r_{ds}$ and is given by

$$r_{ds} = \frac{1}{\left(\frac{\partial I_D}{\partial V_{DS}}\right)_{V_{GS}=K}} = \frac{1}{\lambda I_{DP}} \tag{1.8}$$

The drain-to-source resistance will decrease as $V_{GS}$ increases, leading to higher values of $I_{DP}$.

Another important parameter of the MOSFET is the transconductance which is designated $g_m$ and relates the change in $V_{GS}$ to the resulting change in $I_D$ while holding $V_{DS}$ constant. This parameter is

$$g_m = \left(\frac{\partial I_D}{\partial V_{GS}}\right)_{V_{DS}=K}$$

$$= \frac{\mu W}{L} C_{ox}[V_{GS} - V_T][1 + \lambda(V_{DS} - V_{DSP})] = \frac{2I_D}{V_{GS} - V_T} \tag{1.9}$$

Often $\lambda$ can be taken as zero to give an approximation for transconductance of

$$g_m = \frac{\mu W}{L} C_{ox}(V_{GS} - V_T) \tag{1.10}$$

The active region is generally used for amplifying applications. The parameters $r_{ds}$ and $g_m$ are important in determining amplifier behavior. The constant $\lambda$ can also be related to an Early voltage, similar to that of the BJT.

Another important equation for $g_m$ in longer channel devices is

$$g_m = \sqrt{2\mu C_{ox}(W/L)I_D} \tag{1.11}$$

This equation shows the explicit dependence of transconductance on drain current.

The pMOSFET operates much like the nMOSFET. For the pMOSFET, the threshold voltage is negative, and a negative gate-to-source voltage must be applied to lead to channel conduction. When in the triode or active regions, drain current flows from the drain terminal rather than into the drain terminal as for the nMOS device.

**Practical Considerations**

Submicron devices do not follow the square-law variation of drain current with gate-to-source voltage indicated by Eq. (1.7). Although the physics of this modified behavior is complex, the general trend is the lowering of the exponent of $[V_{GS} - V_T]$ from two toward a value of one. A 0.18-$\mu$ device, for example, may have an exponent of 1.08.

The value of $\lambda$ becomes higher in short-channel devices leading to lower values of $r_{ds}$. Fortunately, the equivalent circuit to be discussed next can apply to any channel length, although the expressions for $g_m$ and $r_{ds}$ will be different, depending on the channel length.

### Practice Problems

**1.3** If $\mu W C_{ox}/2L = 50$ $\mu$A/V$^2$, $V_T = 0.8$ V, $V_{GS} = 2$ V, and $\lambda = 0.02$, calculate $r_{ds}$. *Ans:* 694 k$\Omega$.

**1.4** For the MOSFET of Practice Problem 1.3, evaluate $g_m$. *Ans:* 0.12 mA/V.

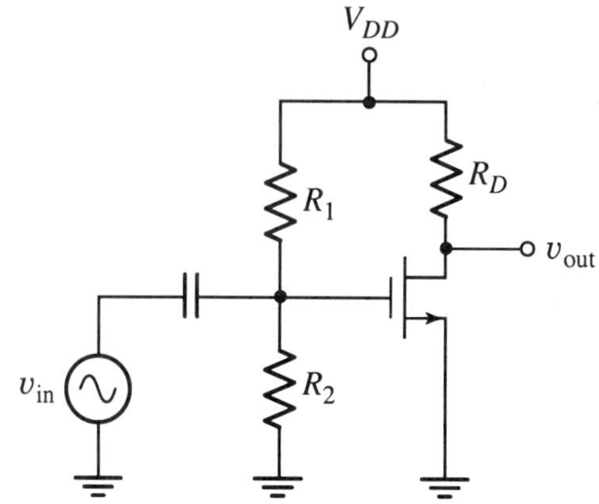


Figure 1.4: A common-source amplifying stage.

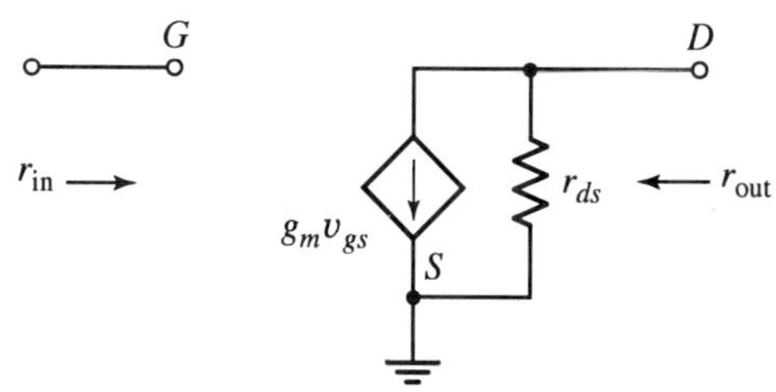


Figure 1.5: Incremental model of the MOSFET.

### 1.3.3 The Midband Incremental MOSFET Model

A discrete component, common-source, resistive load amplifier is shown in Fig. 1.4.

As explained earlier, the MOSFET must be biased into the active region before an input signal is applied. If the output drain-to-source voltage can swing from a minimum of 2 V (at the edge of the triode region) up to 10 V (the cutoff region), the quiescent drain-to-source voltage might be set near 6 V. This would allow the voltage to swing 4 V in either direction from the quiescent value before serious distortion takes place. The desired value of $V_{DSQ}$ is established by setting the proper value of $V_{GSQ}$.

The resistors $R_1$ and $R_2$ are used to establish the necessary gate-to-source quiescent bias voltage, $V_{GSQ}$. These resistors can be very large to avoid loading problems since no gate current is required. Because the gate of the MOSFET draws no current, the bias voltage is established by voltage division of the two bias resistors.

The standard way to analyze this stage is to use the small-signal equivalent circuit of the MOSFET, rather than the graphical method. This model is indicated in Fig. 1.5.

The impedance between the gate and source terminals is represented by an open circuit, reflecting the fact that there is no current flow across the oxide layer between gate and channel. The effect of gate-to-source voltage on the drain current is modeled by the dependent current generator of value $g_m v_{gs}$. The resistance $r_{ds}$ accounts for the incremental resistance between drain and source. This model is an incremental model that applies to the active region when active region values of $g_m$ and $r_{ds}$ are used.

This circuit is the basic incremental model for the MOSFET device in the active region. It is used every time a MOSFET stage is manually analyzed. While the values of $g_m$ and $r_{ds}$ are determined by the quiescent drain current, no dc quantities appear in this model.

The equivalent circuit of the complete amplifier stage is completed by replacing the MOSFET by its model at terminals G, S, and D. All dc power supply voltages are considered to be incremental signal grounds. The resulting circuit is shown in Fig. 1.6.

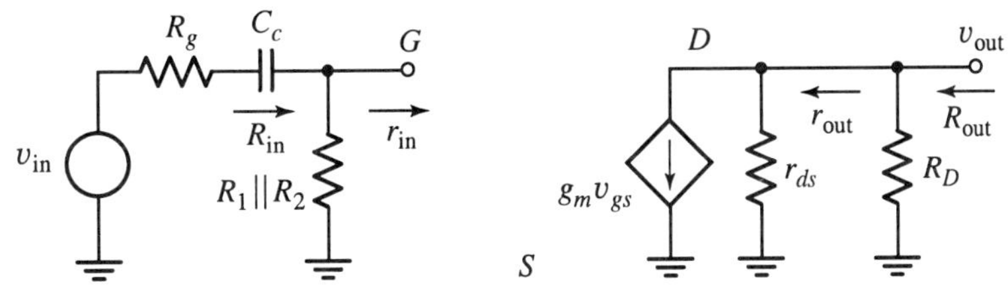


Figure 1.6: The equivalent circuit for the amplifier stage.

The input impedance to the stage is determined by the bias resistors since the gate-to-source impedance of the MOSFET is so high. This input impedance is

$$R_{in} = \frac{R_1 R_2}{R_1 + R_2} \tag{1.12}$$

This value can be made quite high if necessary. Theoretically, the total impedance is this quantity in parallel with $r_{in}$, the gate-to-source impedance, but this value is approximated as an open circuit.

The output impedance of the MOSFET looking into the drain terminal is

$$r_{out} = r_{ds} \tag{1.13}$$

The total output impedance of the amplifier is $r_{ds}$ in parallel with the external drain resistance $R_D$, that is,

$$R_{out} = \frac{r_{ds} R_D}{r_{ds} + R_D} \tag{1.14}$$

The voltage gain from gate to drain can be found by writing the output voltage in terms of the current. This leads to

$$v_{out} = -g_m v_{gs} R_{out} = -g_m v_{gs} \frac{r_{ds} R_D}{r_{ds} + R_D}$$

The gate-to-drain voltage gain is then

$$A_{gd} = \frac{v_{out}}{v_{gs}} = -g_m \times \frac{r_{ds} R_D}{r_{ds} + R_D} \tag{1.15}$$

Ignoring bias considerations, we see that this voltage gain increases as $R_D$ is increased. The maximum gain would then occur for an infinite value of $R_D$. The magnitude of this maximum gain is sometimes designated $A_\mu$ and is expressed as

$$A_\mu = g_m r_{ds} \tag{1.16}$$

The magnitude of voltage amplification of any practical MOSFET stage cannot exceed this value of $A_\mu$.

The overall voltage gain must account for any loading on the signal generator resistance, $R_g$, by the bias resistors. This leads to an overall midband voltage gain of

$$A_{MB} = \frac{v_{out}}{v_{in}} = \frac{-R_B}{R_B + R_g} \times g_m \times \frac{r_{ds} R_D}{r_{ds} + R_D} \tag{1.17}$$

where $R_B$ is equal to the parallel combination of the bias resistors.

The analysis of a MOSFET amplifier stage will be demonstrated by assuming the following values for the stage of Fig. 1.4:

$$R_1 = 50\ \text{k}\Omega,\ R_2 = 15\ \text{k}\Omega,\ R_D = 4\ \text{k}\Omega,\ R_g = 4\ \text{k}\Omega$$

$$g_m = 2.7 \text{ mA/V}, \; r_{ds} = 50 \text{ k}\Omega, \; C_c = 0.1 \,\mu\text{F}, V_{DD} = 10 \text{ V}$$

It is desired to find the midband voltage gain and the lower corner frequency of the circuit.

The midband voltage gain is found from Eq. (1.17) after calculating $R_B$, the parallel combination of $R_1$ and $R_2$. This gives

$$R_B = \frac{50 \times 15}{50 + 15} = 11.5 \text{ k}\Omega$$

The gain is calculated from Eq. (1.17) as

$$A_{MB} = \frac{-11.5}{11.5 + 4} \times 2.7 \times \frac{50 \times 4}{50 + 4} = -7.42 \text{ V/V}$$

The lower corner frequency due to the coupling capacitor is calculated from

$$f_{low} = \frac{1}{2\pi C_c[R_g + R_{in}]} = \frac{1}{2\pi \times 10^{-7} \times 15.5 \times 10^3} = 103 \text{ Hz}$$

Although this analysis is fairly straightforward, an actual design problem is considerably more complex. This is primarily due to the fact that $g_m$ and $r_{ds}$ are both functions of the quiescent operating point. The choice of quiescent drain current is influenced by $R_D$ which may not be known at the start of the design. If $R_D$ is known, the design can proceed as indicated in the preceding paragraphs.

# Example 1.1

Assume that the circuit of Fig. 1.4 has a power supply of $V_{DD} = 8$ V, $R_D = 4$ k$\Omega$, and $R_g = 10$ k$\Omega$. Also assume that the drain current is related to the gate-to-source voltage as in Eq. (1.6), that is,

$$I_D = I_{DP} = \mu \frac{W}{2L} C_{ox} (V_{GS} - V_T)^2$$

If $V_T = 0.8$ V and $\mu W C_{ox}/2L = 0.4 \text{ mA/V}^2$

**(a)** calculate the required value of $V_{GSQ}$ to establish $V_{DSQ}$ at 5 V.

**(b)** Calculate suitable values of $R_1$ and $R_2$ to establish this bias point. Assume that $R_B = R_1 || R_2$ is to equal 100 k$\Omega$.

**(c)** Calculate the small-signal midband voltage gain for this bias point.

**(d)** Calculate the largest peak value of input voltage that can be applied without causing the output to leave the active region.

**Solution (a)** From the load line equation, $V_{DS} = V_{DD} - R_D I_D$. Solving this equation with $V_{DSQ} = 5$ V results in $I_{DQ} = 0.75$ mA. We now substitute this value of quiescent drain current into Eq. (1.6) and solve for $V_{GSQ}$. This results in a value of $V_{GSQ} = 2.17$ V.

**(b)** Since the source terminal of this stage is at 0 V, setting the gate voltage to 2.17 V will establish the proper value of $V_{GSQ}$. This can be accomplished by picking resistor values to satisfy

$$V_{GSQ} = V_{DD} \frac{R_2}{R_1 + R_2}$$

However, in order to result in $R_B = 100\ \text{k}\Omega$, we must also satisfy the equation

$$R_B = \frac{R_1 R_2}{R_1 + R_2} = 100\ \text{k}\Omega$$

Solving these two equations simultaneously yields values of $R_1 = 369\ \text{k}\Omega$ and $R_2 = 137\ \text{k}\Omega$.

**(c)** In order to evaluate the voltage gain, the equivalent circuit of Fig. 1.5 is used. Note that Eq. (1.6) implies a zero value of $\lambda$ and therefore, an infinite value for $r_{ds}$ from Eq. (1.8). We need to find a value for $g_m$ which can be done with the aid of Eq. (1.10). This gives

$$g_m = 2 \times 0.4 \times (V_{GSQ} - V_T) = 0.8 \times (2.17 - 0.8) = 1.096\ \text{mA/V}$$

Using Eq. (1.17), we now calculate the voltage gain as

$$A_{MB} = \frac{-100}{100 + 10} \times 1.096 \times 4 = -3.99\ \text{V/V}$$

**(d)** In order to remain in the active region as the incremental output voltage varies, the stage must avoid cutoff as current decreases and the triode region as current increases. With the given bias point of $V_{DSQ} = 5$ V, a peak voltage of 3 V can be developed in the cutoff direction. This would lead to a cutoff value of $V_{DS} = V_{DD} = 8$ V. We next find how far toward the triode region this output can swing. To do this, we need to find the point where $V_{DS}$, as calculated from the load line equation, equals a value of $V_{DSP}$, the pinchoff value. At pinchoff, we have noted that $V_{DSP} = V_{GSP} - V_T$. We can use the load line equation to write

$$V_{DSP} = 8 - 4I_{DP}$$

Equation (1.6) must also be satisfied, that is,

$$I_{DP} = 0.4(V_{GSP} - 0.8)^2 = 0.4(V_{DSP})^2$$

Solving these equations simultaneously for $V_{DSP}$ leads to a value of 1.94 V for this voltage. The largest peak output voltage swing in the negative direction is the quiescent value less this value, or $5 - 1.94 = 3.06$ V.

The output swing before serious distortion is almost the same in both directions, with the 3-V peak value toward cutoff taken as the limiting value. To relate this to the largest possible peak input voltage, we simply divide by the voltage gain giving

$$|v_{in}| = \frac{3}{3.99} = 0.75\ \text{V}$$

Although an applied voltage of this magnitude or larger would result in serious distortion, even smaller values will result in modest distortion due to the nonlinear relationship between $I_D$ and $V_{GS}$. Distortion decreases as the applied signal magnitude decreases.

---

### 1.3.4 High-Frequency Model of the MOSFET

The high-frequency model of the MOSFET is shown in Fig. 1.7. This model shows capacitance between every possible set of terminals. Making the model even more complex is the fact that many of these capacitors are functions of the voltages across these terminals.

There are basically two kinds of capacitance involved in this model. The first is the parallel plate capacitance that arises from the separation of two conductive materials by a dielectric that is relatively independent of voltage applied. The second is the capacitance associated with the charge included in a depletion region which is highly voltage dependent. Both of these capacitances are functions of dimensions. The capacitances between each set of terminals originate as follows.

**$\mathbf{C_{gs}}$**: In the active or triode regions, this capacitance is made up mainly of parallel plate capacitance. The largest component of this capacitance is that between the gate and the inverted channel. Since the charge that flows through the channel is supplied from the source terminal, this capacitance essentially appears between gate and source. It has been shown that this component of $C_{gs}$ is given by

$$C_{gsch} = \frac{2}{3} W L C_{ox} \tag{1.18}$$

where $W$ is the channel width, $L$ is the channel length, and $C_{ox}$ is the capacitance per unit area due to the oxide layer between gate and substrate.

A second, generally smaller component of $C_{gs}$ is due to the overlap of the gate electrode and the source area. As the source and drain are formed by diffusion or implantation of impurity atoms, these impurities diffuse under the gate by a distance $L_{ov}$. This overlap of the gate and source is much smaller than the channel length $L$ except for very high-frequency, short-channel devices. This capacitance is given by

$$C_{gsov} = W L_{ov} C_{ox} \tag{1.19}$$

**$\mathbf{C_{gd}}$**: The capacitance between the gate and drain terminals is due primarily to overlap capacitance and is generally equal to the gate-to-source overlap capacitance given by Eq. (1.19). While the gate-to-drain overlap capacitance is typically small, it can be multiplied by the Miller effect in the common-source configuration to limit the high-frequency amplification.

**$\mathbf{C_{db}}$**: The drain-to-body capacitance consists of depletion layer capacitance. There is a reverse-biased junction between drain and substrate. This capacitance also includes the depletion layer capacitance between drain and the more heavily doped regions between adjacent MOS devices. This capacitance depends on the voltage from drain to substrate as well as the areas involved.

**$\mathbf{C_{sb}}$**: The source-to-body capacitance is similar to the drain-to-body capacitance except that it includes the depletion layer capacitance from channel to body. The source-to-body component becomes unimportant when both source and body are connected to ground. The channel-to-body component is actually distributed over the length of the channel.

Because of the complexity of the capacitances, it is difficult to manually analyze the high-frequency performance of the MOSFET. Generally, a Spice-based simulation program is used to evaluate and modify MOSFET stages.

It is possible to derive approximate frequency performance in some cases as demonstrated by the following example.

## Example 1.2

For the MOSFET in the circuit of Fig. 1.8, $g_m = 3$ mA/V and $r_{ds} = 63$ kΩ. Assuming that $C_{gs} = 1$ pF and $C_{gd} = 0.1$ pF and that these are the dominant capacitors,

**(a)** calculate the midband voltage gain;

**(b)** calculate the upper corner frequency for the amplifier.

**Solution** Two equivalent circuits for this stage are shown in Fig. 1.9.

**(a)** The midband gain can be calculated simply as

$$A_{MB} = -g_m(r_{ds} \,||\, R_D) = -3 \times 7.1 = -21.3 \text{ V/V}$$

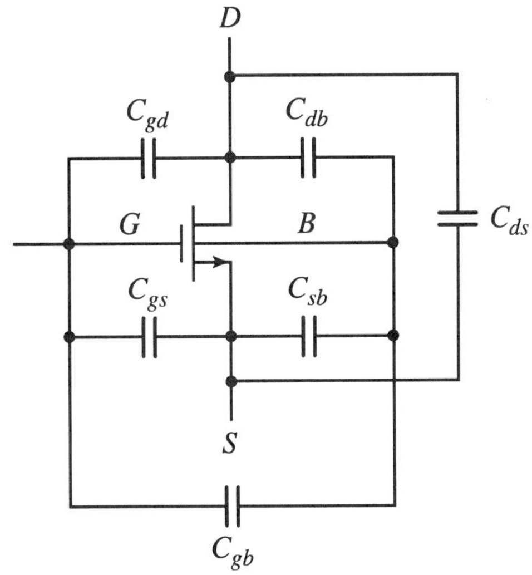


Figure 1.7: The high-frequency MOSFET model.

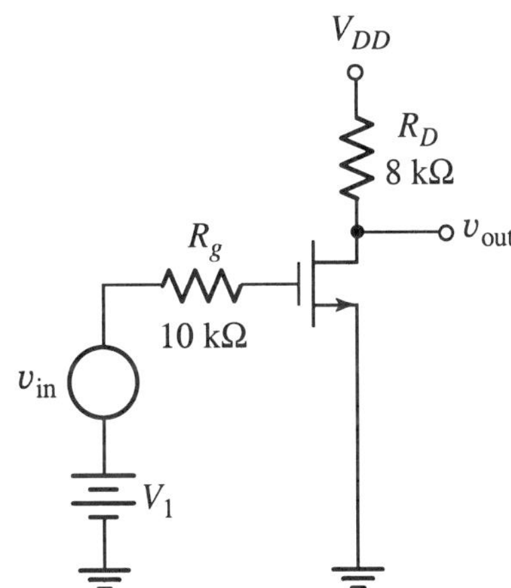


Figure 1.8: Amplifier stage for Example 1.2.

**(b)** There will be two upper corner frequencies for this circuit, one caused by the input loop and the other by the output loop. The Miller effect capacitance reflected to the input loop is

$$C_M = (1 + |A_{gd}|)C_{gd} = 22.3 \times 0.1 = 2.23 \text{ pF}$$

The upper corner frequency due to the input loop is

$$f_{in-high} = \frac{1}{2\pi R_g(C_{gs} + C_M)} = \frac{1}{2\pi \times 10,000 \times 3.23 \times 10^{-12}} = 4.93 \text{ MHz}$$

The upper corner frequency due to the output loop is approximately

$$f_{out-high} = \frac{1}{2\pi C_{gd}(r_{ds} \mid\mid R_D)} = \frac{1}{2\pi \times 10^{-13} \times 7100} = 224 \text{ MHz}$$

Since this latter frequency is so much greater than the input corner frequency, the overall upper corner frequency will be

$$f_{high} = f_{in-high} = 4.93 \text{ MHz}$$

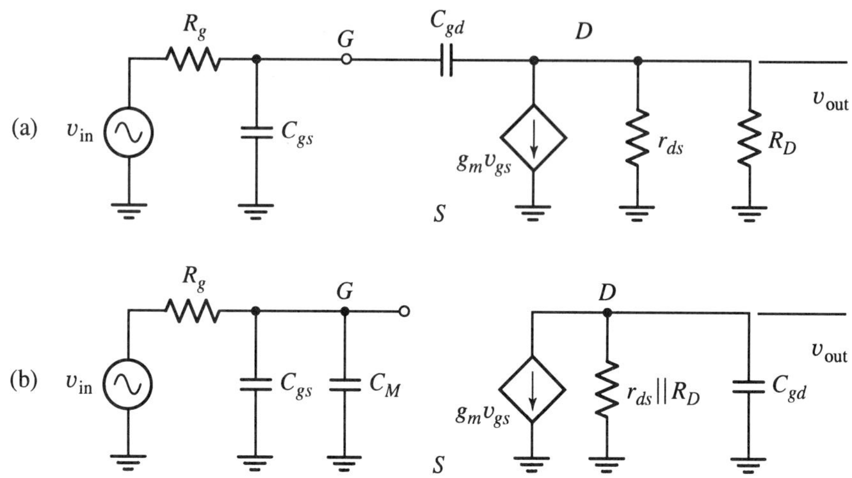


Figure 1.9: (a) An equivalent circuit. (b) After using the Miller effect.

---

## 1.4 MOSFET IC STAGES

**Important Concepts**

**1.** Most MOSFET IC amplifier stages do not use a resistor as the load for the amplifying stage.

**2.** The load is generally created by one or more other MOS devices. These devices are called active loads.

**3.** Active load stages can lead to very high voltage gains.

**4.** The ratio of channel width to channel length, called the aspect ratio, is quite significant in MOSFET circuit design.

### 1.4.1 Active Load Stage

A stage used often in IC design is the active load stage. Typically, the active load is a current source, often based on the current mirror. A simple MOSFET active load stage appears in Fig. 1.10.

There is a compelling reason to use active load stages in IC design, namely, the device that acts as the load for the amplifying stage can present a large incremental load while allowing a reasonable drain current to flow. This large incremental load leads to a high gain for the stage while the dc current can lead to acceptable values of transconductance and bias current. If a fixed resistor with a large value were used to achieve high gain, the dc drop across this element would be prohibitive for normal bias currents.

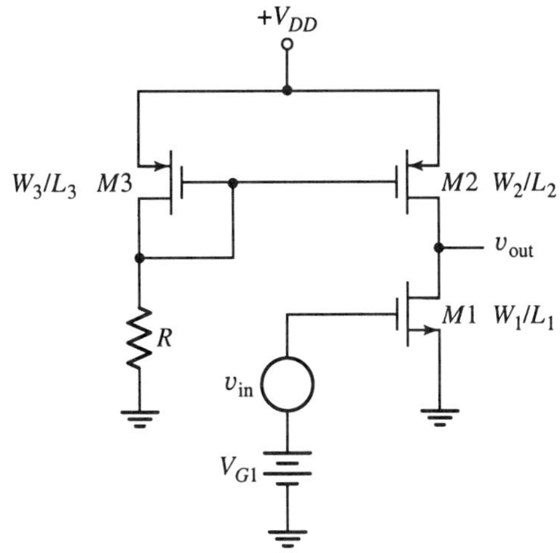


Figure 1.10: A MOSFET stage with an active current source load.

**Practical Considerations**

A simple current mirror might provide 100 $\mu$A of current to an amplifying stage while presenting 60 k$\Omega$ of incremental resistance to the stage. If a simple 60 k$\Omega$ resistance replaced the current mirror, the same incremental resistance would prevail, but the dc voltage drop across the resistor would be

$$V_R = 0.1 \times 60 = 6\ V$$

The dc power supply for many MOSFET IC amplifiers is 5 V or less, thus, using a simple resistor is out of the question.

The difference between an active load and a simple resistor becomes even more apparent when a more complex current mirror is used to produce a resistive load of several hundred kilohms or greater. The dc drop across a several hundred kilohm resistor, conducting a typical bias current, might be over 100 V.

The pMOS current mirror of Fig. 1.10 provides bias current to the amplifying device, $M1$. This device also sees an incremental resistive load of

$$R_{out} = r_{ds1} \,||\, r_{ds2} = \frac{1}{g_{ds1} + g_{ds2}} \tag{1.20}$$

where $g_{ds1} = 1/r_{ds1}$ and $g_{ds2} = 1/r_{ds2}$. The midband voltage gain is now

$$A_{MB} = -g_{m1} R_{out} \tag{1.21}$$

Since the values of $r_{ds}$ can be high for both devices, the voltage gain for a single stage can also be reasonably high.

The output voltage can be driven positive to the point that M2 reaches its pinchoff value for $V_{DS}$. This voltage may be $V_{DD} - V_{eff2} = V_{DD} - 0.4$ V. In the negative direction, the output can be driven to the point that $M1$ reaches pinchoff, which is equal to $V_{eff1} = 0.4$ V.

The output capacitance for the amplifier of Fig. 1.10 is

$$C_{out} = C_{db1} + C_{db2} + C_{gd1} + C_{gd2}$$

giving an upper corner frequency of

$$f_{2o} = \frac{(g_{ds1} + g_{ds2})}{2\pi(C_{db1} + C_{db2} + C_{gd1} + C_{gd2})} \tag{1.22}$$

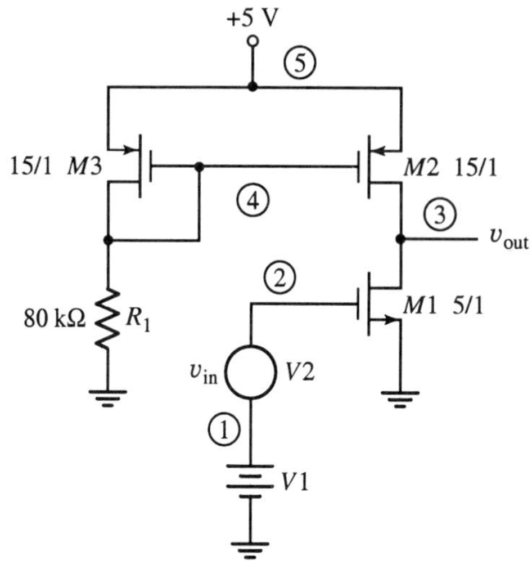


Figure 1.11: Amplifier for Example 1.3.

Again, we emphasize that a load capacitance or input capacitance of a following stage will lower this value.

## Example 1.3

The current mirror of Fig. 1.11 supplies a current of 50 $\mu$A to the amplifying stage. The dc output voltage is adjusted by voltage source, $V_1$, to be 2.4 V. If $g_{m1}$ = 0.19 mA/V, $g_{ds1}$ = 0.95 $\mu$A/V, $g_{ds2}$ = 2 $\mu$A/V, $C_{db1}$ = 11.0 fF, $C_{db2}$ = 32.0 fF, $C_{gd1}$ = 1.5 fF, and $C_{gd2}$ = 4.5 fF,

**(a)** calculate the midband voltage gain, $A_{MB}$;
**(b)** calculate the upper corner frequency, $f_{2o}$.
**(c)** Verify results with a Spice simulation.

**Solution (a)** The output resistance for this stage is

$$R_{out} = \frac{1}{g_{ds1} + g_{ds2}} = \frac{10^6}{2.95} = 339 \text{ k}\Omega$$

Using Eq. (1.21) allows the voltage gain to be found as

$$A_{MB} = -g_{m1}R_{out} = -0.19 \times 10^{-3} \times 339 \times 10^3 = -64.4 \text{ V/V}$$

The output capacitance is

$$C_{out} = C_{db1} + C_{db2} + C_{gd1} + C_{gd2} = 9.0 + 32.0 + 1.5 + 4.5 = 49 \text{ fF}$$

From Eq. (1.22), this gives an upper corner frequency of

$$f_{2o} = \frac{1}{2\pi C_{out}R_{out}} = \frac{1}{2\pi \times 49 \times 10^{-15} \times 339 \times 10^3} = 9.58 \text{ MHz}$$

The Spice netlist file is listed in Table 1.1. Corresponding node and element numbers are shown in Fig. 1.11. This program uses a model for the MOSFET named C5X.LIB.

**Table 1.1 Spice Netlist File for Example 1.3**

```
EX1-3.CIR

R1 0 4 80K
V1 1 0 1.275V
V2 2 1 AC 0.005V
V3 5 0 5V
M1 3 2 0 0 N L=1U W=5U AD=25P AS=25P PD=15U PS=15U
M2 3 4 5 5 P L=1U W=15U AD=75P AS=75P PD=25U PS=25U
M3 4 4 5 5 P L=1U W=15U AD=75P AS=75P PD=25U PS=25U
.AC DEC 10 100 10MEG
.OP
.PROBE
.LIB C5X.LIB
.END
```

The results of this simulation are $A_{MBsim} = -63.6$ V/V and $f_{2o-sim} = 8.53$ MHz, which compare well with the calculated results. A dc scan on the input voltage allows the active region of the output voltage to be evaluated. For this circuit, the active region extends from 0.4 V up to 4.6 V, which also agrees well with theory.

---

**Practical Considerations**

In specifying a MOSFET device for analysis by Spice, the length and width of the channel are first specified. The surface area of the drain and source follow, after which the external perimeters of the drain and source are specified. For device $M1$, the channel length in microns is 1 and the width is 5. The area of the drain is determined by $W_1$ and the layout length of the drain region. For this device, the width is 5 $\mu$, thus the length of the drain region must be 5 in order to result in an area of $5 \times 5 = 25$ square microns. The perimeter used for the drain is not equal to $2W_1 + 2L_D$. Rather, it is the perimeter of the drain region minus the width of the channel. It is $PD = W_1 + 2L_D = 5 + 10 = 15$. Certain capacitances are based on the external perimeter of the associated region, but the capacitance associated with the side of the region that abuts the channel is accounted for in a separate calculation. The same considerations also apply to the source terminal.

### Adding Generator Impedance and Load Impedance

Only the output capacitance was used in developing Eq. (1.22) for upper corner frequency. As mentioned in connection with the diode-connected load stage, two other considerations must be made for the practical circuit. One is the additional capacitance added between output and ground due to the input capacitance of the following stage or from the output pad of an IC chip. The second is the generator resistance or output resistance of the previous stage that drives the input capacitance of this stage. Both input and output loops must often be considered to result in an accurate upper corner frequency.

Let us suppose that the output of the amplifier of Fig. 1.11 is brought to an output pad and pin of an IC chip. The capacitance associated with the pad might exceed 1 pF. If the simulation is repeated with a 1 pF load added, the upper corner frequency drops from 8.35 MHz to 431 kHz, a very large drop.

If this stage were loaded by a comparable stage rather than by 1 pF, the corner frequency might drop by a factor of two rather than a factor of 20. In order to avoid this drop, a buffer stage may be added to drive the 1 pF load without the large reduction in corner frequency. Another possibility is to increase the values of $W$ for all the devices. Although this increases the capacitance, the output resistance decreases and $g_m$ increases resulting in a comparable voltage gain. The effect of the load capacitance on upper corner frequency will now be much less.

As mentioned earlier, the stage of Fig. 1.11 is driven by a perfect voltage generator with zero output resistance. If this stage were driven by an identical stage, the output impedance of the first stage would become the generator resistance for the second stage. To demonstrate the effect of generator resistance on upper corner frequency, a resistance of 339 kΩ is used as a generator resistance for the circuit of Fig. 1.11. This particular value of resistance will be used in succeeding examples for comparison purposes. No external load capacitance is used in this simulation. The simulated upper corner frequency is lowered in this situation from a value of 8.35 MHz to 3.08 MHz.

This value can be calculated by noting that the input circuit will now cause a corner frequency determined by the generator resistance and the input capacitance. The input capacitance equals the sum of $C_{gs1}$, $C_{gb1}$, and the Miller effect capacitance $(1 + |A_{MB}|)C_{gd1}$. From the output file of the simulation, these values are

$$|A_{MB}| = 63.6 \text{ V/V}, \quad C_{gs1} = 13.35 \text{ fF}, \quad C_{gd1} = 1.5 \text{ fF}, \quad C_{gb1} = 0.3 \text{ fF}.$$

The total input capacitance that results is approximately 111 fF. With a value of $R_g = 339$ kΩ for generator resistance, this adds a corner frequency of

$$f_{in-high} = \frac{1}{2\pi C_{in} R_g} = 4.23 \text{ MHz}$$

The amplifier now has an input corner frequency of 4.23 MHz and an output corner frequency of 8.53 MHz. We use an iterative method to calculate the overall upper corner frequency when two single-pole upper corner frequencies make up the amplifier response. This results in a calculated overall 3-dB point or upper corner frequency of $f_{2o-calc} = 3.56$ MHz. This value exceeds the simulated value of 3.08 MHz by 15%.

If a generator resistance of 339 kΩ and a 1-pF load capacitance are both added to the amplifier, the new upper corner frequency is found from simulation to be 399 kHz. Table 1.2 summarizes the results of this simulation.

**Table 1.2 Summary of Simulation Results for the Current-Source Load Stage**

$A_{MBsim} = -63.6$ V/V

| $C_L$, pF | $R_g$, kΩ | $f_{2o-sim}$, MHz |
|---|---|---|
| 0 | 0 | 8.53 |
| 1 | 0 | 0.43 |
| 0 | 339 | 3.08 |
| 1 | 339 | 0.40 |

The active current-source load provides a method to achieve relatively high midband voltage gains. It suffers from poor frequency response when a large capacitive load, much larger than the output capacitance of the stage, is present.

The midband voltage gain given by Eq. (1.21) will be increased if $R_{out}$ is increased. This resistance can be increased by using a more complex current mirror to make the output resistance much larger than the output resistance of the amplifying stage. Whereas this increases the midband voltage gain, the upper corner frequency due to the output loop, calculated from Eq. (1.22), decreases by the same factor. If this corner frequency is the dominant one, midband voltage gain and bandwidth can be exchanged directly by varying the output resistance of the current mirror.

**Practical Considerations**

Although the upper corner frequency of the active load stage is significantly affected by a 1-pF capacitance, it is possible to minimize this problem. The channel width of the output stage can be increased considerably to result in an increase in output capacitance and a corresponding decrease in output resistance. The upper corner frequency with no external load can be approximately equal to that of the smaller device. However, when a 1-pF load capacitance is added, the percentage increase of capacitance is much less for the large device than for the small device. The effect on upper corner frequency is then much less for the larger device.

The effect of channel width, $W$, on the voltage gain can be seen in the following example.

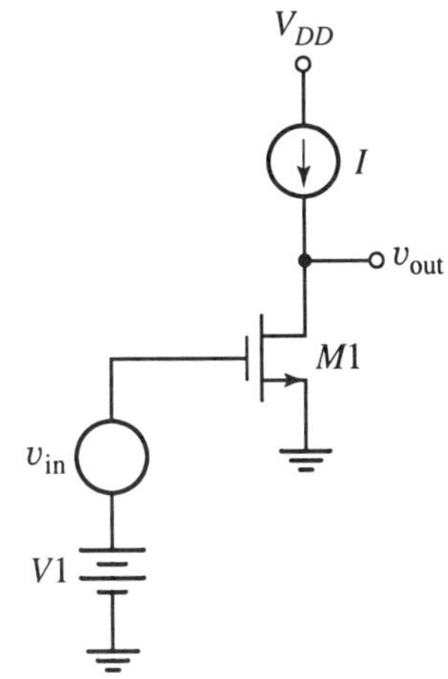


Figure 1.12: Circuit for Example 1.4.

# Example 1.4

In the circuit of Fig. 1.12, the current source generates 100 $\mu A$ of current and has an incremental output resistance of $r_{cs} = 100$ k$\Omega$. The device $M1$ is a $1-\mu$ gate length device with $\mu C_{ox} = 0.06$ mA/V$^2$ and $\lambda = 0.03$ V$^{-1}$. Find the channel width to result in a midband voltage gain of 100 V/V.

**Solution** The midband voltage gain of this stage is

$$A_{MB} = -g_m R_{out}$$

where

$$R_{out} = r_{cs} \mid\mid r_{ds1}$$

The value of $r_{ds1}$ can be calculated by assuming that the drain pinchoff current is approximately equal to 100 $\mu$A. This results in

$$r_{ds1} = \frac{1}{\lambda I_D} = \frac{1}{0.03 \times 0.1} = 333 \text{ k}\Omega$$

Using the 100 k$\Omega$ impedance of the current source, we find the output resistance to be

$$R_{out} = 100 \mid\mid 333 = 76.9 \text{ k}\Omega$$

Recalling that

$$g_m = \sqrt{2\mu C_{ox}(W/L)I_D}$$

we can write the magnitude of the voltage gain as

$$|A_{MB}| = g_m R_{out} = \sqrt{2 \times 0.06 \times (W/L) \times 0.1} \times 76.9 = 100 \text{ V/V}$$

Solving this equation for the ratio of $W/L$ leads to a value of 141 $\mu$ for $W$.

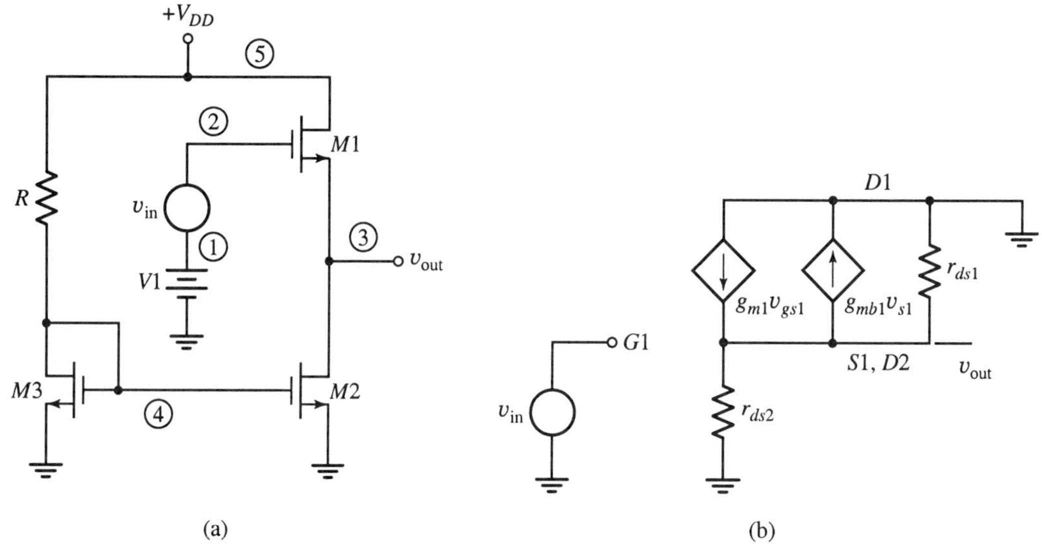


Figure 1.13: (a) A source-follower stage with current-source load. (b) Equivalent circuit.

**Practice Problems**

**1.5** Work Example 1.3 for a generator resistance of $R_g = 50$ kΩ.

**1.6** Work Example 1.3 for a generator resistance of $R_g = 50$ kΩ and an input capacitance to a following stage of $C_{in2} = 100$ fF.

**1.7** The drain current of the current mirror output stage of Fig. 1.11 is increased to 100 $\mu A$. Assuming the capacitances remain constant, calculate the new midband voltage gain and upper corner frequency for the amplifier. *Ans:* $A_{MB} = -45.4$ V/V, $f_{2o} = 19.2$ MHz.

**1.8** Work Example 1.4 if a midband voltage gain of 180 V/V is required. *Ans:* $W = 457$ $\mu$.

## 1.4.2 Source Follower with Active Load

The source follower provides a buffer stage, but the midband voltage gain is low-even less than the value of unity approached by the BJT emitter-follower stage. The bandwidth is quite high for both the emitter-follower and source-follower stages. Figure 1.13 demonstrates a source follower with a current mirror load.

The device M2 presents a resistance of $r_{ds2}$ between the source of $M1$ and ground. In addition, device $M1$ presents a resistance of $r_{ds1}$ in parallel with $1/g_{mb1}$ to the dc power supply which is also ground for incremental signals. We include the body effect in $M1$ ($g_{mb1}$) since the source-to-substrate voltage of this device varies with the output signal. In fact, it equals the output signal.

The circuit of Fig. 1.13(b) is redrawn in Fig. 1.14, and the pertinent parasitic capacitances are added. The current-source strength, $g_{m1}v_{gs1}$, can be written as $g_{m1}(v_{in} - v_{out})$. In the equivalent circuit, this current can be generated by two separate sources as shown in Fig. 1.14. The source terminal now appears at the top of the figure, while the drain terminal is grounded.

The two current sources $g_{m1}v_{out}$ and $g_{mb1}v_{out}$ can be converted to conductances $g_{m1}$ and $g_{mb1}$, respectively. As a voltage appears at S1, the currents through these conductances equal the values that would be generated by the sources. Figure 1.15 shows an alternate equivalent circuit that is used to find the voltage gain as a function of frequency. This circuit results from taking a Thévenin equivalent of the current source, $g_{m1}v_{in}$, and the parallel resistance, $R_{out}$.

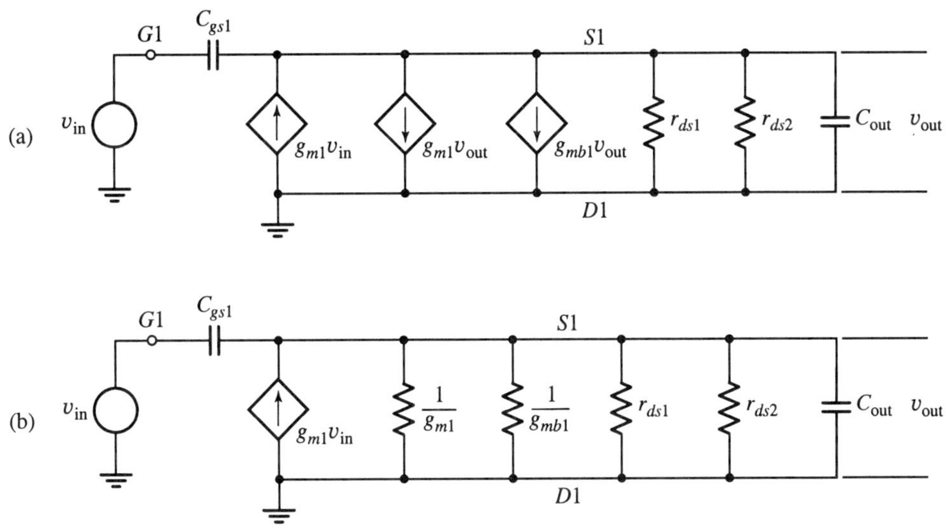


Figure 1.14: (a) The high-frequency source-follower equivalent circuit. (b) Using transconductances for the current sources.

The circuit of Fig. 1.15 is analyzed to find that

$$A(j\omega) = g_{m1}R_{out}\,\frac{1 + j\omega C_{gs1}/g_{m1}}{1 + j\omega(C_{gs1} + C_{out})R_{out}} \tag{1.23}$$

where

$$R_{out} = \frac{1}{g_{m1} + g_{mb1} + g_{ds1} + g_{ds2}} \tag{1.24}$$

and

$$C_{out} = C_{sb1} + C_{db2} + C_{gd2} \approx C_{sb1} + C_{db2} \tag{1.25}$$

The midband gain can be evaluated from Eqs. (1.23) and (1.24) to be

$$A_{MB} = \frac{g_{m1}}{g_{m1} + g_{mb1} + g_{ds1} + g_{ds2}} \tag{1.26}$$

In many submicron process, the value of the denominator of Eq. (1.26) might equal 1.15 to 1.2 times $g_{m1}$. This leads to values of midband gain ranging from about 0.8 to 0.9 V/V.

The transfer function for voltage gain as a function of frequency shows a zero at

$$f_{zero} = \frac{g_{m1}}{2\pi C_{gs1}} \tag{1.27}$$

and a pole at

$$f_{pole} = \frac{g_{m1} + g_{mb1} + g_{ds1} + g_{ds2}}{2\pi(C_{gs1} + C_{out})} \tag{1.28}$$

Typically, the zero frequency is higher than the pole frequency, and the asymptotic frequency response appears as in Fig. 1.16. At high frequencies, the capacitor $C_{gs1}$ feeds through from input directly to output, causing a gain of

$$A_{high} = \frac{C_{gs1}}{C_{gs1} + C_{out}} \tag{1.29}$$

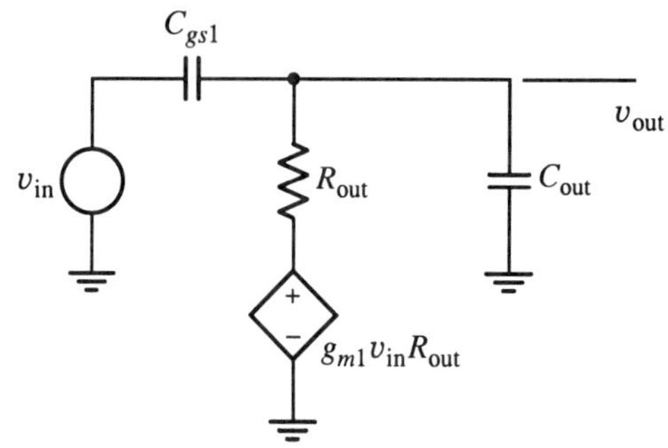


Figure 1.15: Alternate equivalent circuit of the source follower.

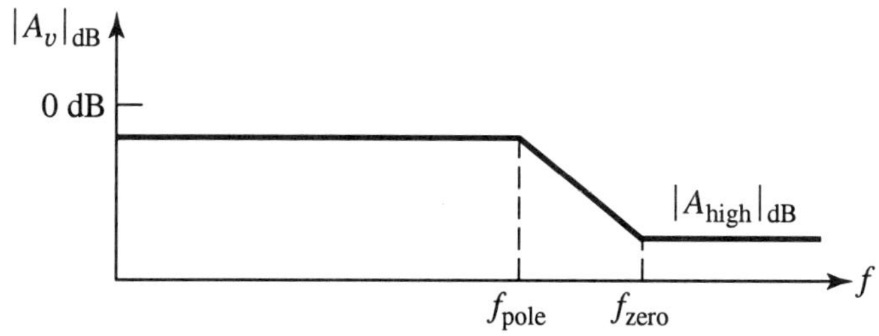


Figure 1.16: Asymptotic frequency response of source follower.

As we consider the large-signal operation of the source follower, we see that the output voltage can be driven positive to a value of $V_{DD} - V_{eff1} - V_{T1}$. The negative voltage can reach $V_{eff2}$ before $M2$ leaves the active region.

# Example 1.5

In the source follower and current mirror of Fig. 1.13, a dc drain current of 50 $\mu$A passes through M1 and M2. The following parameters apply to the devices of the circuit at this bias point: $g_{m1} = 199\ \mu$A/V, $g_{mb1} = 44\ \mu$A/V, $g_{ds1} = 1\ \mu$A/V, $g_{ds2} = 1.6\ \mu$A/V, $C_{gs1} = 7.1$ fF, $C_{sb1} = 12.6$ fF, $C_{db2} = 12.7$ fF, and $C_{gd2} = 1.5$ fF.

Calculate the midband voltage gain, the pole frequency or upper corner frequency, and the zero frequency. Also calculate the high-frequency gain due to capacitor feedthrough of $C_{gs1}$. Simulate this circuit with Spice.

**Solution** The midband voltage gain is calculated from Eq. (1.26) as

$$A_{MB} = \frac{g_{m1}}{g_{m1} + g_{mb1} + g_{ds1} + g_{ds2}} = \frac{199}{245.6} = 0.810 \text{ V/V}$$

The pole frequency which will be near the upper corner frequency is found from Eq. (1.28) to be

$$f_{pole} = \frac{g_{m1} + g_{mb1} + g_{ds1} + g_{ds2}}{2\pi(C_{gs1} + C_{out})} = \frac{245.6 \times 10^{-6}}{2\pi \times 33.9 \times 10^{-15}} = 1.15 \text{ GHz}$$

From Eq. (1.27), the zero frequency is calculated to be

$$f_{zero} = \frac{g_{m1}}{2\pi C_{gs1}} = \frac{199 \times 10^{-6}}{2\pi \times 7.1 \times 10^{-15}} = 4.46 \text{ GHz}$$

The high-frequency gain is

$$A_{high} = \frac{C_{gs1}}{C_{gs1} + C_{out}} = \frac{7.1}{33.9} = 0.209 \text{ V/V}$$

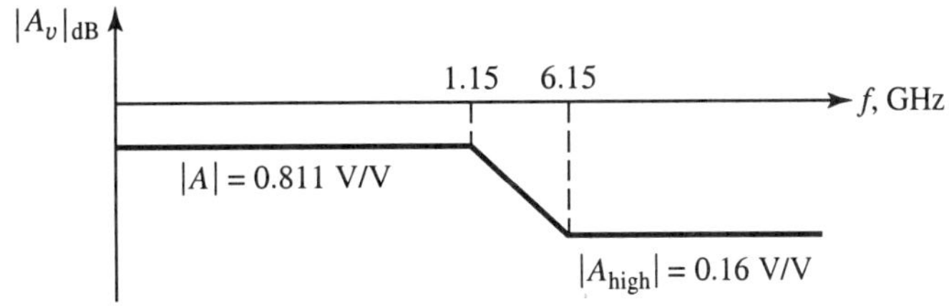


Figure 1.17: Simulated frequency response for source follower.

The Spice simulation file is listed in Table 1.3.

**Table 1.3 Spice File for Example 1.5**

```
EX1-5.CIR

R1 4 5 75K
V1 1 0 3.0V
V2 2 1 AC 0.1V
V3 5 0 5V
M1 5 2 3 0 N L=1U W=5U AD=25P AS=25P PD=15U PS=15U
M2 3 4 0 0 N L=1U W=5U AD=25P AS=25P PD=15U PS=15U
M3 4 4 0 0 N L=1U W=5U AD=25P AS=25P PD=15U PS=15U
.AC DEC 10 100 10000MEG
.OP
.PROBE
.LIB C5X.LIB
.END
```

The frequency response resulting from this simulation is shown in Fig. 1.17.

The simulated midband gain is 0.811 V/V, which is very close to the calculated value of 0.810 V/V. The upper corner frequency from the simulation is 1.15 GHz compared to a calculated value of 1.01 GHz. The simulated zero frequency is 6.15 GHz while the calculated value is 4.46 MHz. The high-frequency gain had a simulated value of 0.16 V/V, while the calculated value was 0.21 V/V.

A dc scan on the input voltage shows an output active region extending from approximately 0.32 to 3.03 V. The lower voltage is approximately $V_{eff2}$, whereas the upper voltage should approximate $V_{DD} - V_{eff1} - V_{T1}$.

---

The upper corner frequency for the source follower is 1.01 GHz compared to 8.53 MHz for the comparable-sized current source load stage of Example 1.4. The primary reason for this is the difference in the output resistance of the two stages. For the current-source load stage, the output resistance is 339 kΩ, whereas the source-follower output resistance is about 4 kΩ. In order to compare this source-follower stage when driven by a signal generator resistance, a 339-kΩ resistance is inserted in the input lead. This simulation leads to an overall upper corner frequency of 101.5 MHz. The corresponding value for the current-source load of Example 1.3 is 3.08 MHz.

With a capacitive load of 1 pF and a generator resistance of 339 kΩ, a Spice simulation shows an upper corner frequency of 37 MHz. Although this value is much lower than the unloaded value, it is considerably more than the corresponding frequency of the current source load stage. This value was 399 kHz. These results are summarized in Table 1.4.

**Table 1.4 Summary of Simulation Results for Source Follower**

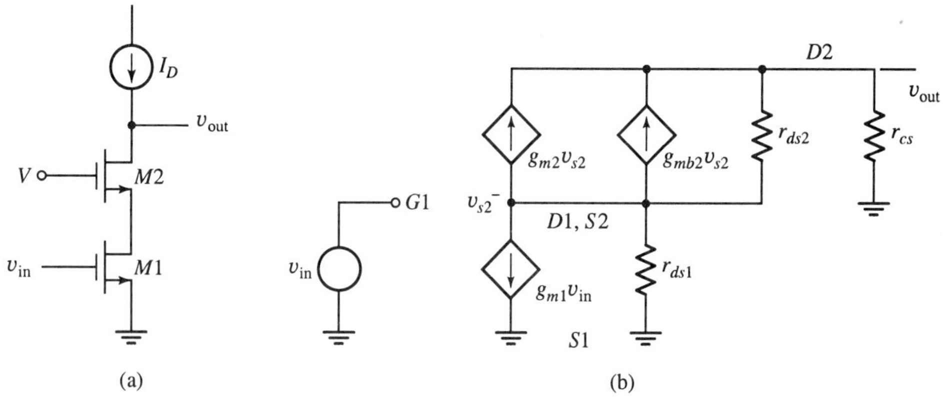


Figure 1.18: (a) The MOSFET cascode amplifier stage. (b) The equivalent circuit.

$$A_{MBsim} = 0.811 \text{ V/V}$$

| $C_L$, pF | $R_g$, kΩ | $f_{2o-sim}$, MHz |
|---|---|---|
| 0 | 0 | 1,150.0 |
| 1 | 0 | 36.6 |
| 0 | 339 | 101.6 |
| 1 | 339 | 37.2 |

**Practical Considerations**

As noted earlier, capacitive loading can be a serious problem for an output stage that connects to an IC pin connected to the input of another circuit. Because of the high output resistance of configurations such as the active load stage, a small capacitance of one pF or less will decrease the upper corner frequency significantly. The source follower stage is often inserted between a high-gain stage and the output pin of a chip to serve as a buffer stage. This buffer presents a rather low capacitance to the preceding stage since there is no Miller multiplication of $C_{gd}$ and, as seen from the simulation, a 1-pF load capacitance can result in a reasonably high upper corner frequency.

**Practice Problems**

**1.9** For the source follower of Fig. 1.13, derive the expression for output resistance looking into the source terminal of $M1$.

**1.10** If $g_{ds1}$ and $g_{ds2}$ are considered negligible for the source follower of Example 1.5, what will the midband voltage gain be? *Ans:* 0.819 V/V.

### 1.4.3 The Cascode Connection

The cascode connection has been used for many years in high-frequency BJT circuits and has become more important in IC MOSFET design in recent years. Figure 1.18(a) shows the basic cascode circuit with a current source load.

The equivalent circuit for the upper device would have a current generator of value $g_{m2}v_{gs2}$. For this connection, $v_{gs2} = -v_{s2}$; thus, this current generator reverses the direction of current flow and removes the negative sign to lead to the value of $g_{m2}v_{s2}$ shown in Fig. 1.18(b).

The resistance $r_{cs}$ is the output resistance of the current source. For a simple source, this value would be $r_{ds}$; for a Wilson mirror, it would be considerably larger. The equivalent circuit can be converted to a Thévenin

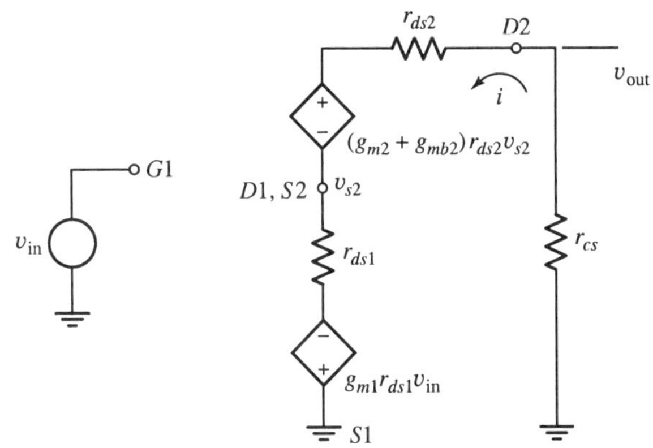


Figure 1.19: An alternate equivalent circuit for the cascode.

circuit as shown in Fig. 1.19.

The output voltage is

$$v_{out} = -ir_{cs}$$

where $i$ is the incremental drain current. Note that the upper voltage source is proportional to $v_{s2}$, the source voltage of $M2$. This voltage can be expressed as

$$v_{s2} = -g_{m1} r_{ds1} v_{in} + ir_{ds1}$$

If the transconductance of $M2$ is added to the body effect transconductance, that is,

$$g_{t2} = g_{m2} + g_{mb2} \tag{1.30}$$

the midband voltage gain can be found to be

$$A_{MB} = -\frac{(g_{m1} r_{ds1} + g_{t2} g_{m1} r_{ds1} r_{ds2}) r_{cs}}{r_{ds1} + r_{ds2} + r_{cs} + g_{t2} r_{ds1} r_{ds2}} \tag{1.31}$$

This expression represents the product of two relatively large gains. As bias current decreases, the $r_{ds}$ terms increase faster than the $g_m$ terms decrease. Gains of several thousand can be obtained with this circuit. However, as $r_{ds}$ increases, the high-frequency response decreases, since the output resistance must drive the parasitic capacitance at the output of the stage.

A differential version of this stage is often used as an input stage in MOSFET op amp chips. A slightly different version of the cascode is the folded cascode that replaces $M2$ by a pMOS device with a drain connection to ground. We will not consider the folded cascode stage.

The upper corner frequency of the cascode can be found from the equivalent circuit shown in Fig. 1.20.

The precise analysis of this circuit is somewhat complex. An accurate value of the upper corner frequency can be found by simulation, but it can be approximated rather easily from the equivalent circuit. We first note that with no signal generator resistance, the input capacitance will not influence the bandwidth. We next observe that the corner frequency due to the capacitance $C_1$ will be affected by the parallel combination of $r_{ds1}$ and the impedance looking into the source of $M2$. This impedance will be much less than $r_{ds1}$ and can be approximated by

$$R_{S2} = \frac{1}{g_{m2} + g_{mb2}}$$

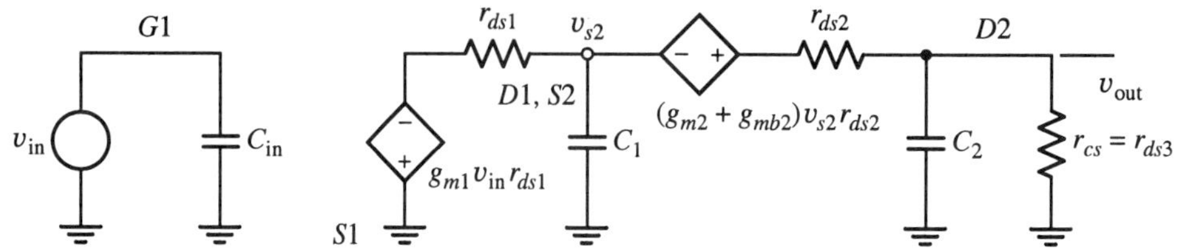


Figure 1.20: High-frequency model of the cascode.

This corner frequency will be quite large, at least compared to that caused by $C_2$. If the current source is a simple current mirror, its output impedance will be $r_{cs} = r_{ds3}$. The resistance seen by $C_2$ is then $r_{ds3} \parallel R_{out2}$ where $R_{out2}$ is the impedance looking into the drain of $M2$. This latter impedance will be quite large since the source resistance of $M2$ equals $r_{ds1}$. This source resistance increases the output resistance of $M2$ significantly, thus, the resistance seen by $C_2$ is approximately $r_{ds3}$. The upper corner frequency of the cascode can then be approximated by

$$f_{2o} \approx \frac{1}{2\pi r_{ds3} C_2} \tag{1.32}$$

where $C_2 = C_{db2} + C_{gd2} + C_{db3} + C_{gd3}$.

# Example 1.6

The cascode circuit of Fig. 1.21 has a dc drain current for all transistors of 50 $\mu$A. This current is supplied by a simple current mirror with $M3$ as the output device. With the bias voltages shown, the parameters are $g_{m1} = 181\ \mu$A/V, $g_{m2} = 195\ \mu$A/V, $g_{ds1} = 5.87\ \mu$A/V, $g_{mb2} = 57.1\ \mu$A/V, $g_{ds2} = 0.939\ \mu$A/V, $g_{ds3} = 3.76\ \mu$A/V, $C_{db2} = 9.8$ fF, $C_{gd2} = 1.5$ fF, $C_{db3} = 40.9$ fF, and $C_{gd3} = 4.5$ fF.

Calculate the midband voltage gain and the approximate upper corner frequency of this cascode stage. Verify these results with a Spice simulation.

**Solution** The midband gain is calculated from Eq. (1.30).

$$A_{MB} = -\frac{(g_{m1} r_{ds1} + g_{t2} g_{m1} r_{ds1} r_{ds2}) r_{ds3}}{r_{ds1} + r_{ds2} + r_{ds3} + g_{t2} r_{ds1} r_{ds2}}$$

$$= -46.8 \text{ V/V}$$

We note that this gain can be approximated by

$$A_{MB} = -g_{m1} r_{ds3} = -\frac{181}{3.76} = -48.1 \text{ V/V}$$

This assumes that the incremental current generated by device $M1$, $g_{m1}v_{in}$, flows through $M2$ to develop the output voltage across $r_{ds3}$. The low value of voltage gain could be increased sharply by increasing the output impedance of the current mirror to something much greater than $r_{ds3}$.

The upper corner frequency is approximated by Eq. (1.32) as

$$f_{2o} = \frac{1}{2\pi r_{ds3} C_2} = \frac{1}{2\pi \times 56.7 \times 10^{-15} \times 266 \times 10^3} = 10.6 \text{ MHz}$$

The Spice netlist is included in Table 1.5.

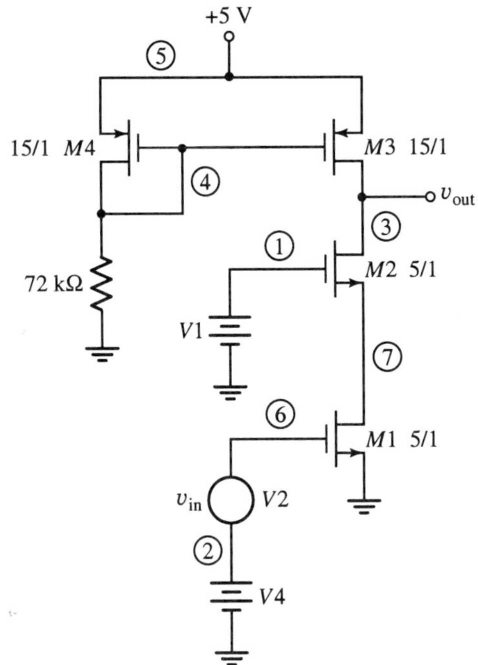


Figure 1.21: Cascode circuit for Example 1.6.

**Table 1.5 Spice Netlist for Example 1.6**

```
EX1-6.CIR

R1 0 4 72K
V1 1 0 2.1V
V2 2 6 AC 0.005V
V3 5 0 5V
V4 2 0 1.3V
M1 7 6 0 0 N L=1U W=5U AD=25P AS=25P PD=15U PS=15U
M2 3 1 7 0 N L=1U W=5U AD=25P AS=25P PD=15U PS=15U
M3 3 4 5 5 P L=1U W=15U AD=75P AS=75P PD=25U PS=25U
M4 4 4 5 5 P L=1U W=15U AD=75P AS=75P PD=25U PS=25U
.AC DEC 10 100 100MEG
.OP
.PROBE
.LIB C5X.LIB
.END
```

The results of the simulation are $A_{MBsim} = -46.9$ V/V and $f_{2o-sim} = 9.65$ MHz. The calculated gain is very close to the simulated value, while the upper corner frequency is calculated to be about 10% higher than the simulated value.

---

In order to demonstrate the effect of current on gain and bandwidth, the dc drain current of the circuit is dropped from 50 $\mu$A to 10.7 $\mu$A. For this situation, the simulated results are $A_{MBsim} = -94.1$ V/V and $f_{2o-sim} = 2.52$ MHz. The gain has approximately doubled while the bandwidth has dropped by a factor of about 4.

The effect on voltage gain can be seen from the approximation

$$A_{MB} = -g_{m1} r_{ds3}$$

Since $g_m$ varies as the square root of drain current and $r_{ds}$ varies approximately inversely with current, the voltage gain at a new current, $I_{D2}$, compares to the voltage gain at a current of $I_{D1}$ as

$$A_{MB}(I_{D2}) = \sqrt{\frac{I_{D1}}{I_{D2}}}\, A_{MB}(I_{D1})$$

For this circuit, the new gain at 10.7 $\mu$A would be approximated as

$$A_{MB}(10.7) = \sqrt{\frac{50}{10.7}}\, A_{MB}(50) = 2.16 \times (-46.8) = -101 \text{ V/V}$$

The bandwidth decrease is due to the inverse increase in output resistance when drain current is lowered.

If a signal generator resistance is present, a second corner frequency is added at the input due to $C_{in}$ of Fig. 1.20. When a 339-k$\Omega$ generator resistance is added to the simulation of Table 1.5, the new upper corner frequency is 8.95 MHz. This represents a relatively small change from the original value of 9.65 MHz. This implies that $C_{in}$ is rather small. This is one of the advantages of the cascode connection, that is, the input capacitance is small since the Miller effect is minimized by a small voltage gain from gate to drain of M1.

If a load capacitance of 1 pF is added, but no generator resistance is present, the upper corner frequency is lowered to 552 kHz. Adding the generator resistance to this circuit leads to no further reduction of the upper corner frequency.

A summary of performance for the circuit of Fig. 1.21 is included in Table 1.6.

**Table 1.6 Summary of Simulation Results for Cascode Amplifier**

$A_{MBsim} = -46.9$ V / V

| $C_L$, pF | $R_g$, k$\Omega$ | $f_{2o-sim}$, MHz |
|---|---|---|
| 0 | 0 | 9.65 |
| 1 | 0 | 0.55 |
| 0 | 339 | 8.95 |
| 1 | 339 | 0.55 |

**Practice Problem**

**1.11** Rework Example 1.6 if the current from the mirror is changed to 75 $\mu$A. *Ans:* $A_{MB} = -39.3$ V/V, $f_{2o} = 15.9$ MHz.

## 1.5 THE BJT

**Important Concepts**

**1.** The hybrid-$\pi$ circuit is often used to analyze the common-emitter configuration of the BJT.

**2.** Knowledge of key DC and AC equations allows a designer to calculate gains and impedances pertaining to the BJT amplifier.

This device is the transistor that began the semiconductor era. It is still important in many applications. The improved heterojunction transistor, which operates similarly to the standard BJT, can amplify signals at

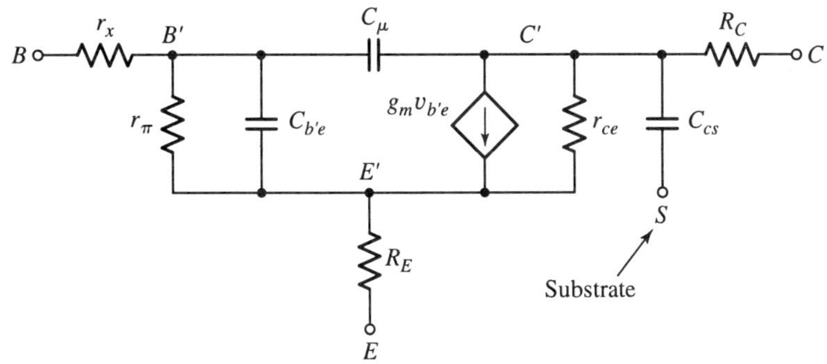


Figure 1.22: The small-signal hybrid-$\pi$ equivalent circuit.

much higher frequencies than can the MOSFET. Many high-frequency communication applications use the BJT for this reason.

For smaller devices, the BJT also has a higher output current capability than the MOSFET, allowing simpler interfacing to higher current systems. In high power applications, the BJT can exhibit less distortion as well as a higher value of transconductance than the power MOSFET. These and other considerations continue to keep the BJT at the forefront of the analog circuits field, even though the MOSFET dominates the personal computer area.

## 1.5.1 The Hybrid-$\pi$ Equivalent Circuit

This important model of the transistor has been used for years and can be easily extended to the Gummel-Poon model used in many versions of Spice. The model is attractive because each element relates directly to some physical effect that contributes to the operation of the device. This is not true of the latest models of the MOSFET which use pages of equations in an attempt to accurately model that device.

Figure 1.22 shows the small-signal equivalent circuit for the BJT. This model is used for high-frequency analysis while all capacitors are assumed to be open circuits for dc to midband frequencies. The elements of the model are as indicated in Table 1.7.

## 1.5.2 Some Important Equations for the BJT

**DC Equations:** In the active region, the following equations are valid.

$$I_E = I_C + I_B \tag{1.33}$$

This assumes that $I_E$ is positive when leaving the *npn* device and that both $I_C$ and $I_B$ are positive when entering the device.

$$I_C = \alpha_0 I_C + I_{co} \tag{1.34}$$

$$I_C = \beta I_B + (\beta + 1) I_{co} \tag{1.35}$$

The current $I_{co}$ is the reverse leakage current of the base-collector junction. For low power silicon devices, $I_{co}$ is generally negligible.

**Table 1.7 Equivalent Circuit Parameters for the BJT**

| Element | Name | Comment |
|---|---|---|
| $r_x$ | Base resistance | Base contact resistance plus base region resistance |
| $r_\pi$ | Base-emitter resistance | $(\beta+1)r_e$ where $r_e = kT/qI_E$ |
| $C_{b'e}$ | Base-emitter capacitance | Diffusion capacitance plus junction capacitance Diffusion capacitance dominates |
| $C_\mu$ | Collector-base junction capacitance | $C_\mu = C_0/(\phi - V_{BC})^{0.33}$ |
| $g_m$ | Transconductance | $g_m = \alpha/r_e$ |
| $r_{ce}$ | Collector-emitter resistance | $r_{ce} = V_{AF}/I_C$ Forward Early voltage over collector current at saturation |
| $R_E$ | Emitter contact resistance | Typically negligible (2 Ω) |
| $R_C$ | Collector contact resistance | Typically negligible (50 Ω) |
| $C_{cs}$ | Collector-substrate capacitance | Neglected in manual analysis |

The dc current gain from emitter to collector is

$$\alpha_0 = \frac{I_C}{I_E} \tag{1.36}$$

As long as $I_{co}$ is negligible, the dc current gain from base to collector is

$$\beta_0 = \frac{I_C}{I_B} \tag{1.37}$$

**AC Equations at Midband Frequencies:** For small signals, the following equations are useful when the device is operating in the active region.

$$\Delta I_C = \alpha \Delta I_E \text{ or } i_c = \alpha i_e \tag{1.38}$$

$$\alpha = \frac{\Delta I_C}{\Delta I_E} \text{ or } \alpha = \frac{i_c}{i_e} \tag{1.39}$$

Typically, $\alpha_0 = \alpha$.

$$\Delta I_C = \beta \Delta I_B \text{ or } i_c = \beta i_b \tag{1.40}$$

$$\beta = \frac{\Delta I_C}{\Delta I_B} \text{ or } \beta = \frac{i_c}{i_b} \tag{1.41}$$

Generally, we will assume that $\beta_0 = \beta$ for ease of calculation.

Several useful relationships can be summarized using the circuit of Fig. 1.23.

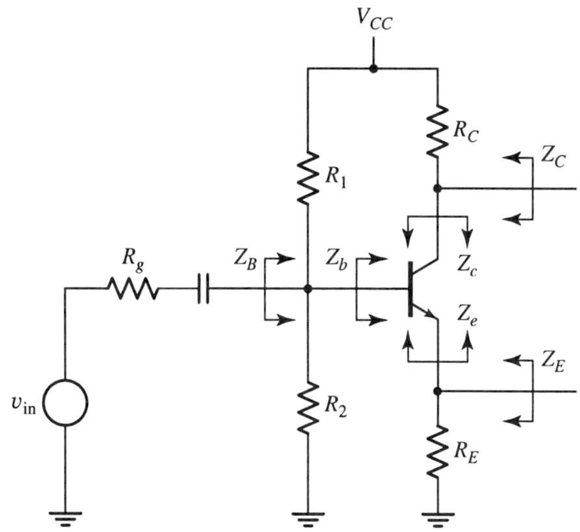


Figure 1.23: A BJT stage.

**General Equations**

$$\beta = \frac{i_c}{i_b} \qquad i_c = \beta i_b \qquad g_m = \frac{\alpha}{r_e}$$

$$r_e = \frac{26}{I_E\ (\text{mA})} \qquad r_{ce} = \frac{V_A}{\beta I_B} \qquad V_{BE(on)} = 0.6\text{ V}$$

**DC Bias Equations**

$$V_{Th} = \frac{R_2}{R_1 + R_2} V_{CC}$$

$$R_{Th} = R_1 || R_2$$

$$I_B = \frac{V_{Th} - 0.6}{R_{Th} + (\beta + 1) R_E}$$

**Impedances for Small-Signal Operation**

Looking into the base:

$$Z_b = (\beta + 1)(R_E + r_e)$$

Overall input impedance:

$$Z_B = Z_b || R_{Th}$$

Looking into the emitter:

$$Z_e = r_e + \frac{R_g || R_{Th}}{\beta + 1} \approx r_e \approx \frac{1}{g_m}$$

Overall output impedance at the emitter:

$$Z_E = R_E || Z_e$$

Looking into the collector:

If $R_E$ is bypassed with a capacitor

$$Z_c = r_{ce}$$

If $R_E$ is unbypassed

$$Z_c = r_{ce}\left(1 + \frac{\beta R_E}{(\beta+1)r_e + R_g + R_E}\right)$$

This value is generally much larger than $r_{ce}$.

Overall output impedance at the collector:

If $R_E$ is bypassed with a capacitor

$$Z_C = r_c || R_C$$

If $R_E$ is unbypassed

$$Z_C \approx R_C$$

Voltage gain from base to collector

$$A_{bc} = -\frac{\beta R_C}{(\beta+1)(r_e + R_E)} = -\frac{g_m R_C}{R_E + 1/g_m} = -\frac{R_C}{r_e + R_E}$$

Voltage gain from base to emitter

$$A_{be} = \frac{R_E}{r_e + R_E} = \frac{R_E}{R_E + 1/g_m}$$

A thorough understanding of these equations allows BJT circuit analysis or design problems to be easily solved.

# Example 1.7

In the emitter-bias circuit of Fig. 1.23, $R_C = 10$ kΩ, $R_E = 1$ kΩ, and $V_{CC} = 20$ V. Given that $\beta = 80$ and $V_{BE} = 0.6$ V, find $R_1$ and $R_2$ to set $V_{CQ}$ at the center of the active region.

**Solution** We will first assume that $R_2 = 10R_E$ is a condition imposed on the circuit by stability requirements. This gives a value of $R_2 = 10$ kΩ for this example. We now must find the desired value of $I_{BQ}$.

1. The center of the active region must be located. We calculate this value to be

$$V_{CQ} = 10\left(1 + \frac{1}{11}\right) = 10.9 \text{ V}$$

2. The required collector current and base current can now be found to give this value of $V_{CQ}$:

$$I_{CQ} = \frac{V_{CC} - V_{CQ}}{R_C} = \frac{20 - 10.9}{10} = 0.91 \text{ mA}$$
$$I_{BQ} = \frac{I_{CQ}}{\beta_0} = \frac{0.91}{80} = 0.0114 \text{ mA}$$

3. We now select $R_1$ to give the calculated value of $I_{BQ}$. By noting in Fig. 1.23 that $I_1 = I_{BQ} + I_2$, we can find the necessary value of $R_1$:

$$I_2 = \frac{V_{BQ}}{R_2} = \frac{I_{EQ}R_E + V_{BE}}{R_2} = \frac{(I_{BQ} + I_{CQ})R_E + 0.6}{R_2}$$
$$= \frac{(0.92) \times 1 + 0.6}{10} = 0.152 \text{ mA}$$

Then

$$I_1 = I_{BQ} + I_2 = 0.011 + 0.152 = 0.163 \text{ mA}$$

Since the voltage across $R_1$ is $V_{CC} - V_{BQ}$, then

$$R_1 = \frac{V_{CC} - V_{BQ}}{I_1} = \frac{20 - 1.52}{0.163} = 113 \text{ k}\Omega$$

**Practice Problems**

**1.12** Rework Example 1.7 with $R_2 = 10$ kΩ and a desired value of $V_{CQ} = 14$ V. *Ans:* $R_1 = 147$ kΩ.

**1.13** Rework Example 1.7 with $R_2 = 10$ kΩ and a desired value of $V_{CQ} = 7$ V. *Ans:* $R_1 = 88$ kΩ.

# Example 1.8

Assume that the circuit of Fig. 1.23 is biased such that $I_E = 1$ mA. Given that $R_C = 5$ kΩ, and $\alpha = 0.99$, find the input impedance, output impedance, and voltage gain of the amplifier.

**Solution** From the equivalent circuit, we see that the input impedance is

$$r_{in} = (\beta + 1)r_e \tag{1.42}$$

This results in a value of

$$r_{in} = 26(100) = 2600 \ \Omega$$

The voltage gain is

$$A_{MB} = \frac{v_{out}}{v_{in}} = -\frac{\alpha R_C}{r_e} = -190 \text{ V/V}$$

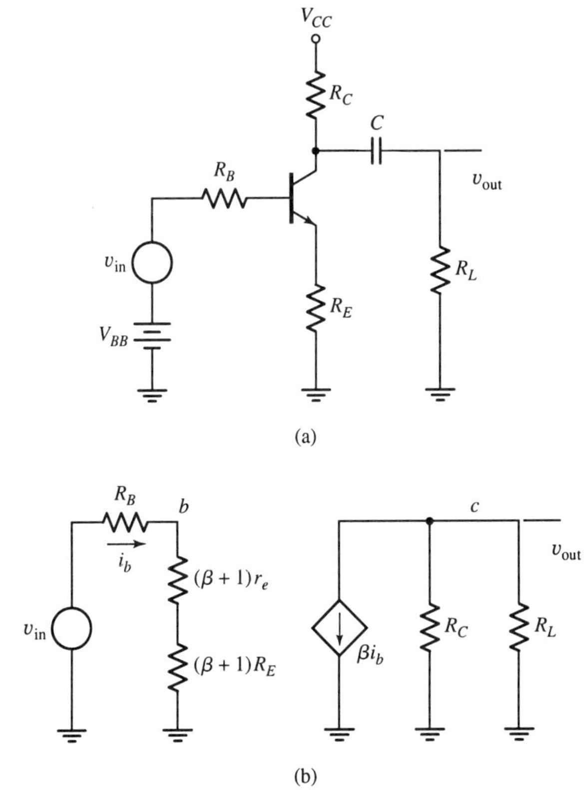


Figure 1.24: Common-emitter stage with external resistors. (a) Actual circuit. (b) Equivalent circuit.

The output impedance is again 5 kΩ, neglecting $r_{ce}$, but the current gain in the common-emitter stage is quite high. This value is found by taking the ratio of the output or collector current to input or base current. This ratio is equal to $\beta$.

---

Because the common-emitter amplifier has a high current gain along with a high voltage gain, it is the most well-used stage of the three possible configurations.

**Practice Problems**

**1.14** If an input signal source with a resistance of 1 kΩ drives the stage of Example 1.8, what is the overall voltage gain? *Ans:* - 137 V/V.

**1.15** If an input signal source with a resistance of 1 kΩ drives the stage of Example 1.8, what value of $R_C$ is needed to result in an overall midband voltage gain of - 90? *Ans:* 3.27 kΩ.

# Example 1.9

The circuit of Fig. 1.24 is biased such that $I_E = 0.5$ mA. The BJT has a value of $\beta = 180$. If $R_B = 4$ kΩ, $R_E = 500\ \Omega$, $R_C = 8$ kΩ, $R_L = 2$ kΩ, and $V_{CC} = 12$ V, calculate the midband voltage gain of the circuit.

Can the absolute value of voltage gain be increased to 6 by modifying the values of $R_C$ and $R_E$ without changing the dc emitter current? If so, select proper values.

**Solution** The calculation of the voltage gain is done rather easily. The effective collector load is $R_{eff} = 1.6$ kΩ. The dynamic emitter resistance is

$$r_e = \frac{26}{0.5} = 52\ \Omega$$

With these values the voltage gain is

$$A_v = \frac{-180 \times 1.6}{4 + 181 \times 0.552} = -2.77\ \text{V/V}$$

This part of the example is an analysis problem. The second part is more closely related to design.

The equation for voltage gain shows that in order to increase the absolute value of gain, $R_C$ can be increased, $R_E$ can be decreased, or a combination of both actions can be effected. There is no single correct solution to this problem. It can be seen, however, that the voltage gain varies directly with $R_{eff}$. Attempts to increase $R_{eff}$ by increasing $R_C$ meet with little success. The maximum value of $R_{eff}$ is 2 kΩ occurring when $R_C$ increases to infinity. Not only does $R_{eff}$ increase little with increasing $R_C$, but the BJT would near saturation if $R_C$ were increased toward 20 kΩ. Changing $R_E$ appears to be a good approach to satisfying the voltage gain requirement. Assuming that $R_{eff}$ remains constant, the equation for voltage gain can be solved for that value of $R_E$ that leads to the correct gain. This results in

$$\begin{aligned} R_E &= \frac{\beta \times R_{eff}}{(\beta + 1) \times |A_v|} - \frac{R_B}{(\beta + 1)} - r_e \\ &= \frac{180 \times 1.6}{181 \times 6} - \frac{4}{181} - 0.052 = 0.191\ \text{k}\Omega \end{aligned}$$

---

**Practice Problems**

**1.16** Calculate the voltage gain from input to the collector for the stage of Fig. 1.23. Assume that $I_E = 1$ mA, $\beta = 100$, $R_S = 1$ kΩ, $R_1 \| R_2 = 6$ kΩ, and $R_C = 4$ kΩ. *Ans:* - 25.4 V/V.

**1.17** Repeat Practice Problem 1.16 if $\beta$ is changed to 200 while all other parameters remain the same. Compare the gain to that of Practice Problem 1.16 *Ans:* - 26.3 V/V, $\beta$ increases 100% while gain only changes by 3.5%.

# 1.6 BJT IC STAGES

**Important Concepts**

**1.** A simple *pnp* current mirror (source) can act as the active load for an *npn* common-emitter stage. The load resistance equals the parallel combination of output resistances of the mirror stage and the gain stage.

**2.** An *npn* current mirror stage (sink) can act as the load for an *npn* emitter follower stage.

**3.** The cascode stage minimizes Miller effect capacitance at the input terminal, thereby improving frequency response.

The current mirror serves as an active load for several important BJT IC amplifier stages to be considered in the following paragraphs.

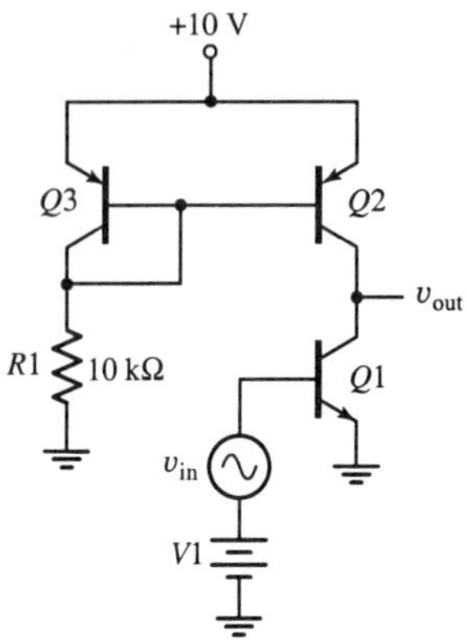


Figure 1.25: A common-emitter stage with current mirror active load.

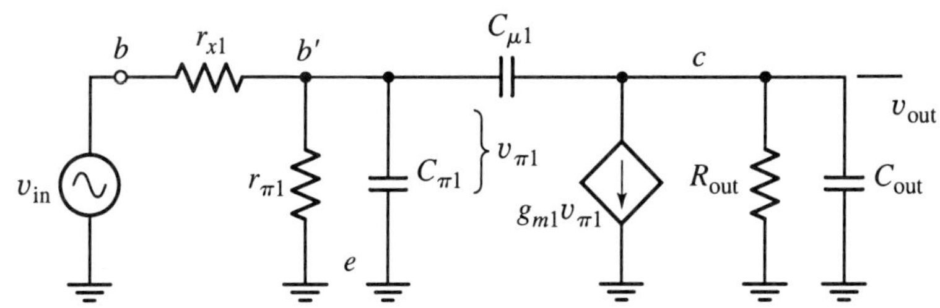


Figure 1.26: Equivalent circuit of the amplifier in Fig. 1.25.

### 1.6.1 The Current Mirror Load

A rather simple configuration for an amplifying stage is shown in Fig. 1.25. In this stage, the output impedance of the current mirror is not large enough to be negligible. Thus, the analysis will have to account for this element.

In an earlier section, the high frequency response of a discrete circuit was considered. Typically, this value was determined by the input circuit, including the Miller effect capacitance. The collector load resistance in a discrete stage is usually small enough that the output circuit does not affect the upper corner frequency. In the circuit of Fig. 1.25, as in most IC amplifier stages, the output resistance is very high compared to the discrete stage. For this circuit, the output resistance of the amplifier consists of the output impedance of $Q2$ in parallel with that of $Q1$. This value will generally be several tens of kΩ.

The equivalent circuit of the amplifier of Fig. 1.25 is indicated in Fig. 1.26.

The value of $R_{out}$ is

$$R_{out} = r_{o1} \parallel r_{o2} = r_{ce1} \parallel r_{ce2} \tag{1.43}$$

The capacitance in parallel with $R_{out}$ is approximately

$$C_{out} = C_{\mu 1} + C_{\mu 2} + C_{cs1} + C_{cs2} \tag{1.44}$$

In this equation, $C_{\mu 1}$ and $C_{\mu 2}$ are the collector-to-base junction capacitances, and $C_{cs1}$ and $C_{cs2}$ are the collector-to-substrate capacitances of the respective transistors. If no generator resistance is present, $C_{\mu 1}$ will also appear in parallel with the output terminal and ground. When $R_g$ is present, we will still approximate the output capacitance with the same equation, although feedback effects between the output and the bases of $Q1$ and $Q2$ actually modify the value slightly.

The midband gain is easy to evaluate as

$$A_{MB} = \frac{-\beta_1 R_{out}}{r_{x1} + r_{\pi 1}} \tag{1.45}$$

The upper corner frequency is now more difficult to evaluate than that of the discrete circuit which often has a low value of collector load resistance. In the discrete circuit, the input loop generally determines the overall upper corner frequency of the circuit. Although the Miller effect will be much larger in the IC stage, lowering the upper corner frequency of the input loop, the corner frequency of the output loop will also be smaller due to the large value of $R_{out}$. Both frequencies may influence the overall upper corner frequency of the amplifier.

The calculation of upper corner frequency begins by reflecting the bridging capacitance, $C_\mu$, to both the input and the output. The value reflected to the input side, across terminals $b'$ and $e$, is

$$(1 - A_{b'c1})C_{\mu 1} \tag{1.46}$$

as in the discrete circuit amplifier. Thus, the total input capacitance in parallel with $r_{\pi 1}$ is

$$C_{in} = C_{b'e1} + (1 - A_{b'c1})C_{\mu 1} \tag{1.47}$$

The major component of $C_{b'e1}$ is the diffusion capacitance, $C_\pi$, of $Q1$.

The upper corner frequency resulting from the input circuit of this stage is

$$f_{in-high} = \frac{1}{2\pi C_{in} R_{eq}} \tag{1.48}$$

where $R_{eq} = r_{x1} \parallel r_{\pi 1}$.

The upper corner frequency resulting from the output side of the stage is

$$f_{out-high} = \frac{1}{2\pi C_{out} R_{out}} \tag{1.49}$$

An example will demonstrate how to calculate the overall upper corner frequency for this two-pole response.

# Example 1.10

Assume that the circuit of Fig. 1.25 is biased so that the collector currents of $Q1$ and $Q2$ have a magnitude of 1.14 mA. The parameters for $Q1$ are $\beta = 160$, $r_{x1} = 10\ \Omega$, $r_{ce1} = 68\ k\Omega$, $C_{b'e1} = C_{\pi 1} = 20$ pF, and $C_{\mu 1} = 2.1$ pF. For device $Q2$, the necessary parameters are $r_{ce2} = 21$ kΩ and $C_{\mu 2} = 3.1$ pF. Each device has a value of $C_{cs1} = C_{cs2} = 2.5$ pF. In this circuit, the power supply is 10 V and $R1 = 10$ kΩ.

Calculate the midband voltage gain and the upper corner frequency for this amplifier stage. Do a Spice simulation using 2N3904 (*npn*) and 2N3905 (*pnp*) transistors.

**Solution** The midband voltage gain can be calculated from Eq. (1.45) after evaluating $r_{\pi 1}$ and $R_{out}$. These resistances are

$$r_{\pi 1} = (\beta + 1)r_{e1} = 161 \times \frac{26}{1.14} = 3672\ \Omega$$

and

$$R_{out} = r_{ce1} \parallel r_{ce2} = 68 \parallel 21 = 16\ \text{k}\Omega$$

The midband gain is then

$$A_{MB} = -\frac{\beta R_{out}}{r_{x1} + r_{\pi 1}} = -\frac{160 \times 16,000}{10 + 3672} = -695\ \text{V/V}$$

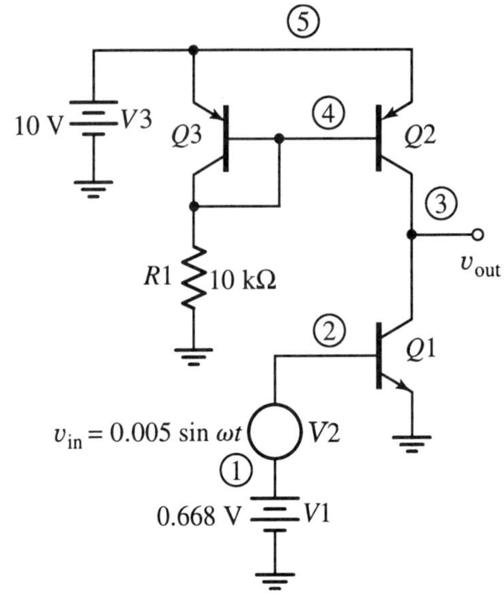


Figure 1.27: Schematic for Spice simulation.

The upper corner frequency is found from a consideration of the two poles caused by the input circuit and the output circuit. The corner frequency of the input circuit is

$$f_{in-high} = \frac{1}{2\pi C_{in} R_{eq}} = \frac{1}{2\pi(C_{\pi 1} + [1 - A_{b/c}]C_{\mu 1})(r_{x1} \mid\mid r_{\pi 1})}$$

$$= \frac{1}{2\pi \times 1481 \times 10^{-12} \times 10} = 10.7 \text{ MHz}$$

Since $r_{\pi 1} >> r_{x1}$, the value of 10 Ω was used for $R_{eq}$. In addition, the midband gain was used to approximate $A_{b/c}$.

The corner frequency of the output circuit is

$$f_{out-high} = \frac{1}{2\pi C_{out} R_{out}}$$

$$= \frac{1}{2\pi \times (2.1 + 3.1 + 2.5 + 2.5) \times 16{,}000} = 975 \text{ kHz}$$

The input corner frequency is much higher than the output corner frequency; consequently, the latter value approximates the overall corner frequency. This gives a value of $f_{2o} = 975$ kHz.

The schematic for a Windows Spice simulation is shown in Fig. 1.27, with node and element numbers added. The Spice netlist file used to simulate this circuit is shown in Table 1.8.

**Table 1.8 Spice Netlist File for Example 1.10**

```
EX1-10.CIR

R1 0 4 10K
V1 1 0 0.668V
V2 2 1 AC 0.005V
V3 5 0 10V
Q1 3 2 0 0 Q2N3904
Q2 3 4 5 5 Q2N3905
Q3 4 4 5 5 Q2N3905
.AC DEC 100 100 1G
.OP
.PROBE
.LIB BIPOLAR.LIB
.END
```

Note that the connections for the BJT correspond to collector node, base node, emitter node, and substrate node. Normally, the substrate for a *pnp* device connects to the positive power supply voltage. The substrate for the *npn* device connects to the most negative power supply rail. This is often the ground terminal.

The results of the simulation are $A_{MBsim} = -685$ V/V and $f_{2o-sim} = 900$ kHz. The calculated and simulated values of midband voltage gain agree within 2%, while the upper corner frequency values are within 10%.

---

In order to evaluate the effects of a signal generator resistance and a larger capacitive load, a 10 kΩ resistance was inserted in the base lead of $Q1$. This resulted in a simulated midband gain of −186 V/V and an upper corner frequency of 39 kHz. If a 10-pF capacitor is placed across the output and no generator resistance is used, the voltage gain returns to the value of −685 V/V, but the upper corner frequency is lowered to 480 kHz. Adding the generator resistance while the capacitor loads the output gives $A_{MB} = -186$ V/V and $f_{2o} = 37.8$ kHz. The comparison of calculated and simulated values for the different loading conditions is given in Table 1.9.

**Table 1.9 Summary of Results for the Common-Emitter Current-Source Load Stage**

| $C_L$, pF | $R_g$, kΩ | $A_{MBcal}$, V/V | $A_{MBsim}$, V/V | $f_{2o-cal}$, kHz | $f_{2o-sim}$, kHz |
|---|---|---|---|---|---|
| 0 | 0 | −696 | −685 | 975 | 900 |
| 10 | 0 | −696 | −685 | 480 | 493 |
| 0 | 10 | −187 | −186 | 40 | 39 |
| 10 | 10 | −187 | −186 | 39 | 38 |

The calculated results are designated $A_{MBcal}$ and $f_{2o-cal}$ while the simulated values are designated $A_{MBsim}$ and $f_{2o-sim}$. The task of calculating values for Table 1.9 is left to the student. Adding a 10-kΩ generator resistance lowers the upper corner frequency by a large factor, approximately 25, while adding 10 pF to the output lowers this frequency by a factor of about two.

**Practical Considerations**

For a BJT with a high value of current gain-bandwidth product, $f_t$, the current source load stage has a relatively low upper corner frequency. In Example 1.10, the value of $f_t$ for these BJTs is 300 - 400 MHz. The resulting upper corner frequency is just 975 kHz. This type of stage is used in op amp chips that will ultimately have a very low upper corner frequency. When used in an amplifier, the op amp chip will use feedback to improve the overall upper corner frequency. Thus, the low value of $f_{2o}$ for an individual gain stage is not significant in these applications.

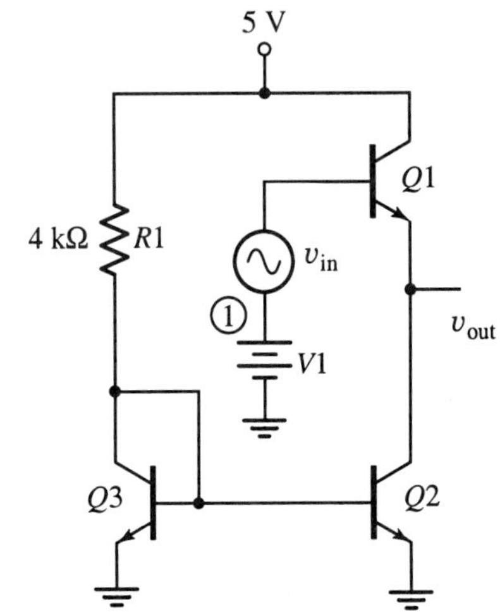


Figure 1.28: An emitter follower.

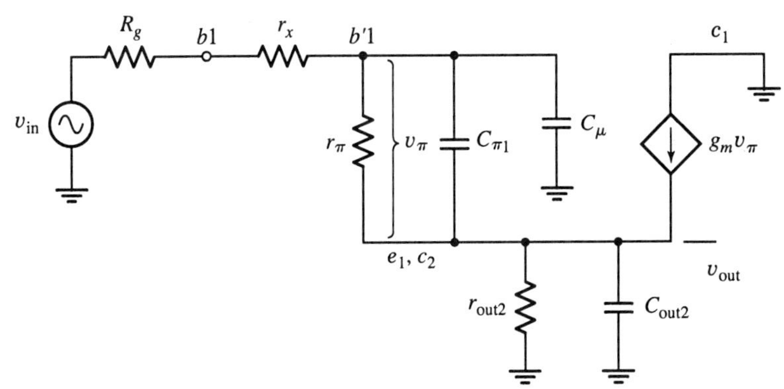


Figure 1.29: Equivalent circuit for the emitter follower.

**Practice Problem**

**1.18** Calculate the value of $R_g$ that lowers the upper corner frequency of the circuit in Example 1.11 to 100 kHz. Assume a 10-pF load capacitance. Calculate the midband voltage gain. *Ans:* $R_g = 1.43\text{ k}\Omega$, $A_{MB} = -501\text{ V/V}$.

## 1.6.2 The Emitter Follower

The emitter-follower stage can be used to minimize the adverse effect of a generator resistance on frequency response. Although this stage has a voltage gain near unity, it can be driven by a higher voltage gain stage, while the emitter follower can drive a low-impedance load. A typical stage is shown in Fig. 1.28.

The output stage of the *npn* current mirror, $Q2$, serves as a high impedance load for the emitter follower, $Q1$. An equivalent circuit that represents the emitter follower of Fig. 1.28 is indicated in Fig. 1.29.

For this circuit, $g_{m1} = \alpha_1 / r_e \approx 1/r_e$.

This circuit can be analyzed to result in a voltage gain of

$$A = \frac{\frac{C_{\pi 1}}{\Pi C(R_g + r_{x1})}\left(j\omega + \frac{g_{m1}r_{\pi 1}+1}{r_{\pi 1}C_{\pi 1}}\right)}{-\omega^2 + bj\omega + d} \tag{1.50}$$

where

$$\Pi C = C_{out2}C_{\mu 1} + C_{out2}C_{\pi 1} + C_{\pi 1}C_{\mu 1} \tag{1.51}$$

$$b = \frac{1}{\Pi C}\left(\frac{C_{\pi 1} + C_{out2}}{R_g + r_{x1}} + \frac{C_{\pi 1} + C_{\mu 1}(1 + g_{m1}r_{ce2})}{r_{ce2}} + \frac{C_{out2} + C_{\mu 1}}{r_{\pi 1}}\right) \tag{1.52}$$

and

$$d = \frac{R_g + r_{x1} + r_{\pi 1} + r_{ce2}(g_{m1}r_{\pi 1} + 1)}{\Pi C(R_g + r_{x1})r_{\pi 1}r_{ce2}} \tag{1.53}$$

Note that the output capacitance of $Q2$ can be approximated as the sum of $C_{\mu 2}$ and $C_{cs2}$.

The midband voltage gain is found from Eq. (1.50) by letting $\omega \to 0$. This value is

$$A_{MB} = \frac{(1 + g_{m1}r_{\pi 1})r_{ce2}}{R_g + r_{x1} + r_{\pi 1} + (1 + g_{m1}r_{\pi 1})r_{ce2}} \tag{1.54}$$

This gain is very near unity for typical element values.

The bandwidth is more difficult to calculate because the response has one zero and two poles. The zero for the circuit of Fig. 1.29 is typically larger than the lowest frequency pole. If these frequencies canceled, the larger pole would determine the corner frequency. Since they do not cancel, the overall upper corner frequency is expected to be smaller than the larger pole frequency. An accurate calculation can be made from Eq. (1.50) when the parameters are known.

## Example 1.11

The emitter-follower circuit of Fig. 1.28 is biased so that $I_{C1} = 1.08$ mA. The value of $\beta_1$ is 155, whereas the ohmic base resistance is 10 Ω. The collector-base depletion capacitance for both $Q1$ and $Q2$ is 2.5 pF as is also the collector to substrate capacitance. The value of $f_t$ is approximately 300 MHz and the Early voltage is 75 V for the transistors.

Calculate

**(a)** the midband voltage gain of the circuit,

**(b)** the approximate upper corner frequency of the circuit.

Simulate the operation of this stage using Spice to find the voltage gain and upper corner frequency.

**Solution** The value of $r_e$ is found as

$$r_e = \frac{26}{I_E} \approx \frac{26}{1.09} = 24\ \Omega$$

Based on this value, the diffusion capacitance or base-to-emitter capacitance can be found as

$$C_\pi = \frac{1}{2\pi r_e f_t} = \frac{1}{2\pi \times 24 \times 3 \times 10^8} = 22\ \text{pF}$$

The output resistance of $Q2$ is calculated from

$$r_{ce2} \approx \frac{V_A}{I_C} = \frac{75}{1.09} = 68.8\ k\Omega$$

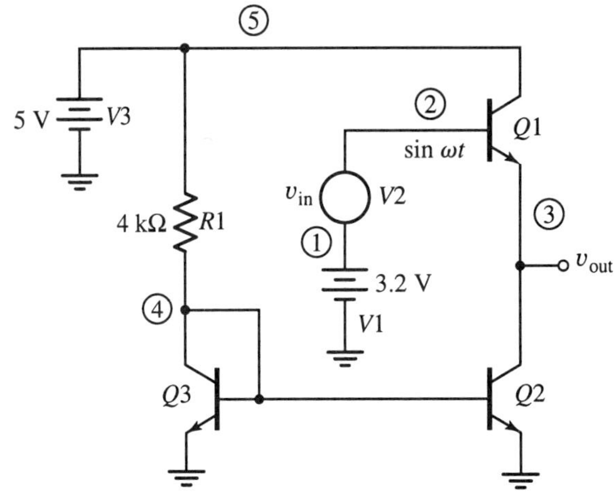


Figure 1.30: Schematic for simulation of the emitter follower.

This equation neglects the voltage $V_{CEQ2}$. The value of $r_{\pi 1}$ is $(\beta + 1)r_e = 3744\ \Omega$. All element values in the equivalent circuit of Fig. 1.29 are now known.

Substituting element values into Eq. (1.50) results in

$$A = \frac{1.239 \times 10^{10} \left(1 + j\frac{\omega}{1.89 \times 10^9}\right)}{-\omega^2 + j\omega 1.581 \times 10^{10} + 2.348 \times 10^{19}}$$

The midband gain of this circuit is $A_{MB} = 0.9993$ V/V. The zero frequency is 301 MHz. The two pole frequencies are 264 MHz and 2.28 GHz. The gain expression can also be written as

$$A = \frac{0.9993 \left(j\frac{f}{301\ MHz} + 1\right)}{\left(j\frac{f}{264\ MHz} + 1\right)\left(j\frac{f}{2.28\ GHz} + 1\right)}$$

Using iterative methods shows the upper corner frequency to be 1.67 GHz. Although the voltage gain is only unity, the upper corner frequency is much higher than the common-emitter amplifier with a current mirror load.

The schematic for simulation is shown in Fig. 1.30, with node and element numbers added.

The Spice netlist file is shown in Table 1.10.

**Table 1.10 Spice Netlist File for Example 1.11**

```
EX1-11.CIR

R1 5 4 4K
V1 1 0 3.2V
V2 2 1 AC 1V
V3 5 0 5V
Q1 5 2 3 0 Q2N3904
Q2 3 4 0 0 Q2N3904
Q3 4 4 0 0 Q2N3904
.AC DEC 100 100 10G
.OP
.PROBE
.LIB BIPOLAR.LIB
.END
```

The results of the simulation are $A_{MBsim} = 0.9993$ V/V and $f_{2o-sim} = 2.56$ GHz. Although the midband voltage gain compares well to the calculated value, the simulated upper corner frequency is somewhat higher than the calculated value.

---

It is again useful to consider the effects of adding a large generator resistance or a load capacitance to the emitter-follower stage. Table 1.11 summarizes the results of simulations for different combinations of source resistance and load capacitance.

**Table 1.11 Summary of Results for the Emitter-Follower Stage**

| $C_L$, pF | $R_g$, kΩ | $A_{MBcal}$, V/V | $A_{MBsim}$, V/V | $f_{2o-cal}$, MHz | $f_{2o-sim}$, MHz |
|---|---|---|---|---|---|
| 0 | 0 | 0.9993 | 0.9993 | 1670 | 2560 |
| 10 | 0 | 0.9996 | 0.9993 | 420 | 420 |
| 0 | 10 | 0.9982 | 0.9975 | 6.56 | 6.61 |
| 10 | 10 | 0.9987 | 0.9975 | 6.85 | 6.87 |

Note that the insertion of a 10-kΩ generator resistance lowers the upper corner frequency more than the addition of the 10-pF load capacitance. Note also that the addition of the 10 pF capacitance to the circuit that includes a 10 kΩ generator resistor leads to a higher upper corner frequency rather than a lower value. This results from a shift in the lower pole frequency to a value that exceeds that of the zero frequency when the load capacitance is added. The added value of $C_L$ when no generator resistance is present does not have the same effect.

Surprisingly, the two poles may become complex, depending on element values. When this occurs, the frequency response can exhibit a peak and the step response can exhibit ringing.

**Practical Considerations**

The emitter follower, like the source follower, is often used as a buffer to interface between a high voltage gain stage and a low-impedance load. The emitter follower loads the preceding stage only slightly, but provides a high current to the load with an accompanying high upper corner frequency.

**Practice Problem**

**1.19** An emitter follower has a midband voltage gain of 0.99. The voltage gain has a zero at 500 MHz, one pole at 200 MHz, and another pole at 1.6 MHz. Find the upper corner frequency of the voltage gain. *Ans:* 236 MHz.

## 1.6.3 The Cascode Amplifier Stage

One problem associated with the common-emitter stage using an active load is the Miller effect. This stage has a high voltage gain from base to collector. The circuit of Fig. 1.24, with the values of Example 1.11 has an inverting voltage gain, $A_{MB}$, that approaches −700 V/V. The base-collector junction capacitance is multiplied by $(1 + |A_{MB}|)$ and reflected to the input loop. This capacitance adds to the diffusion capacitance from point $b'$ to point $e$ and decreases the upper corner frequency to a relatively small value.

The cascode amplifier stage of Fig. 1.31 minimizes the capacitance reflected to the input.

In this circuit, the input capacitance is composed primarily of the diffusion capacitance of $Q1$. The gain from base-to-collector of $Q1$ is quite low since the collector load of this device consists of the impedance looking into the emitter of $Q2$. This impedance is approximately equal to the base-emitter diode resistance of $Q2$, which is

$$r_{e2} = \frac{26}{I_{E2}}$$

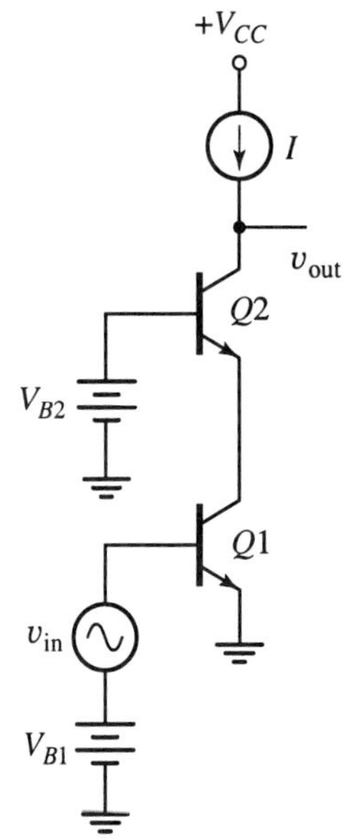


Figure 1.31: A cascode stage.

The upper device passes the incremental signal current of $Q1$ to its collector and develops a large voltage across the current-source impedance. There is no Miller multiplication of capacitance from the input of $Q2$ (emitter) to the output (collector) since the gain is noninverting and negligible capacitance exists between emitter and collector. Thus, the cascode stage essentially eliminates Miller effect capacitance and its resulting effect on upper corner frequency. A high-frequency equivalent circuit of this stage is shown in Fig. 1.32.

The resistance $R$ includes any generator resistance and the base resistance, $r_{x1}$ of $Q1$. The resistance $r_{cs}$ is the output resistance of the current source. The output capacitance is the sum of $C_{\mu 2}$, $C_{cs2}$, and any capacitance at the current source output. The resistance $r_{out2}$ can be quite large since $Q2$ sees a large emitter resistance looking into the collector of $Q1$. This emitter load leads to negative feedback that increases the output resistance of $Q2$.

The midband voltage gain is calculated from the equivalent circuit of Fig. 1.32 after eliminating the capacitors. This gain is found rather easily by noting that the input current to $Q1$ is

$$i_{b1} = \frac{v_{in}}{R + r_{\pi 1}} \tag{1.55}$$

This current will be multiplied by $\beta_1$ to become collector current in $Q1$. This current also equals the emitter current of $Q2$. The emitter current of $Q2$ is multiplied by $\alpha_2$ to become collector current of $Q2$. The output voltage is then

$$v_{out} = i_{c2} \times R_3 \tag{1.56}$$

where $R_3 = r_{out2} \,||\, r_{cs}$. This resistance can be very large if the current-source resistance, $r_{cs}$, is large. The value of $r_{out2}$ will be high since the emitter of $Q2$ sees a resistance of $r_{ce1}$. Combining this information results in a midband voltage gain of

$$A_{MB} = \frac{-\beta_1 \alpha_2 R_3}{R + r_{\pi 1}} \tag{1.57}$$

If no generator resistance is present and if $R_3 = 100$ kΩ, this midband voltage gain might exceed 5000 V/V.

The gain as a function of frequency can be found as

$$A = A_{MB} \frac{1}{1 + j\omega C_{\pi 1}(r_{\pi 1} \,||\, R)} \frac{1}{1 + j\omega r_{e2} C_{\pi 2}} \frac{1}{1 + j\omega R_3 C_{out}} \tag{1.58}$$

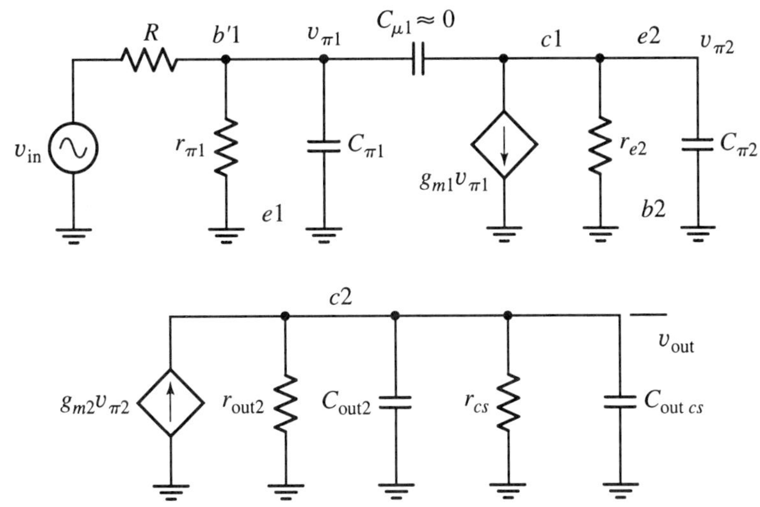


Figure 1.32: Equivalent circuit of the cascode stage.

Typically, the corner frequency of the second frequency term in Eq. (1.58), that is,

$$f_2 = \frac{1}{2\pi r_{e2} C_{\pi 2}}$$

is much higher than that of the first term,

$$f_{in-high} = \frac{1}{2\pi (r_{\pi 1} \mid\mid R) C_{\pi 1}}$$

This is especially true if $R$ is large compared to $r_{\pi 1}$. In this case, since $r_{e1} \approx r_{e2}$ as a result of equal emitter currents, then $r_{e2} \ll (\beta + 1) r_{e1}$. For hand analysis of the cascode circuit, the second term in the expression for gain is often neglected.

The gain can then be written as

$$A = A_{MB} \frac{1}{1 + j\frac{f}{f_{in-high}}} \frac{1}{1 + j\frac{f}{f_{out-high}}} \tag{1.59}$$

where

$$f_{out-high} = \frac{1}{2\pi C_{out} R_3} \tag{1.60}$$

and $f_{in-high}$ was defined previously.

The capacitance $C_{out}$ is the sum of the current-source output capacitance and the output capacitance of $Q2$, giving

$$C_{out} = C_{out2} + C_{outcs}$$

If a current mirror with output stage $Q3$ generates the collector bias current for $Q1$ and $Q2$, the output capacitance is

$$C_{out} = C_{\mu 2} + C_{\mu 3} + C_{cs2} + C_{cs3}$$

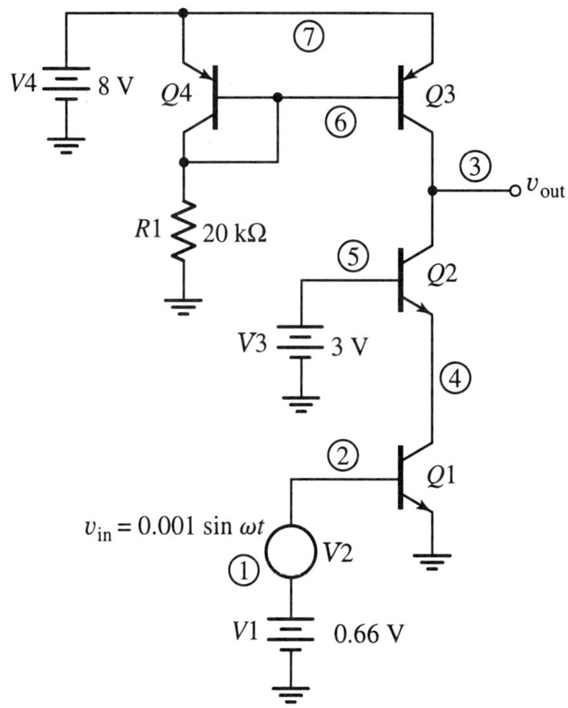


Figure 1.33: Schematic for simulation of cascode amplifier.

## Example 1.12

The cascode circuit of Fig. 1.33 is driven by a current mirror with an output current of 0.38 mA. Devices $Q1$ and $Q2$ have values of $\beta = 140$ and $r_x = 10\Omega$. The capacitor values are $C_{\pi 1} = C_{\pi 2} = 10.8$ pF, $C_{\mu 1} = C_{\mu 2} = C_{cs2} = 2.5$ pF, and $C_{\mu 3} = C_{cs3} = 5$ pF. The output impedance of the current mirror is 52.2 kΩ.

Calculate the midband voltage gain and the upper corner frequency of the circuit. Do a Spice simulation and compare the measured to the simulated results.

**Solution** The value of $r_\pi$ for $Q1$ and $Q2$ is calculated to be

$$r_\pi = (\beta + 1)r_e = 141 \times \frac{26}{0.38} = 9647\ \Omega$$

Assuming that $r_{out2}$ is very large, the value of $R_3$ is approximated by $r_{cs} = 52.2$ kΩ. From Eq. (1.57), the midband voltage gain is

$$A_{MB} = \frac{-\beta_1 \alpha_2 R_3}{r_{\pi 1}} \approx \frac{-\beta_1 r_{cs}}{r_{\pi 1}} = \frac{-141 \times 52.2}{9.647} = -763\ \text{V/V}$$

The corner frequency of the input stage is

$$f_{in-high} = \frac{1}{2\pi(r_{\pi 1} \,||\, R)C_{\pi 1}} = \frac{1}{2\pi \times 10 \times 10.8 \times 10^{-12}} = 1.47\ \text{GHz}$$

The upper corner frequency due to the output circuit is

$$f_{out-high} = \frac{1}{2\pi R_3 C_{out}} = \frac{1}{2\pi \times 52,200 \times 15 \times 10^{-12}} = 203\ \text{kHz}$$

In this equation, the output capacitance was taken as

$$C_{out} = C_{\mu 2} + C_{\mu 3} + C_{cs2} + C_{cs3} = 2.5 + 5 + 2.5 + 5 = 15 \text{ pF}$$

The overall upper corner frequency is $f_{2o} = 203$ kHz.

The Spice netlist file for this circuit is shown in Table 1.12 using the node and element numbers of Fig. 1.33.

**Table 1.12 Spice Netlist File for Example 1.12**

```
EX1-12.CIR

R1 6 0 20K
V1 1 0 0.66V
V2 2 1 AC 0.001V
V3 5 0 3V
V4 7 0 8V
Q1 4 2 0 0 Q2N3904
Q2 3 5 4 0 Q2N3904
Q3 3 6 7 7 Q2N3905
Q4 6 6 7 7 Q2N3905
.AC DEC 100 100 100Meg
.OP
.PROBE
.LIB BIPOLAR.LIB
.END
```

The results of the simulation are $A_{MBsim} = -758$ V/V and $f_{2o-sim} = 205$ kHz.

---

Additional calculations and simulations were done using a 10-pF load capacitor and/or a 10-kΩ generator resistance. These results are summarized in Table 1.13.

**Table 1.13 Summary of Results for the Cascode Stage**

| $C_L$, pF | $R_g$, kΩ | $A_{MBcal}$, V/V | $A_{MBsim}$, V/V | $f_{2o-cal}$, kHz | $f_{2o-sim}$, kHz |
|---|---|---|---|---|---|
| 0 | 0 | −763 | −758 | 203 | 205 |
| 10 | 0 | −763 | −758 | 121 | 123 |
| 0 | 10 | −372 | −368 | 201 | 202 |
| 10 | 10 | −372 | −368 | 121 | 123 |

We observe that the insertion of a 10-kΩ generator resistance has little effect on the upper corner frequency. This is to be expected as a result of the minimization of the Miller effect. The input capacitance is small enough that it has little effect on the overall upper corner frequency even with the larger generator resistance in the input loop. In the active load stage considered earlier, insertion of a 10-kΩ generator resistance lowered the upper corner frequency by a factor of about 25.

**Practice Problem**

**1.20** If a very large generator resistance is inserted in the circuit of Example 1.12, what is the lower limit on $f_{in-high}$? *Ans:* 1.53 MHz.

## SUMMARY

**1.** The MOSFET is an important amplifying element in electronic circuit design.

**2.** The incremental or small-signal equivalent circuit of the MOSFET is used to find the voltage gain, the input impedance, and the output impedance of the amplifier circuit.

**3.** High-frequency performance of the MOSFET is limited by parasitic capacitances between terminals and from terminal to substrate.

**4.** MOSFET IC amplifiers use active loads, often formed by current mirrors to replace the resistive loads of discrete circuits.

**5.** The BJT can also serve as the basis for an electronic amplifier. A small-signal model is again used to evaluate important parameters of an amplifying stage.

**6.** Current mirror loads are often used as active loads for this device when implemented on an IC chip.

**7.** For both MOSFET and BJT circuits, simulation is used as a design aid, particularly for IC design.

## PROBLEMS

### Section 1.3.1 The Triode Region

**1.1** If $\mu W C_{ox}/L = 0.4$ mA/V$^2$ and $V_T = 0.8$ V in Eq. (1.1),

**(a)** Calculate the drain current for $V_{GS} = 2$ V and $V_{DS} = 0.5$ V.
**(b)** Calculate the pinchoff drain current.
**(c)** What is the minimum value of $V_{DS}$ that leads to pinchoff?

### Section 1.3.2 The Active Region

**1.2** For the MOSFET circuit shown, $\mu W C_{ox}/2L = 80\ \mu$A/V$^2$, $V_T = 0.9$ V, and $\lambda = 0$.

(a) What value must $V_{GS}$ have to bring the device from the active region to the edge of the cutoff region?

(b) What value must $V_{GS}$ have to bring the device from the active region to the edge of the triode region?

**1.3** In the circuit of Problem 1.2, identify the region in which the MOSFET exists for the following conditions:

**(a)** $V_{GS} = 2.84$ V, $V_{DS} = 6$ V
**(b)** $V_{GS} = 3.27$ V, $V_{DS} = 3$ V
**(c)** $V_{GS} = 3.46$ V, $V_{DS} = 2$ V

**1.4** The threshold voltage for a MOSFET is 0.86 V. In which region does the device exist for

**(a)** $V_{GS} = 2$ V, $V_{DS} = 1$ V
**(b)** $V_{GS} = 1.8$ V, $V_{DS} = 1.8$ V
**(c)** $V_{GS} = 1.5$ V, $V_{DS} = 4$ V

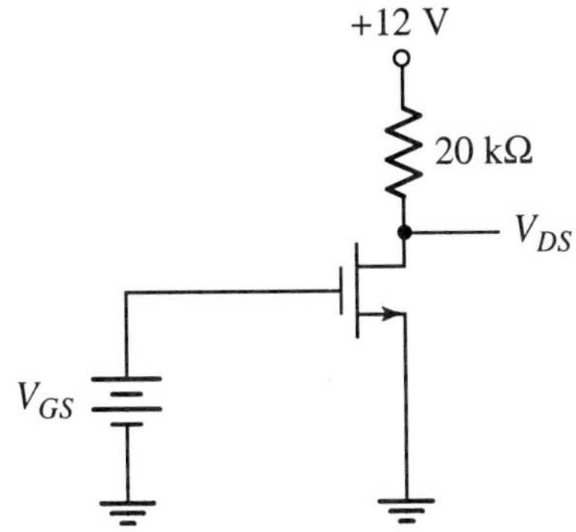


**Figure P1.2**

**1.5** A MOSFET with $V_T = 1$ V has the following measured data taken:

| $V_{GS}$, V | $V_{DS}$, V | $I_D$, $\mu$A |
|---|---|---|
| 2 | 1 | 80 |
| 2 | 8 | 91 |

Calculate $\lambda$ for this device.

**1.6** With $V_{GS} = 2.4$ V, calculate $r_{ds}$ at $V_{DS} = 4$ V and again at $V_{DS} = 8$ V using $r_{ds} = 1/\lambda I_{DP}$ and compare to the approximate values at these voltages given by $r_{ds} = 1/\lambda I_D$. Use values of $\mu W C_{ox}/2L = 0.5$ mA/V$^2$, $\lambda = 0.025$/V, and $V_T = 0.8$ V.

**1.7** If $g_m = 3.2$ mA/V at $I_D = 120$ $\mu$A, calculate $g_m$ when $I_D = 820$ $\mu$A.

**1.8** If $r_{ds} = 80$ k$\Omega$ at $I_D = 500$ $\mu$A, estimate $r_{ds}$ when $I_D = 100$ $\mu$A.

**1.9** If $\mu W C_{ox}/2L = 0.4$ mA/V$^2$, $\lambda = 0.025$/V, and $V_T = 0.8$ V calculate $r_{ds}$ at $V_{DS} = 4$ V for $V_{GS}$ values of 2 V and 4 V.

**1.10*** For the conditions of Problem 1.9, calculate the two values of $g_m$. Compare these values to those found by assuming $\lambda = 0$.

### Section 1.3.3 The Midband Incremental MOSFET Model

**1.11** A MOSFET is used in a common-source configuration with a drain resistance of $R_D = 8$ k$\Omega$. If the drain-to source resistance is $r_{ds} = 70$ k$\Omega$, what must the value of $g_m$ be to result in a midband voltage gain of $-18$ V/V? If the quiescent drain current is 1 mA, what is the value of $\mu C_{ox} W/2L$?

**1.12** If the quiescent drain current in Problem 1.11 is decreased to 0.1 mA, what are the new values of $g_m$ and $r_{ds}$? What is the new value of midband voltage gain?

**1.13*** A device with $\mu W C_{ox}/2L = 0.4$ mA/V$^2$, $\lambda = 0.025$/V, and $V_T = 0.8$ V is used in a common-source amplifier stage with $V_{GS} = 2.5$ V. If the power supply is 8 V and $R_D = 4$ k$\Omega$, calculate the voltage gain from gate to drain. Compare this result to that found by assuming $\lambda = 0$.

**1.14*** Repeat Problem 1.13 if the power supply is changed to 12 V and the load resistance is changed to 6 k$\Omega$.

**1.15*** For the amplifier of Problem 1.13, calculate the minimum and maximum possible output voltages before serious distortion is encountered (triode region and cutoff).

**1.16*** For the amplifier of Problem 1.14, calculate the minimum and maximum possible output voltages before serious distortion is encountered (triode region and cutoff).

**D1.17*** For the amplifier shown, assume that $\mu W C_{ox}/2L = 0.2$ mA/V$^2$, $\lambda = 0$/V, and $V_T = 1$ V. The quiescent drain voltage is to be in the range of 4 V to 9 V and the quiescent drain current should be in the range of 0.1 mA to 4 mA. Select $R_D$ to achieve the maximum possible midband voltage gain magnitude. Calculate $A_{MB}$.

**1.18*** For the amplifier shown, $g_m = 2$ mA/V and $r_{ds} = 60$ kΩ at the quiescent bias point. Find the midband voltage gain of the amplifier.

**1.19*** Assume that $g_m = 3$ mA/V, $r_{ds} = 40\ k\Omega$, and $V_T = 0.8$ V for the MOSFET of the circuit.

**(a)** Calculate the quiescent drain current, $I_{DQ}$, neglecting the effects of $\lambda$.

**(b)** Calculate the quiescent drain-to-source voltage, $V_{DSQ}$.

**(c)** Draw the incremental equivalent circuit.

**(d)** Calculate the midband incremental voltage gain, $A_{MB} = \frac{v_{out}}{v_{in}}$.

**D1.20*** For the MOSFET shown, $V_T = 1$ V, $\lambda = 0$, and $\mu C_{ox} W/2L = 0.1$ mA/V$^2$. Select $R_D$ and $R_1$ to result in a midband voltage gain of -4 V/V and $V_{DSQ} = 7$ V.

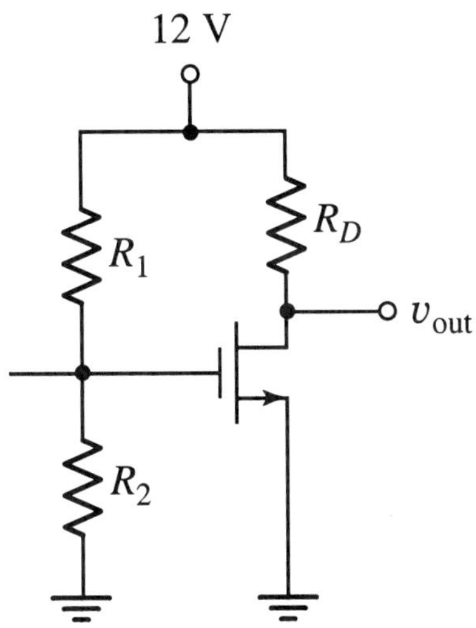


**Figure P1.17**

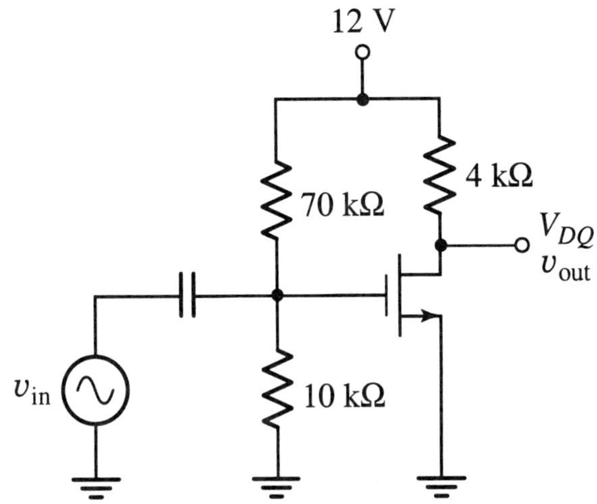


**Figure P1.19**

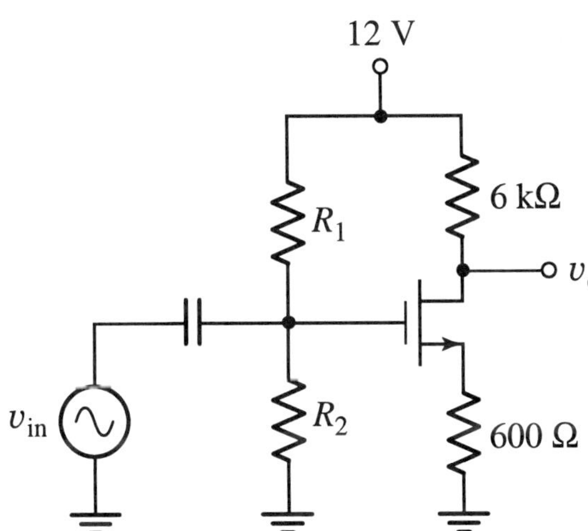


**Figure P1.18**

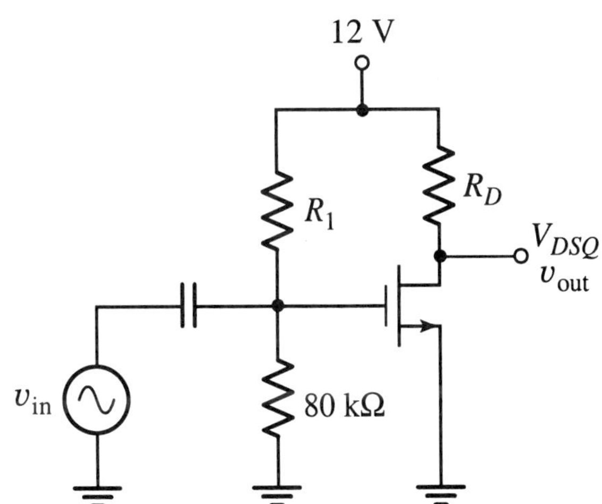


**Figure P1.20**

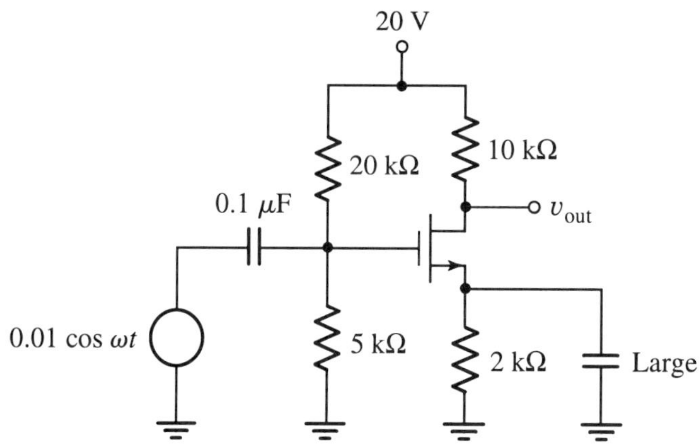


**Figure P1.21**

**1.21*** For the amplifier stage shown, $\mu W C_{ox}/2L = 75\ \mu\text{A/V}^2$, $V_T = 1$ V, and $\lambda = 0$.

(a) Calculate $V_{DQ}$ and $V_{SQ}$.
(b) Calculate the ac output voltage at midband frequencies.
(c) Find the lower corner frequency.

**1.22*** Repeat Problem 1.21 if $\lambda = 0.02$/V.

**D1.23** A MOSFET has values of $\mu C_{ox} W/2L = 0.75\ \text{mA/V}^2$, $V_T = 1$ V, and $\lambda = 0$. Design an amplifier to have a midband voltage gain of - 10 V/V. Assume a resistance of 5 kΩ replaces the 2-kΩ resistance between source terminal and ground and is bypassed by a large capacitor at midband frequencies. The power supply voltage is 12 V.

**1.24*** For the MOSFET in the circuit, $\frac{\mu W C_{ox}}{2L} = 2\ \text{mA/V}^2$, $\lambda = 0.02$/V, and $V_T = 0.8$ V. Calculate the midband voltage gain of the stage.

**1.25** Repeat Problem 1.24 if all parameters remain the same except that $\lambda = 0$.

**1.26*** For the amplifier shown, $\frac{\mu W C_{ox}}{2L} = 3\ \text{mA/V}^2$, $\lambda = 0.02$/V, and $V_T = 1.0$ V.

(a) Calculate $V_{DSQ}$ when $V_{GSQ} = 1.5$ V.

(b) Calculate the midband voltage gain for the stage.

**1.27** Repeat Problem 1.26 if all parameters remain the same except that $\lambda = 0$.

**1.28D*** For the MOSFET in the circuit, $\frac{\mu W C_{ox}}{2L} = 2.1\ \text{mA/V}^2$, $\lambda = 0.02$/V, and $V_T = 0.94$ V. Select $R_1$ to result in a midband gain of $A_{MB} = -25$ V/V $\pm 5\%$.

### Subsection 1.3.4 High-Frequency Model of the MOSFET

**1.29** Assume that $g_m = 3$ mA/V, $r_{ds} = 100$ kΩ, $C_{gs} = 1$ pF, $C_{gd} = 0.5$ pF, and $V_{DSQ} = 6$ V for the MOSFET of the circuit. The input voltage is a positive-going 20-mV step function.

(a) Calculate the time constant, $\tau$, and sketch the output voltage with time showing values of voltage levels.

**(b)** If the circuit is now driven with a small sinusoidal input signal, what is the upper corner frequency of the circuit, $f_{high}$?

**(c)** How do $\tau$ and $f_{high}$ relate to each other?

**Section 1.4.1 Active Load Stage**

**1.30** In the active load stage of Fig. 1.11, the current from the mirror is increased from 50 $\mu$A to 100 $\mu$A. What would you now expect the midband voltage gain to be? Explain your calculations.

**1.31** In the circuit of Fig. 1.11, a resistance of value 400 k$\Omega$ is placed in series with the input signal generator. Calculate the value of load capacitance that causes the input upper corner frequency to equal the output upper corner frequency. What is the overall upper corner frequency? Use the capacitance values from the circuit simulation.

**1.32** In the circuit of Fig. 1.11, a resistance is placed in series with the input signal generator. Calculate the value of this resistor to cause the input upper corner frequency to equal the output upper corner frequency. Assume a zero value for load capacitance. Use the capacitance values from the circuit simulation. What is the overall upper corner frequency?

**D1.33*** In the circuit shown, the current source has infinite impedance and can be adjusted to supply whatever current is required to keep $M1$ in the active region as $V_{GS}$ is changed. If $\lambda = 0.03/\text{V}$, $V_T = 1.02$ V, and $\mu C_{ox}W/2L = 0.2\ \text{mA/V}^2$,

**(a)** Calculate $A_{MB}$ when $V_{GS} = 2$ V.

**(b)** Find the value of $V_{GS}$ required to double this voltage gain. Assume that $V_{DS} = 4$ V for both cases.

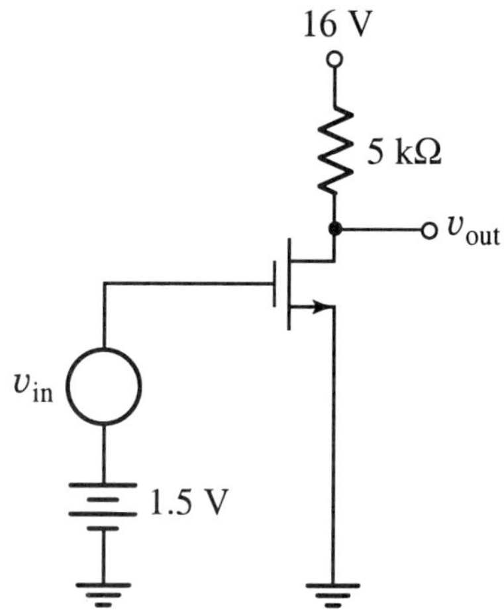


**Figure P1.24**

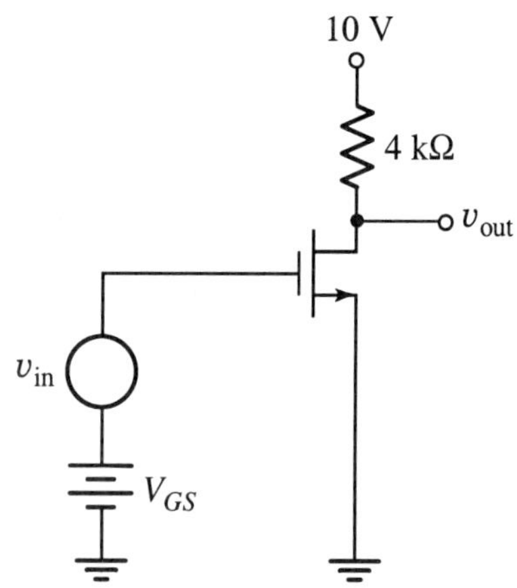


**Figure P1.26**

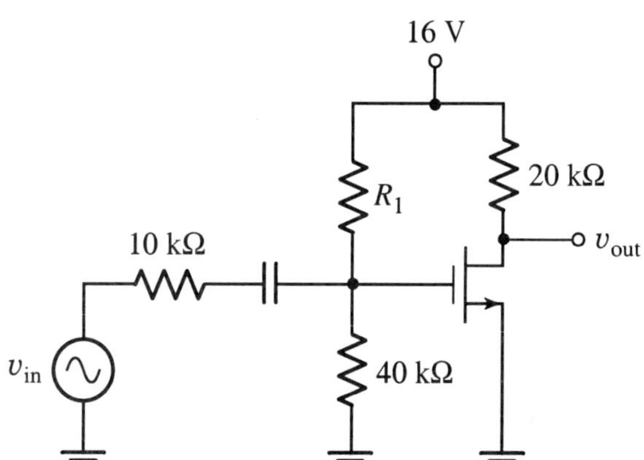


**Figure P1.28**

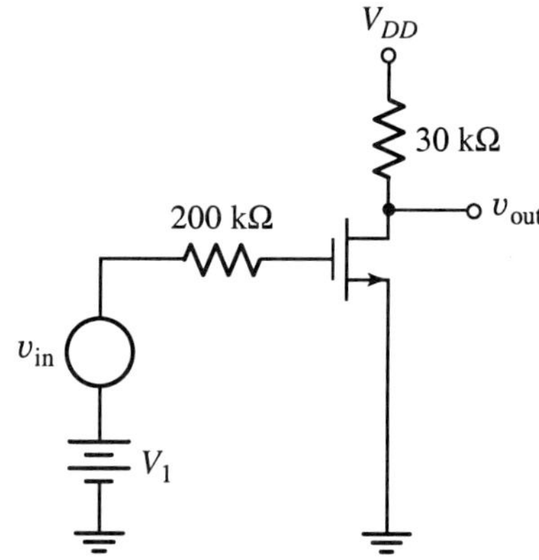


**Figure P1.29**

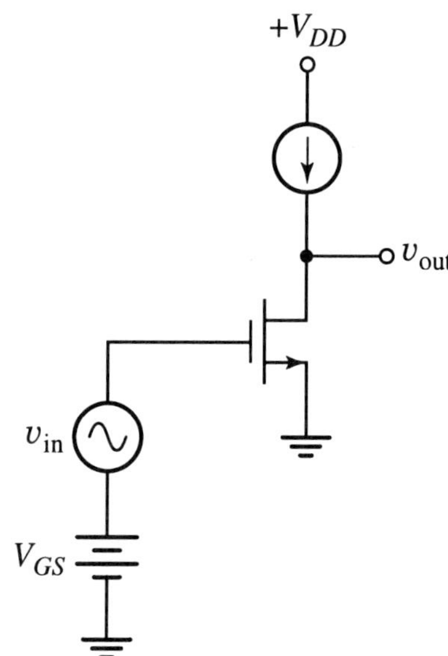


**Figure P1.33**

**D1.34** If the current source in Problem 1.33 is fixed at a value of 100 $\mu$A, what must $V_{GS}$ be to result in $V_{DS} = 4$ V? What is the value of $A_{MB}$ for this bias?

**1.35** Rework Example 1.5 for a required midband voltage gain of -76 V/V.

**1.36** In the circuit of Example 1.5, the drain current is lowered to 50 $\mu$A. Approximate the new value of midband voltage gain. Explain this approximation.

**D1.37** Using devices with $L = 1\ \mu$, $|V_T| = 1$ V, $\lambda = 0.02$/V, and $\mu C_{ox} = 0.06$ mA/V$^2$, design an active load stage with a simple current mirror load to have a midband gain of -120 V/V.

**Section 1.4.2 Source Follower with Active Load**

**1.38** Rework Example 1.5 if all parameters remain the same except $g_m$, which becomes $g_{m1} = 300$ mA/V.

**1.39** Find the overall upper corner frequency of the source follower of Example 1.5 if a load capacitance of 0.6 pF is added.

**1.40*** Find the overall upper corner frequency of the source follower of Example 1.5 if a load capacitance of 0.6 pF is added and a generator resistance of 100 k$\Omega$ is present. Calculate the midband voltage gain.

**Section 1.4.3 The Cascode Connection**

**1.41** In Example 1.6, using the cascode circuit of Fig. 1.21, the current from the mirror is increased from 50 $\mu$A to 100 $\mu$A. What would you now expect the midband voltage gain to be? Explain your calculations.

**Section 1.5 The BJT**

**D1.42** Select $R_1$ for the emitter-bias circuit of Fig. 1.23 with $R_C = 8.2$ k$\Omega$, $R_E = 1$ k$\Omega$, $R_2 = 20$ k$\Omega$, and $V_{CC} = 12$ V. Assume that $\beta = 100$, and $V_{BE} = 0.7$ V. Select $R_1$ to set $V_{CQ}$ at the midpoint of the active region. Find the maximum symmetrical peak-to-peak output voltage that can be obtained before saturation or cutoff occurs.

**D1.43** The emitter-bias circuit of Fig. 1.23 has $R_C = 8.2$ kΩ, $R_E = 1$ kΩ, $V_{CC} = 20$ V, $\beta = 100$, and $V_{BE} = 0.7$ V. Choose reasonable values of $R_1$ and $R_2$ to set $V_{CQ}$ to the midpoint of the active region.

**D1.44** An emitter-bias circuit (see Fig. 1.23) having $R_C = 6.8$ kΩ, $V_{CC} = 12$ V, and $R_E = 750$ Ω is to be capable of delivering a sinusoidal output signal of 3-V amplitude. Given that a BJT is used with $\beta = 60$ and $V_{BE} = 0.7$ V, find the upper and lower limits on $V_{CQ}$ to satisfy the design. Choose reasonable values of $R_1$ and $R_2$ that will result in the lower of these two values.

**1.45** Calculate the midband voltage gain of the circuit shown if $\alpha = 0.99$ and $I_E = 2$ mA. Calculate the input impedance of the stage. What will the maximum peak-to-peak output signal be when cutoff is reached?

**1.46** Repeat Problem 1.45 if $I_E$ is changed to 1 mA.

**1.47** If $\beta = 100$, calculate the midband voltage gain and sketch the output voltage including the dc component for the circuit shown.

**1.48** If $V_{BE(on)} = 0.6$ V, calculate the midband voltage gain for the circuit shown.

### Section 1.6.1 The Current Mirror Load

**1.49** For the circuit shown, $\beta_1 = 200$, $\beta_2 = 100$, $|V_{A1}| = 60$ V, $|V_{A2}| = 40$ V, and $V_{BE1} = -V_{BE2} = -V_{BE3} = 0.7$ V. Assume that the quiescent output voltage is 5 V. Calculate the midband voltage gain of the circuit.

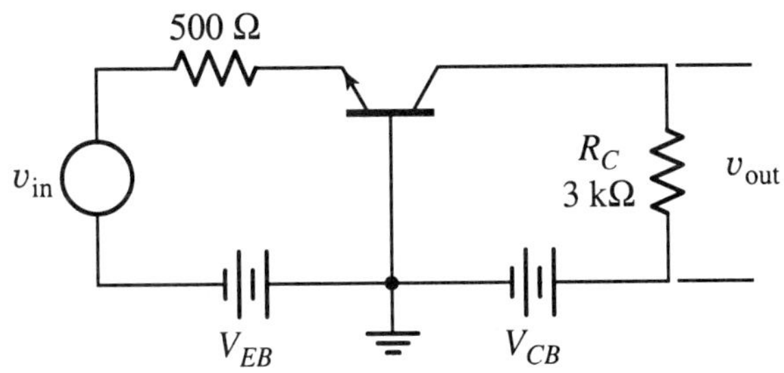


**Figure P1.45**

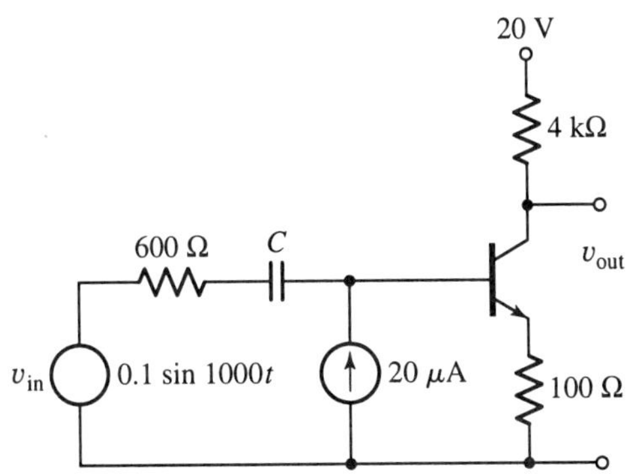


**Figure P1.47**

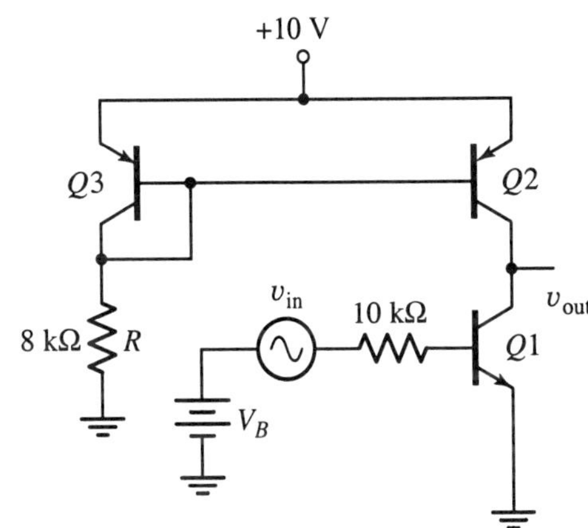


**Figure P1.49**

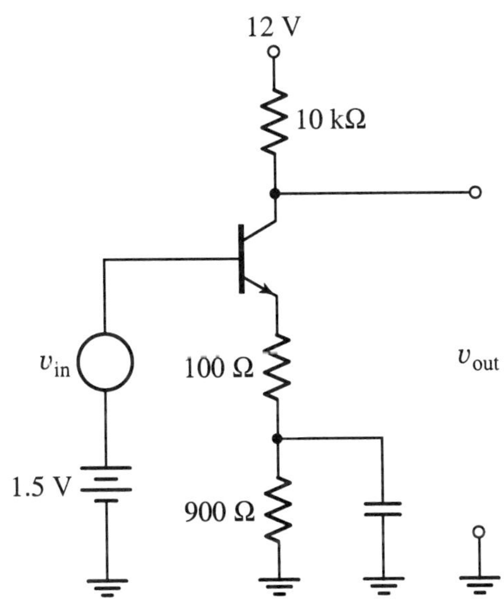


**Figure P1.48**

**1.50** If $R$ of the current mirror in Problem 1.49 is changed to 20 kΩ and $V_B$ is adjusted to keep $V_{CQ1} = 5$ V, calculate the midband voltage gain of the circuit.

**1.51** If the simple current mirror of Problem 1.49 is replaced by a Wilson current mirror with an output impedance of 420 kΩ, calculate the midband voltage gain of the stage. Assume that the output current of the mirror remains equal to the value in Problem 1.49.

**1.52** In the circuit of Problem 1.49 at the bias point used, the capacitances of $Q1$ are $C_\pi = 20$ pF, $C_{\mu 1} = 2$ pF, $C_{\mu 2} = 4$ pF, $C_{cs1} = 3$ pF, and $C_{cs2} = 3.5$ pF. Calculate the upper corner frequency of the circuit.

**1.53** Repeat Problem 1.52 if the source resistance is changed to 1 kΩ. Assume that $r_x = 50\ \Omega$.

**D1.54*** In Problem 1.52, how large can a load capacitance be to result in a 10% reduction in bandwidth compared to the case of no load capacitance?

### Subsection 1.6.2 The Emitter Follower

**1.55** The three transistors of Fig. 1.28 are identical, with $V_{BE} = 0.68$ V, $\beta = 180$, $r_x = 100\Omega$, $r_\pi = 2.4\ k\Omega$, and $r_{out} = 40$ kΩ. If a 100-kΩ generator resistance is inserted in series with $v_{in}$, keeping the bias current constant, what is the midband voltage gain?

**1.56** In Problem 1.55, the capacitors $C_\mu = 3$ pF, $C_{\pi 1} = 30$ pF, and $C_{out2} = 5$ pF. Calculate the upper corner frequency of the voltage gain.

### Section 1.6.3 The Cascode Amplifier Stage

**1.57** In Fig. 1.33, the resistor $R1$ is changed from 20 kΩ to 10 kΩ. Calculate the new midband voltage gain. Assume that the output resistance of the current source, $r_{cs}$, changes from 52.2 kΩ to 38 kΩ.

# Chapter 2

# FUNDAMENTAL POWER AMPLIFIER STAGES

Many discrete and integrated amplifiers are intended for low-power applications; these circuits are used primarily for voltage amplification. Although power is also amplified by these voltage amplifiers, typical output power levels fall in the milliwatt to hundreds of milliwatt range.

Many requirements have arisen for amplifiers that produce higher levels of output power. Medium-power amplifiers, having from 1 to 10 W of output power are common in radio, portable CD players, and tape recorder circuits. High-power stages may deliver from 20 to 300 W to a load in the case of public address systems or musical instrument amplifiers. Of course, many stereo systems, whether mobile or stationary, may drive a speaker with 200 to 300 W of power.

Low-power amplifiers can be constructed from discrete transistor stages or from op amp chips. Often such stages are also used prior to the power amplifying stages in a power amplifier. Medium-power amplifiers can also be made from discrete elements or integrated circuit chips. Integrated circuit chips that include an entire amplifier with a 5- to 10-W power output are available. Another option allows a voltage amplifier to be constructed with op amps followed by a power buffer chip. A third alternative is the hybrid circuit that implements an amplifier on a chip, which is then used to drive discrete transistor output stages. These amplifiers can provide power approaching the 100-W level. For high-power amplifiers, either this hybrid approach or the discrete transistor amplifier must be used. The physical volume of the integrated circuit is simply too small to dissipate the thermal energy associated with these high-power amplifiers. When power devices are constructed to be sufficiently large to conduct the necessary thermal energy, the size benefit of the integrated circuit cannot be realized.

This chapter will consider fundamental amplifiers and power output stages that can be used for medium- and high-power output circuits. The next chapter will discuss advanced power stages that offer higher efficiencies in the same range of power outputs.

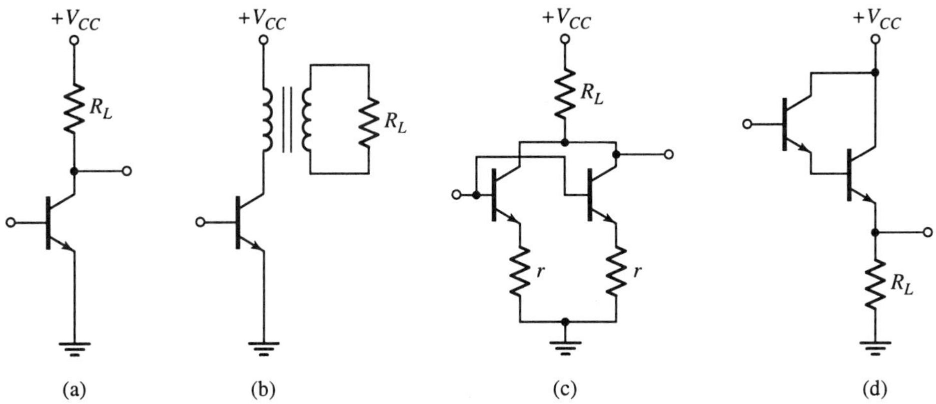


Figure 2.1: Class-A stages: (a) resistive load; (b) transformer load; (c) parallel output; (d) Darlington connection.

## 2.1 CLASSIFICATION OF POWER STAGES

**Important Concepts**

**1.** Power stages are traditionally classified as class-A, class-B, or class-C stages. This classification is based on the portion of the waveform period over which the stage conducts output current.

**2.** Power stages can have resistive loads or can use transformers to couple the output signal to the load.

**3.** Parallel stages can be used to produce higher levels of output current or power.

Several types of power output stages can be used in amplifiers. In general, the first few stages of a power amplifier will make up a voltage amplification section while the last few stages are used primarily to boost the power level. We will consider only the last stage of the amplifier because this stage is so important in developing output power.

A class-A stage conducts output current continuously during the complete cycle of a periodic input signal. This stage operates as a linear stage at all times; that is, the output signal developed is always directly proportional to the input signal. Later, we shall see that this stage, though less complex than the class-B or class-C stage, is less efficient and delivers less power to the load under comparable conditions than do the class-B and class-C stages. Examples of class-A stages are shown in Fig. 2.1.

The transformer-coupled stage can be more efficient than the resistive stage in delivering power to the load. The parallel stage must have a small resistance, $r$, placed in each emitter lead to ensure that the current of the load is distributed equally between the two transistors. If too much current were to flow through one of these transistors, the added drop across $r$ would decrease the forward bias to limit the current flow. More than two stages can be paralleled in very high-current applications.

Class-B operation requires two output stages. One stage conducts output current only when the input signal moves in the positive direction from its zero level. The other stage conducts current only when the input signal moves in the negative direction from the zero level. The currents from both stages are summed and used to create a voltage that is an amplified version of the input signal. For a sinusoid or periodic input signal with half-wave symmetry, one stage conducts for exactly one-half cycle and the other stage conducts during the alternate half-cycle. It is sometimes advantageous to cause each stage to conduct more than one-half cycle, but less than the full cycle. This mode of operation is called class-AB. Class-B operation is relatively efficient and allows high levels of power output. Figure 2.2 shows two examples of class-B stages.

Class-C operation occurs when a stage conducts during less than a half-cycle of the periodic input signal.

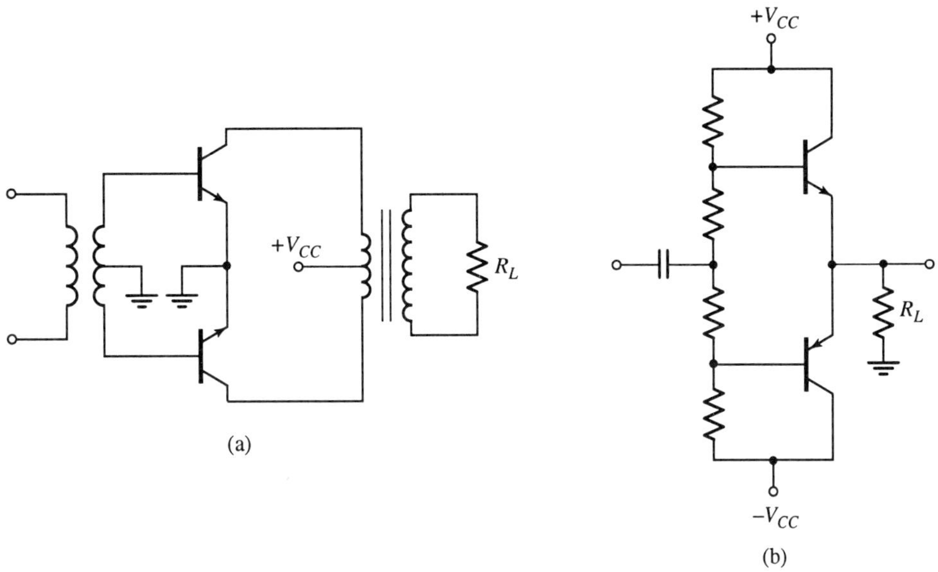


Figure 2.2: Class-B stages: (a) class-B push-pull; (b) complementary emitter followers.

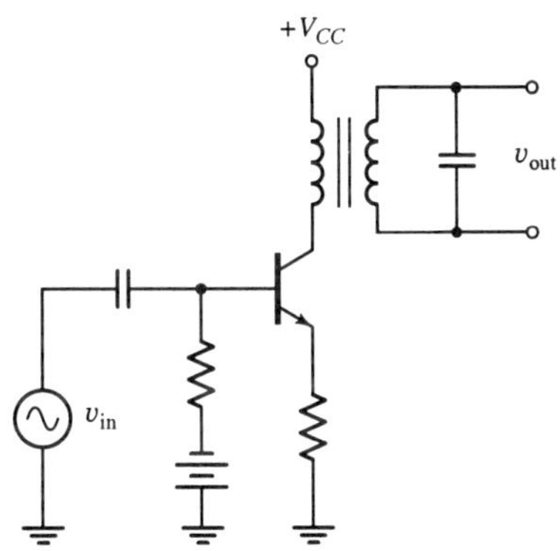


Figure 2.3: Class-C stage.

This type of stage is generally only used to amplify a single frequency or a narrow band of frequencies. The current conducted by the output stage flows through a resonant circuit and excites an output voltage at the resonant frequency. Although the output current is nonsinusoidal, the output voltage is a sinusoid. Class-C operation can be very efficient, approaching a maximum theoretical efficiency of 100%. This circuit can only amplify the signals that fall within the passband range of frequencies of the tuned circuit. The class-C stage is quite useful in narrowband applications such as radio frequency production and transmission. It cannot be used in wideband applications such as high-fidelity audio amplifiers. A class-C amplifier is shown in Fig. 2.3. In this circuit the transformer inductance serves as the resonating inductance.

In addition to the various types of amplifier previously mentioned, there is a two-state or switching amplifier that achieves amplification by rapidly switching an inductive load between a positive and negative voltage. The inductance acts as a low-pass element and rejects high-frequency components of the rectangular waveform while passing the average value. This amplifier uses a pulse-width modulating stage that creates a rectangular signal with a duty cycle that varies in accordance with the input signal to be amplified. The load of the output stage then demodulates this signal, reproducing the input signal at a higher power level. It will later be shown that a transistor dissipates little power if saturated or cut off. The power output stage of the two-state amplifier can deliver a great deal of power to the load while experiencing little power loss by continually operating in either the cutoff or saturation region. A theoretical efficiency of 100% can be obtained just as in the case of a class-C amplifier. However, the two-state amplifier can be used for a broader range of audio signals and is not limited to narrowband frequency operation as is the class-C stage. This stage is often called a class-D amplifier. Practical efficiencies of 95% have been reported [2].

Integrated circuit amplifiers delivering tens of watts are often built in this configuration. Although some audio applications use class-D to achieve very large output powers with little dissipation of the output stage, for example, 200 W of power to the load with only 10 W of output stage dissipation, there are also disadvantages in applying this class. The high switching speeds required in the class-D stage typically generate noise spikes that can interfere with adjacent circuit operation. In addition, the harmonic distortion of a well-designed class-B stage is superior to that of the class-D stage [2].

This chapter will discuss the class-A, class-B, and class-C amplifier stages and the following chapter will consider some modifications to the class-B stage as well as the class-D stage and the popular class-G stage [4]. This latter stage is built around the premise that voice and music generate average power levels that are much lower than the occasional instantaneous peak power levels. Output stage dissipation can be minimized by using a low-voltage power supply to handle the average power levels and providing higher-voltage supplies only when the instantaneous power is large.

When a transistor is used as a power output stage, it is the external circuit that determines the amount of power actually dissipated by the transistor. The later sections of this chapter will deal with determining the actual power output of the transistor when employed in various configurations. Before one can choose the proper transistor for a given output, however, one must know the allowable dissipation of the device. The allowable dissipation of a transistor is a function of the construction of the device and the thermal properties of the surrounding medium. The following section will discuss the practical calculation of the allowable device dissipation. Once the allowable dissipation is known, the circuit must be designed to ensure that the actual dissipation never exceeds this allowable dissipation.

## 2.2 ALLOWABLE DISSIPATION

**Important Concepts**

**1.** The construction of the transistor along with the thermal environment determines the amount of power that the transistor can safely dissipate.

**2.** The maximum junction temperature is a major factor that limits the allowable transistor dissipation.

Several factors limit the power that the transistor can safely dissipate. For example, there is a maximum value of the collector-to-base breakdown voltage. There is also a maximum value of current that can flow, limited by cross-sectional area of bonded leads and physical dimensions of the transistor. Both of these factors indirectly influence the maximum power dissipation of the transistor.

One of the most significant limitations on dissipation is the junction temperature. As the temperature rises, several potentially dangerous effects may occur. First, the solder or alloys used in the transistor can be softened or even melted. Second, the impurity profiles in the doped regions can be affected if elevated temperatures exist for long periods of time. A third result of higher temperatures is the increase in collector leakage current. In power transistors, this current doubles for an incremental increase in temperature of 7 to 10°C. The leakage current, $I_{co}$, at temperature $T_2$ can be related to $I_{co}$ at temperature $T_1$ by

$$I_{co}(T_2) = I_{co}(T_1) \times 2^{(T_2-T_1)/T_K}$$

where $T_K$ can range from 7°C to 10°C. For a temperature increase of 100°C, the minimum factor of increase in $I_{co}$ is 1024 ($2^{100/10}$), whereas the maximum factor is approximately 20,000 ($2^{100/7}$). This marked increase in $I_{co}$ can lead to increased power dissipation which leads to an increased temperature followed by a further increase in $I_{co}$. In some cases, this feedback effect is large enough to cause thermal runaway and destroy the transistor.

This effect is minimized by placing a resistance in the emitter to decrease the forward bias on the base-emitter junction as current increases.

In other situations, the leakage current can approach the value of the quiescent collector current. Since leakage current is not controlled by the applied input signal, its effects can severely limit the amplifying properties of the stage. For these reasons, the manufacturer places a limit on the maximum allowable junction temperature of the device.

The maximum junction temperature is therefore an important quantity that limits the power a transistor can deliver. The junction temperature will be determined by the power being dissipated by the transistor, the thermal conductivity of the transistor case, and the heat sink that is being used. The collector junction is the point at which most power is dissipated; hence, it is this junction that concerns us here.

Basically, manufacturers specify the allowable dissipation of a transistor in two ways. One way is to specify the maximum junction temperature along with the thermal resistance between the collector junction and the exterior of the case. Although this method is very straightforward, it incorrectly implies that the allowable power increases indefinitely as the transistor is cooled to lower temperatures. Actually, there is a maximum limit on the allowable dissipation of the transistor which is reflected by the second method of specification. This method shows a plot of allowable power dissipation versus the temperature of the mounting base. Quite often this plot shows power dissipation versus ambient temperature, where an infinite heat sink is used. However, if the transistor could be mounted on an infinite heat sink, the ambient temperature would equal the mounting base temperature; thus both plots convey the same information. This second method indicates the maximum allowable power dissipation, in addition to the maximum junction temperature.

The maximum limit on power dissipation ensures that chip temperature differentials do not become excessive as excess power is dissipated in collector regions. It also minimizes the possibility of excessive collector currents in typical applications.

### 2.2.1 Thermal Resistance

The thermal resistance of a material is defined as the ratio of the temperature difference across the material to the power flow through the material. This assumes that the temperature gradient is linear throughout. The symbol $\theta$ is used for thermal resistance, and

$$\theta = \frac{\Delta T}{P} = \frac{\text{Temperature difference across conductor}}{\text{Power flowing through conductor}} \tag{2.1}$$

An analogy can be made between the thermal circuit and electrical circuit. If temperature corresponds to voltage, power flow to current flow, and thermal resistance to electrical resistance, Eq. (2.1) is a statement of Ohm's law. The analogy does not end here, since it is obvious that $\theta$ will be increased if the material is made longer and will be decreased if the cross-sectional area is made larger. We could then write $\theta$ as

$$\theta = \rho \frac{L}{A}$$

where $\rho$ is the thermal resistivity, $L$ is the length of the path, and $A$ is the cross-sectional area perpendicular to the power flow.

Let us consider the thermal circuit shown in Fig. 2.4 of a transistor surrounded by free air. Here, $\theta_{JM}$ is the thermal resistance from collector to mounting base and $\theta_A$ is the thermal resistance of that portion of air in contact with the mounting base; $T_A$ is the temperature of air far away from the transistor; $T_M$ is the mounting base temperature; and $T_J$ is the collector junction temperature. The power $P$ will be determined by the electrical circuit that includes the transistor; $P$, in turn, will determine the temperatures $T_J$ and $T_M$. The temperature $T_J$ can be written as

$$T_J = T_A + P(\theta_{JM} + \theta_A) \tag{2.2}$$

Again the analogy to the electrical case is noted. The voltage on one side of a resistance is equal to the voltage on the other side plus the $IR$-rise across the resistance. Equation (2.2) shows that as more power is

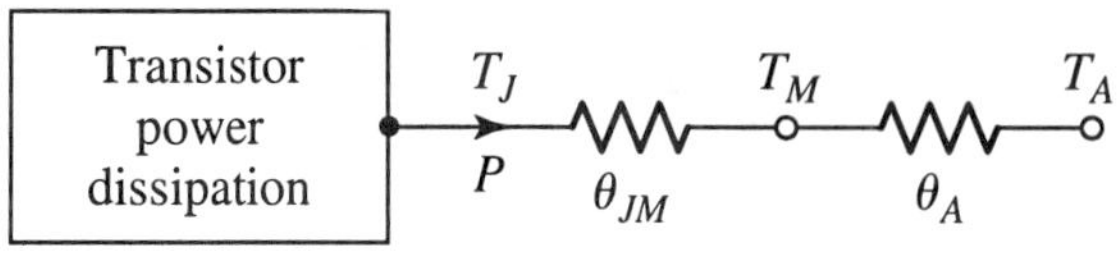


Figure 2.4: Transistor dissipating power in free air.

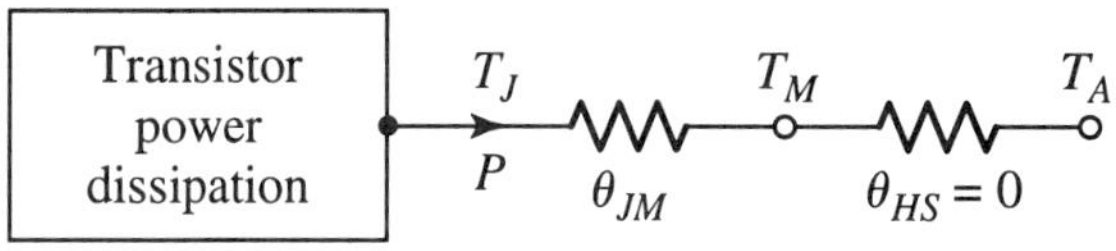


Figure 2.5: Transistor mounted on infinite heat sink.

dissipated by the transistor, $T_J$ must rise.

For high-power transistors $\theta_A$ is usually many times greater than $\theta_{JM}$. If $\theta_{JM}$ were 1°C/W, then $\theta_A$ might be 5 to 20°C/W. Of course, $\theta_A$ depends on the area of the transistor case.

We next consider the circuit of Fig. 2.5 that shows the transistor mounted on an infinite heat sink.

The heat-sink temperature will equal the ambient temperature and, since an infinite power can be absorbed by the heat sink without raising its temperature, $T_M$ will also equal $T_A$. The temperature $T_J$ can now be written as

$$T_J = T_A + P\theta_{JM} \tag{2.3}$$

For a given power dissipation, $T_J$ will be much less when an infinite heat sink is used than when no heat sink is used. In practice, an infinite heat sink is not obtainable. The values of $\theta_{HS}$ presented between the transistor case and free air might range from 0.4°C/W for air-cooled systems to 2°C/W for flat vertical-finned aluminum heat sinks to 8°C/W for cylindrical heat sinks that slide over the transistor. The flat vertical-finned heat sinks are the most commonly used. The manufacturer of the heat sink will specify the thermal resistance of that particular heat sink. Often a long heat sink will be used to accommodate several transistors on the same sink. When this is done, two points should be remembered.

**1.** Appropriate space should be allotted for each transistor to allow the heat to be conducted away from the transistor properly. The manufacturer will usually state the distance to be allowed between adjacent transistors.

**2.** Since the collector is in contact with the case in power transistors, the case is always at the same electrical potential as the collector. The heat sink is conductive and therefore an insulating washer must be used since it has a high electrical resistance and a low thermal resistance.

There are two methods of calculating the allowable power dissipation of the transistor, as indicated in the following subsections.

### Practice Problems

**2.1** If $\theta_{JM} = 1.5$°C/W and $\theta_A = 6$°C/W and the transistor dissipates 8 $W$ of power, calculate the junction temperature if the ambient air temperature is 30°. *Ans:* 90°.

**2.2** If the maximum junction temperature of the device in Practice Problem 2.1 is 140°C, what is the maximum power that can be dissipated? *Ans:* 14.7 W.

### 2.2.2 Method 1: Specification of $T_{Jmax}$ and $\theta_{JM}$

The power flowing from the transistor can be expressed as

$$P_T = \frac{T_J - T_A}{\theta_{JM} + \theta_i + \theta_{HS}}$$

where $\theta_i$ is the thermal resistance of the insulating washer, if used. This equation results from solving Eq. (2.2) replacing $\theta_A$ by $\theta_i + \theta_{HS}$.

If the manufacturer specifies a maximum junction temperature and $\theta_{JM}$, the maximum power that can be conducted away from the transistor can be found by letting $T_J = T_{J\text{max}}$. This value is

$$P_{T\text{max}} = \frac{T_{J\text{max}} - T_A}{\theta_{JM} + \theta_i + \theta_{HS}} \tag{2.4}$$

If a higher power is dissipated by the transistor, the actual junction temperature, $T_J$, will exceed the allowed maximum value, $T_{J\text{max}}$.

# Example 2.1.

Given that $T_{J\text{max}} = 100°\text{C}$, $\theta_{JM} = 1°\text{C/W}$, $\theta_i = 0.5°\text{C/W}$ and $\theta_{HS} = 2.5°\text{C/W}$, find the maximum power that can be dissipated by the transistor when the highest ambient temperature is $T_A = 25°\text{C}$.

**Solution** Equation (2.4) gives

$$P_{T\text{max}} = \frac{100 - 25}{1.0 + 0.5 + 2.5} = 18.75\ \text{W}$$

If the transistor could be mounted on an infinite heat sink so that $\theta_i + \theta_{HS} = 0$, the amount of power that could be dissipated would be

$$P_{Tmax} = \frac{100 - 25}{1.0} = 75\ \text{W}$$

It is interesting to compare these two figures. For a practical heat sink, the transistor can handle 18.75 W; for an infinite heat sink, the transistor is capable of 75 W. The manufacturer will specify the transistor as a 75-W device. However, this figure cannot be reached, although if one uses expensive forced air or liquid cooling, the figure might be approached.

### 2.2.3 Method 2: Graphical Determination of Allowable Dissipation

Most manufacturers of power transistors supply a graph of allowable power dissipation versus transistor mounting-base temperature, as shown in Fig. 2.6. This graph is sometimes called allowable dissipation versus ambient temperature using an infinite heat sink. If this is the case, $\theta_i + \theta_{HS} = 0$, and $T_A = T_M$. It can be seen from Eq. (2.4) that when the mounting-base temperature equals the maximum junction temperature, no power can be allowed to flow. From Fig. 2.6, the intersection of the curve with the $P = 0$ line occurs at $T_{J\text{max}}$. One would expect the graph to be continuous, but the manufacturer limits the maximum power of the device by leveling the curve for all values below $T_1$, which is usually 25°C for germanium and from 25 to 50°C for silicon.

We can evaluate $\theta_{JM}$ from the curve, since

$$\theta_{JM} = \frac{T_J - T_M}{P}$$

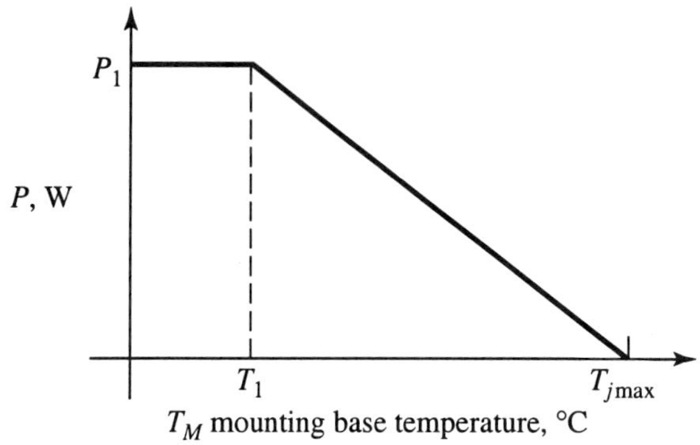


Figure 2.6: Allowable dissipation vs. $T_M$.

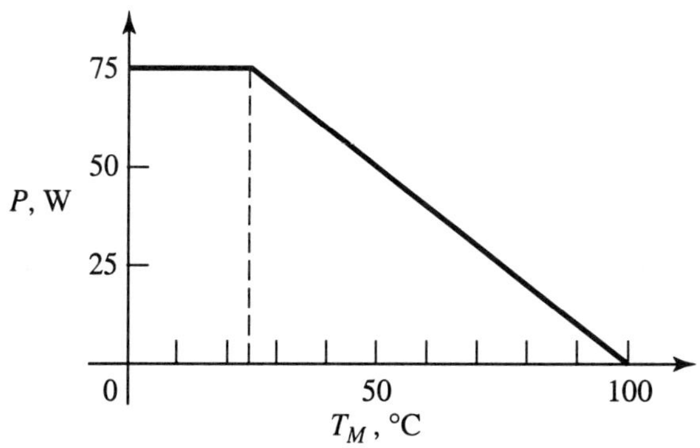


Figure 2.7: Power dissipation curve.

The curve gives the maximum power that can be dissipated at any $T_M$; hence, $T_J$ must be the maximum junction temperature. The equation becomes

$$\theta_{JM} = \frac{T_{J\max} - T_M}{P}$$

and any $P$ and corresponding $T_M$ falling on the linear portion of the curve can be used. Quite often, the point $(P_1,\ T_1)$ will be used for convenience, giving

$$\theta_{JM} = \frac{T_{Jmax} - T_1}{P_1} \tag{2.5}$$

# Example 2.2.

If $\theta_i = 0.5°\text{C/W}$ and $\theta_{HS} = 2.5°\text{C/W}$, and the maximum ambient temperature is to be 25°C, find the allowable dissipation of a transistor whose thermal characteristics are those given in Fig. 2.7.

**Solution** If $\theta_{JM}$ is evaluated along with $T_{J\max}$, the first method can again be used. From the curve we see that $T_{J\max} = 100°C$. From Eq. (2.5) and the curve we can write

$$\theta_{JM} = \frac{100 - 25}{75} = 1°\text{C/W}$$

Equation (2.4) can now be solved to give

$$P_{T\max} = 18.75\ \text{W}$$

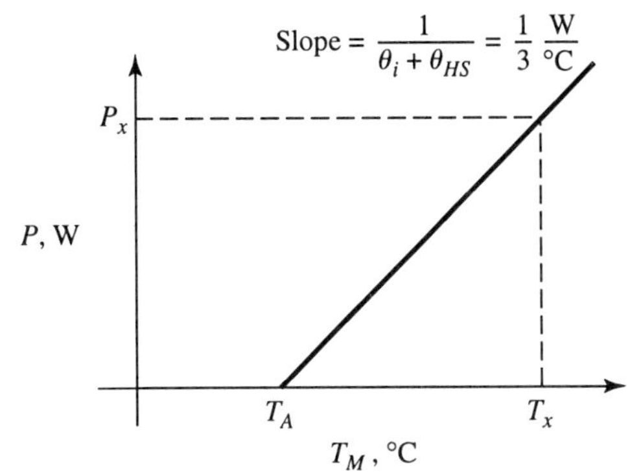


Figure 2.8: Power conducted through $(\theta_i + \theta_{HS})$ vs. $T_M$.

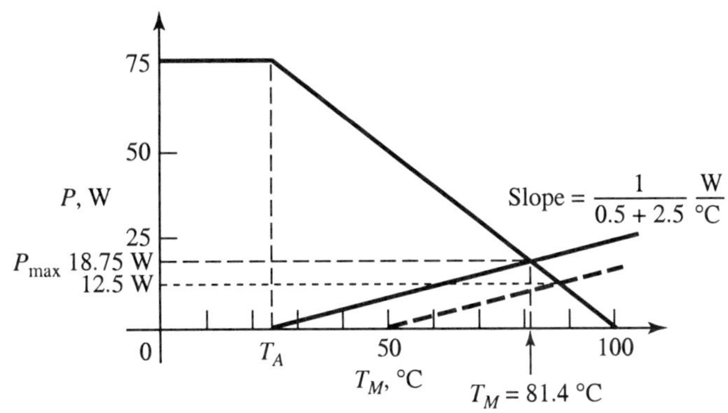


Figure 2.9: Graphical solution for $P_{T\max}$.

---

This same answer can also be obtained graphically. Consider a plot of $(\theta_i + \theta_{HS})$ versus $T_M$, as shown in Fig. 2.8. Since one side of the insulating washer is in contact with the transistor case, the temperature at that point will be $T_M$. The thermal path then travels through the washer and the heat sink to the air at ambient temperature. The slope of the curve in Fig. 2.8 is equal to the reciprocal of $(\theta_i + \theta_{HS})$.

When the thermal circuit of the transistor is in equilibrium, the power flowing through $\theta_{JM}$ must equal the power flowing through $(\theta_i + \theta_{HS})$. At some particular $T_M$, the curves of Figs. 2.7 and 2.8 will show the same power flow. This is the simultaneous solution of the equations represented by the two graphs. Figure 2.9 shows this solution, obtained by superimposing the two graphs. The point of intersection is the only point which will cause the same power to flow through $\theta_{JM}$ as that flowing through $(\theta_i + \theta_{HS})$. The graphical technique is a very simple one to apply. The fact that $T_A$ has the same value as $T_1$ in this example is merely a coincidence. Another problem might give all the same parameters, but this circuit is to work in an ambient temperature of 50°C. This condition is shown by the dotted line on Fig. 2.9, where it can be seen that only 12.5 W could be dissipated in this case.

**Practice Problems**

**2.3** Rework Example 2.1 if $T_{J\max} = 160°\text{C}$. *Ans:* 33.75 W.

**2.4** Rework Example 2.2 if $T_{J\max} = 160°\text{C}$ and $T_1 = 40°\text{C}$. Note that $\theta_{JM}$ is different from the value

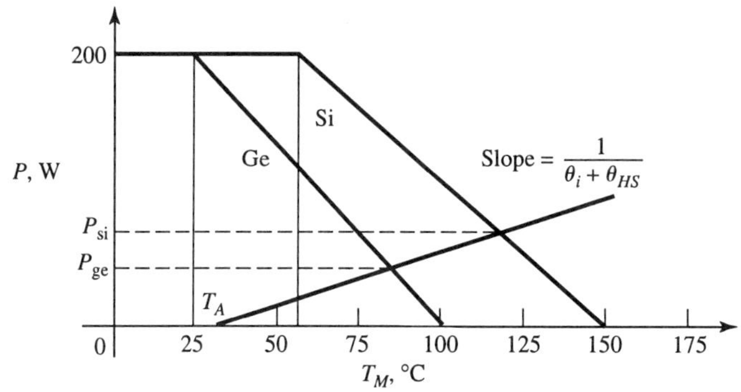


Figure 2.10: Comparison between silicon and germanium.

in Example 2.1. *Ans:* 32.1 W.

### 2.2.4 Power Transistors

The maximum junction temperature of a silicon transistor is greater than that of a germanium transistor. The power area is one in which the germanium transistor is still used. For silicon $T_{J\max}$ will usually fall in the range of 150°C to 200°C. The corresponding value for germanium is normally near 100°C. In addition, $T_1$ is often higher for silicon. Figure 2.10 shows a comparison between silicon and germanium for a 200-W transistor. It should also be apparent that as $T_A$ is lowered, the allowable dissipation increases. This demonstrates why forced air, liquid cooling, or a refrigeration unit improves the performance.

Quite often, if more power is needed, many output transistors can be paralleled to deliver the required power. This method is usually the cheapest way to obtain high-power outputs, since power transistor prices are now quite reasonable. New processes and improvements in old processes continue to extend the performance of power transistors. Many types are available for low-frequency work that are advertised as 200-W transistors. Work is being done at the high-frequency end of the spectrum, and transistors that work up to frequencies above 100 MHz can deliver over 100 W of power. However, at present these units are quite expensive.

For switching applications, the capacitance of the thermal circuit must be considered. This subject will not be covered here. In cases where the transistor switches through a region of excessive dissipation, the thermal capacitance is an important consideration. For class-A, class-B, and class-C amplifiers operating at frequencies above 20 Hz, only the resistance of the thermal circuit need be considered.

### 2.2.5 Power Dissipation Review

Before proceeding to the calculation of actual transistor dissipation, it is appropriate to review certain aspects of power dissipation in electronic circuits.

1. **We are normally interested only in average dissipation.**

   The thermal inertia of physical devices such as semiconductors or heat sinks is such that the effect on temperature is a function of average dissipation. Instantaneous power dissipation generally changes so rapidly that the thermal circuit is only affected by the average value of this instantaneous power. Instantaneous power is used only to calculate average power.

**2. The product of voltage and current at any time gives the instantaneous power at that time.**

This product would be used primarily to find average power.

**3. If voltage and current are constant values, the average power equals the instantaneous power.**

If $V = 3$ V and $I = 30$ mA, the instantaneous power dissipation is

$$p = 3 \times 30 = 90 \text{ mW}$$

Since this power is constant at all times, the average power dissipation is also $P = 90$ mW. This situation may correspond to a calculation of quiescent power dissipation of a transistor.

**4. If both V and I are periodic with the same period, the average power is calculated by integrating the instantaneous power over one cycle and dividing by the period.**

Since integration is more complex than multiplication, electrical engineers have invented a better way to calculate the average power using only multiplication. The root-mean-square or RMS value of a voltage or a current can be used to avoid repeating the integration. The average power for waveforms having an RMS value is given by $P = V_{RMS} \times I_{RMS}$. This formula is based on the integration of instantaneous power over a cycle and the creation of a *fictitious* quantity called the RMS value that forces the product of the two RMS values to result in the same value derived from integration. Use of RMS values to calculate average power dissipation is popular for sinusoidal or other periodic waveforms having no dc component. We note that the average power is *not* equal to the product of the average values of the voltage and current waveform in this case.

This is one of the few times RMS values are used. Note that the product of RMS voltage and RMS current is *average* power–not RMS power. We rarely encounter a situation in power amplifiers where both current and voltage are pure ac signals.

**5. If voltage is constant (dc), but current is not, the average dissipation can be calculated as $P = V \times I_{AV}$.**

Suppose the current drawn from a 12-V power supply is

$$i = 40 + 10 \sin \omega t \text{ mA}$$

The instantaneous power varies as

$$p = 12 \times (40 + 10 \sin \omega t) = 480 + 120 \sin \omega t \text{ mW}$$

The average dissipation can be found by averaging this instantaneous power over one complete cycle of the sinusoid. However, a sinusoid has zero average value for any integer multiple of periods, so the average power is $P = 480$ mW. A key point is that this value can be easily calculated from

$$P = V_{DC} \times I_{AV}$$

In this example, $I_{AV}$ is just 40 mA. This may correspond to a calculation of source power delivered to the collector of a transistor. Collector current is often the sum of the quiescent current and the sinusoidal signal current. Thus, it has a dc component and an ac component. The average current equals the dc component.

**6. Both voltage and current have dc and ac components.**

For example, if $v(t) = V_{CQ} - V \sin \omega t$ and $i(t) = I_{CQ} + I \sin \omega t$, which is the case for the collector of an output stage, the power is calculated by

$$P = V_{CQ} \times I_{CQ} - \frac{VI}{2}$$

where $V$ is the magnitude of the ac voltage and $I$ is the magnitude of the ac current. This result will be derived in the next section.

**Practice Problems**

**2.5** A 12-V power supply delivers a half-wave rectified current having a peak value of 500 mA. How much average power does this source deliver? Note that the average value of a half-wave rectified signal with peak value $I_p$ is $I_p/\pi$. *Ans:* 1.91 W.

**2.6** The voltage across a device is given by $v = V_{CQ} - V \sin \omega t = 6 - 2 \sin \omega t$ V. The current through the device is $i = I_{CQ} + I \sin \omega t = 0.6 + 0.2 \sin \omega t$ A. Calculate the average power dissipated by the device. *Hint:* Integrate instantaneous power over one cycle and divide by the period. *Ans:* 3.40 W.

## 2.3 CLASS-A POWER OUTPUT STAGES

**Important Concepts**

**1.** The class-A, resistive-load stage has a maximum circuit efficiency of 25% when used as an amplifier of a sinusoidal signal.

**2.** The transistor dissipates maximum power under quiescent conditions. As the AC output signal increases, transistor dissipation decreases.

**3.** A transformer can be inserted between the output transistor and the load to increase the maximum circuit efficiency to 50% for a sinusoidal output signal.

### 2.3.1 Resistive Stages

According to the maximum-power transfer theorem, a source will deliver maximum power to the load if the load resistance equals the source resistance. This theorem applies only to linear operation. In the case of a common-emitter stage as shown in Fig. 2.11, the output circuit can be represented by a current source and an impedance of relatively large value. For a class-A circuit the transistor must operate in the linear region and one might be tempted to apply the maximum-power transfer theorem to select the optimum load impedance. This would result in a load resistance of value $R_L = r_{out}$. If this resistance were used in a power stage that allows the collector current to reach its maximum value, the voltage developed across the load would tend to be very large. Of course, saturation would occur under these conditions and the maximum collector voltage would be greatly exceeded. Thus, the assumption of linear operation would be invalid. We cannot apply linear theory to this case because of the nonlinear constraints imposed on the problem by the transistor.

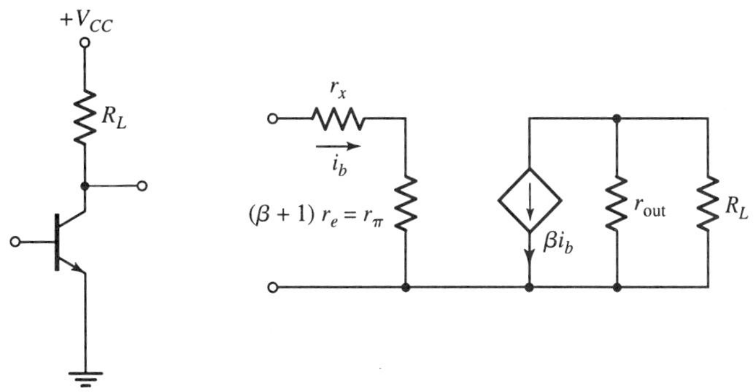


Figure 2.11: Common-emitter power stage.

**DC Power**

In the class-A stage of Fig. 2.11(a) to which the equivalent circuit of Fig. 2.11(b) applies, the maximum load current that can flow through the load equals the maximum collector-current rating of the transistor, $I_{\max}$. The maximum voltage across the load resistance is limited by the power supply to $V_{CC}$. If we are interested in delivering maximum dc power to the load, we must cause the maximum voltage across the load to occur simultaneously with maximum load current. We can select a load resistance to result in the satisfaction of these conditions if

$$R_{LM} = \frac{V_{CC} - V_{CE(\text{sat})}}{I_{\max}} \approx \frac{V_{CC}}{I_{\max}} \tag{2.6}$$

where $V_{CE(\text{sat})}$ is the saturation voltage of the transistor corresponding to a collector current of $I_{\max}$. Often $V_{CE(\text{sat})}$ will be small enough to be neglected, with little loss of accuracy.

The maximum dc power dissipated by $R_{LM}$ occurs when maximum current flows through this load resistance. This power is

$$P_R = (V_{CC} - V_{CE(\text{sat})})I_{\max}$$

which is independent of the output impedance of the transistor.

When the transistor is delivering maximum power to the load, very little power is dissipated by the transistor. For maximum load power, $I_C = I_{\max}$, and the transistor power is

$$P_T = I_{\max}V_{CE(\text{sat})} \approx 0$$

When $I_C = 0$, no power is being delivered to the load and the transistor voltage is maximum at $V_{CE} = V_{CC}$. The transistor dissipation is

$$P_T = 0 \times V_{CC} = 0$$

When $I_C = I_{\max}/2$, then

$$V_{CE} = \frac{V_{CC}}{2} + \frac{V_{CE(\text{sat})}}{2}$$

and the transistor dissipation is

$$P_T = \frac{I_{\max}}{2}\left(\frac{V_{CC} + V_{CE(\text{sat})}}{2}\right) = \frac{1}{4}(I_{\max}V_{CC} + I_{\max}V_{CE(\text{sat})})$$

If $V_{CE(\text{sat})}$ is assumed to be zero, maximum power is dissipated by the transistor when $I_C = I_{\max}/2$ and $V_{CE} = V_{CC}/2$. The similarity between this result and that predicted by the maximum-power transfer theorem

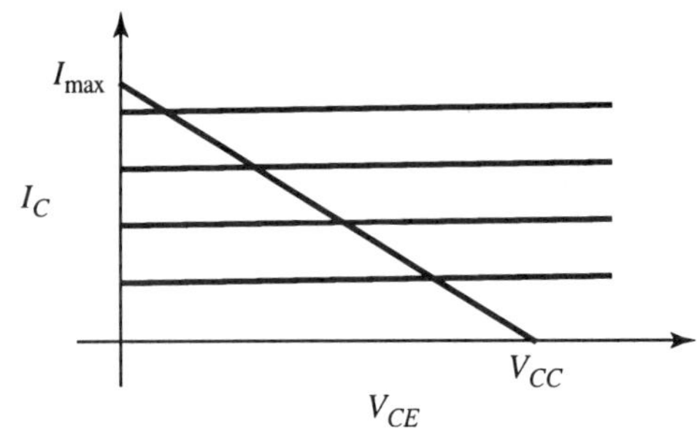


Figure 2.12: V - I characteristics and load line.

is readily apparent. That is, the dissipation of the transistor is maximum when one-half of the dc source voltage appears across it.

We conclude that the transistor delivers maximum power to the load when saturated, and that this power depends on the maximum allowable collector current. Furthermore, we note that the transistor dissipation is very low when the transistor is saturated. As $I_C$ decreases (reducing the power delivered to the load), $P_T$ increases to a maximum value, which occurs at $I_C = I_{\max}/2$.

### AC Power-Ideal Circuit

For this discussion, we will assume that the saturation voltage of the transistor is $V_{CE(\text{sat})} = 0$ V and that the only resistance in the circuit is that of the load. The ac considerations are slightly different from those of the dc case. With a given power supply voltage $V_{CC}$, the greatest peak-to-peak voltage that can be delivered to a resistive load is equal to $V_{CC}$, and the quiescent voltage must be chosen as $V_{CEQ} = V_{CC}/2$ for this output voltage to occur. Minimizing the value of $R_L$ will maximize the power delivered to the load. However, the minimum value of $R_L$ is determined by the collector current rating, as shown in Fig. 2.12.

The minimum value of $R_L$ that will not result in excessive collector current is

$$R_L = \frac{V_{CC}}{I_{\max}}$$

If the quiescent voltage is $V_{CEQ} = V_{CC}/2$, the collector current will be $I_{CQ} = I_{max}/2$, and the ac power that can be delivered to the load is

$$P_{ac} = \frac{V_{\text{peak}}^2}{2R_L} = \frac{V_{CC}^2}{8R_L} \tag{2.7}$$

The circuit efficiency of an amplifier is a measure of the effectiveness with which the stage converts the dc input power to ac output power. This efficiency is defined as

$$\eta = \frac{P_{\text{ac}}}{P_S} \times 100 \tag{2.8}$$

where $P_S$ is the power delivered by the dc power supply. Using this equation for the resistive load stage gives

$$\eta = \frac{V_{\text{peak}}^2/2R_L}{V_{CC}I_{CQ}} \times 100$$

Both ac power and source power depend on $I_{CQ}$. If $I_{CQ}$ is less than $I_{\max}/2$, the peak voltage swing that can occur before cutoff is $V_{\text{peak}} = I_{CQ}R_L$. Thus, the circuit efficiency becomes

$$\eta = \frac{I_{CQ}R_L}{2V_{CC}} \times 100 \text{ for } I_{CQ} \le \frac{I_{\max}}{2}$$

The efficiency is seen to increase as $I_{CQ}$ increases up to the point where $I_{CQ} = I_{\max}/2$. Above this value of current, the peak voltage swing is limited by saturation of the transistor. As $I_{CQ}$ increases above $I_{\max}/2$, the maximum possible peak voltage swing decreases and the input source power increases. Therefore we conclude that maximum efficiency occurs when $I_{CQ} = I_{\max}/2$. This value of efficiency is

$$\eta_{\max-\text{ideal}} = \frac{V_{CC}^2/8R_L}{V_{CC} \times \frac{I_{\max}}{2}} \times 100 \tag{2.9}$$

Since $I_{\max} = V_{CC}/R_L$, $\eta_{\max-\text{ideal}} = 25\%$ for the ideal class-A stage.

We note that the quiescent collector current resulting in maximum possible efficiency is the same current that allows maximum output ac power. Unfortunately, this current also results in maximum transistor dissipation, as we have seen earlier. The transistor dissipation with no input signal is

$$P_T = V_{CC}^2/4R_L \tag{2.10}$$

which is twice the ac power that can be delivered to the load.

# Example 2.3.

It is desired to drive a $10 - \Omega$ load using a power supply of 20 V. Select a transistor that will allow maximum efficiency to be obtained.

**Solution** The maximum possible ac power output is

$$P_{\text{ac}} = \frac{V_{CC}^2}{8R_L} = \frac{400}{80} = 5 \text{ W}$$

The transistor must dissipate twice this power under no signal conditions, so $P_T = 10$ W. The maximum collector current will be

$$I_{\max} = \frac{V_{CC}}{R_L} = \frac{20}{10} = 2 \text{ A}$$

The transistor selected must handle at least 2 A of collector current and dissipate at least 10 W of power.

The figure of 25% efficiency is never obtained in the practical case. The presence of a finite saturation voltage, $V_{CE(\text{sat})}$, and leakage current, $I_{co}$, other resistors in the collector-emitter loop, and the fact that distortion will be excessive if large voltage swings occur, limit the efficiency of the stage. Of course, an emitter resistance will be necessary in any class-A power output stage to prevent thermal runaway. If unbypassed by a capacitance, this emitter resistance will also reduce nonlinear distortion in the output due to negative feedback. Values of circuit efficiency ranging from 10 to 15% can typically be obtained, depending on the amount of allowable distortion and the resistance in the loop in addition to the load. A rule of thumb suggests that the emitter resistance used to avoid thermal runaway should drop about 0.5 V at the quiescent emitter current. This value is

$$R_E = \frac{0.5}{I_{EQ}} \approx \frac{0.5}{I_{CQ}}$$

### Transistor Dissipation-Ideal Circuit

The power dissipated by the transistor decreases as the ac output voltage increases. This can be seen by considering the product of collector current and collector voltage over a complete cycle. If the current is expressed as $i = I_{CQ} + I \sin \omega t$, the corresponding voltage will be $v = V_{CQ} - V \sin \omega t$. The voltage across the transistor decreases as the collector current increases, hence, the negative sign on the ac voltage. The transistor power is

$$P_T = \frac{1}{2\pi} \int_0^{2\pi} (V_{CQ} - V \sin \omega t)(I_{CQ} + I \sin \omega t)\, d(\omega t)$$

If $\omega t = \theta$,

$$\begin{aligned} P_T &= \frac{1}{2\pi}\left[\int_0^{2\pi} V_{CQ}I_{CQ}\,d\theta + \int_0^{2\pi} V_{CQ}I\,\sin\,\theta\,d\theta\right] \\ &- \frac{1}{2\pi}\left[\int_0^{2\pi} I_{CQ}V\,\sin\,\theta\,d\theta + \int_0^{2\pi} VI\,\sin^2\,\theta\,d\theta\right] \end{aligned}$$

The second and third integrals in the preceding equation have zero value when evaluated over a full cycle. Thus, the equation for transistor dissipation reduces to

$$P_T = V_{CQ}I_{CQ} - \frac{VI}{2} = P_{TQ} - \frac{VI}{2} \tag{2.11}$$

This equation indicates that no matter where the transistor is biased, the dissipation is maximum when $V = I = 0$, or with no ac output signal. As the ac signal increases, $P_{TQ}$ decreases by the amount $VI/2$.

In selecting the transistor, the worst-case condition or the no-signal power is the important consideration. This assumes that the amplifier will experience periods when the input signal is zero. The allowable dissipation of the device must exceed $V_{CQ}I_{CQ}$, or

$$P_{T\max} \geq V_{CQ}I_{CQ} \tag{2.12}$$

When $V = V_{CQ}$ and $I = I_{CQ}$, maximum power is delivered to the load while the transistor dissipates minimum power. The dissipation in this case is $V_{CQ}I_{CQ}/2$. The source power is constant with output signal and equals the sum of transistor dissipation and load power. It follows that as load power increases from zero to $V_{CQ}I_{CQ}/2$ or $V_{CC}^2/8R_L$, transistor dissipation decreases from $V_{CQ}I_{CQ}$ or $V_{CC}^2/4R_L$ to $V_{CC}^2/8R_L$.

Returning to Example 2.3, we can apply Eq. (2.11) to calculate the various power relationships of the circuit. We will assume that the quiescent collector current is set at 1 A to result in a value of $V_{CQ} = 10$ V. Since a voltage drop of 10 V appears across both the load resistance and the transistor, the power dissipated by each of these elements is 10 W, when no ac signal is present. The total source power, neglecting bias power, is

$$P_S = 20 \text{ V} \times 1 \text{ A} = 20 \text{ W}$$

which equals the sum of transistor and resistance dissipation.

If an ac source is now applied to create a peak output voltage of 4 V, a corresponding peak ac collector current of 0.4 A is required. Using these values of V and I in Eq. (2.11) results in a transistor dissipation of

$$P_T = 10 \times 1 - \frac{4 \times 0.4}{2} = 9.2 \text{ W}$$

The transistor dissipation has decreased under signal conditions. Since the average value of collector current is unchanged, the input source power remains constant with output signal at 20 W. The decrease in transistor dissipation is offset by an increase in resistor dissipation with output signal. The resistor is now dissipating 10.8 W of power. The resistor power can be expressed as

$$P_R = V_R I_{CQ} + \frac{VI}{2} \tag{2.13}$$

As output signal is increased up to the maximum of 10 V peak, the transistor dissipation decreases from its quiescent level of 10 W to 5 W. The resistor dissipation increases from its quiescent level of 10 W to 15 W while the dc source power remains constant at 20 W.

Equation (2.11) remains valid even when transformer coupling to the load resistance is used, as discussed in the next subsection.

### The Nonideal Resistive Stage

In general, a resistance will be added in the emitter circuit to prevent thermal runaway in higher power circuits as mentioned earlier. This will result in an additional dc dissipation in this resistance, $R_E$. Occasionally, this resistance is bypassed by a capacitor while it is often unbypassed to reduce nonlinear distortion in the stage.

When an unbypassed emitter resistance is added to the circuit of Fig. 2.11, the active region is reduced in size. It can be shown that the midpoint collector voltage of the active region is now

$$V_{C\text{mid}} = \frac{V_{CC}}{2}\left[1 + \frac{R_E}{R_E + R_L}\right]$$

If $V_{CQ}$ is selected to equal $V_{C\text{mid}}$ to maximize the output voltage swing before reaching cutoff or saturation, the quiescent collector current must be

$$I_{CQ} = \frac{V_{CC} - V_{C\text{mid}}}{R_L} = \frac{V_{CC}}{2[R_E + R_L]} \qquad (2.14)$$

The maximum peak current that can now be delivered to the load resistance equals $I_{CQ}$, and the corresponding power is

$$P_{\text{acmax}} = \frac{I_{CQ}^2 R_L}{2} = \frac{V_{CC}^2 R_L}{8[R_E + R_L]^2} \qquad (2.15)$$

The dc source power delivered to the circuit is

$$P_S = V_{CC} I_{CQ} = \frac{V_{CC}^2}{2[R_E + R_L]} \qquad (2.16)$$

This leads to a maximum circuit efficiency of

$$\eta_{\max} = \frac{P_{\text{acmax}}}{P_S} = 0.25 \times \frac{R_L}{R_E + R_L} \quad \text{for an unbypassed } R_E \qquad (2.17)$$

The presence of $R_E$ leads to a reduction of the maximum ideal circuit efficiency by the factor $R_L/[R_E + R_L]$. Of course, distortion and a finite value of saturation voltage will further reduce the actual circuit efficiency achieved in a practical circuit.

If the emitter resistance is bypassed, the theoretical efficiency can be increased. This derivation is left to the student and will probably appear on the next exam.

### Practice Problems

**2.7** A common-emitter stage with $R_C = 24\ \Omega$, $R_E = 0\ \Omega$, and $V_{CC} = 16$ V is driven with a total collector current of $i_c = (300 + 200 \sin \omega t)$ mA. Calculate the signal power delivered to $R_C$. *Ans:* 0.48 W.

**2.8** Calculate the power dissipation of the transistor in Practice Problem 2.7. *Ans:* 2.16 W.

**2.9** If $i_c = 300$ mA (no-signal condition), what is the power dissipation of the transistor? *Ans:* 2.64 W.

## 2.3.2 Class-A Transformer-Coupled Stage

A resistive load dissipates both dc power and ac power. The ac signal is the signal of interest while the dc power dissipation only leads to lower efficiency. If the dc load could be removed from the collector circuit, the dc power

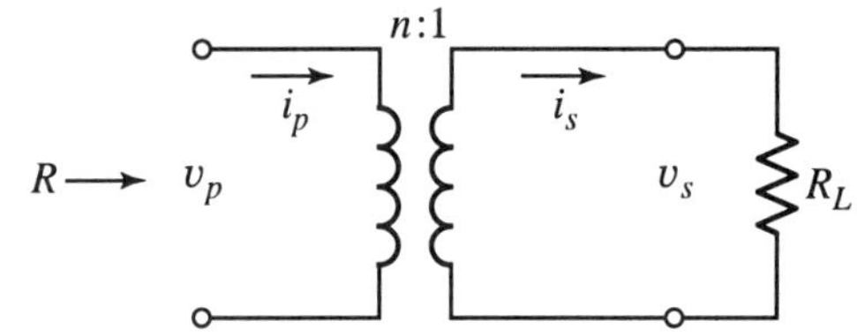


**An ideal transformer**

loss could be avoided. Removal of the dc load can be accomplished by inserting a transformer primary in the collector circuit with the load placed in the transformer secondary. With this arrangement, no dc current flows through the load and the efficiency increases.

**Practical Considerations**

For an ideal transformer with a secondary load of $R_L$, the equivalent resistance seen at the primary terminals is $R = n^2 R_L$. This results from a consideration of the currents and voltages of the accompanying figure.
Since there is no power loss in the ideal transformer, the power into the primary must equal the power out of the secondary. As voltage from primary to secondary steps down by a factor of $n$, the secondary current must step up by this factor of $n$ over the primary current. The load resistance forces the secondary voltage and current to be related by

$$\frac{v_s}{i_s} = R_L$$

Since $v_p = nv_s$ and $i_p = i_s/n$, we can write

$$R = \frac{v_p}{i_p} = \frac{nv_s}{i_s/n} = n^2 \frac{v_s}{i_s} = n^2 R_L$$

Transformer action only takes place for changing inputs. For the following discussions, we note that the transformer does not pass a dc signal and the relations between primary and secondary variables apply only to ac quantities.

### The Ideal Transformer-Coupled Stage

Figure 2.13 shows a transformer-coupled stage. The load line construction of Fig. 2.14 assumes that no emitter resistance is included for protection against thermal runaway and that the ideal transformer has zero dc resistance. The ac resistance seen looking into the primary of the transformer is $n^2 R_L$.

The equation for the dc load line is

$$V_{CE} = V_{CC} - I_C R_{\text{dc}} = V_{CC}$$

where $R_{\text{dc}}$ is the dc resistance around the collector-emitter loop. When there is no dc resistance in the loop, this equation represents a vertical line at $V_{CE} = V_{CC}$.

The ac load line relates the ac voltage to the ac current, and since the transistor voltage decreases as the current increases, the slope of the ac load line is negative and equal to

$$\text{slope} = \frac{\Delta I}{\Delta V} = -\frac{1}{R_{\text{ac}}} \tag{2.18}$$

If the magnitude of ac current is decreased to zero, the output voltage must correspond to the quiescent level. Therefore, the ac load line intersects the dc load line at the quiescent point.

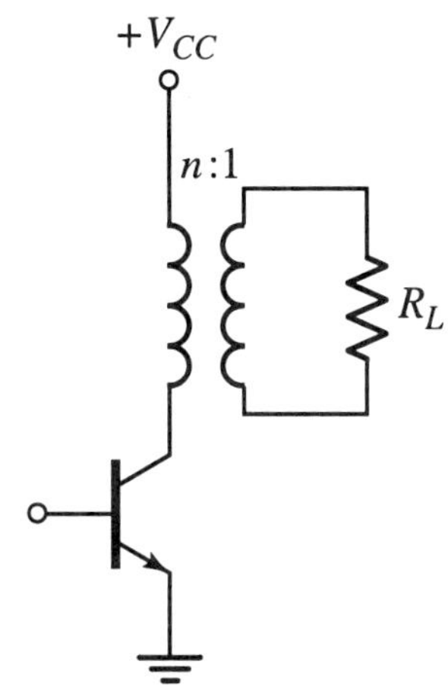


Figure 2.13: Transformer-coupled stage.

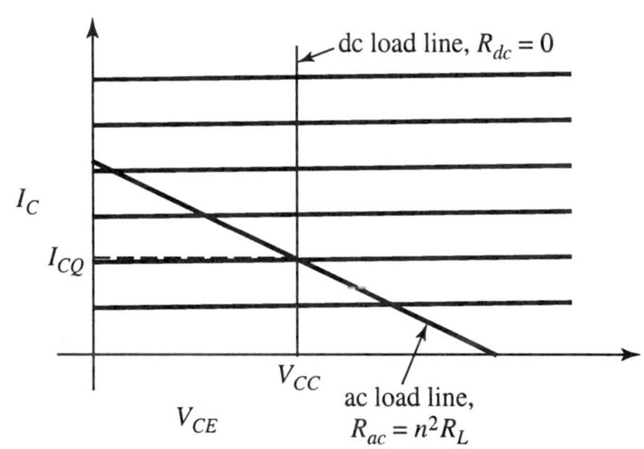


Figure 2.14: Load lines for circuit of Fig. 2.13.

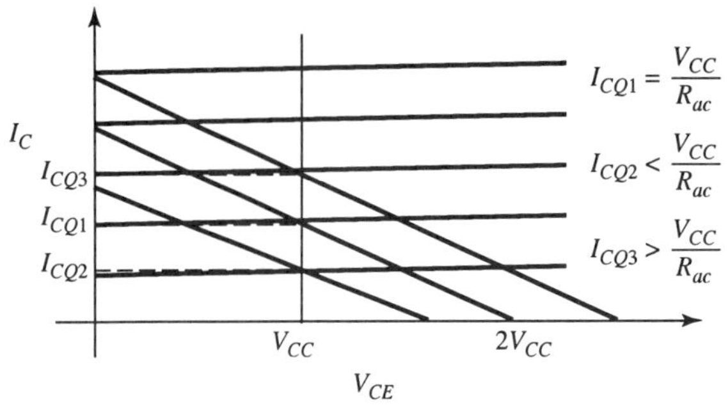


Figure 2.15: Load lines for various values of $I_{CQ}$.

The maximum symmetrical voltage swing that can occur has a value of 2 $V_{CC}$ peak-to-peak. At first thought, it may seem strange that the ac voltage can swing more positive than the power supply voltage. However, this is always possible when the ac resistance in the collector circuit is greater than the dc resistance. It can be shown that the maximum efficiency and maximum ac output voltage occur when the quiescent current is

$$I_{CQ} = \frac{V_{CC}}{R_{\text{ac}}} \tag{2.19}$$

Figure 2.15 shows the results of using different values of $I_{CQ}$.

For values of $I_{CQ}$ less than $V_{CC}/R_{\text{ac}}$, the peak voltage swing is less than $V_{CC}$ and is limited by cutoff. As $I_{CQ}$ is increased beyond $V_{CC}/R_{\text{ac}}$, the voltage can swing to a peak voltage of $V_{CC}$, but the efficiency drops off, since the input source power increases with $I_{CQ}$. The greatest ac output power is

$$P_{\text{ac}} = \frac{V_{CC}^2}{2R_{\text{ac}}} \tag{2.20}$$

The maximum ideal circuit efficiency occurs when $I_{CQ} = V_{CC}/R_{\text{ac}}$ and is then

$$\eta_{\text{max-ideal}} = \frac{V_{CC}^2/2R_{\text{ac}}}{V_{CC}^2/R_{\text{ac}}} \times 100 = 50\% \text{ for ideal class} - \text{A stage}$$

The circuit efficiency is increased over the resistive load case because the load resistor is removed from the dc circuit and no longer dissipates dc power.

In both cases, the element values and quiescent values can be chosen without referring to the V-I characteristics of the transistor. They have been shown only to demonstrate some pertinent points.

**Practical Considerations**

Only on a rare occasion would an amplifier be designed to amplify a single sinusoidal signal. In general, the amplifier would be designed for audio signals, including voice and music. These signals are largely composed of harmonic sinusoids. Yet the calculations of efficiency for power stages are based on a single sinusoidal input signal.

The justification for this is that the analysis is intractable for the complex waveforms of voice and music. Furthermore, since they are composed of sums of sinusoids, it seems reasonable to use a single component for the analysis. Although overall circuit efficiency will be different for each type of waveform, the sinusoidal efficiency gives a good approximation to that of the audio waveform.

### The Nonideal Transformer-Coupled Stage

Some of the nonideal effects that will cause the actual design to differ from the theoretical design are:

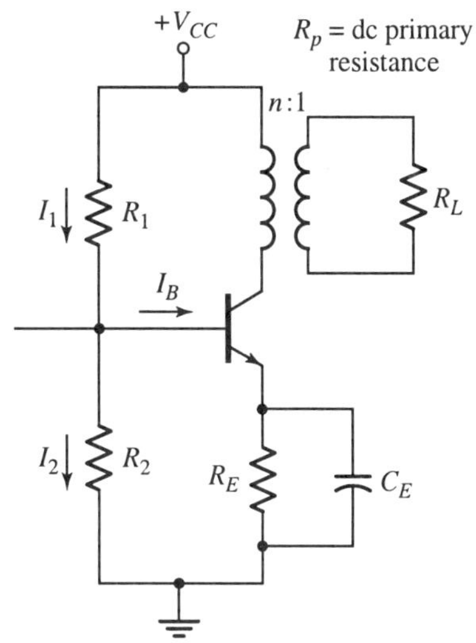


Figure 2.16: A practical transformer-coupled stage.

1. $V_{CE(\text{sat})} > 0$.

2. The transformer will have a dc resistance, in addition to some ac losses.

3. Some dc power must be supplied to the bias circuit.

4. A resistor will be inserted in the emitter lead to avoid thermal runaway.

The dc resistance of the transformer primary must be known along with the efficiency of the transformer in order to design the output stage. The efficiency of the transformer is defined as the ratio of power output to the power applied to the primary (not including the power dissipated by the dc winding resistance):

$$n_t = P_{\text{out}}/P_{\text{in}}$$

Usually, the transformer parameters must be measured, since the manufacturer does not specify these values.

The problem of thermal runaway is especially prevalent in germanium power transistors, which exhibit higher leakage currents at room temperature than comparable silicon transistors. On the other hand, the leakage current in silicon increases more rapidly with temperature, so both types of power transistor must be protected. A bypassed emitter resistance is typically used to prevent thermal runaway. The additional drop across $R_E$ as $I_E$ increases tends to reduce the forward bias of the emitter-base junction and limits the increase of current.

Figure 2.16 shows a practical transformer-coupled stage with biasing and an emitter resistance to prevent thermal runaway. If maximum output power and maximum circuit efficiency are desired, the quiescent point should be set to allow an equal voltage swing on both sides of this point. Since the peak value of current swing is $I_{CQ}$, the ac condition for maximum output is

$$V_{CEQ} = R_{\text{ac}} I_{CQ} \tag{2.21}$$

Although dc variables appear in Eq. (2.21), it is an ac equation that relates the peak voltage that can exist to peak current that can flow.

The second required equation is that for the dc load line, which is

$$V_{CEQ} = V_{CC} - I_{CQ} R_{\text{dc}} \tag{2.22}$$

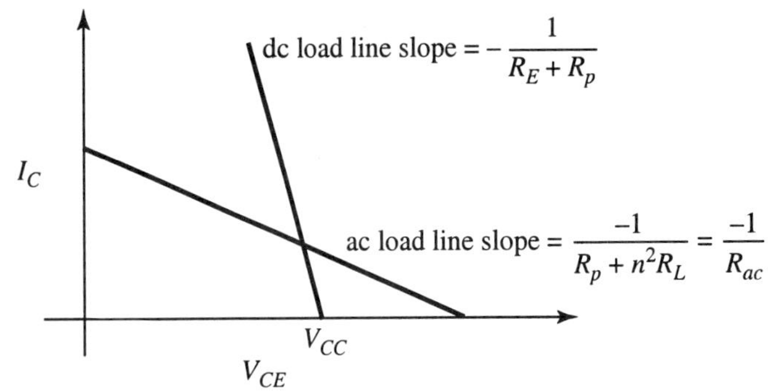


Figure 2.17: Load lines.

Solving these two equations simultaneously for $I_{CQ}$ gives

$$I_{CQ} = \frac{V_{CC}}{R_{\text{ac}} + R_{\text{dc}}} \text{ (for maximum symmetrical output)} \tag{2.23}$$

Figure 2.17 shows the load lines for this situation.

With a value of quiescent current less than or equal to this maximum value, the maximum power that can be delivered to the load is

$$P_{\text{ac}} = n_t \frac{I_{CQ}^2 n^2 R_L}{2} \tag{2.24}$$

The maximum circuit efficiency occurs when $I_{CQ}$ is set to the value given by Eq. (2.23). Using this value of $I_{CQ}$ gives the maximum output power:

$$P_{\text{ac}-\max} = n_t \left( \frac{V_{CC}}{R_{\text{ac}} + R_{\text{dc}}} \right)^2 \times n^2 \frac{R_L}{2} \tag{2.25}$$

The source power for this value of $I_{CQ}$, neglecting bias network current, is

$$P_S = V_{CC} I_{CQ} = V_{CC} \left( \frac{V_{CC}}{R_{\text{ac}} + R_{\text{dc}}} \right) \tag{2.26}$$

Taking the ratio of $P_{\text{ac}}$ to $P_S$ for this case gives

$$\eta_{\max} = n_t \frac{n^2 R_L}{2(R_{\text{ac}} + R_{\text{dc}})} \text{ for finite } R_{\text{dc}} \text{ in output circuit} \tag{2.27}$$

It is obvious that this efficiency will be reduced from 50% by the transformer efficiency and by the factor

$$\frac{n^2 R_L}{(R_{\text{ac}} + R_{\text{dc}})}$$

An example will demonstrate the considerations in a nonideal class-A stage.

## Example 2.4

In the circuit of Fig. 2.16, $R_1 = 760\ \Omega$, $V_{CC} = 20\ V$, $R_2 = 100\ \Omega$, $R_L = 4\ \Omega$, $R_E = 10\ \Omega$, $C_E = 100\ \mu\text{F}$, the turns ratio is 4:1, the transformer efficiency is $n_t = 0.9$, the dc primary resistance is 10 Ω, and the silicon transistor has a $\beta_0 = 30$ and $V_{BE(\text{on})} = 0.8$ V.

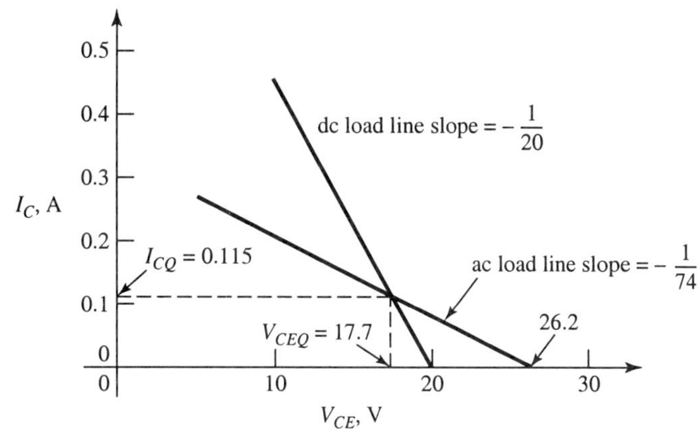


Figure 2.18: Load lines for Example 2.4.

**(a)** Find the greatest output power that can be delivered to the load at the given bias point without saturating or cutting off the transistor; and determine the circuit efficiency.

**(b)** Find the no-signal power dissipated by the transistor.

**(c)** Find the bias point that gives maximum possible circuit efficiency, and calculate this efficiency.

**Solution (a)** We must first determine $I_{CQ}$. This can be done by taking a Thévenin equivalent of the bias network. This value is found to be $I_{CQ} = 115$ mA. The load lines are shown in Fig. 2.18.

These lines use values of $R_{\text{dc}} = 20\ \Omega$ and $R_{\text{ac}} = n^2 R_L + R_p = 74\ \Omega$. It is obvious that a symmetrical sinusoid will be limited by cutoff rather than saturation. This means that $I_{CQ}$ is less than the value given by Eq. (2.23) and that output power can be calculated from Eq. (2.24). This power is found to be

$$P_{ac} = \frac{0.9 \times (0.115)^2 \times 64}{2} = 0.381 \text{ W}$$

The source power, including the bias power, is

$$P_S = V_{CC}(I_{CQ} + I_1) = 20 \times 0.139 = 2.78 \text{ W}$$

The circuit efficiency is

$$\eta = \frac{P_{\text{ac}}}{P_S} \times 100 = \frac{0.381}{2.78} \times 100 = 14\%$$

**(b)** The no-signal power dissipated by the transistor is

$$P_T = V_{CEQ} I_{CQ} = 17.7 \times 0.115 = 2.04 \text{ W}$$

The voltage $V_{CEQ}$ is found from the dc load line or the corresponding equation; that is,

$$V_{CEQ} = V_{CC} - I_{CQ} R_{\text{dc}} = 20 - 0.115 \times 20 = 17.7 \text{ V}$$

**(c)** The bias point that gives the highest circuit efficiency is calculated by using Eq. (2.23):

$$I_{CQ} = \frac{20}{94} = 0.213 \text{ A}$$

Using this collector current results in a maximum output power of

$$P_{\rm ac} = \frac{0.9 \times (0.213)^2 \times 64}{2} = 1.31 \text{ W}$$

The source power is

$$P_S = 20(0.213 + 0.021) = 4.67 \text{ W}$$

The circuit efficiency is now

$$\eta_{\rm max} = \frac{1.31}{4.67} \times 100 = 28.1\%$$

The circuit efficiency of 14% is higher than the actual value that could be obtained in part (a). Distortion would limit the maximum output swing, causing $\eta$ to be somewhat lower. However, the optimum bias point of part (c) would lead to a higher circuit efficiency, though not as high as calculated, since the output voltage amplitude would again be limited in order to minimize distortion.

---

**Practical Considerations**

If the output stage of Example 2.4 had been biased higher than 213 mA, say to $I_{CQ} = 280$ mA, the output would be limited by saturation. This would be a poor choice of bias since $I_{CQ}$ exceeds the maximum efficiency point. The peak output voltage can be calculated by determining the peak current that causes saturation. The quiescent voltage across the transistor is

$$V_{CEQ} = 20 - 0.28 \times 20 = 14.4 \text{ V}$$

The peak value of current to bring the transistor to saturation is

$$\Delta I = \frac{V_{CEQ}}{R_{\rm ac}} = \frac{14.4}{74} = 0.195 \text{ A}$$

The peak output voltage to the transformer primary is

$$v_p = \Delta I n^2 R_L = 0.195 \times 64 = 12.5 \text{ V}$$

The output power is

$$P_{\rm ac} = \frac{0.9 \times 12.5^2}{2 \times 64} = 1.09 \text{ W}$$

This value is less than the power output when biased for maximum efficiency. Furthermore, the source power is greater for this bias point, leading to an even lower efficiency. We conclude that a bias point higher than that calculated from Eq. (2.23) should not be used.

**Practical Considerations**

An interesting and practical problem occurs when the allowable transistor dissipation limits the bias current to values lower than that given by Eq. (2.23). For example, if the allowable dissipation of the transistor of Example 2.4 is $P_{T\max} = 2.5$ W, a collector current of 0.213 A cannot be used. This optimum value would result in a transistor dissipation of 3.35 W. In order to limit the transistor dissipation to 2.5 W, the following equation is solved for $I_{CQ}$:

$$P_{T\max} = P_T = I_{CQ}V_{CEQ} = I_{CQ}(V_{CC} - I_{CQ}R_{dc}) \tag{2.28}$$

For the circuit of Example 2.4 this equation becomes

$$2.5 = 20I_{CQ} - 20I_{CQ}^2$$

Solving this quadratic equation for $I_{CQ}$ results in

$$I_{CQ} = 0.854\ A,\ 0.146\ \text{A}$$

Either value of quiescent current will limit the transistor dissipation to 2.5 W. However, we concluded earlier that $I_{CQ}$ should never exceed the optimum value given by Eq. (2.23) of 0.213 A. Thus, the lower value of $I_{CQ}$ is always chosen.
For an $I_{CQ}$ value of 146 mA, the value of $V_{CEQ}$ is calculated to be 17.1 V. The product of these quantities gives a transistor dissipation of 2.5 W.

**Practice Problems**

**2.10** Calculate the quiescent power dissipated by the transistor in part (c) of Example 2.4. *Ans:* 3.35 W.

**2.11** If the maximum allowable power dissipation of the transistor in Example 2.4 is 2.8 W, what is the maximum power that can be delivered to the load? *Ans:* 0.9 W.

## 2.4 CLASS-B POWER OUTPUT STAGES

**Important Concepts**

1. The class-B stage dissipates little or no quiescent power.
2. The class-B stage offers higher values of circuit efficiency than class-A stages.

A transformer-coupled, class-B stage is shown in Fig. 2.19.

The voltage of the secondary windings of the input transformer will appear between points $a$ and $b$. For the half-cycle when point $a$ is positive with respect to point $b$ (assuming a sinusoidal input), $T_1$ will be in its active region and will amplify this signal, while $T_2$ remains off. On the alternate half-cycle, point $b$ will be positive with respect to point $a$, and $T_2$ will amplify this half-cycle, while $T_1$ is shut off. Since $I_{C1}$ travels in the opposite direction to $I_{C2}$ through the output transformer primary windings, the two currents will cause opposite polarity signals in the secondary windings; when $I_{C2}$ flows, the alternate negative-going waveform will appear. This arrangement wherein both sides of the output transformer are driven is called a double-ended amplifier as opposed to the single-ended class-A transformer stage previously discussed. It is also classified as a push-pull stage since one transistor drives the transformer in one direction, while the other transistor drives the transformer in the

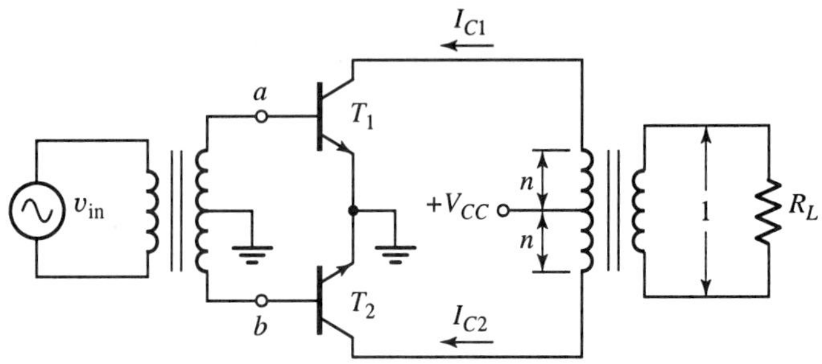


Figure 2.19: A class-B, push-pull stage.

opposite direction.

### Advantages of the Class-B Stage

The class-B stage has the following advantages and disadvantages compared to the class-A stage.

**1.** The class-B stage has very small no-signal power dissipation in $T_1$ and $T_2$. The class-A stage causes maximum power dissipation of the transistor under no-signal conditions.

**2.** If transistors with the same power rating are used for class-A and class-B stages, the class-B stage will deliver more power to the load than will the class-A stage.

**3.** Push-pull stages have less distortion than single-ended stages if designed properly.

**4.** The class-B stage has a higher theoretical maximum for circuit efficiency. This value is $n_{\max} = 78.5\%$.

### Disadvantages of the Class-B Stage

The advantages of the class-B stage come with a price. The following lists some of the resulting disadvantages.

**1.** The typical class-B stage requires two center-tapped transformers.

**2.** Transistors that have reasonably similar characteristics are required for the class-B stage.

**3.** The reduction of crossover distortion in class-B stages requires additional circuitry. Crossover distortion occurs because of the small base-to-emitter voltage that is required before collector current will flow.

## 2.4.1 Analysis of Class-B Transformer Stages

We will first consider an ideal class-B amplifier.

### An Ideal Class-B Amplifier

The ideal analysis of the stage begins with several unrealistic assumptions. However, after this analysis is done, it is relatively easy to extend the results to the practical case. The simplifying assumptions made are as follows.

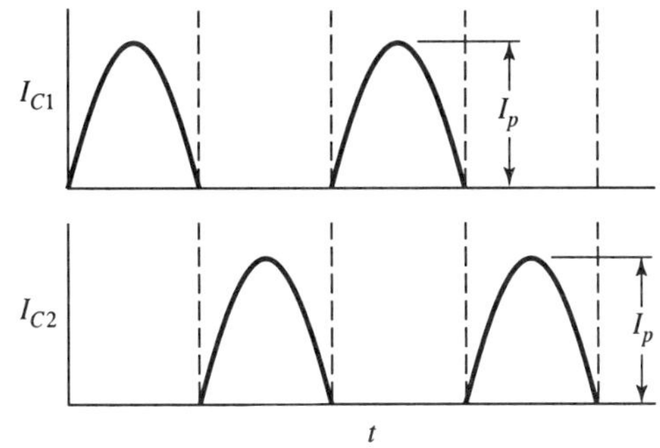


Figure 2.20: Collector current waveforms.

**1.** The transistors turn on with negligible drop across the base-emitter junction, that is, $V_{BE(on)} = 0$ V.

**2.** There is no dc resistance in the output loop implying no dc transformer resistance and no emitter resistance to avoid thermal runaway.

**3.** The transformer efficiency is 100%.

For a sinusoidal input signal, the ideal current waveforms $I_{C1}$ and $I_{C2}$ will be half-wave rectified signals as shown in Fig. 2.20.

Each transistor drives collector current through one-half of the total number of windings of the output transformer primary. If the total turns ratio is $2n:1$, the ac load seen by each collector is $n^2 R_L$. The primary is subjected to a complete sine wave of current of peak value $I_p$. One-half of the sinusoidal current travels upward through the transformer and through transistor $T_1$. The other half of the sinusoid travels downward through the transformer and through transistor $T_2$. The total power delivered to the transformer secondary is

$$P_{\text{ac}} = \frac{I_p^2 n^2 R_L}{2} \tag{2.29}$$

The maximum amplitude that $I_p$ can reach is that value which causes saturation of the transistor. For the ideal case, this corresponds to a peak voltage drop of $v_p = V_{CC}$ across the active half of the primary windings, causing zero volts across the transistor. This maximum value of current is then

$$I_{p-\max} = \frac{V_{CC}}{n^2 R_L} \tag{2.30}$$

The maximum power to the load is found by combining Eqs. (2.29) and (2.30) to give

$$P_{\text{ac}-\max} = \frac{V_{CC}^2 \times n^2 R_L}{2n^4 R_L^2} = \frac{V_{CC}^2}{2n^2 R_L} \tag{2.31}$$

The input source power is found by taking the product of the power supply voltage and the dc current flowing into the circuit. The current flowing from the source is the sum of $I_{C1}$ and $I_{C2}$, which is a full-wave rectified waveform. A half-wave of current flows through $T_1$ from the source; then a half-wave, delayed by 180°, flows from the source through $T_2$. The dc source current is just the average value of this full-wave rectified waveform, or

$$I_{\text{dc}} = \frac{2I_p}{\pi}$$

The dc source power is

$$P_S = V_{CC} I_{\text{dc}} = \frac{2V_{CC}I_p}{\pi} \tag{2.32}$$

The greatest power drain will occur for the maximum value of $I_p$, so

$$P_{S-\max} = \frac{2}{\pi} \frac{V_{CC}^2}{n^2 R_L} \tag{2.33}$$

Equation (2.32) shows that the source power depends on the magnitude of the output signal. In the class-A stage this was not the case. Only the quiescent current affects the source power in the class-A stage; therefore, this power is independent of the magnitude of the output signal.

The circuit efficiency in percent for the ideal circuit is calculated from Eqs. (2.29) and (2.32) as

$$\eta = \frac{(I_p^2 \times n^2 R_L)/2}{(2V_{CC}I_p)/\pi} \times 100 = \frac{\pi}{4} \times n_t \times \frac{I_p n^2 R_L}{V_{CC}} \times 100 \tag{2.34}$$

The maximum efficiency occurs when maximum output power is present, or when $I_p = I_{p-\max}$, so from Eq. (2.30),

$$\eta_{\max} = \frac{\pi}{4} \times \frac{n^2 R_L}{n^2 R_L} \times 100 = \frac{\pi}{4} \times 100 \tag{2.35}$$

For the ideal case, the maximum circuit efficiency is $\eta_{\max} = 78.5\%$.

### A Practical Class-B Amplifier

In a practical stage, the transformer will present dc primary resistance and a transformer efficiency less than unity. The power transistors will require several tenths of a volt to begin conduction, and an emitter resistance will be used to avoid thermal runaway and reduce distortion.

**Practical Considerations**

An emitter resistor could be placed between each emitter terminal and ground. By noting that each transistor conducts when the other is off, we see that a single emitter resistance can be shared. The two emitter terminals are connected, and the emitter resistance appears between these emitters and ground. This resistance cannot be bypassed by a capacitor as in the case of the class-A stage. Since each emitter conducts current into the emitter resistance, a full-wave rectified voltage waveform is developed across this resistance. If a capacitor is placed in parallel, it will charge toward the average voltage level, creating a dc voltage at the emitters that varies with amplitude of the input signal. This would cause each transistor current to be something different than a half-wave signal, and distortion of the output would therefore result.

The effects of $V_{BE(\text{on})}$ can be reduced significantly and will not affect the circuit efficiency. The effects on circuit efficiency due to resistances in the output loop and transformer efficiency can be included in the analysis rather easily. The power to the load can now be expressed as

$$P_{\text{ac}} = \frac{n_t \times I_p^2 n^2 R_L}{2} \tag{2.36}$$

The maximum amplitude that $I_p$ can reach before saturation of the transistor is

$$I_{p-\max} = \frac{V_{CC}}{R_{\text{ac}}} \tag{2.37}$$

where $R_{\text{ac}}$ includes all resistance of the output loop. This value can be expressed as

$$R_{\text{ac}} = n^2 R_L + \frac{R_p}{2} + R_E \tag{2.38}$$

In this equation $R_p$ is the total dc resistance across the primary.

The maximum power to the load is found by combining Eqs. (2.36) and (2.37) to give

$$P_{\text{ac-max}} = \frac{n_t \times V_{CC}^2 \times n^2 R_L}{2R_{\text{ac}}^2} \tag{2.39}$$

The dc source power is

$$P_S = V_{CC} I_{\text{dc}} = \frac{2V_{CC} I_p}{\pi} \tag{2.40}$$

Again, the greatest power drain will occur for the maximum value of $I_p$, so

$$P_{S-\text{max}} = \frac{2}{\pi}\frac{V_{CC}^2}{R_{\text{ac}}} \tag{2.41}$$

The maximum efficiency occurs when maximum output power is present, or when $I_p = I_{p-\text{max}}$, so from Eq. (2.37),

$$\eta_{\text{max}} = \frac{\pi}{4} \times n_t \times \frac{n^2 R_L}{R_{\text{ac}}} \times 100 = \frac{\pi}{4} \times n_t \times \frac{n^2 R_L}{n^2 R_L + R_p/2 + R_E} \times 100 \tag{2.42}$$

This efficiency will be less than the ideal value of 78.5% owing to the presence of $R_E$, $R_p$, and a transformer efficiency less than one. It will also suffer from the limitation that the collector signal amplitude will never reach its maximum possible level due to nonlinearities in the voltage as saturation is approached.

### Transistor Dissipation

The power dissipated by the transistors in the class-B stage can be found by considering a power balance of all dissipations. The input power to the circuit will be accounted for by the dissipation of both transistors, the ac power delivered to the transformer primary, and the power dissipated by the resistances of the collector-emitter circuit. This can be expressed as

$$P_S = \frac{P_{\text{ac}}}{n_t} + 2P_T + P_R \tag{2.43}$$

where $P_R$ is the resistive losses and $P_T$ is the power dissipated by each transistor. Since we can evaluate $P_S$, $P_R$, and $P_{\text{ac}}$, the transistor dissipation can be found from

$$P_T = \frac{P_S - P_R - P_{\text{ac}}/n_t}{2} \tag{2.44}$$

The ac power delivered to the transformer plus the power dissipated by the dc resistances is

$$\frac{P_{\text{ac}}}{n_t} + P_R = \frac{I_p^2}{2}(n^2 R_L + R_p/2 + R_E) = \frac{I_p^2}{2} R_{\text{ac}}$$

Returning to Eq. (2.43), we can now write

$$P_T = I_p \left( \frac{V_{CC}}{\pi} - \frac{I_p R_{\text{ac}}}{4} \right) \tag{2.45}$$

A plot of this transistor dissipation as a function of peak collector current is shown in Fig. 2.21.

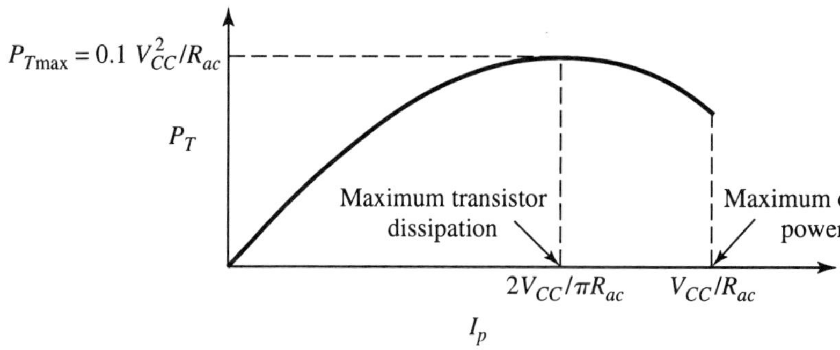


Figure 2.21: Transistor dissipation in a class-B stage.

Maximum power output to the load occurs when $I_p = V_{CC}/R_{\text{ac}}$. For this value of peak current, the transistor dissipation is calculated from Eq. (2.45) to be

$$P_T = \frac{0.068V_{CC}^2}{R_{\text{ac}}} \tag{2.46}$$

when

$$I_p = \frac{V_{CC}}{R_{\text{ac}}}$$

In order to do *worst-case design*, we must find the greatest possible power dissipation of each transistor and design our thermal environment to exceed this value. The power dissipated by the transistor does not reach its greatest value when maximum output power is being delivered to the load. From Fig. 2.21 it is seen that the power $P_T$ reaches its greatest value when $P_{\text{ac}}$ is less than the maximum value. In order to find this maximum dissipation, Eq. (2.45) can be differentiated with respect to $I_p$, and the result can be set equal to zero. Setting

$$\frac{dP_T}{dI_p} = \frac{V_{CC}}{\pi} - \frac{I_p R_{\text{ac}}}{2} = 0$$

results in a value of $I_p = 2V_{CC}/\pi R_{\text{ac}}$. For this value of $I_p$, $P_T$ is maximum and is calculated by substituting this current value into Eq. (2.45) to get

$$P_{T-\max} = \frac{0.10V_{CC}^2}{R_{\text{ac}}} \tag{2.47}$$

when

$$I_p = \frac{2}{\pi}\frac{V_{CC}}{R_{\text{ac}}}$$

This point is shown in Fig. 2.21. When selecting transistors for this stage, Eq. (2.47) is often used, since this is the maximum possible $P_T$ that can occur for a sinusoidal output.

### Distortion in Class-B Stages

If the stages are perfectly matched, and if $T_2$ turns on at the exact time that $T_1$ turns off, and conversely, the even harmonic distortion will be very low. Physically, this is easy to visualize. If there is even harmonic distortion in the waveform due to $T_1$, the same distortion will be present in $T_2$, but will be out of phase by 180° of the fundamental frequency. This 180° of the fundamental frequency corresponds to phase shifts of integral multiples of 360° for the higher harmonics (360° for second, 720° for fourth, etc.). Since the voltage induced in the transformer by $T_2$ will be in the opposite direction to that induced by $T_1$, the even harmonics will cancel in the load. Theoretically, this fact is explained by noting that the waveform will possess half-wave symmetry, and a waveform of this type will not have even harmonics. Because perfectly matched conditions are impossible to achieve in practice, some even harmonic distortion will be present in the output signal.

A very-high odd harmonic-distortion content will occur if *crossover distortion* is allowed to take place. This distortion arises from the nonzero values of $V_{BE(\text{on})}$ for the transistors. Since the base-emitter junction must be forward biased a few tenths of a volt before appreciable collector current will flow, the center portion of the input signal will not be amplified unless proper precautions are taken. The effects of this offset voltage that is required for conduction are shown in Fig. 2.22.

In order to reduce crossover distortion, the transistors can be forward biased enough to overcome this offset voltage. For germanium, the forward bias should be 0.2 to 0.5 V; for silicon, the voltage should be 0.5 to 1.2 V. The circuits in Fig. 2.22 can be used to bias the base voltages to the proper level.

The diode can be biased so that the diode voltage equals or slightly exceeds the offset voltage of the transistors. Usually, a very small collector current is caused to flow to ensure conduction by the transistors for the full half-cycle. If more no-signal current is made to flow in the transistors, more no-signal dissipation results. This condition can reduce distortion since nonlinearities are more prevalent near cutoff and is used in low-distortion applications. It is referred to as class-AB operation.

# Example 2.5.

In the circuit of Fig. 2.23, the total dc primary resistance is 10 Ω (5 Ω on each side of center tap) and the transformer efficiency is $n_t = 0.9$. Assuming that $V_{CE(sat)} = 0$ and $I_{co} = 0$,

**(a)** find the maximum output power before clipping of the waveform occurs;

**(b)** find the circuit efficiency for the condition of part (a);

**(c)** find the maximum transistor dissipation that could occur in this circuit.

**Solution (a)** The reflected ac load between center tap and each side of the primary is $n^2 R_L = 32\ \Omega$. When each transistor conducts, the circuit is as shown in Fig. 2.24. The peak value of current that can occur is

$$I_{p-\max} = \frac{20}{42} = 0.476 \text{ A}$$

The maximum power delivered to the load is found by using Eq. (2.39):

$$P_{\text{ac}-\max} = \frac{0.9(20)^2 \times 32}{2 \times (42)^2} = 3.27 \text{ W}$$

**(b)** The source power when $I_{p-\max}$ is present is

$$P_{S-\max} = \frac{2}{\pi} \times \frac{(20)^2}{42} = 6.07 \text{ W}$$

The bias power can also be calculated here as

$$P_{\text{bias}} = I_1 \times V_{CC} = \frac{20 - 0.5}{3.9 \times 10^3} (20) = 0.1 \text{ W}$$

The total source power is

$$P_{S-\text{total}} = 6.07 + 0.1 = 6.17 \text{ W}$$

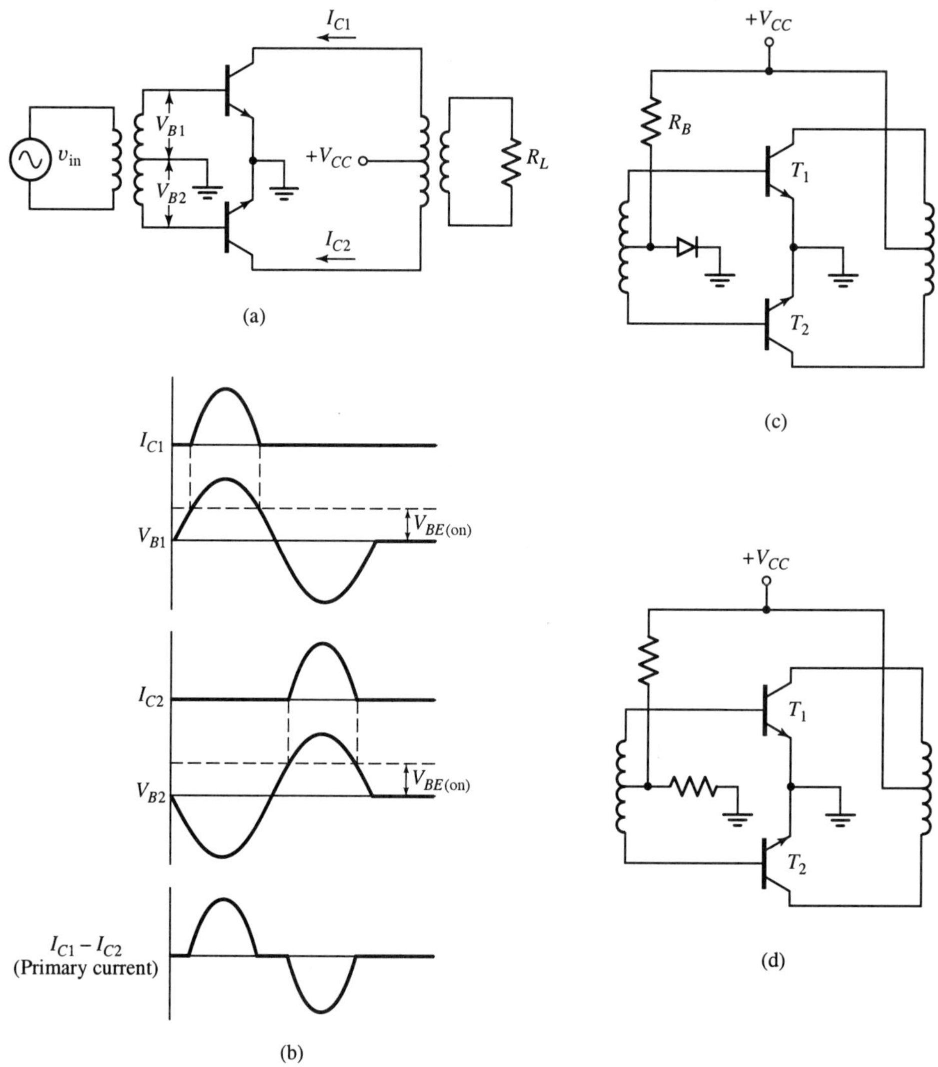


Figure 2.22: Distortion effects and reduction of distortion in class-B: (a) actual circuit with no bias; (b) resulting waveforms; (c) diode bias; (d) resistor bias.Class-B stages: (a) class-B push-pull; (b) complementary emitter followers.

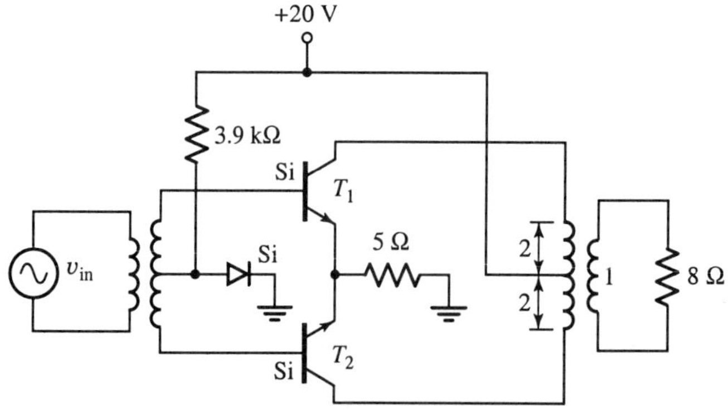


Figure 2.23: Circuit for Example 2.5.

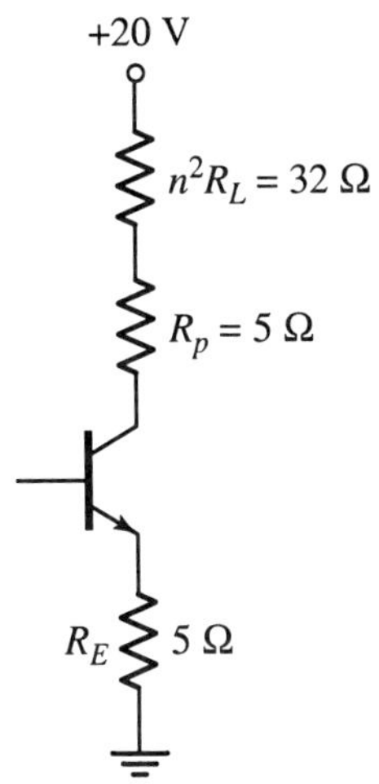


Figure 2.24: Equivalent circuit of each stage.

The circuit efficiency is then

$$\eta_{\max} = \frac{3.27}{6.17} \times 100 = 53\%$$

**(c)** The transistor power will be maximum at a current of less than $I_{pmax}$. Using Eq. (2.47), we find that the greatest transistor dissipation is

$$P_{T\max} = \frac{(0.1) \times (20)^2}{42} = 0.95 \text{ W}$$

The allowable distortion will of course limit $\eta_{\max}$ to something less than the value of 53% calculated.

---

**Practice Problems**

**2.12** Calculate the ratio of maximum required allowable transistor dissipation to the power delivered to the load in Example 2.5. Compare this ratio to that for part (c) of Example 2.4, the class-A stage. *Ans:* For class-B: 29.1%, for class-A: 256%.

**2.13** Rework Example 2.5(a) if $n^2R_L = 50\ \Omega$ *Ans:* 2.78 W.

## 2.5 CLASS-C POWER OUTPUT STAGES

**Important Concepts**

**1.** Class-C stages can approach 100% circuit efficiency, typically achieving 95% efficiency in practical applications.

**2.** The class-C amplifier uses a tuned circuit load and can only amplify frequencies within the range of the passband of this resonant circuit.

**3.** This amplifier finds application in narrowband communication applications and cannot be used in wideband applications such as a high-fidelity amplifier for music.

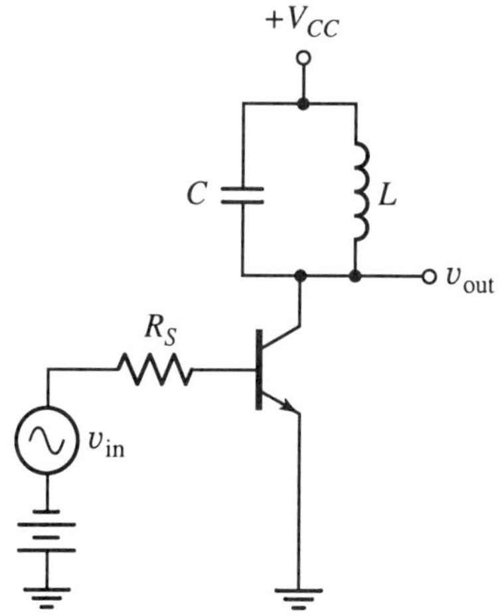


Figure 2.25: A class-C stage.

Class-C amplifiers are used for amplification of a single frequency or a very narrow frequency band. In class-B output stages, the output current ideally flows for exactly 180° of the sinusoidal waveform. If the duration of the output current flow is less than one half-cycle, class-C operation is taking place. The periodic output current generates a sinusoidal output voltage by flowing through a resonant circuit tuned to the fundamental frequency or one of the harmonic components.

As the conduction angle of output current approaches zero degrees, the circuit efficiency approaches 100%. Unfortunately, the output power also tends toward zero in this instance. Some compromise between good efficiency and high power output is necessary under normal conditions with a resulting typical efficiency of 90%. Vacuum-tube amplifiers produce output powers exceeding 50,000 W in some applications such as AM radio transmission. Transistor stages generally are limited to hundreds of watts of output power although parallel stages can increase this capability considerably. These class-C stages find their greatest use in harmonic generation and radio frequency production.

A simple class-C amplifier is shown in Fig. 2.25. The resistance $R_S$ is much greater than the input impedance to the transistor. The input voltage, collector current, and collector voltage waveforms are shown in Fig. 2.26.

Since collector current is periodic, it will contain a fundamental component of current and higher harmonics in addition to a dc or average value. The tank circuit or parallel resonant circuit will ideally present zero impedance to dc and all harmonics except the fundamental. A resistance $R_p$ will be presented to the fundamental frequency. This resistance is related to the circuit $Q$ and other tuned-circuit parameters as derived in Chapter 4. The fundamental frequency component of current will develop an ac voltage across the tuned circuit that is in proportion to this component. The ac output voltage is given by

$$v_{\text{out}} = -R_p I \cos \omega t \tag{2.48}$$

where $I$ is the magnitude of the fundamental current component. For an ideal tank circuit, there will be no distortion of the voltage waveform since load impedance exists only for the fundamental frequency. In practice, the higher components may appear in the voltage waveform due to nonzero impedances presented to harmonics by the tank circuit. Some class-C stages are driven from cutoff through the active region to the saturation region. Additional distortion is introduced as the transistor output impedance drops drastically when saturated. If harmonic generation is required, the tank circuit is tuned to the desired frequency and develops a voltage only at this particular frequency.

To calculate the input power, we must find the dc collector current that flows. Output power can be found only if the magnitude of the fundamental frequency component is found. For transistor stages, the collector-current waveform closely approximates a portion of a sinusoid. A Fourier analysis of this waveform yields the two quantities of interest.

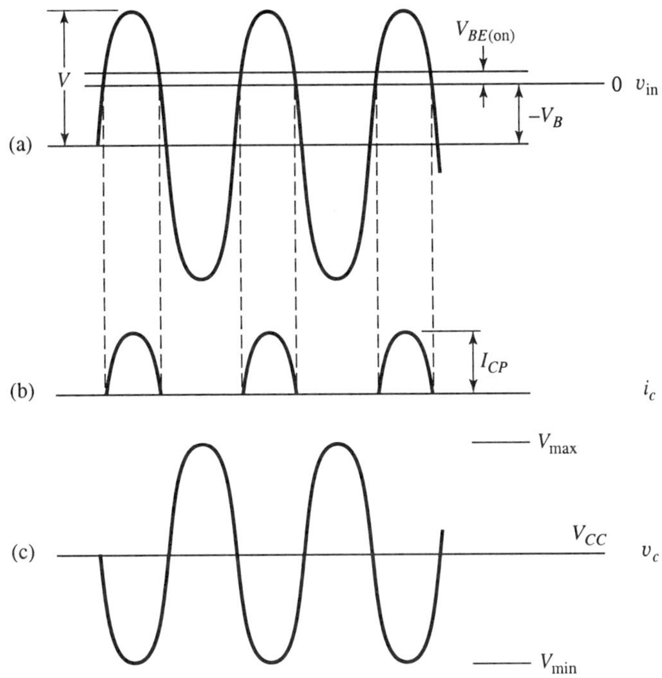


Figure 2.26: (a) Input voltage. (b) Collector current. (c) Collector voltage.

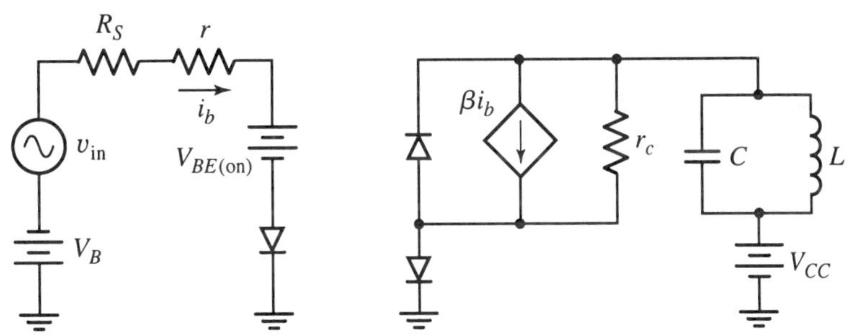


Figure 2.27: Equivalent circuit of class-C stage.

The equivalent circuit for the class-C stage can be found by using the large-signal transistor model shown in Fig. 2.27. Base current will flow only when the input voltage given by $V \cos \omega t$ exceeds $V_B + V_{BE(on)}$. This current can be written as

$$i_b = \frac{V \cos \omega t - (V_B + V_{BE(\text{on})})}{R_S} \quad (\text{for } i_b > 0) \tag{2.49}$$

assuming that the input capacitor appears as a short circuit to this frequency. The collector current given by $\beta i_b$ is shown in Fig. 2.28. The peak value of collector current that flows is

$$I_{CP} = \frac{\beta[V - (V_B + V_{BE(\text{on})})]}{R_S} \tag{2.50}$$

Collector current can be expressed as

$$i_c = \frac{\beta[V \cos \omega t - (V_B + V_{BE(\text{on})})]}{R_S} \tag{2.51}$$

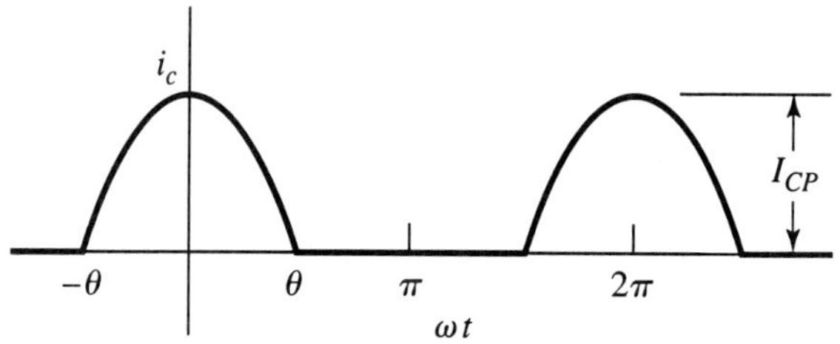


Figure 2.28: Collector-current waveform.

for values of $i_c$ greater than zero. For values of $t$ that result in a zero or negative value of Eq. (2.51), the collector current is zero. The angle $\theta$ can be found by noting that

$$\cos\,\theta = \frac{V_B + V_{BE\text{on}}}{V} \tag{2.52}$$

Equation (2.51) is then valid for times that satisfy

$$-\theta < (\omega t - 2n\pi) < \theta$$

where $n = 0, 1, 2, ...$ and

$$\theta = \cos^{-1}\left[\frac{V_B + V_{BE(\text{on})}}{V}\right] \tag{2.53}$$

A Fourier analysis of the collector-current waveform leads to the values of dc and fundamental frequency components. The dc value of collector current is

$$I_{\text{dc}} = \frac{1}{2\pi}\int_{-\theta}^{\theta}\frac{\beta V}{R_S}[\cos\,\omega t - \cos\,\theta]\,d(\omega t) = \frac{\beta V}{\pi R_S}[\sin\,\theta - \theta\,\cos\,\theta] \tag{2.54}$$

The fundamental component has a magnitude given by

$$I = \frac{1}{\pi}\int_{-\theta}^{\theta}\frac{\beta V}{R_S}[\cos\,\omega t - \cos\,\theta]\,\cos\,\omega t d(\omega t) = \frac{\beta V}{\pi R_S}\left[\theta - \frac{\sin 2\,\theta}{2}\right] \tag{2.55}$$

The ac output voltage is then expressible from Eq. (2.48) as

$$v_{\text{out}} = \frac{-\beta V R_p}{\pi R_S}\left[\theta - \frac{\sin 2\theta}{2}\right]\cos\,\omega t \tag{2.56}$$

In the preceding equations we have neglected $r_{\text{out}}$, the output resistance of the transistor. In the simple circuit shown, this may not be acceptable inasmuch as $R_p$, the resonant resistance of the tank circuit, may be quite large. In practical applications, transformer mismatching or neutralization of the circuit will generally be used, especially at higher frequencies. Both of these schemes minimize the effect of $r_{\text{out}}$; thus, it seems reasonable to neglect this element in the present discussion. If it is not negligible, it can be placed in parallel with $R_p$ to result in a modified value for this parameter.

We are now in a position to calculate the various power relationships of the circuit. The dc input power from the source is

$$P_S = V_{CC}I_{\text{dc}} = \frac{\beta V_{CC}V}{\pi R_S}[\sin\,\theta - \theta\,\cos\,\theta] \tag{2.57}$$

The angle of conduction, which equals $2\theta$, can be controlled by the bias voltage $V_B$. As this value approaches the magnitude $V$ of the input signal, the conduction angle approaches zero. The input power in this case also approaches zero, as does the output power.

The power dissipated by the transistor equals the source power minus output power. The output power is given by squaring the RMS value of output voltage and dividing by $R_p$, or

$$P_{\text{ac}} = \frac{\beta^2 V^2 R_p}{2\pi^2 R_S^2}\left[\theta - \frac{\sin\,2\theta}{2}\right]^2 \tag{2.58}$$

The transistor dissipation is

$$P_T = \frac{\beta V_{CC} V}{\pi R_S}\left[\sin\,\theta - \theta\,\cos\,\theta\right] - \frac{\beta^2 V^2 R_P}{2\pi^2 R_S^2}\left[\theta - \frac{\sin\,2\theta}{2}\right]^2 \tag{2.59}$$

It is difficult to apply the power expressions to design without imposing some additional constraints. Thus far we have not discussed methods of selecting $\theta$, $R_p$, $V_{CC}$, or $V$; consequently, additional information must be considered to specify these values. The power supply voltage is generally chosen from limitations imposed by the transistor or the power supply. The impedance $R_p$ is a function of the load that must be driven and the quality of the tuned circuit. The maximum output voltage and power that can be delivered to the load occur when the magnitude of the output voltage equals $V_{CC}$, that is, $|v_{\text{out}}| = V_{CC}$. The product of $R_p$ and the fundamental component must then equal $V_{CC}$, or from Eq. (2.56)

$$V_{CC} = \frac{\beta V R_p}{\pi R_S}\left[\theta - \frac{\sin\,2\theta}{2}\right] \tag{2.60}$$

Using this relationship allows the maximum power output to be found from Eq. (2.58). This value is

$$P_{ac-max} = \frac{\beta V_{CC} V}{2\pi R_S}\left[\theta - \frac{\sin\,2\theta}{2}\right] = \frac{V_{CC}^2}{2R_p} \tag{2.61}$$

The circuit efficiency for this case is

$$\eta = \frac{\beta V_{CC} V 2\pi R_S\left[\theta - \frac{\sin\,2\theta}{2}\right]}{\frac{\beta V_{CC} V}{\pi R_S}\left[\sin\,\theta - \theta\,\cos\,\theta\right]} = \frac{\theta - \frac{\sin\,2\theta}{2}}{2\,\sin\,\theta - 2\theta\,\cos\,\theta} \tag{2.62}$$

When $\theta$ equals $\pi/2$, the angle of conduction is 180° and the amplifier operates in the class-B mode. The circuit efficiency calculated from Eq. (2.62) is 78.5%, which agrees with previous results. As $\theta$ approaches zero, the limit of Eq. (2.62) is indeterminate. Applying L'Hopital's rule indicates that

$$\lim_{\theta\to 0} \frac{\theta - \frac{\sin\,2\theta}{2}}{2\,\sin\,\theta - 2\theta\,\cos\,\theta} = 1$$

An efficiency of 100% is approached as $\theta$ tends toward zero. However, the output power also tends toward zero resulting in a practical lower limit on $\theta$. Figure 2.29 shows plots of efficiency and output power as a function of $\theta$. The graph of output power is based on the assumption that the amplitude of input voltage $V$ is constant as $\theta$ is varied. This can be accomplished by varying the bias voltage $V_B$. The development of maximum output power requires the output voltage magnitude to equal $V_{CC}$. As $\theta$ decreases, the fundamental frequency component of collector current decreases in magnitude; thus the resonant tank circuit impedance must be increased to develop the maximum output signal. As $\theta$ approaches zero, a very small fundamental current component is present. Consequently, an extremely large value of $R_p$ is required to develop the maximum output voltage. It is reasonable to expect little output power under this condition. A reasonable compromise between circuit efficiency and output power often puts $\theta$ in the range of 60° to 80°, giving a conduction angle of 120° to 160°.

In applications with a variable input voltage $V$, the output voltage will depend on both $V$ and $\theta$. Equation (2.53) indicates that $\theta$ changes as $V$ is varied. Thus, it is difficult to calculate the maximum dissipation of the transistor. We can calculate this power when maximum output power is being delivered to the load.

The power dissipation of the transistor for maximum output power is found from Eq. (2.59) to be

$$P_{T-\text{maxout}} = \frac{\beta V_{CC} V}{\pi R_S}\left[1 - \frac{\theta}{2} + \frac{\sin\,2\theta}{4}\right] \tag{2.63}$$

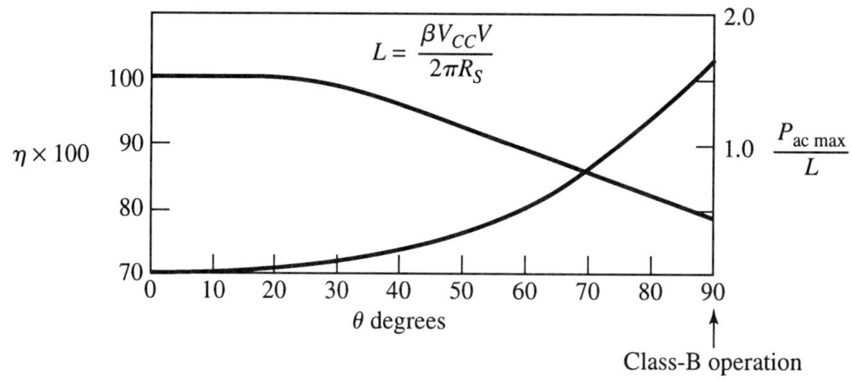


Figure 2.29: Circuit efficiency and output power as a function of $\theta$.

Maximum dissipation by the transistor does not occur simultaneously with maximum output. To calculate the maximum dissipation as $\theta$ and $V$ are varied, one could set the partial derivatives of $P_T$ with respect to these variables equal to zero. This would result in values of $\theta$ and $V$ that maximize $P_T$. Following this procedure, however, leads to a value of $\theta$ that is greater than 90°. The largest value of $P_T$ for the practical case can be found by setting $(\partial P_T/\partial V) = 0$ and allowing $\theta$ to equal 90°. For this case, $V$ is found to be

$$V = \frac{4}{\pi} \frac{V_{CC} R_S}{\beta R_p} \tag{2.64}$$

Using these values of $\theta$ and $V$ in Eq. (2.63) leads to a maximum dissipation of

$$P_{T\max} = \frac{2}{\pi^2} \frac{V_{CC}^2}{R_P} = 0.2 \frac{V_{CC}^2}{R_p} \tag{2.65}$$

This value can be used as a worst-case design value on which to base the allowable transistor dissipation.

One particular mode of operation that is of considerable interest relates to the class-C modulator stage. When the tuned circuit is tuned to the fundamental frequency and if the load $R_p$ is great enough to cause saturation of the transistor during the positive peaks of fundamental collector current, the output voltage will reach a peak-to-peak value of 2 $V_{CC}$. An increase in collector current will not increase this upper limit of output voltage if the tank circuit has a high-$Q$ value. This can be seen by noting that the peak swing of output voltage in the negative direction is $V_{CC}$ volts, from the quiescent level of $V_{CC}$ down to zero volts at saturation. The tank circuit offers zero impedance to all harmonics of the fundamental; thus, a voltage across this network can only be developed at the fundamental frequency. A perfect sinusoid at this frequency must be the resultant output. The swing in the positive direction must then be $V_{CC}$ volts. In practice, a small impedance at the higher harmonics may exist, and a modest amount of distortion will appear in the waveform.

As the base drive increases beyond that necessary for maximum output voltage, the collector-current waveform distorts considerably. Although current distortion is severe, the output variable of interest is voltage, which is almost free of distortion as explained earlier.

Later we will see that a class-C collector modulator operates under saturation-limited conditions. The output voltage is then determined by the value of the power supply voltage. To change the amplitude of the ac output voltage or carrier signal, the power supply voltage is modulated, resulting in a proportional change in magnitude in the carrier.

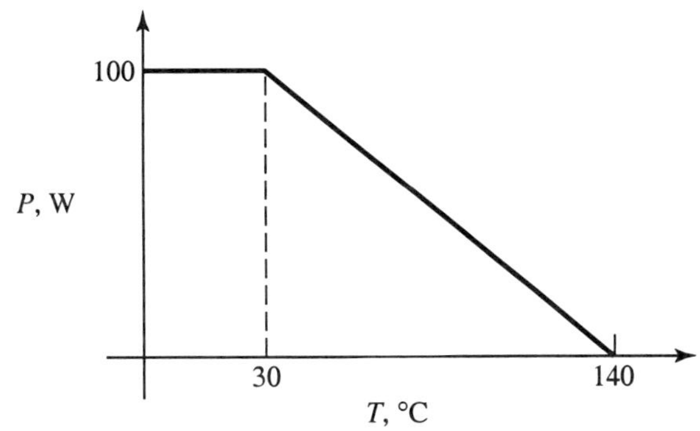


**Figure P2.1**

## SUMMARY

**1.** Traditional power stages are classified as class-A, class-B, or class-C stages, depending on the portion of the waveform period over which the stage conducts output current.

**2.** Class-A stages may have low distortion, but are somewhat limited in the amount of output power that can be produced.

**3.** Class-B stages can produce more output power than class-A stages, but care must be taken to limit distortion.

**4.** Class-C stages can be efficient and can produce very high levels of output power. Unfortunately, these stages function only over a very narrow band of frequencies and cannot be used for general-purpose amplifiers.

## PROBLEMS

### Sections 2.2.1-2.2.4 Allowable Dissipation

**2.1** The thermal characteristics of a power transistor are shown in the accompanying figure. Find the maximum allowable power dissipation for a maximum ambient temperature of 40°C given that the heat sink and insulating washer present a total thermal resistance of

(**a**) 2°C/W;

(**b**) 4°C/W.

(**c**) Calculate the allowable dissipation when no heat sink is used given that the thermal resistance between transistor case and air is 10°C/W.

**2.2** Repeat Problem 2.1 assuming that refrigeration is used to maintain an ambient temperature of −20°C.

**2.3** Calculate the maximum ambient temperature possible in Problem 2.1(a) to allow a transistor dissipation of 100 W.

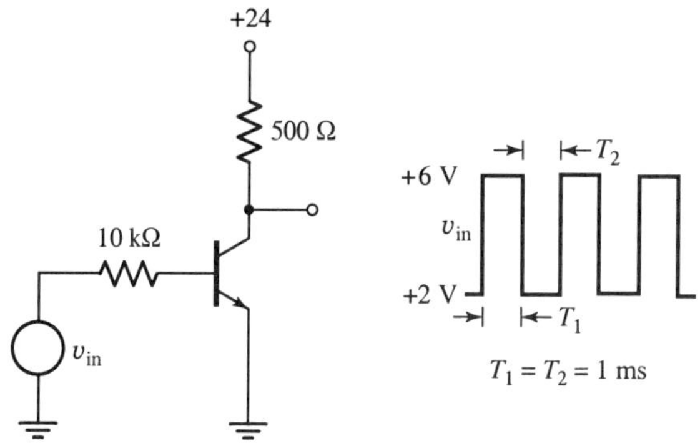


**Figure P2.10**

**2.4** A power stage must dissipate 20 W. If a transistor with thermal characteristics, as shown in the figure for Problem 2.1, is to be used and the maximum ambient temperature is limited to 30°C, what is the maximum usable value of thermal resistance of the heat sink and insulating washer?

**2.5** Repeat Problem 2.4 given that the transistor must dissipate 10 W.

**2.6** A low-power transistor is rated as a 750-mW transistor (infinite heat sink). The specifications state that above 25°C the rating should be reduced 10 mW/°C. If the thermal resistance between the case and free air is 200°C/W, what is the maximum allowable dissipation for an ambient temperature of 30°C? What is the maximum allowable junction temperature?

**2.7** Repeat Problem 2.6, given that a heat sink and insulating washer having a total thermal resistance of 100°C/W are used.

**2.8** Repeat Problem 2.6 for an ambient temperature of 0°C.

### Section 2.2.5 Power Dissipation Review

**2.9*** A circuit is connected between node $a$ and ground. Calculate the average power dissipation for the following situations. The current given is into node $a$.

**(a)** $v_a = 12$ V; $i = 0.1$ A.

**(b)** $v_a = 6 - 4 \sin \omega t$ V; $i = 0.1 + 0.07 \sin \omega t$ A.

**(c)** $v_a = 10$ V; $i = 0.1 + 0.07 \sin \omega t$ A.

**(d)** $v_a = 10$ V; $i$ is a repetitive rectangular wave that equals 1 A for 3 ms, goes to 1.5 A for 1 ms, returns to 1 A, and repeats the period.

**(e)** $v_a = 4 \sin \omega t$ V; $i = 3 \sin \omega t$ A.

### Section 2.3.1 Class-A Resistive Stages

**2.10*** Given that $\beta_o = 50$ and $V_{BE} = 0.6$ V find the power dissipated by the transistor shown in the figure. Assume that the switching time of the transistor is zero. If the transistor thermal characteristics correspond to those given in Problem 2.6, what is the maximum thermal resistance that the heat sink and insulating washer can have to prevent the destruction of the transistor? Assume a maximum ambient temperature of 30°C.

**2.11*** Repeat Problem 2.10, assuming that $T_1$ of the input waveform equals 4 ms and $T_2$ equals 1 ms.

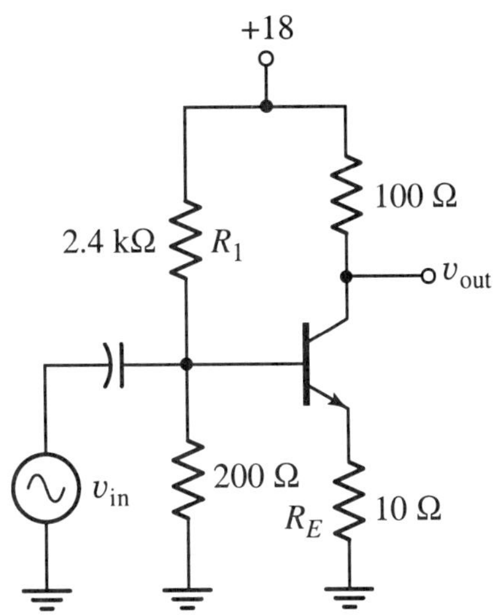


**Figure P2.16**

**2.12** Repeat Problem 2.10 given that $T_1$ of the input waveform equals 1 ms and $T_2$ equals 4 ms.

**2.13** Calculate the power dissipated by the 500-Ω resistor in Problem 2.10. Calculate the power delivered to the circuit by the dc source.

**2.14** Calculate the power dissipated by the 500-Ω resistor in Problem 2.11. Calculate the power delivered to the circuit by the dc source.

**2.15** Calculate the power dissipated by the 500-Ω resistor in Problem 2.12. Calculate the power delivered to the circuit by the dc source.

**2.16*** The transistor in the circuit shown in the figure has a current gain of $\beta_0 = 60$, and $V_{BE} = 0.6$ V.

**(a)** Calculate the no-signal power dissipated by the transistor.

**(b)** Calculate the maximum ac power that can be delivered to the load before cutoff or saturation occurs.

**(c)** Calculate the power dissipated by the transistor when maximum ac power output is present (assume an undistorted output signal).

**(d)** Calculate the power delivered by the dc source.

**2.17*** Repeat Problem 2.16 given that $R_1$ is changed to 1.5 kΩ.

**2.18*** Repeat Problem 2.16 assuming that $R_E$ is bypassed with a capacitor.

**2.19*** Repeat Problem 2.16 given that $R_1$ is changed to 1.5 kΩ and $R_E$ is bypassed with a capacitor.

**2.20*** The transistor used in the circuit has the thermal characteristics shown.

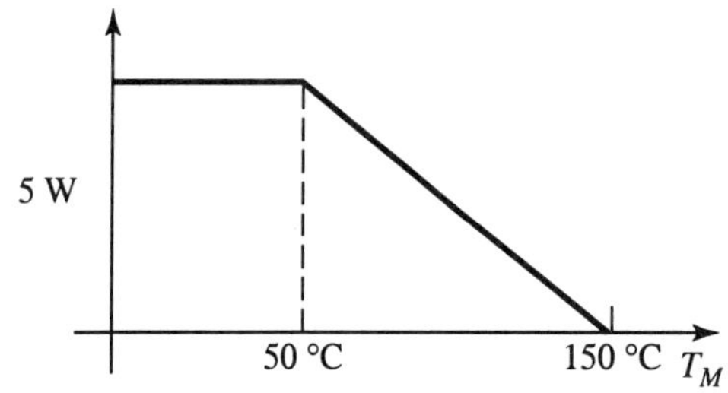


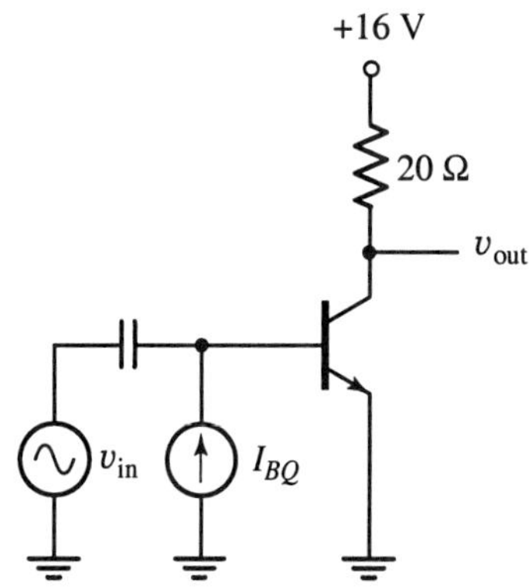


**Figure P2.20**

**(a)** If a heat sink and insulating washer are used with

$$\theta_{HS} + \theta_i = 30°\text{C/W}$$

and the maximum ambient temperature is 30°C, what is the allowable dissipation of the transistor?

**(b)** If there were no limit on allowable transistor dissipation, what is the maximum ac power that could be delivered to the 20-Ω load?

**(c)** Imposing the limit on allowable dissipation found in part (a), what is the maximum possible ac power that can be delivered to the load without exceeding $T_{J\max}$?

### Section 2.3.2 Class-A Transformer-Coupled Stage

**2.21** For the circuit shown in the figure,

**(a)** Assuming that the bias point can be adjusted and that the saturation voltage is zero, what is the maximum power that can be delivered to the ideal transformer in the circuit shown in the figure?

**(b)** In order to deliver the maximum power to the load, what value of quiescent collector current is required?

**(c)** What is the no-signal transistor dissipation at the bias point calculated in part (b)?

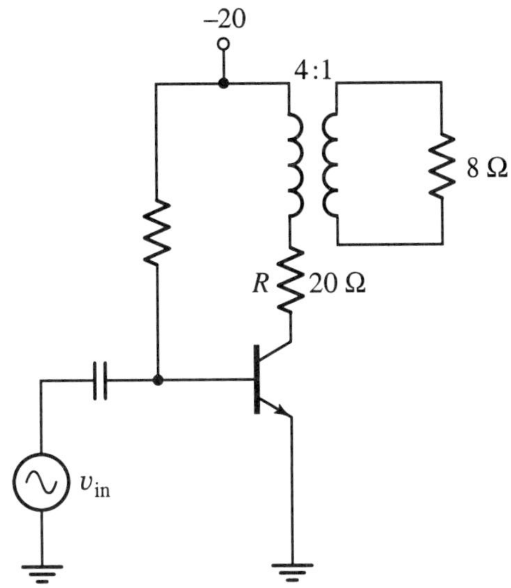


**Figure P2.21**

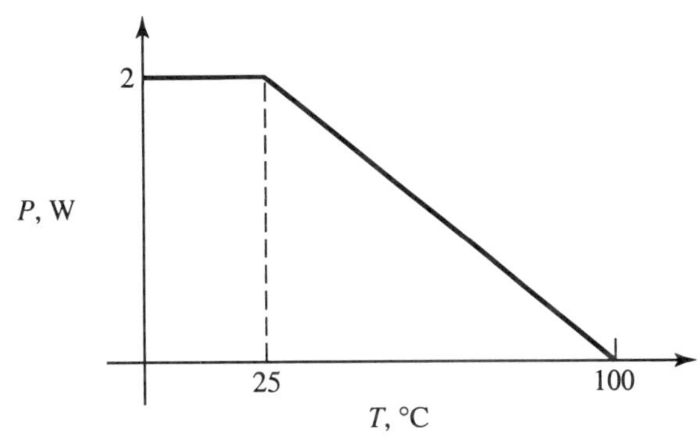


**Figure P2.23**

**Fig. P2.21**

**2.22** Repeat Problem 2.21 given that a 12-V power supply is used and that $R$ is 10 Ω.

**2.23*** If the transistor dissipation curve corresponds to that shown in the figure, and $\theta_i + \theta_{HS} = 15°\text{C/W}$, what is the maximum power that can be delivered to the load in the circuit of Problem 2.21 without exceeding the allowable transistor dissipation? Assume a maximum ambient temperature of 25°C.

**2.24** Repeat Problem 2.23 for $\theta_i + \theta_{HS} = 10°\text{C/W}$ and a maximum ambient temperature of 40°C.

**2.25** Work parts (a) and (b) of Example 2.4 for $\beta_0 = 100$.

**2.26** In Example 2.4, assume that the transistor has a $T_{J\max} = 100°\text{C}$ and $\theta_{JM} = 7.5°\text{C/W}$. What is the maximum value of $\theta_i + \theta_{HS}$ that will allow maximum power to be delivered to the load without destroying the transistor? The maximum ambient temperature will be limited to 20°C by forced-air cooling.

**D2.27** A transistor is rated as a 10-W transistor. The maximum junction temperature is 150°C and $\theta_{JM} =$

12°C/W. The transistor is used in a class-A stage, as shown in Fig. 2.16, with $R_E = 5\ \Omega, V_{CC} = +12$ V and a 4:1 transformer ratio. Given that $\theta_i + \theta_{HS} = 10$°C/W, and the maximum ambient temperature is to be 50°C,

**(a)** find the allowable dissipation of the transistor;

**(b)** determine the collector-base junction temperature when the transistor is dissipating the maximum allowable power;

**(c)** calculate the value of load resistance that allows maximum power to be delivered to the load (assuming an ideal transformer);

**(d)** find the maximum power that can be delivered to the load.

**D2.28** Repeat Problem 2.27 given that the transformer turns ratio is 2:1.

**D2.29** In the circuit of Fig. 2.16, $R_L = 8\ \Omega$, $V_{CC} = +20$ V, $R_E = 10\ \Omega$, and the dc resistance of the transformer primary is 10 Ω. Find the turns ratio $n$ that will cause maximum power to be delivered to the load, given that the maximum collector current is 200 mA. If the transformer efficiency is 0.9, what is the maximum power delivered to $R_L$?

**D2.30*** In the circuit of Fig. 2.16, $R_L = 8\ \Omega$, $R_2 = 100\ \Omega$, $V_{CC} = +40$ V, $R_E = 10\ \Omega$, $R_p = 20\ \Omega$, and the transformer turns ratio is 4 while the transformer efficiency is 0.9. The transistor has $\beta_0 = 40$, $V_{BE} = 0.7$ V, $\theta_{JM} = 4.5$°C/W, $T_{J\max} = 130$°C, and a maximum power dissipation of 20 W; $\theta_i + \theta_{HS} = 10$°C/W, and the ambient temperature will never exceed 50°C.

Find:

**(a)** the greatest power that can be delivered to the load without exceeding transistor dissipation;

**(b)** the quiescent collector current to deliver this power;

**(c)** $R_1$ to give this value of collector current;

**(d)** the circuit efficiency at this point;

**(e)** the no-signal transistor dissipation;

**(f)** the required value of $\theta_i + \theta_{HS}$ such that maximum circuit efficiency can be obtained;

**(g)** the maximum circuit efficiency;

**(h)** the no-signal transistor dissipation for the maximum circuit efficiency situation;

**(i)** the quiescent collector current for maximum circuit efficiency.

**D2.31*** Repeat Problem 2.30 given that the transformer turns ratio is 6:1.

**D2.32** It is desired to deliver at least 4 W of power to the 8-Ω load of the circuit. Three ideal transformers are available with turns ratios of $n_1$ =1:1, $n_2$ =2:1, and $n_3$ =3:1.

**(a)** Which turns ratio or ratios will satisfy the power specification? Support your answer with appropriate calculations.

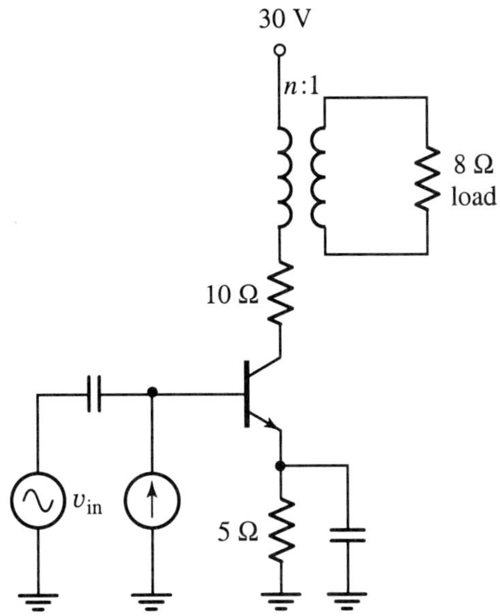


**Figure P2.32**

**(b)** What is the maximum power that can be delivered to the load if a turns ratio of 4:1 is used?

**(c)** What value of quiescent collector current should be used to achieve the power of part (b)?

**(d)** Calculate the necessary transistor dissipation to achieve the power of part (b).

**Fig. P2.32**

**D2.33*** Assume that an ideal transformer is used and that the $\beta$ of the BJT in the circuit is 80.

**(a)** If $I_{BQ} = 3$ mA, calculate the maximum power that can be delivered to the 16-Ω load before serious distortion is encountered.

**(b)** Calculate the circuit efficiency for the conditions of part (a).

**(c)** Calculate the no-signal power dissipation of the transistor for $I_{BQ} = 3$ mA.

**(d)** Calculate the necessary value of $I_{BQ}$ to achieve maximum efficiency.

**(e)** Calculate the maximum circuit efficiency for the conditions of part (d).

**(f)** Calculate the transistor power dissipation for the bias of part (d) when maximum power is being delivered to the load.

**D2.34*** Assume that the transformer of the circuit shown is ideal.

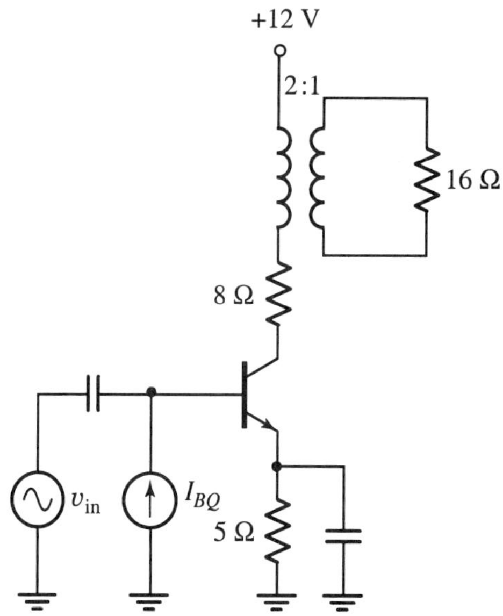


**Figure P2.33**

**(a)** Calculate $R_{\text{ac}}$ for the output loop.

**(b)** Calculate $R_{\text{dc}}$ for the output loop.

**(c)** Calculate $I_{CQ}$ for maximum circuit efficiency.

**(d)** Calculate $\eta_{\text{max}}$, the maximum circuit efficiency that can be obtained, neglecting $I_{BQ}$.

**(e)** Calculate the transistor dissipation when the peak value of ac collector current is 100 mA when using the $I_{CQ}$ of part (c).

**(f)** Calculate the no-signal transistor dissipation for the bias of part (c).

**(g)** If the allowable BJT dissipation is $P_{\text{ALLOW}} = 2$ W, what is the highest circuit efficiency that can be obtained before serious distortion occurs?

**D2.35*** Assume that $\beta = 50$ for the BJT of the circuit and the transformer is ideal.

**(a)** What is the value of $R_{\text{dc}}$ in the output loop?

**(b)** What is the value of $R_{\text{ac}}$ in the output loop?

**(c)** What is the value of $I_{CQ}$?

**(d)** What is the value of $V_{CEQ}$?

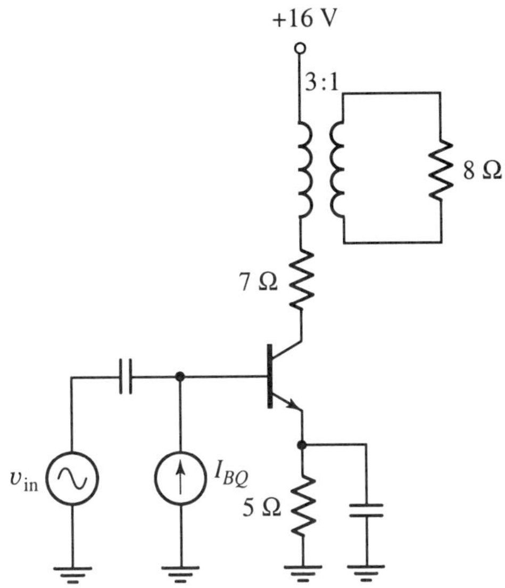


**Figure P2.34**

**(e)** If $v_{\text{in}}$ can be adjusted to any value, what is the maximum power that can be delivered to the 8-Ω load before encountering serious distortion for this bias point?

**(f)** To what value would you change $I_{BQ}$ in order to deliver the greatest possible power to the load?

**(g)** What is the maximum power that can be delivered to the load for the conditions of part (f)?

**Section 2.4.1 Analysis of Class-B Transformer Stages**

**2.36** Rework Example 2.5 for the case where $R_E = 10\ \Omega$ and the transformer turns ratio is doubled. Assume that $R_p = 20\ \Omega$ for the total dc transformer primary resistance.

**D2.37** Design a class-B amplifier to deliver 30 W to an 8-Ω load. Assume that a 12-V power supply exists and allow for a transformer dc primary resistance of 20 Ω (total) and $n_t = 0.9$. Specify a reasonable value for $\theta_i + \theta_{HS}$ assuming that the transistor thermal characteristics are as given in the figure for Problem 2.23 and that the maximum ambient temperature is 20°C.

**D2.38** Repeat Problem 2.37 given that the amplifier is to deliver 3 W to the load.

**2.39** Plot the transistor dissipation as a function of peak collector current $I_p$ for a class-B stage with a sinusoidal input. Assume $V_{CC} = 20$ V and $R_{\text{ac}} = 100\ \Omega$.

**2.40** Rework Example 2.5, given that the reflected ac load between center tap and each side of the primary is $n^2 R_L = 64\ \Omega$.

**2.41** Assume an ideal output transformer and negligible crossover distortion for the class-B stage shown. The peak value of the output voltage across the 8-Ω load is 4 V.

**(a)** Calculate the dissipation for each BJT.

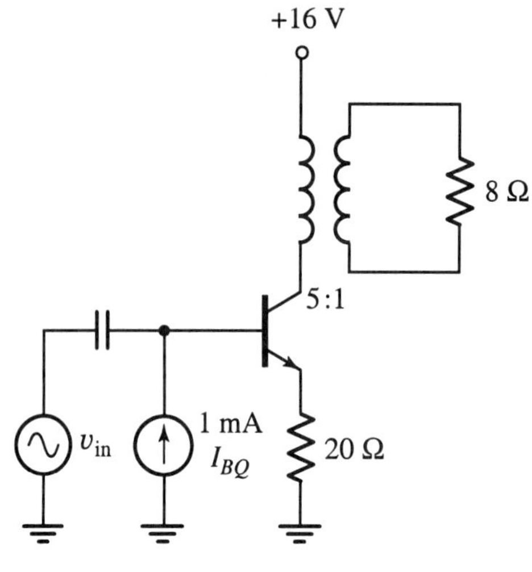


**Figure P2.35**

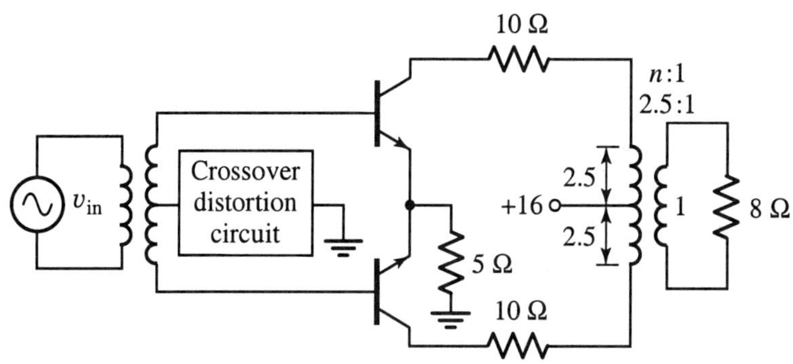


**Figure P2.41**

**(b)** Calculate the circuit efficiency.

**2.42** Assume ideal transformers and negligible crossover distortion for the class-B stage shown. The input $v_{in}$ can be adjusted to any value.

**(a)** How much power can be delivered to the load before saturation of the BJTs is reached?

**(b)** What is the worst-case transistor dissipation for this circuit?

### Section 2.5 Class-C Power Output Stages

**2.43*** A transistor can dissipate 10 W with the heat sink being used. If the power supply voltage for the stage is 30 V, $V_B = 1$ V, $\beta = 100$, $V_{BEon} = 0.5$, V and $V_{CE(\text{sat})} = 0$ V,

**(a)** find the minimum value of $R_p$ that can be used;

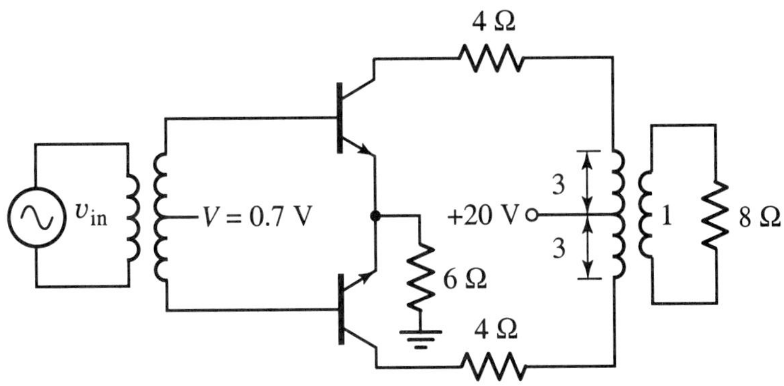


**Figure P2.42**

**(b)** calculate the maximum output power that can be delivered to $R_p$;

**(c)** calculate the values of $\theta$, $V$, and $R_S$ to deliver this power;

**(d)** find the circuit efficiency of the stage.

## REFERENCES

1. Comer, D. J. and Comer, D. T., "Control voltage determines multivibrator pulse width," *Electronics,* Vol. 40, No. 16, August 1967, pp. 115-116.

2. Duncan, B., *High Performance Audio Power Amplifiers.* Jordan Hill, Oxford, UK: Newnes, 1996.

3. Mohan, N., Undeland, T. M., and Robbins, W. P., *Power Electronics,* 2nd Ed. New York: John Wiley and Sons, 1995.

4. Self, D., *Audio Power Amplifier Design Handbook.* Jordan Hill, Oxford, UK: Newnes, 2000.

# Chapter 3

# ADVANCED POWER AMPLIFICATION

Chapter 2 developed design and analysis methods for the basic power stages that have served as workhorses of the field for many years. This chapter will consider more recent developments of power stages including those used in IC amplifier design. Before discussing these stages, some background theory will be developed.

## 3.1 AMPLIFIER SPECIFICATIONS

**Important Concepts**

1. Two important specifications in power amplifiers are power gain and total harmonic distortion.
2. The third-order intercept point is also used to characterize a power amplifier.

Bandwidth and voltage gain have been defined and discussed in connection with low-power stages. The parameters to be discussed next are often used to specify power amplifier performance.

### 3.1.1 Power Gain

The previous chapter did not mention power gain of the output stages considered. This may appear to be somewhat unusual, especially since the entire chapter is concerned with power amplification. The justification for deleting these calculations is that the actual design of the power stage can be based on voltage gain calculations that have been developed earlier. If, for example, 5 W of power are to be delivered to a 10-$\Omega$ load, the output voltage and load are established. If the input voltage and source impedance are known, the necessary voltage gain and impedance requirements can be accounted for in the amplifier design. Thus, power gain is related indirectly to these considerations.

Although there are many definitions of power gain, we will deal with the ratio of output power delivered to the load to the power reaching the amplifier input. Power gain can be calculated from the diagram shown in

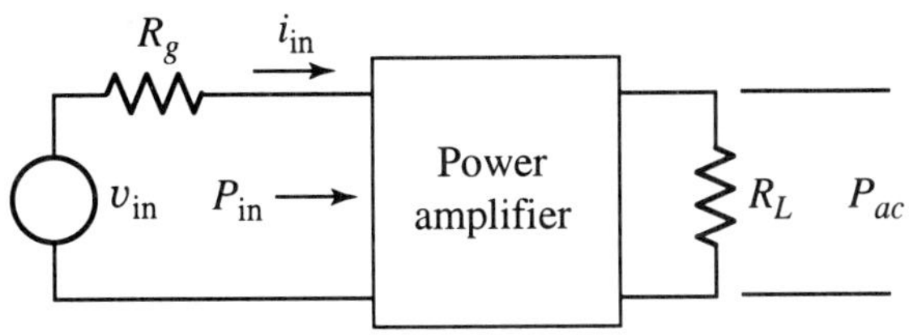


Figure 3.1: Power amplifier.

Fig. 3.1. The power gain is defined as

$$G_P = \frac{P_{ac}}{P_{in}} \tag{3.1}$$

This gain can be related to voltage gain for sinusoidal waveforms since

$$P_{ac} = \frac{[v_{out}]^2}{R_L}$$

and

$$P_{in} = (i_{in})^2 R_{in} = \left[\frac{v_{in}}{R_{in} + R_g}\right]^2 R_{in}$$

In these expressions, $i_{in}$, $v_{in}$, and $v_{out}$ are taken as rms quantities.

If $A_v$ is the loaded voltage gain of the circuit, that is,

$$A_v = \frac{v_{out}}{v_{in}}$$

then Eq. (3.1) can be written as

$$G_p = \left(\frac{v_{out}}{v_{in}}\right)^2 \frac{(R_{in} + R_g)^2}{R_L R_{in}} = (A_v)^2 \frac{(R_{in} + R_g)^2}{R_L R_{in}} \tag{3.2}$$

Thus, power gain can easily be related to voltage gain and impedance levels. For the amplifier in which $R_{in}$ is much larger than $R_g$, Eq. (3.2) can be written as

$$G_p = \left(\frac{v_{out}}{v_{in}}\right)^2 \frac{R_{in}}{R_L} = (A_v)^2 \frac{R_{in}}{R_L} \tag{3.3}$$

Let us assume that an ideal source drives the class-B amplifier of Example 2.5. The turns ratio of the input transformer is 1:1 between the primary and each side of the secondary to center tap. Losses in this transformer are negligible. If we assume that the secondary voltage of the output transformer can be related to the primary voltage by

$$v_s = \frac{v_p \sqrt{n_t}}{n}$$

where $n$ is the ratio of one-half the total primary to secondary turns and $n_t$ is the transformer efficiency, we can easily calculate the voltage gain. The magnitude of the voltage gain is

$$|A_v| = \frac{\beta n^2 R_L}{r_x + r_\pi + (\beta + 1) R_E} \frac{\sqrt{n_t}}{n}$$

If $\beta = 50$, $r_x = 20\ \Omega$, and the linearized value of $r_\pi$ is taken to be 10 Ω, the value of voltage gain is $|A_v| = 2.66$. The input impedance to the input transformer is 285 Ω, giving a power gain of

$$G_p = (2.66)^2 \frac{285}{8} = 253\ W/W$$

Although the voltage gain is rather low in power output stages, the power gain may be high as a result of the difference between input and load impedance levels. These stages typically have a high current gain.

**Practical Considerations**

In general, the voltage gain of a power output stage is not a key specification of the stage. A voltage amplifier will often precede the output power stages to provide the necessary voltage gain, while the power stage will provide a high current gain along with a moderate voltage gain. Many amplifying systems use a low-noise preamplifier followed by a voltage amplifier that drives the power output stages. The overall power gain of an amplifier with high voltage gain and high current gain can be quite large. The power gain is expressed in dB, noting that for power the definition of gain in dB is

$$G_{P-\mathrm{dB}} = 10 \log \frac{P_{\mathrm{ac}}}{P_{\mathrm{in}}}$$

A typical power gain for an amplifier that couples a microphone to a 10-W speaker is 80 dB.

**Practice Problem**

**3.1** A microphone with 10-kΩ resistance drives an amplifier with an input impedance of 50 kΩ. If the amplifier has a loaded voltage gain of $A_v = 100$ V/V when driving a 50-Ω load, what is the power gain of the amplifier? *Ans:* 71.6 dB.

### 3.1.2 Distortion in Power Stages

There are two major causes of nonlinear distortion in BJT stages. The first is the change of current gain $\beta$ with changes in $I_C$ or $V_{CE}$. If an undistorted base current enters the device, the output current is distorted as a result of the change in current gain as the output quantities vary. This effect can also be explained in terms of the output characteristics which show unequal spacing as $I_C$ or $V_{CE}$ is changed. Obviously, if the output signal amplitude is limited, less distortion occurs. In power stages, very large output-current changes are prevalent, and higher distortion levels are to be expected.

The second important cause of distortion arises from the nonlinearity of the base-emitter characteristics of the transistor. When a voltage source drives a common-emitter stage with little series resistance, the base current can be quite distorted. This current varies with voltage in the same way that diode current varies with diode voltage. It is possible to use this input distortion to partially offset the output distortion; however, feedback techniques are most often used in distortion reduction.

In an amplifier circuit, the output signal should ideally contain only those frequency components that appear in the input signal. As mentioned previously, the nonlinear nature of amplifying devices introduces extraneous frequencies in the output signal that are not contained in the input signal. These signals are referred to as harmonic distortion. One measure of the amplifier's performance is called the *total harmonic distortion* (THD).

If a sinusoidal signal is applied to the amplifier input, the output will also contain a major component of this signal at the fundamental frequency, with amplitude designated $v_f$. In addition, there will be smaller harmonic components in the output signal. These amplitude values will be designated $v_2$, $v_3$, $v_4$,...,$v_n$,... Fortunately, in engineering applications, the harmonics decrease in amplitude as frequency increases. Thus, perhaps only the second or third harmonic is large enough to effect the total harmonic distortion. The THD can defined as a

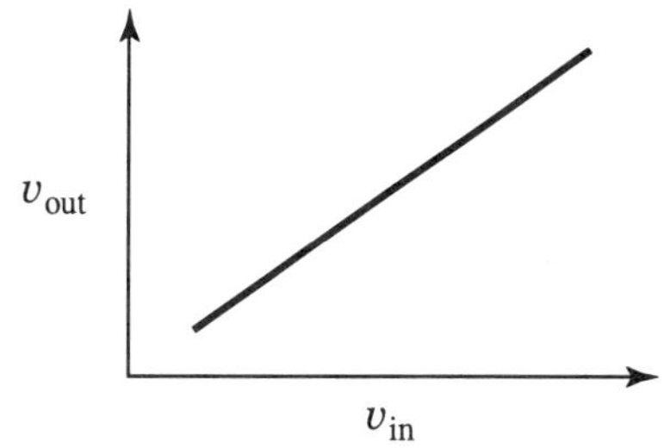


Figure 3.2: Transfer characteristics of a distortion-free amplifier.

percentage by

$$\text{THD} = \frac{\sqrt{v_2^2 + v_3^2 + v_4^2 + \cdots}}{v_f} \times 100 \tag{3.4}$$

In dB, this parameter becomes

$$\text{THD} = 10 \log \left( \frac{v_2^2 + v_3^2 + v_4^2 + \cdots}{v_f^2} \right) \tag{3.5}$$

The THD of an amplifier is a major specification used to characterize the amplifier performance.

The MOSFET also introduces nonlinear distortion due to the nonlinear relationship between input (gate) voltage and output (drain) current. This can readily be seen from the equation for transconductance

$$g_m = \sqrt{2\mu C_{\text{ox}}(W/L)I_D}$$

The voltage gain of the common-source stage is proportional to $g_m$, which in turn depends on $I_D$. As output voltage changes, this change in $g_m$ is driven by changes in $I_D$, resulting in nonlinear behavior.

A perfect, distortion-free amplifier would have transfer characteristics that form a straight line as shown in Fig. 3.2.

Unfortunately, as we have mentioned, the two most popular devices used for amplification do not have linear transfer functions. The BJT with no emitter degeneration has a collector current that relates to base-emitter voltage given by

$$I_C = \beta K_1 e^{qV_{BE}/kT}$$

where $k$ is Boltzmann's constant. The MOSFET has a drain current expressed as

$$I_D = K_2[V_{GS} - V_T]^m$$

where $V_T$ is the threshold voltage and $m$ is an exponent that varies between one and two, depending on the channel length. These functions are highly nonlinear, and only small variations of $V_{BE}$ or $V_{GS}$ can be applied to approximate linear behavior. An emitter resistance or emitter degeneration can be added to make the circuit more linear at the expense of reduced gain. This amounts to feedback to improve the nonlinear distortion. Another method of reducing distortion is to use several cascaded stages to achieve a very high gain; then feedback is applied around the amplifier to decrease the distortion.

In some instances, distortion is desirable. For example, the MOSFET mixer uses the square-law relationship between $V_{GS}$ and $I_D$ resulting in multiplication of two signals applied to the gate. The distortion in a mixer is high, but this is a desired result. A MOSFET amplifier, on the other hand, should minimize the nonlinearities between applied voltage and output voltage.

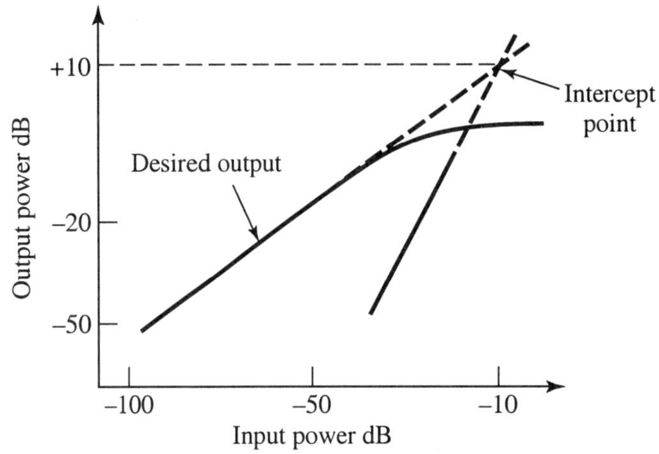


Figure 3.3: Amplifier output power vs. input power.

### Intercept Points

The unwanted harmonic distortion of an amplifier and the desired harmonic distortion of a mixer are functions of output signal size. As the signal becomes larger, the linear output will increase, and so will the distortion. In fact, a component due to harmonic distortion will increase more rapidly than the linear component.

Typically, an amplifier with a small output signal, compared to the size of the active region, will have a large linear output component or fundamental component relative to the harmonic components. As signal size increases, the size of the second or third harmonic may become as large as the fundamental component. Generally, this will not occur before the output signal reaches the extremes of the active region, consequently it would be impossible to measure the size of the output signal that results in a harmonic component that is equal in amplitude to the fundamental component.

The concept of the intercept point allows this output to be approximated. A plot, often in dB, is made of the input power versus the output power of the fundamental component. This plot is a straight line with a slope of unity for smaller output signals. As the output approaches the edges of the active region, the output power increases much more slowly than does the input power. Finally, the output power is limited to some fixed value as the output signal size is limited by the active region boundaries. Figure 3.3 shows this variation. Often, this plot is in terms of dBm or power relative to a 1 mW reference.

For small signals, each harmonic component is small. The third harmonic is plotted in Fig. 3.3 in addition to the fundamental component. As input power is increased, the third harmonic power may increase at three times the rate of the fundamental power. Before the input power is high enough to significantly limit the fundamental power, a straight-line extension is added to the fundamental power plot and the third harmonic power plot. The point where these two extensions intersect is called the *third-order intercept* point (TOI). Note that this point falls beyond the actual output power that can be achieved by the amplifier. The TOI is specified in terms of the output power along the linear line at this point of intersection. For example, in Fig. 3.3, the TOI is +10 dBm. The higher the intercept point, the more linear is the amplifier.

A second-order intercept can also be defined in much the same way as the TOI is defined. For communication amplifiers, the TOI is generally more important than the second-order intercept because the third-order term can result in a frequency that falls within the desired passband of the amplifier. For a mixing circuit, the second-order intercept is important, and mixing will be more efficient for smaller values of this intercept point.

### Basic Concepts of Harmonic Distortion

The mathematics involved in dealing with linear systems is much simpler than the mathematics required for nonlinear systems. For this reason, electrical engineers often assume linear operation of devices and systems and ignore the nonlinear effects to the extent possible. As electrical engineers analyze amplifier performance, they often assume the devices are linear. This assumption is valid only when the output signals do not contain significant distortion.

In fact, small-signal operation should be defined as output signals that are small enough so that the distortion contained in the signal is insignificant to the application at hand. For a high-fidelity amplifier, the distortion may be more important than for a circuit designed to amplify a telephone signal. We should recognize that all devices have nonlinear characteristics that become more significant as the output signal becomes larger.

Let us assume that an amplifier is constructed whose output contains a linear term, a second-order term, and a third-order term. The output voltage is expressed as

$$v_{\text{out}} = av_{\text{in}} + bv_{\text{in}}^2 + cv_{\text{in}}^3 \tag{3.6}$$

Incidentally, for a good amplifier, the coefficient $a$ should be much larger than $b$ or $c$. For a good mixer, $b$ should be maximized. Now suppose that the input signal consists of two components of different frequencies, but equal magnitude; that is,

$$v_{\text{in}} = A \cos 2\pi f_1 t + A \cos 2\pi f_2 t$$

Using various trigonometric identities, we find that the output signal contains the components listed in Table 3.1.

**Table 3.1 Signal and Distortion Components**

| Due to | Output Components |
|---|---|
| linear | $f_1$<br>$f_2$ |
| 2nd order | dc<br>$f_1 + f_2$<br>$2f_1$<br>$2f_2$ |
| 3rd order | $f_1$<br>$f_2$<br>$3f_1$<br>$3f_2$<br>$2f_1 - f_2$<br>$2f_1 + f_2$<br>$2f_2 - f_1$<br>$2f_2 + f_1$ |

In an amplifier, the output terms should be determined by the linear coefficient, $a$, and should contain frequencies of $f_1$ and $f_2$. The terms introduced by the second-order term are at frequencies of dc, $f_1 + f_2$, $2f_1$ and $2f_2$. These are unwanted components, but can generally be filtered out of the output signal. Except for the dc term, the frequencies are often outside the range of frequencies of interest.

The third-order term also introduces components with the desired frequencies of $f_1$ and $f_2$. If a bandpass filter is used to pass only these components, the output will still contain unwanted components at these frequencies due to the third-order term. Furthermore, depending on the two frequencies, one of the components at frequencies $2f_1 - f_2$ or $2f_2 - f_1$ introduced by the third-order term might also be near the frequencies $f_1$ or $f_2$. The TOI point is important in amplifier applications because of the possible frequency components introduced near the desired frequencies. In most applications, a signal that contains several frequencies might be applied to the amplifier, resulting in many unwanted terms.

For a mixer, the sum or difference of the two frequencies is important. In this case, the second-order term is the important term that results in the correct output frequencies. The TOI is an oft-used specification for power stages as well as high-frequency amplifiers.

## Example 3.1

An output signal is expressed as

$$v_{\text{out}} = 9.6 \sin 6283t + 0.4 \sin 12,566t + 0.16 \sin 18,849t$$

when the input power is -40 dB. Assuming the third harmonic increases three times as fast as the fundamental with increasing input power, calculate the TOI of this signal.

**Solution** The power in the fundamental component of the output signal is higher than the power in the third harmonic by

$$10 \log \frac{9.6^2}{0.16^2} = 35.6 \text{ dB}$$

The dB change in the fundamental will equal the change in input power while the dB change in the third harmonic will be three times that of the input power change. Thus, we can write

$$P_f + \Delta P_{\text{in}} = P_3 + 3\Delta P_{\text{in}}$$

where $P_f$ is the fundamental power when $P_{\text{in}} = -40$ dB, $P_3$ is the third harmonic power when $P_{\text{in}} = -40$ dB, and $\Delta P_{\text{in}}$ is the increase in input power. Noting that $P_f - P_3$ has been found to be 35.6 dB, we solve this equation to find that

$$\Delta P_{\text{in}} = 17.8 \text{ dB}$$

The increase of input power required to make the third harmonic power equal the fundamental power is 17.8 dB. Consequently, the TOI is

$$\text{TOI} = -40 + 17.8 = -22.2 \text{ dB}$$

**Practice Problem**

**3.2** For the signal defined in Example 3.1, calculate the magnitude of the fundamental voltage and the third harmonic voltage when $P_{\text{in}}$ is -50 dB. *Ans:* $v_f = 3.04$ V, $v_3 = 0.00506$ V.

## 3.2 TRANSFORMERLESS CLASS-B POWER STAGES

**Important Concepts**

**1.** The complementary output stage can be used to eliminate the transformer from the fundamental class-B amplifier.

**2.** The complementary stage can be tailored for implementation as an IC by adding a distortion reduction circuit and short-circuit protection to the output stage.

**3.** The bridge output stage can be used to drive an output in both directions when powered by single dc voltage without requiring a transformer.

**4.** The class-G stage delivers very high levels of peak power to the load with reduced transistor dissipation.

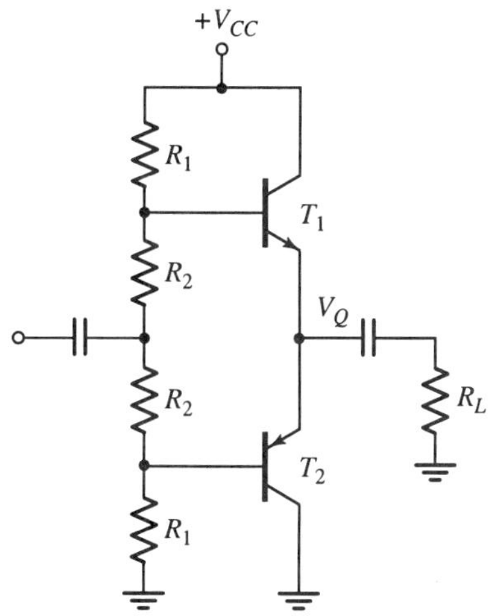


Figure 3.4: Complementary emitter followers.

The class-B stage was introduced by the transformer-coupled stage of Fig. 2.19. However, transformers present several problems in high-quality amplifiers such as audio high-fidelity or hi-fi amplifiers. They are heavy, the core can saturate at high current levels causing waveform distortion, and poor transformer efficiency will lower the overall circuit efficiency of the class-B stage.

Since the 1960s, transformers have been used less and less in audio stages to the point that it is rare to encounter a transformer-coupled stage in high-quality audio amplifiers. There are lower-quality applications that may use transformer coupling, but newer methods of output stage design depend on good power BJTs or MOSFETs.

In lower-quality (higher distortion), high-power output stages, heat sinks and transformers are often used in fundamental configurations. Since these stages require large heat sinks to dissipate thermal energy, it may not be imperative to minimize amplifier size or weight. The use of transformers is not uncommon in these applications. On the other hand, if size or weight is a consideration, elimination of a transformer can result in a smaller stage. In addition, if power dissipation of the transistors can be reduced while delivering specified power to the load, the amplifier size and weight can be further decreased. In this instance, transistor and heat-sink size can be reduced.

This section will consider modifications to the class-B stage that eliminate transformers, enhance efficiency, or improve performance in some other way. Some of these stages can be implemented on IC chips for moderate output powers. Class-D stages will also be introduced.

### 3.2.1 Class-B Complementary Emitter Followers

It is possible to design a class-B amplifier without transformers if *complementary power transistors* (an em *npn* device and a matched *pnp* device) are used. A simple push-pull, complementary emitter-follower stage is shown in Fig. 3.4. The resistors $R_1$ and $R_2$ are chosen to set $V_Q$ at a value equal to one-half the power supply voltage. The base of $T_1$ should be a few tenths of a volt more positive than $V_Q$, while the base of $T_2$ is a few tenths lower. This slight forward bias on each transistor minimizes the crossover distortion.

The positive half-cycle of an input sinusoid leads to conduction by $T_1$, while the alternate half-cycle leads to conduction by $T_2$. This power amplifier results in a current gain, but has a voltage gain of approximately unity. Any voltage amplification required can by done be earlier stages at a lower power level before reaching the complementary emitter followers.

Capacitor coupling to the load is often not desirable because a very large capacitance is required to effectively couple the signal to a low resistance load down to reasonable frequencies. If use of a coupling capacitor is inappropriate in a given application, the circuit of Fig. 3.5 can be utilized. The circuit requires two voltage

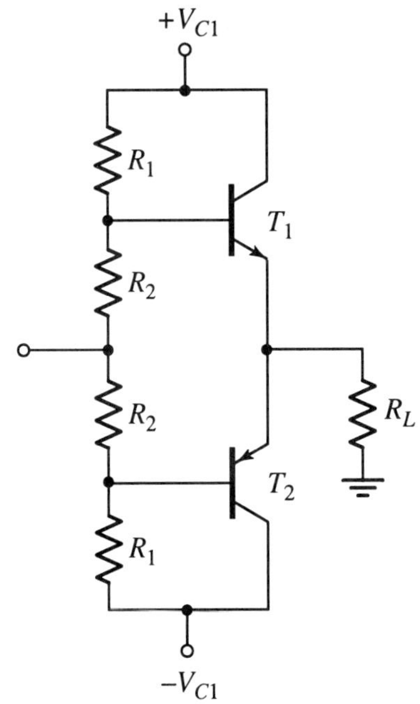


Figure 3.5: Complementary emitter followers with two dc sources.

sources, but gives performance similar to the previous circuit when each source equals $V_{CC}/2$. The input signal can now be dc-coupled if the average value is zero volts.

The input power from the dc voltage sources for the circuit of Fig. 3.5 is

$$P_S = \frac{2V_{C1}I_p}{\pi} \tag{3.7}$$

for a peak load current of $I_p$. The load power is

$$P_{\text{ac}} = \frac{I_p^2 R_L}{2} \tag{3.8}$$

The power delivered by the source must equal the output power plus the power dissipated by the two transistors. Individual transistor dissipation is then

$$P_T = \frac{(P_S - P_{\text{ac}})}{2} = \frac{V_{C1}I_p}{\pi} - \frac{I_p^2 R_L}{4} \tag{3.9}$$

As in the transformer-coupled class-B stage, the maximum transistor dissipation can again be found by differentiating $P_T$ with respect to $I_p$ and equating the result to zero. This peak dissipation occurs for a peak collector current of

$$I_p = \frac{2V_{C1}}{\pi R_L} \tag{3.10}$$

and is

$$P_{T\max} = \frac{V_{C1}^2}{\pi^2 R_L} = \frac{0.1V_{C1}^2}{R_L} \tag{3.11}$$

This value of transistor dissipation is the worst-case design value, which must be exceeded by the allowable transistor dissipation.

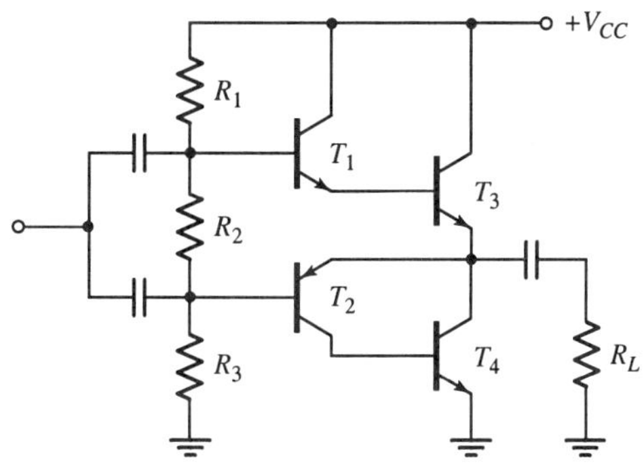


Figure 3.6: A quasi-complementary symmetry class-B stage.

Maximum output power to the load is

$$P_{\text{ac-max}} = \frac{V_{C1}^2}{2R_L} \tag{3.12}$$

and the corresponding transistor dissipation is

$$P_T = \frac{0.068V_{C1}^2}{R_L} \tag{3.13}$$

Maximum circuit efficiency occurs at maximum output power and again has a theoretical value of $\eta_{\max} = 78.5\%$.

**Practical Considerations**

In practice, small emitter resistors will be used to avoid thermal instability and decrease nonlinear distortion. This fact, along with distortion problems and transistor mismatch, limit the practical maximum circuit efficiency to something lower than 78.5%. The emitter resistors are selected to be less than 10% of the load resistor, if possible.

**Practice Problem**

**3.3** A 5-Ω resistance is inserted in series with each emitter in the circuit of Fig. 3.5. If 12-V power supplies are used and the saturation voltage of the transistors is 0.5 V, calculate the maximum circuit efficiency for the circuit when driving a 60-Ω load. Neglect bias power. *Ans:* 69.5%.

### 3.2.2 The Quasi-Complementary Symmetry Class-B Stage

Other types of complementary circuits can be used to advantage for class-B stages. A quasi-complementary symmetry amplifier is shown in Fig. 3.6. Transistors $T_1$ and $T_3$ make up a Darlington pair or double emitter follower. These stages amplify the positive-going portion of the input signal. Transistors $T_2$ and $T_4$ form an arrangement that amplifies the negative-going portion of the waveform. When the signal at the base of $T_2$ swings in the negative direction, the emitter must follow. Since the emitter of $T_2$ is connected to the load, current must be pulled through the load to force the voltage at this point to follow the input signal. The necessary current is conducted primarily by $T_4$ rather than $T_2$. As the emitter and collector currents of $T_2$ increase, the base drive to $T_4$ rises. This increased base drive of $T_4$ leads to the desired increase of collector and load current. Hence the appropriate load voltage is established as a result of the emitter-follower action of $T_2$, while the large amount of current necessary to drive the load is handled by $T_4$.

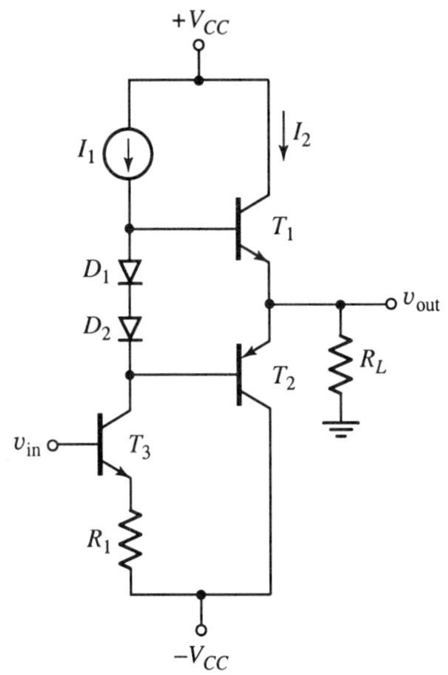


Figure 3.7: Class-B IC stage.

This circuit was originally developed in response to the difficulty in finding *pnp* power transistors that were truly complementary to the better *npn* power device. This problem has been minimized in recent years as *pnp* fabrication methods have improved. In the arrangement of Fig. 3.6, $T_1$ and $T_2$ can be matched, whereas the high-power stages, $T_3$ and $T_4$, can both be *npn* transistors.

### 3.2.3 Integrated Amplifier Class-B Stages

Although some of the circuits previously considered can be integrated, there are other versions of the class-B stage that are more popular in IC amplifiers. Figure 3.7 shows one such stage.

The current source $I_1$ consists of an IC current-source stage to provide a small bias current through diodes $D_1$ and $D_2$. A small quiescent collector current $I_2$ is necessary to reduce crossover distortion at the output. As $v_{\text{in}}$ swings positive the input to $T_2$ goes negative, turning $T_2$ on while shutting $T_1$ completely off. A negative-going waveform at the amplifier input drives the bases of $T_1$ and $T_2$ positive to shut $T_2$ off and pass the signal to the output through $T_1$.

The amplifier can be made *short-circuit proof* by limiting the output current that can flow if the output terminal is accidentally shorted to one of the supplies. Figure 3.8 indicates the additional circuitry required for this purpose.

The emitter-base junctions of transistors $T_4$ and $T_5$ are driven by the voltage drops across the resistances $R_{E1}$ and $R_{E2}$. Under normal operating conditions, these voltages are too small to turn $T_4$ and $T_5$ on; thus, circuit operation is unaffected. If the output is short-circuited to the negative supply voltage, serious damage to device $T_1$ could result if transistor $T_4$ were not present. A large voltage would appear across $T_1$ and the base-emitter junction would be forward biased resulting in a high value of emitter current. The excessive power dissipation could destroy $T_1$ if the output were to be shorted to the negative supply or even to ground. When this occurs for the circuit of Fig. 3.8, $T_4$ becomes sufficiently forward biased to divert the base-current drive from $T_1$. The maximum current that can flow in the output circuit is then limited to $V_{BE4}/R_{E1}$. Typical maximum currents for the short-circuit case range from 10 mA to 50 mA for modern IC amplifiers.

The $V_{BE}$ multiplier circuit often replaces the two diodes in Figs. 3.7 and 3.8 to get better cancellation of the crossover voltage of the output devices. This circuit appears in Fig. 3.9.

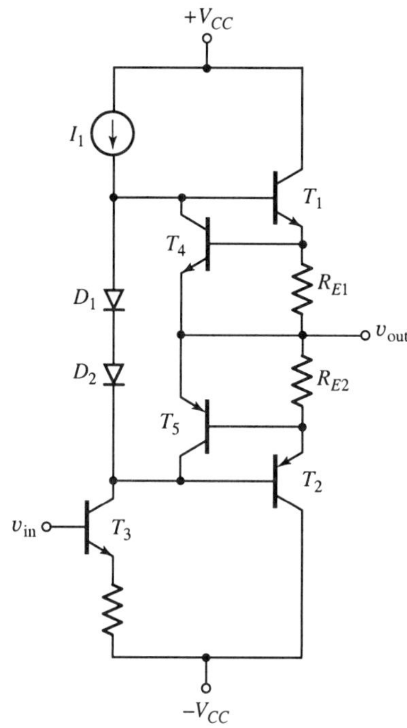


Figure 3.8: Short-circuit protection.

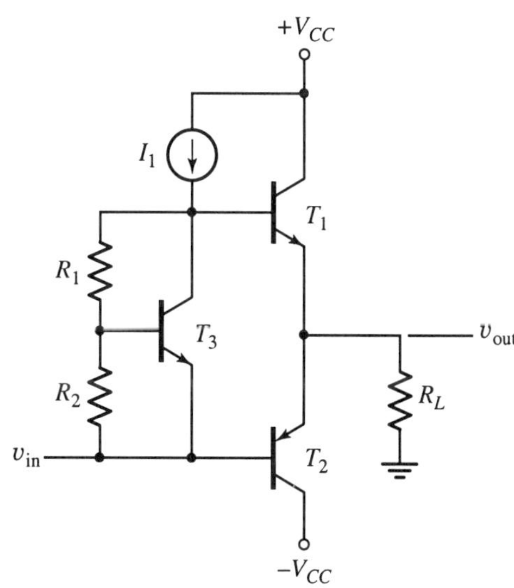


Figure 3.9: Class-B stage using a $V_{BE}$ multiplier.

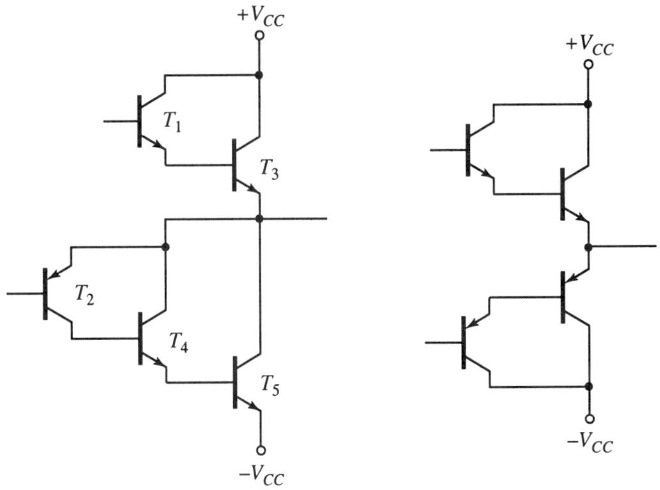


Figure 3.10: Darlington output stages.

If negligible base current flows in $T_3$, the voltage across $R_1$ and $R_2$ is

$$V_{CE3} = (1 + R_1/R_2)V_{BE3} \tag{3.14}$$

This voltage is found by noting that the drop across $R_2$ is constrained to be $V_{BE3}$, and this value must also equal

$$\frac{R_2}{R_1 + R_2} V_{CE3}$$

The voltage $V_{CE3}$ is used to eliminate crossover distortion and can be adjusted by the ratio of the two resistors. Whereas the absolute values of individual resistors cannot be accurately determined in standard IC fabrication processes, the ratio of two resistors can be determined to the required accuracy.

There are several other output configurations based on the complementary emitter follower. A popular one is the Darlington output stage modified for IC amplifiers as shown in Fig. 3.10. The current gain of this stage is very high, approximately equal to $\beta^2$. If high-gain *pnp* devices are available, a Darlington pair similar to the upper *npn* pair can replace devices $T_2$, $T_4$, and $T_5$.

For the Darlington pairs, a larger difference of input bias voltage must be provided to the inputs of the respective pairs due to the larger voltage drop between each input and output which is now 2 $V_{BE(on)}$ instead of just $V_{BE(\text{on})}$. The $V_{BE}$ multiplier circuit can be designed to generate this increased bias voltage.

**Practice Problem**

**3.4** The $V_{BE}$ multiplier is used to produce a voltage drop of 2.5 V with a value of $R_2 = 5$ kΩ (see Fig. 3.9). Assuming that $V_{BE} = 0.68$ V, calculate the necessary value of $R_1$. *Ans:* $R_1 = 13.4$ kΩ.

### 3.2.4 Additional Class-B Output Configurations

Emitter followers can be used to drive a high-current speaker in the complementary configuration. This configuration exhibits no voltage gain in the output stages, although current gain can be quite high. The common-emitter stages of Fig. 3.11 make up an alternate configuration that can result in voltage gain as well as current gain.

We are more familiar with the configuration of part (a) which shows a load in each collector, but the conducting wire from point *a* to point *b* conducts positive current from *Q*1 for one half-cycle, and then conducts

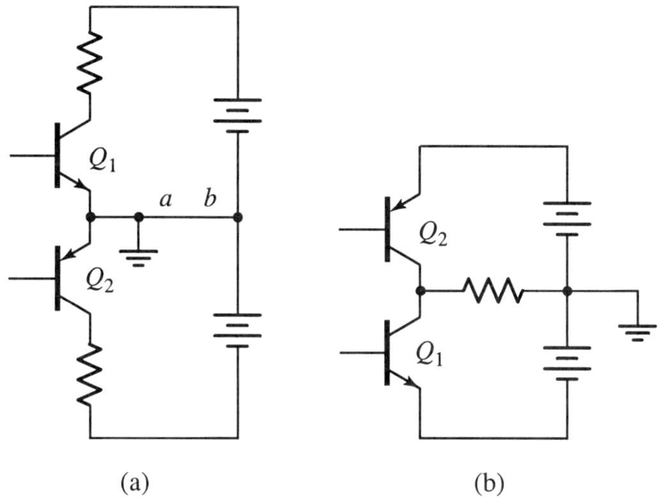


Figure 3.11: (a) The common-emitter output stage with dual loads. (b) The common-emitter output stage with a single load.

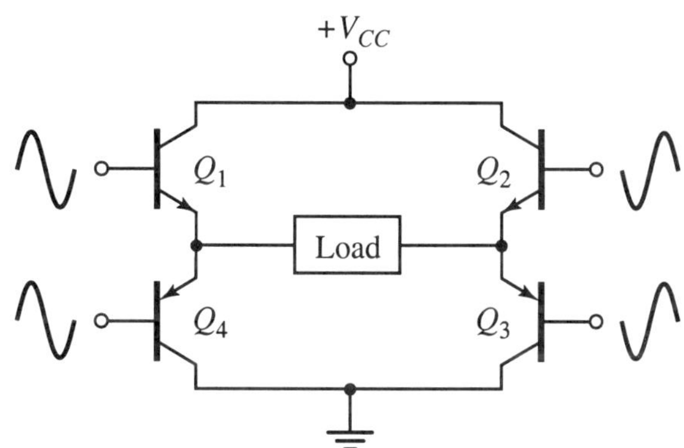


Figure 3.12: A bridge output stage.

negative current from $Q2$ during the second half cycle. Thus, a single load can be placed in this path and will develop a complete sinusoidal output voltage for an applied sinusoidal input voltage. Figure 3.11(b) shows a modified architecture that results in push-pull operation. Although this stage exhibits voltage gain as well as current gain, it requires a complex bias circuit to function properly.

### Bridge Output Stages

A popular high-power output stage for driving an electric motor in both directions with a single power supply was adapted to high-power audio amplifiers in the 1960s and 1970s. This configuration is shown in Fig. 3.12.

In this circuit, two legs of the transistor bridge circuit drive current through the load in one direction while the other two legs drive current in the opposite direction. For example, $Q1$ and $Q3$ are turned on as the input signal goes positive. Note that the positive-going signal is applied to $Q1$ while an inverted signal is applied to $Q3$. As the input goes negative, $Q2$ turns on and an inverted input also drives $Q4$ on.

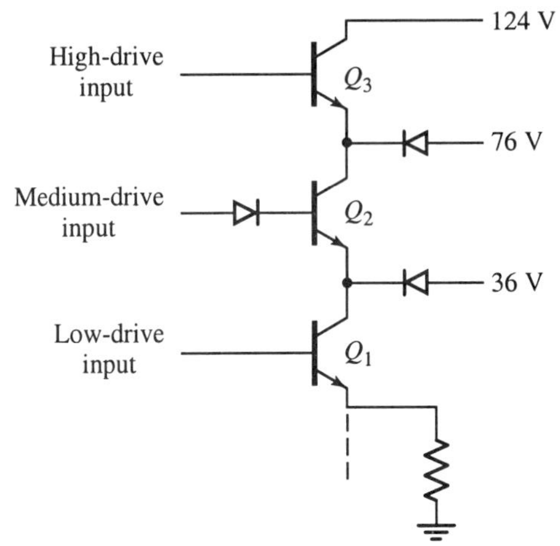


Figure 3.13: The positive half of a class-G output stage.

### Class-G Output Stages

Speech and music signals are very dynamic. The average amplitude of audio is a small fraction of the occasional large peak voltages that occur. In order for a conventional class-B stage to deliver the large peak voltage without saturation, the power supply must be large enough to accommodate the largest voltage and current peaks. The power delivered from the dc source for a complementary emitter-follower stage was previously derived as

$$P_S = \frac{V_{CC} I_P}{\pi} \tag{3.15}$$

The maximum transistor dissipation was found to be

$$P_{T\max} = \frac{0.025 V_{CC}^2}{R_L} \tag{3.16}$$

where $R_L$ is the load resistance.

In a given audio amplifier, the power supply voltage may be 124 V to supply the maximum peak current needed. Very large transistor dissipations and very large source powers are needed to supply current from such a high-voltage source. The class-G stage and its variations can be used to cut down the power dissipation while allowing large peak voltages to drive the load.

One version of the class-G stage is partially shown in Fig. 3.13.

For most of the time, the output voltage will be less than 36 V, and the current to the load is supplied through device $Q1$ from the 36-V supply. The collector of $Q1$ will be at 36 - 0.7 = 35.3 V. The emitter of $Q2$ is also at this potential, whereas the base of $Q2$ is less than this voltage. Consequently $Q2$ is in the off state. Thus, no current flows from the 76-V or the 124-V supplies. From Eq. (3.15), the source input power is determined by the average source current and the 36-V source voltage.

As the input signal increases above 35-36 V, the transistor $Q2$ turns on. There is a dc voltage-level shift between the base of $Q1$ and the base of $Q2$ to turn $Q2$ on at the proper input level. As the base of $Q2$ goes more positive, the emitter of this device is pulled higher and reverse biases the diode connected to the 36-V supply. Now all current comes from the 76-V supply until the input exceeds this value, turning on $Q3$ and turning off $Q2$. Only on very large peaks does the 124-V source supply current to the load. The 76-V source supplies current for an intermediate output voltage amplitude and the 36-V source supplies current for smaller amplitudes.

The design of the input circuit to drive these three transistors while making smooth transitions, without introducing excess distortion, is rather involved and will not be considered here.

For the greatest percentage of time, the input dc source dissipation is proportional to 36 V times the average current of the half-wave audio signal. For a smaller time, the input dissipation is proportional to 76 V times the average current, and for only small fractions of time, when large voltage peaks are present, the input dissipation is proportional to 124 V times the average current. Thus, large peak voltages and peak powers can be amplified when needed, but average input power to the circuit and transistor dissipation are minimized.

# Example 3.2

A class-G amplifier uses power supplies of ±120 V, ±80 V, and ±40 V. For a certain input signal, the average currents from these dc sources are $I_{120} = 100$ mA, $I_{80} = 250$ mA, and $I_{40} = 2$ A. These currents represent the total average values supplied by both positive and negative supplies. Calculate the total power delivered from the sources. Compare this to the value that would be delivered if only the $\pm 120 - V$ sources were used.

**Solution** The powers delivered from the six dc sources are

$$P_{S120} = 120 \times 0.1 = 12 \text{ W}, \quad P_{S80} = 80 \times 0.25 = 20 \text{ W}, \quad P_{S40} = 40 \times 2 = 80 \text{ W}.$$

The total power is then

$$P_{ST} = P_{S120} + P_{S80} + P_{S40} = 112 \text{ W}$$

If supplied by the ±120-V power supplies only, the average current will be

$$I_T = 0.1 + 0.25 + 2 = 2.35 \text{ A}$$

The source power is then

$$P_{S120} = 2.35 \times 120 = 282 \text{ W}$$

A conventional class-B stage with ±120-V supplies would require 282 W whereas the class-G amplifier would need only 112 W. Implied in this result is the fact that the individual transistor dissipations are also correspondingly smaller. Offsetting this power advantage is the need for complex drive circuitry, six transistors rather than two, and several high-current diodes.

**Practical Considerations**

The Carver Cube Model M400A amplifier introduced in 1981, weighed under 10 pounds and occupied 6.8 cubic inches. This amplifier delivered up to 500 W of power output when driven with music signals. Only during large instantaneous peak output signals does the amplifier dissipate maximum power. For a large percentage of the time, power is delivered from a lower supply voltage that minimizes amplifier dissipation.

There are many other variations in the audio power amplifier field. The class-G can also operate in a parallel mode rather than in the serial mode of Fig. 3.13.

## 3.3 PULSE-WIDTH MODULATED AUDIO POWER AMPLIFIERS

**Important Concepts**

**1.** The class-D amplifier is a pulse-width modulated stage with information carried in the average value of the rectangular waveform.

**2.** Efficiencies near 90% are common for wideband signals in the audio range. Power outputs of audio stages in the 400 to 600 W range are possible.

**3.** The switching frequencies of 200 to 500 kHz tend to produce electromagnetic interference and other noise components.

The integrated circuit offers many advantages over the conventional, lumped-element circuit in a number of applications. Chief among these advantages are digital networks, low-power amplifiers, and low-power, subminiature communications systems. On the other hand, applications requiring large capacitors, inductors, transformers, or high-power capability rarely utilize the integrated circuit. A direct, integrable replacement for a large inductor or transformer has yet to be found, while large capacitors cannot be integrated due to volume requirements.

A power amplifier requires that the output stage or stages dissipate a certain fraction of the power delivered to the circuit from the source. For class-A resistive stages, the maximum circuit efficiency is 25%. The device in this case dissipates twice the power delivered to the load, or one-half of the total power delivered by the dc power supply. For a 5-W output signal, the output transistor must be capable of dissipating 10 W. In a class-B stage the situation is improved somewhat as a result of obtainable efficiencies of 50-60%. A 5-W output signal now results in a 4-W to 5-W device dissipation. The heat generated by the output stages must be conducted away from the device by means of a high thermal conductance path. The minimum size of an integrated circuit is now limited by thermal conductance path. Not only does the small integrated circuit experience difficulty in conducting heat from the device, but the reliability of the unit is adversely affected when the chip temperature is elevated. It is then desirable to develop more efficient amplifiers that do not require transformers or large capacitors.

### 3.3.1 The Concept of Pulse-Width Modulation

A pulse-width modulation scheme known for many years appears to be quite appropriate for integrated circuit power amplifiers. This type of amplification has been used successfully in high-power transistor amplifiers to deliver several hundred watts of audio frequency power to a load. Integrated circuit amplifiers have been reported with efficiencies approaching 90% at 10-W output and a frequency response from 20 Hz to 20 kHz. Many IC chips are available that drive power BJTs or MOSFETs, delivering 30 to 50 W to a load. Larger discrete circuits report audio amplifiers based on the class-D stage that deliver 600 W per stereo channel [2]. This type of amplifier is often used in *low-end* car radios.

Pulse-width modulation (PWM) is used to reduce the power dissipated by the transistor while delivering a large power to the load. The varying load signal is applied by means of output devices that switch between on and off states. The resulting load voltage has a rectangular waveform that contains an average or dc value dependent on the duty cycle. In addition, the load voltage would also contain several ac components-however, these unwanted components can be filtered before reaching the load.

Any periodic waveform can be represented by a Fourier series consisting of a dc component (if present), a fundamental frequency component, and higher harmonics of the fundamental frequency. A rectangular wave switching between $+V$ and $-V$ that remains positive for $t^+$ seconds is said to have a *duty cycle* of

$$d = \frac{t^+}{T} \tag{3.17}$$

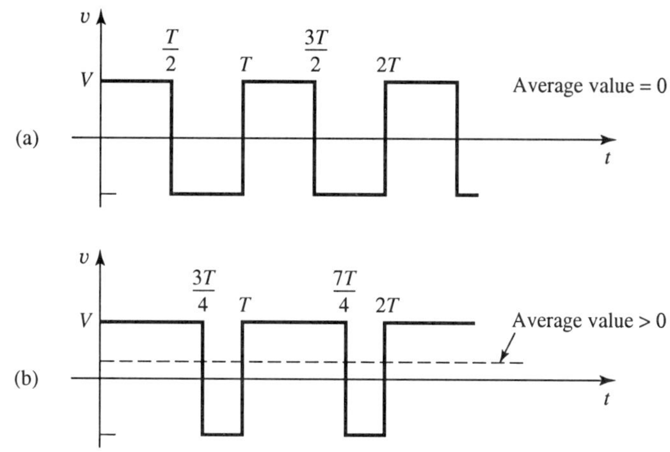


Figure 3.14: (a) Rectangular waveform with 50% duty cycle. (b) Rectangular waveform with 75% duty cycle.

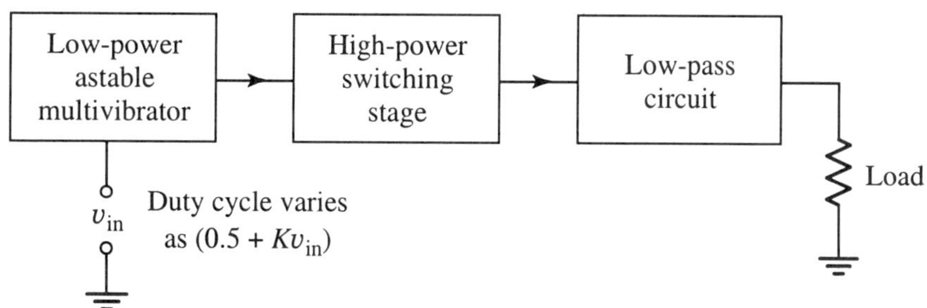


Figure 3.15: Pulse-width modulated amplifier.

where $T$ is the repetition period of the waveform. A rectangular waveform with a 50% duty cycle contains no dc or average value, but a change of duty cycle will give the waveform an average value as shown in Fig. 3.14. If the period, $T$, remains constant as the duty cycle is varied, the average value of a rectangular waveform of amplitude $V$ is

$$V_{av} = V(2d - 1) \tag{3.18}$$

The average value varies directly with $d$.

The Fourier coefficients of the ac components also vary as $d$ is changed and new frequencies may be introduced, but in general, these components are of little interest to us as they can be easily eliminated. If, for example, the repetition frequency is 200 kHz, all ac components of the waveform will be greater than this value and far out of the audio range. Now if we vary the duty cycle sinusoidally at some low frequency, the average value will also vary sinusoidally. Mathematically, we can express this by saying that if $d = 0.5 + k \sin \omega t$, then the average value is

$$V_{av} = 2kV \sin \omega t \tag{3.19}$$

The waveform with variable duty cycle can be filtered by a low-pass filter to eliminate all frequencies above $\omega$. The result is a low-frequency output sinusoid with an amplitude that varies in the same manner as the duty cycle. A block diagram of PWM amplification is shown in Fig. 3.15.

The voltage control circuit must have a 50% duty cycle when the input signal is at zero volts. As $v_{in}$ becomes nonzero, the duty cycle varies proportionally. The high-power switching stage amplifies this rectangular wave and applies the output to a lowpass circuit that allows only the changes in average value to pass to the load. The output signal is then proportional to the input signal, but can be at a much higher power level than the input signal. The power output stage may dissipate only 5 to 10% of the power delivered to the load.

The major advantage of PWM is that the output transistors need operate in only two states to produce the rectangular waveform: either fully on or fully off. In saturation we know that the very small voltage drop across a transistor leads to very low power dissipation. A very small dissipation is also present when the transistor is cut off. If switching times were negligible, no device power loss would occur during the transition between states. Actually, there is a finite switching time and this leads to an increased total dissipation of the output stages. Still, the efficiency figures for the class-D amplifier are very high, as reported earlier. This leads to higher possible power outputs for lumped-element circuits and to smaller chip areas for integrated PWM amplifiers.

The stages can be direct-coupled to the load, which eliminates the necessity of capacitors. Nonlinear distortion can be less than that of class-B stages, and matching of transistors is unnecessary.

On the other hand, the disadvantages of this amplifier ultimately dictate the limits of usefulness of the PWM scheme. The upper frequency response is limited to a small fraction of the switching frequency. The operating frequency of power transistors generally decreases with higher power ratings. It follows that the upper corner frequency of the amplifier may be lower for higher power transistors. Furthermore, a low-pass filter is required to eliminate the unwanted frequency components of the waveform. An inductance or capacitance will often be required for the filter circuit, although active filtering methods can be applied to reduce the values of any reactive elements used. The generation of radio frequencies or electromagnetic interference by the switching circuits can also present problems in certain applications.

**Practice Problems**

**3.5** A rectangular waveform swinging between +12 V and -12 V has a duty cycle of 0.65. What is the dc content of the waveform? *Ans:* 3.6 V.

**3.6** Repeat Practice Problem 3.5 if the duty cycle is changed to 0.32. *Ans:* -4.32 V.

### 3.3.2 Class-D Output Stages

In addition to compound emitter followers, the power output stages can be designed in several arrangements. Figure 3.16 shows two possible configurations.

The diodes appearing across the output transistors are present to protect the transistors against inductive voltage surges. If the filter is inductive, the current reversals that occur over short switching times generate very large voltage spikes, unless the protective diodes are used.

In the push-pull circuit, the low-power, pulse-width modulated input turns $T_1$ and $T_3$ on when the signal is at its maximum value. Transistors $T_2$ and $T_4$ are off at this time, and current is forced through the load. When the signal switches to the minimum value, $T_1$ and $T_3$ go off, while $T_2$ and $T_4$ turn on to pull current through the load.

Figure 3.16(b) shows a bridge circuit that can drive a floating load with a single power supply. When the input signal reaches its maximum value, $T_1$ and $T_3$ are on while $T_2$ and $T_4$ are held off. The input signal is inverted and applied to the bases of $T_5$ and $T_6$. This inverted signal is at its minimum value during the time when the normal input is maximum; thus, $T_5$ and $T_7$ will be off while $T_6$ and $T_8$ are on. Current will leave the collector of $T_3$, flow through the load, and enter the collector of $T_8$. When the input assumes the most negative value, $T_1$, $T_3$, $T_6$ and $T_8$ turn off while $T_2$, $T_4$, $T_5$, and $T_7$ turn on. Current now leaves the collector of $T_7$, flows through the load, and enters the collector of $T_4$. During this period, the load current flows in the opposite direction to that flowing when the input is maximum. The load current then reverses each time the input signal makes a transition.

We will not discuss the filter in detail, but will note that for many applications a large inductance in series with a resistive load is sufficient for filtering purposes. The corner frequency is selected to slightly exceed the highest frequency to be amplified. Higher frequency components, for example those near the fundamental

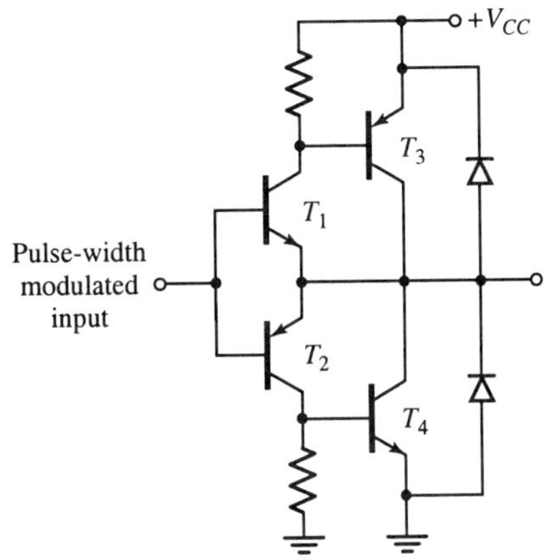


(a)

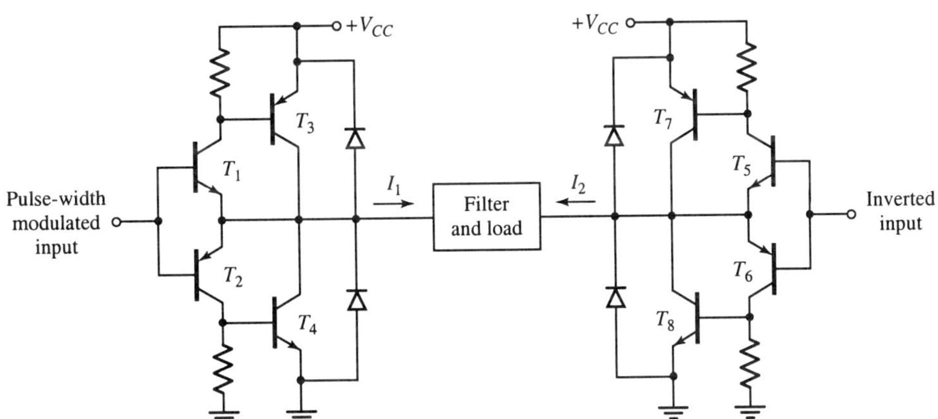


(b)

Figure 3.16: (a) Push-pull output stage. (b) Bridge output.

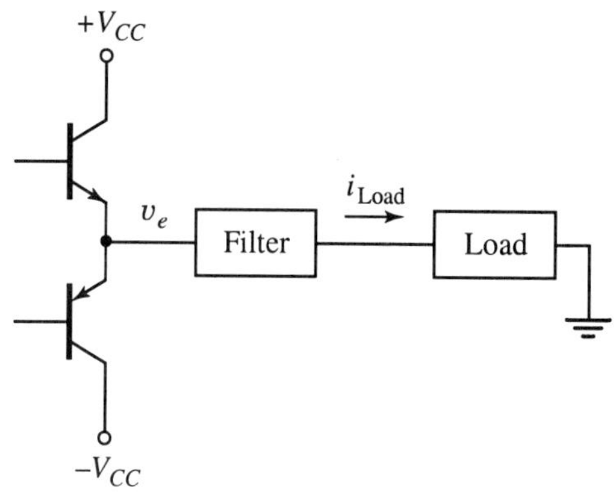


Figure 3.17: An ideal class-D power stage.

frequency, will be attenuated considerably. In some applications, such as motor control or high-output audio systems, the load serves as a filter of high frequencies since these particular loads cannot respond to the switching frequencies.

### Calculation of Class-D Circuit Efficiency

The ideal circuit efficiency of the class-D stage is 100% and, unlike the class-C stage, can deliver maximum power at maximum efficiency. An idealized model of the class-D stage using complementary emitter followers is depicted in Fig. 3.17.

Maximum efficiency can be calculated if we make the following assumptions.

1. The filter offers infinite impedance to the switching frequency and higher harmonics while offering no impedance to the signal frequency.

2. The transistors switch between saturation and cutoff with zero switching time.

3. The saturation voltage from collector to emitter is zero for the transistors.

If no input signal is present, the duty cycle will be 50%. The voltage appearing at the emitters as input signal is varied is shown in Fig. 3.18 along with load current. This figure does not show the large difference between switching frequency and signal frequency that exists in the practical circuit.

For a 50% duty cycle, a Fourier analysis of the voltage waveform at the emitters gives

$$v_e = \frac{4V_{CC}}{\pi}\left[\sin\,\omega_s t - \frac{1}{3}\,\sin\,3\omega_s t + \frac{1}{5}\,\sin\,5\omega_s t - \cdots\right] \tag{3.20}$$

where $\omega_s$ is the switching frequency. Because the impedance at these frequencies is assumed infinite, no current flows through the load and no current is supplied from the power supply. The quiescent dissipation of the amplifier is zero.

When a sinusoidal input signal is applied, the duty cycle is modulated sinusoidally about the quiescent value of 50%. If the amplitude of the output voltage is $v_1$, the load current is $i_1\,\sin\,\omega t$. This current is supplied from the two power supplies through the transistors. The maximum output voltage and current occur when $v_1 = V_{CC}$. Load power is then

$$P_{\text{ac}} = \frac{V_{CC}^2}{2R_L} \tag{3.21}$$

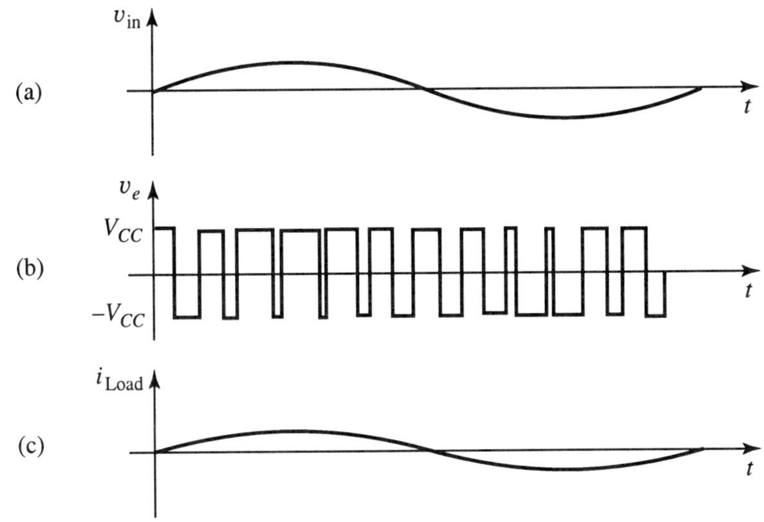


Figure 3.18: (a) Input voltage. (b) Voltage at emitters. (c) Load current

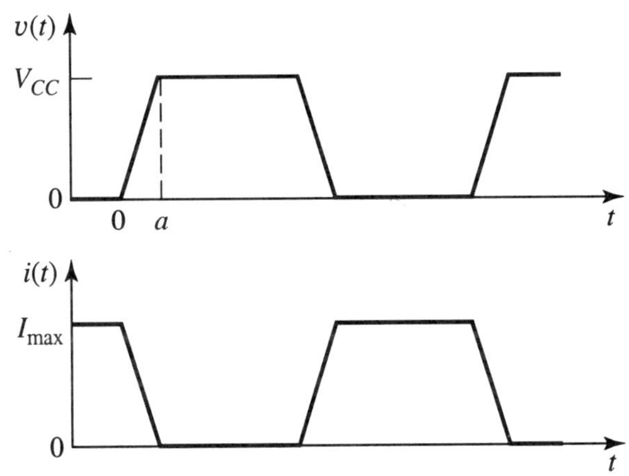


Figure 3.19: Linear switching transitions.

Since there is no power dissipation in the filter or the transistors, the power delivered by the source must also equal this value, to result in 100% circuit efficiency.

We can extend this ideal circuit one step toward the practical amplifier by calculating the approximate transistor dissipation taking place due to finite rather than zero switching times. Whereas an accurate analysis of the switching waveform is quite complex, a good approximation results when a linear voltage transition is assumed, as indicated in Fig. 3.19.

The transitions of the emitter voltage of each transistor from 0 to $V_{CC}$ volts and from $V_{CC}$ to 0 volts are assumed to take place in $a$ seconds. As the emitter voltage increases from 0 to V volts, the voltage across the transistor decreases from V to 0 volts. At the same time, the current increases from 0 to $I_{\max}$ in $a$ seconds. At the peak of the maximum output voltage, the maximum current is

$$I_{\max} = \frac{V_{CC}}{R_L}$$

The voltage across the transistor from $t = 0$ to $t = a$ can be expressed as

$$v(t) = \frac{t}{a}\, V_{CC}$$

while the decreasing current can be written as

$$i(t) = I_{\max} - \frac{t}{a}\, I_{\max} = \frac{V_{CC}}{R_L}\left[1 - \frac{t}{a}\right]$$

The instantaneous power dissipated by a transistor over this transition is

$$p(t) = v(t)i(t) = \frac{V_{CC}^2}{R_L}\left(\frac{t}{a} - \frac{t^2}{a^2}\right)$$

Since each transistor makes two transitions in one period of the switching frequency, the average power dissipation of a transistor can be found by averaging the instantaneous power over one period, $T$. This can be done by

$$P_T = \frac{2}{T}\int_0^a \frac{V_{CC}^2}{R_L}\left(\frac{t}{a} - \frac{t^2}{a^2}\right) dt = \frac{a}{T}\frac{V_{CC}^2}{3R_L} \tag{3.22}$$

The total transistor dissipation of both transistors is twice this value, or

$$2P_T = \frac{a}{T}\frac{2V_{CC}^2}{3R_L} \tag{3.23}$$

The total power supplied from the dc power supplies is the sum of the transistor dissipation and the power delivered to the load. An example will demonstrate a typical circuit efficiency for the class-D stage.

## Example 3.3

Evaluate the transistor power dissipation for a class-D circuit with linear switching transitions of 5% of the total period. Assume that 24-V power supplies are used to drive a load resistance of 50 Ω. Calculate the circuit efficiency for the circuit.

**Solution** The total transistor dissipation is calculated from Eq. (3.23) as

$$2P_T = \frac{a}{T}\frac{2V_{CC}^2}{3R_L} = \frac{0.05 \times 2 \times 24^2}{3 \times 50} = 0.384 \text{ W}$$

The power to the load is

$$P_{\text{ac}} = \frac{V_C C^2}{2R_L} = \frac{24^2}{100} = 5.76 \text{ W}$$

The circuit efficiency can be written as

$$\eta = \frac{P_{\text{ac}}}{P_{\text{ac}} + 2P_T} = \frac{V_{CC}^2/2R_L}{V_{CC}^2/2R_L + 0.05\,\frac{2V_{CC}^2}{3R_L}} = 93.8\%$$

The power dissipation of each transistor is only 0.192 W, and the power delivered to the load is 5.76 W. The load power is 30 times the transistor dissipation. This is a vast improvement over a resistive load, class-A stage that can deliver a maximum load power that is just one-half the transistor dissipation!

---

In addition to transistor power dissipation arising from finite switching times, the nonzero saturation voltage also introduces power dissipation. During the time of peak current to the load, one transistor remains saturated

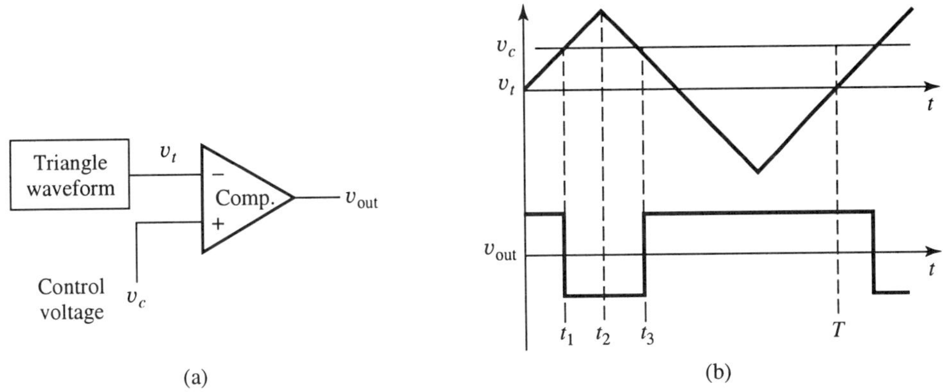


Figure 3.20: (a) A PWM waveform generator. (b) Waveforms.

for almost the entire switching period while the other transistor is on for a very small time. A second effect of the saturation voltage is that the peak value of load voltage is reduced to $V_{CC} - V_{CES}$. The maximum dissipation due to a finite value of $V_{CES}$ can be expressed as

$$P_{TS} = I_{\max}V_{CES} = \frac{V_{CC} - V_{CES}}{R_L}V_{CES} = \frac{V'_{CC}}{R_L}V_{CES} \tag{3.24}$$

where $V'_{CC} = V_{CC} - V_{CES}$. For values of $V_{CES} = 0.4$ V and $V_{CC} = 24$ V, the total circuit efficiency for the class-D stage of Example 3.3 becomes

$$\eta = \frac{P_{\text{ac}}}{P_{\text{ac}} + 2P_T + P_{TS}} = \frac{{V'_{CC}}^2/2R_L}{{V'_{CC}}^2/2R_L + 0.05\frac{2{V'_{CC}}^2}{3R_L} + \frac{V'_{CC}}{R_L}V_{CES}} = 90.9\%$$

Other sources of dissipation originate from the finite rather than infinite impedance of the filter to the switching frequencies, the finite rather than zero impedance of the filter to the signal frequency, and power dissipation in the protective diodes.

**Practice Problems**

**3.7** Recalculate the circuit efficiency from Example 3.2 if $a$ is 10% of the total period. *Ans:* 88.2%.

**3.8** Recalculate the circuit efficiency from Example 3.2 if the power supplies are changed to 12-V sources. *Ans:* 93.8%.

### 3.3.3 Generating the PWM Drive Signal

A conceptually simple method of generating a PWM waveform is shown in Fig. 3.20 [3].

The triangular waveform moves linearly between values of $+V_p$ and $-V_p$ with a total period of $T$ seconds. The time that the comparator generates a negative output, $t^-$, is $2(t_2 - t_1)$. Since these times are given by

$$t_1 = \frac{v_c}{V_p}\frac{T}{4}$$

and

$$t_2 = \frac{T}{4}$$

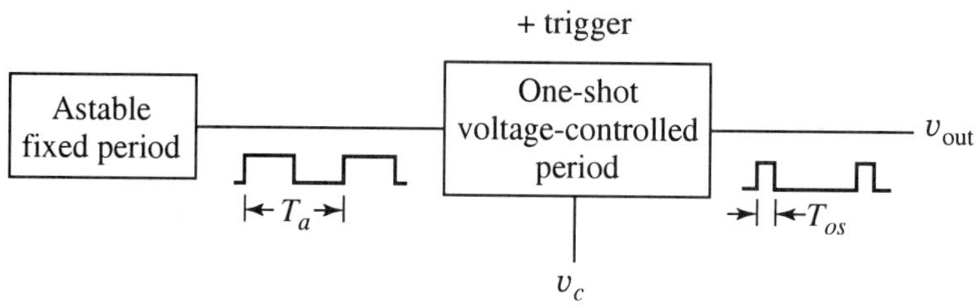


Figure 3.21: Multivibrator method of PWM waveform generation.

the value of $t^-$ can be found as

$$t^- = \frac{T}{2}\left(1 - \frac{v_c}{V_p}\right) \tag{3.25}$$

The time that the comparator output is positive is

$$t^+ = T - t^- = \frac{T}{2}\left(1 + \frac{v_c}{V_p}\right) \tag{3.26}$$

If the output of the comparator now is applied to a power stage such as that of Fig. 3.17, the average voltage applied to the load is

$$V_{av} = \frac{Vt^+ - Vt^-}{T} \tag{3.27}$$

where $V$ is the amplitude of the load voltage. Substituting Eqs. (3.25) and (3.26) into Eq. (3.27) results in

$$V_{av} = \frac{V}{V_p} v_c \tag{3.28}$$

Both $V_p$ and $V$ are constants, therefore, the average output voltage is proportional to the applied control voltage. This signal is the input signal to be amplified, giving an output voltage that is directly proportional.

The duty cycle of the output waveform is

$$d = \frac{t^+}{t^- + t^+} = \frac{\frac{T}{2}\left(1 + \frac{v_c}{V_p}\right)}{\frac{T}{2}\left(1 - \frac{v_c}{V_p}\right) + \frac{T}{2}\left(1 + \frac{v_c}{V_p}\right)} = 0.5 + \frac{v_c}{V_p}$$

Returning to Eq. (3.18), we see that

$$V_{av} = V(2d - 1)$$

Substituting the expression for duty cycle into this equation again results in Eq. (3.28).

Conceptually, the method of Fig. 3.20 is simple and is often used to generate the PWM drive signal. It does, however, require the generation of a triangular waveform with good linearity.

An alternative method of generating a PWM input waveform uses an astable multivibrator along with a voltage-controlled one-shot multivibrator. This method eliminates the need for a sawtooth generator. Figure 3.21 shows the arrangement of these multivibrators.

The key to this method is the one-shot circuit with a period that varies linearly with the control voltage. One possible configuration for this circuit is indicated in Fig. 3.22 [1].

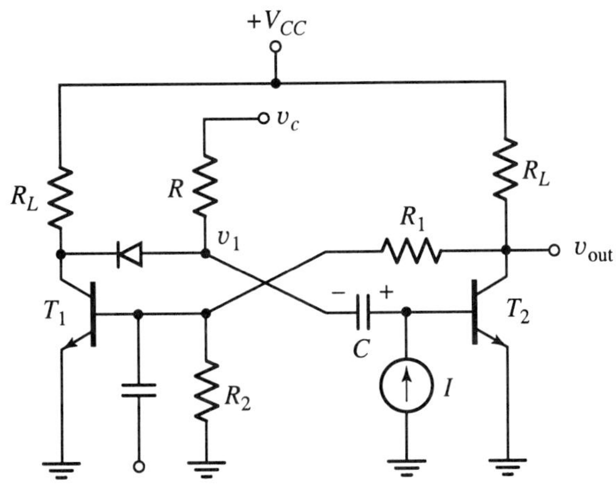


Figure 3.22: A linear voltage control one-shot.

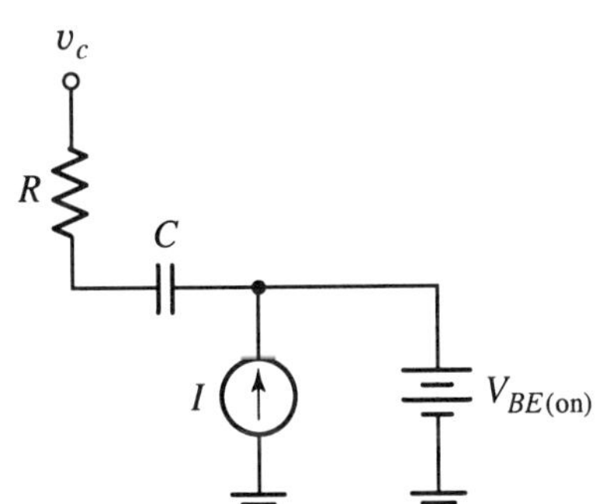


Figure 3.23: Equivalent circuit when $T_2$ is on.

When transistor $T_2$ is on and $T_1$ is off, the capacitor circuit appears as in Fig. 3.23.

The current $I$ saturates the transistor, keeping the voltage at the base of $T_2$ equal to $V_{BE(\mathrm{on})}$. The voltage on the other side of the capacitor will equal $v_c$, the control voltage. This assumes that the circuit has existed in this state for over three times the charging time constant, which is

$$\tau_c = RC$$

The diode D will be reverse biased if $V_{CC}$ is larger than $v_c$.

When the trigger pulse turns $T_1$ on, the collector voltage is pulled to $V_{CE(\mathrm{sat})}$ and $v_1$ is now equal to $V_{CE(\mathrm{sat})} + V_D$ where $V_D$ is the forward-biased drop of the diode. As $v_1$ instantaneously changes from $v_c$ to $V_{CE(\mathrm{sat})} + V_D$, the capacitor cannot change voltage instantaneously. The base of $T_2$ is then pulled from $V_{BE(\mathrm{on})}$ to

$$V_{B2} = -v_c + V_{BE(on)} + V_{CE(sat)} + V_D$$

As $V_{B2}$ changes to this value, the one-shot period begins and $V_{C2}$ switches to a value near $V_{CC}$. The actual value of $v_{\mathrm{out}}$ is now

$$\frac{R_1}{R_1 + R_L} V_{CC}$$

With $T_2$ off, the current generator charges the capacitor, causing a linear voltage rise with a slope of

$$\frac{dV}{dt} = \frac{I}{C} \tag{3.29}$$

When the base voltage of $T_2$ reaches $V_{BE(\mathrm{on})}$, this transistor saturates and the period ends. Since $dV/dt$ is constant and can be written as $\Delta V/T_{os}$, where $T_{os}$ is the period of the one-shot, Eq. (3.29) can be solved for this period. This gives

$$T_{os} = \Delta V \frac{C}{I} = [v_c - V_{CE(\mathrm{sat})} - V_D] \frac{C}{I} \tag{3.30}$$

If $v_c$ is much greater than $V_{CE(\mathrm{sat})} + V_D$, the period is approximated by

$$T_{os} = \frac{v_c C}{I} \tag{3.31}$$

Equation (3.31) reflects a linear variation of the one-shot period. However, this is an approximation of the actual period since the saturation voltage and diode voltage were neglected. A modification of the control voltage circuit as shown in Fig. 3.24 can make Eq. (3.31) quite accurate. If the resistance $R_d$ is chosen to be much smaller than $R$, the voltage $v_1$ now reaches a value of

$$v_1 = v_c + V_{D1}$$

when $T_1$ is off. As $T_1$ is turned on, $v_1$ is again pulled down to a value of $V_{CE(\mathrm{sat})} + V_D$. The value of $V_{B2}$ is now

$$V_{B2} = -v_c - V_{D1} + V_{BE(\mathrm{on})} + V_{CE(\mathrm{sat})} + V_D$$

This base voltage must change to $V_{BE(\mathrm{on})}$ to end the period; thus,

$$\Delta V = v_c + V_{D1} - V_{CE(\mathrm{sat})} - V_D$$

The diode voltage $V_{D1}$ can be used to offset the voltage $V_{CE(\mathrm{sat})} + V_D$, resulting in a $\Delta V$ of

$$\Delta V = v_c$$

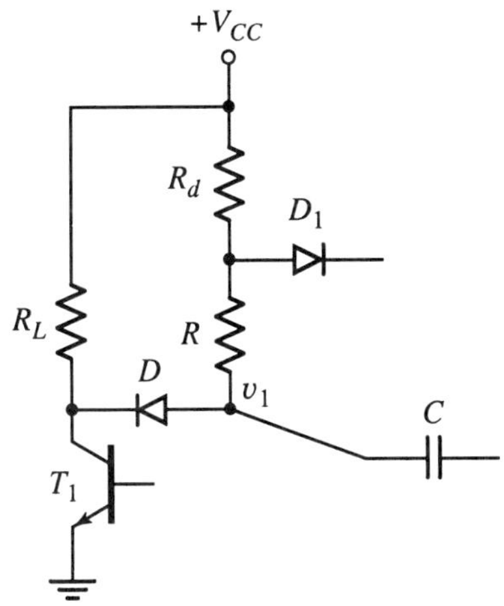


Figure 3.24: Modification of control circuit.

With this modification, Eq. (3.31) becomes rather accurate. If a constant-period astable is used to trigger the one-shot, the on duty cycle is given by

$$\frac{t_{\text{on}}}{T_a} = \frac{T_{os}}{T_a} = \frac{v_c C}{T_a I} \tag{3.32}$$

where $T_a$ is the astable period. In this PWM circuit, the control voltage consists of the ac signal to be amplified plus a dc signal that results in a 50% duty cycle when the ac signal is zero.

**Practice Problems**

**3.9** The circuit of Fig. 3.20 uses a value of $V_p = 10$ V. If the output is driven to the extreme values of +12 V and -12 V, calculate the control voltage to cause a dc output of 4 V. *Ans:* 3.33 V.

**3.10** How much current is needed to charge a 200-pF capacitor at a rate of 600,000 V/s? *Ans:* 120 $\mu$A.

---

## SUMMARY

1. Power gain is often limited by the total harmonic distortion in the output signal of a power amplifier.

2. Class-B stages can be constructed without a transformer. Such stages are required for IC amplifiers.

3. The class-G stage delivers high levels of peak power with little transistor dissipation.

4. Class-D stages use pulse-width modulation to achieve high circuit efficiencies.

## PROBLEMS

### Section 3.1.1 Power Gain

**3.1** An amplifier has an unloaded voltage gain of 26 V/V, an output impedance of 200 Ω, and an input impedance of 2 kΩ. If this amplifier couples a 500-Ω source to a 100-Ω load, calculate the overall voltage gain and the power gain.

**D3.2** An amplifier requires a power gain of 28 dB as it couples a source impedance of 4 kΩ to a 100-Ω load. If the input impedance of the amplifier is 6 kΩ and the output impedance is 50 Ω, what must be the loaded voltage gain of the amplifier? What must the unloaded voltage gain of the amplifier be to satisfy this design?

### Section 3.1.2 Distortion in Power Stages

**3.3** A sinusoidal signal of frequency 1 kHz is applied to a power amplifier. The output consists of a signal expressed as

$$v_{\text{out}} = 9.6 \text{ sin } 6283t + 0.4 \text{ sin } 12,566t + 0.16 \text{ sin } 18,849t \text{ V}$$

Calculate the THD of the output signal. Express the answer in both percent and dB.

**3.4** Repeat Problem 3.3 if the fundamental component has a magnitude of 12.4 V, the second harmonic has a magnitude of 0.8 V, and the third harmonic has a magnitude of 0.56 V.

**3.5*** An output voltage is expressible as

$$v_{\text{out}} = 4[3 + v_{\text{in}}]^2$$

Calculate the second-order intercept point if the output load resistance is 100 Ω.

### Section 3.2.1 Class-B Complementary Emitter Followers

**3.6** The output stage of Fig. 3.5 uses power supplies of 20 V to drive a 20-Ω load resistance. If 4-Ω resistors are placed in series with each emitter for thermal protection, calculate the maximum circuit efficiency of the circuit neglecting bias power. Calculate the maximum transistor dissipation for a sinusoidal input signal.

**3.7** Repeat Problem 3.6 if the power supplies are changed to 32 V.

**3.8** Repeat Problem 3.6 if the load resistance is changed to 16 Ω.

**D3.9** It is desired to deliver a 10-W signal to a 16-Ω resistive load using the circuit of Fig. 3.5. Select an appropriate value of $V_{C1}$ and find the minimum allowable dissipation of the transistors, assuming the transistors are ideal and $R_E = 0$.

**D3.10** Repeat Problem 3.9 if 4-Ω resistors are inserted in series with each emitter.

**D3.11** Repeat Problem 3.9 if 4-Ω resistors are inserted in series with each emitter and 15 W must be delivered to the load.

### Section 3.2.3 Integrated Amplifier Class-B Stages

**D3.12** A $V_{BE}$ multiplier is to be used to create a 2.9-V drop across its output. The multiplier is driven with a current source of 400 $\mu$A. The transistor has an emitter current of

$$I_E = 10^{-13} e^{V_{BE}/0.26} \text{ mA}$$

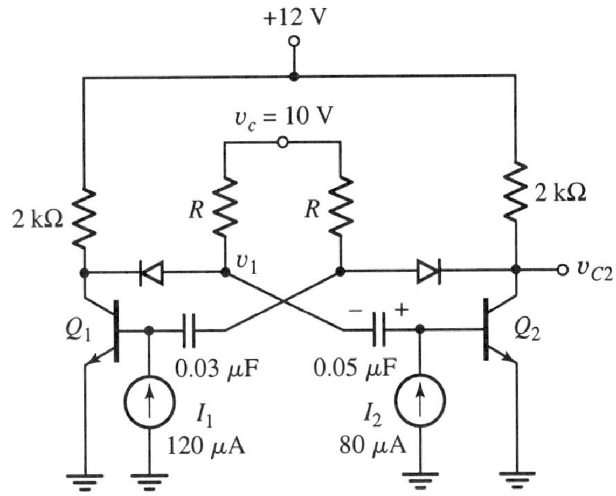


**Figure P3.19**

Select values of $R_1$ and $R_2$ for this circuit.

**3.13*** A Darlington pair is used to drive a 10-Ω load to a peak voltage of 6 V. If $\beta$ for the input transistor is 120 and $\beta$ for the output transistor is 45, calculate the required peak base current for the input transistor.

### Section 3.3.2 Class-D Output Stages

**3.14** A class-D output stage switches between +12 V and −12 V. When the output is −8.7 V, what is the duty cycle of the signal?

**3.15*** For the circuit of Fig. 3.17, $V_{CC} = 16$ V and $V_{CES} = 0.3$ V. If the load resistance is 20 Ω, what is the maximum sinusoidal power that can be delivered to the load? What is the power delivered to the load when the duty cycle is 72%?

**3.16** Repeat Problem 3.15 if the load resistance is changed to 32 Ω.

### Section 3.3.3 Generating the PWM Drive Signal

**3.17** If $V_p = 10$ V in Fig. 3.20, find $v_c$ to result in a 60% duty cycle, that is, $t^+/T = 0.6$.

**3.18** Repeat Problem 3.17 for a duty cycle of 28%.

**3.19*** Assume that $\beta = 100$, the forward-biased diode voltage is zero, and $V_{BE(\text{on})} = V_{CE(\text{sat})} = 0$ V in the circuit shown.

(a) Show the waveforms for both $V_{B2}$ and $V_{C2}$ for one full cycle, beginning with the time that $Q1$ turns on. Show important voltage values and times.

(b) What is the minimum value of current for $I_1$ or $I_2$ to cause saturation of the transistors at the proper times?

## REFERENCES

1. Comer, D. J. and Comer, D. T., "Control voltage determines multivibrator pulse width," *Electronics,* Vol. 40, No. 16, Aug. 1967, pp. 115-116.

2. Duncan, B., *High Performance Audio Power Amplifiers.* Jordan Hill, Oxford, UK: Newnes, 1996.

3. Mohan, N., Undeland, T. M., and Robbins, W. P., *Power Electronics,* 2nd Ed. New York: John Wiley and Sons, 1995.

# Chapter 4

# WIDEBAND AMPLIFIERS

The term *wideband amplifier* is often used synonymously with the term *high-frequency amplifier*. A high-frequency amplifier can be a wideband amplifier, but so also can an amplifier that is designed to work at audio frequencies. A wideband amplifier is defined as one with a gain-magnitude frequency response that is flat over most of the bandwidth. If bandwidth is measured in terms of the upper and lower 3-dB points (corner frequencies), $f_2$ and $f_1$, the ratio of $f_2$ to $f_1$ determines whether or not the amplifier is wideband.

Figure 4.1 demonstrates the frequency responses for an audio amplifier and for a narrowband or tuned amplifier. For the audio amplifier, the upper 3-dB frequency is 20 kHz and the lower corner frequency is 20 Hz. The ratio of upper to lower corner frequency is 1000. The narrowband circuit has an upper corner frequency of 462 kHz and a lower corner frequency of 448 kHz. The ratio here is only 1.03. A ratio of 10 or above is generally considered a wideband response.

This chapter will first consider wideband design after developing some applicable theory. Following wideband amplifier design, we will turn our attention to narrowband or tuned stage designs that are quite important in communication circuits.

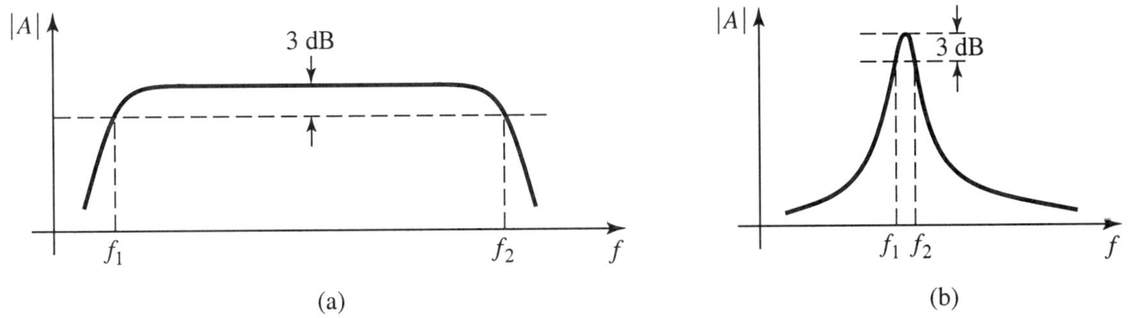


Figure 4.1: (a) Wideband amplifier response. (b) Narrowband response.

## 4.1 WIDEBAND SINGLE STAGES

**Important Concepts**

**1.** Single-stage amplifiers based on the op amp have a figure of merit called the gain-bandwidth product or $GBW$ of the stage.

**2.** The $GBW$ of the noninverting stage equals that of the op amp, whereas the $GBW$ of the inverting stage may be smaller than that of the op amp.

**3.** A constant $GBW$ stage allows well-developed theory to be applied to multistage amplifier design.

**4.** The $GBW$ of transistor stages may not be as useful in design as in the case of op amp stages.

In this section, we will review some of the basic frequency limitations of the MOSFET, the BJT, and the op amp that are typically treated in a basic electronics textbook. Useful gain or bandwidth equations will not be rederived, but will be referenced as they are applied. The major limitation on high frequency amplification with the MOSFET and BJT is the inadvertent formation of parasitic capacitances resulting from the geometry of the fabricated device and the connections to other devices or bonding pads and external pins. Some of these parasitic capacitances are relatively constant with voltage, whereas others, such as depletion capacitances, depend on voltage across the capacitance.

For the common-source MOSFET and the common-emitter BJT stages, two upper critical frequencies may exist. One is due to the input circuit that accounts for the Miller effect and another is due to the output capacitance in parallel with the output resistance. In some situations, both critical frequencies or poles will influence the overall upper 3-dB frequency. In other cases, the lowest of the two poles may determine the overall frequency. If the upper frequency is more than five times the lower frequency, the upper frequency can be ignored.

In discrete BJT multistage amplifiers, a high overall gain can be achieved with several stages, each with only a modest gain. The stages can be capacitively coupled and need only a small load resistance to achieve the modest gain. In general, the poles caused by the input circuits will determine the overall upper corner frequency. The small load resistances lead to very high-frequency, often negligible, poles for the output circuits of each stage.

Small-channel MOSFETs are typically reserved for IC chip design. Very little discrete MOSFET design is done with low-power devices. In wideband IC amplifiers, for both BJT and MOSFET designs, the general philosophy has been to use high-gain, low-bandwidth amplifiers utilizing negative feedback to achieve a higher bandwidth. In these high-gain stages, the load resistance may be a large active load that leads to a low corner frequency. In such a case, the output pole or both poles must be considered to find the overall corner frequency of the stage.

Because MOSFET devices for high-frequency work are quite small, the parasitic capacitances are also very small. These small on-chip capacitances result in good high-frequency amplifiers, but may seriously degrade in bandwidth when the output signal is connected to some off-chip circuit. The output pad and input capacitance to the off-chip circuit may offer picofarads of capacitance, whereas the internal capacitance is often measured in femtofarads. A buffer stage is often required to interface the output stage of the chip to the input of the following circuit.

The BJT does not generally suffer as severely from this bandwidth degradation as does the MOSFET stage. The larger dimensions of the BJT result in larger capacitances and lower resistances in the on-chip stages. Additional output pad and loading circuit capacitance may represent only a small fraction of the total capacitive load for the BJT output circuit whereas it may dominate the MOSFET circuit.

Another problem suffered by common-source and common-emitter stages is that of the Miller effect. When high-gain stages are used, even the very small gate-to-drain capacitance of the MOSFET becomes significant. For the BJT, the base-to-collector capacitance may dominate the frequency response. For this reason, the cascode connection has become prominent in IC design. This stage minimizes the Miller effect, but requires an additional device. For IC design, an extra device may be a small price to pay for a higher performance stage.

Although the op amp consists of many transistor stages, it is often considered a single-stage device. Most general-purpose op amps exhibit a single-pole rolloff of the frequency response over several decades of frequency. Most of these devices use *dominant-pole compensation*, resulting in a rather low value of upper 3-dB frequency. Negative feedback is then used to exchange gain and bandwidth to achieve a useful value of bandwidth.

### 4.1.1 The Op Amp Single-Stage Amplifier

Two common and important specifications of an amplifier are overall midband gain and overall bandwidth. An amplifier may require a gain of 800 V/V and a bandwidth extending from 20 Hz to 20,000 Hz. Although a single op amp stage may achieve a voltage gain of 800 V/V, this gain cannot be achieved simultaneously with a bandwidth of 20 kHz. If op amps are to be used, multiple stages are required to satisfy the specifications.

The gain-bandwidth product is a figure of merit that can be used in the design of multistage amplifiers. The gain-bandwidth product ($GBW$) of an amplifier is defined as the product of the magnitude of midband gain and the 3-dB bandwidth of the amplifier. The units of $GBW$ can be either Hz or rad/s, depending on the units used to express bandwidth. For an op amp with the frequency response of Fig. 4.2, the midband gain is $10^5$ and the bandwidth is 10 Hz resulting in $GBW_{oa} = 10^6$ Hz.

There is a $GBW$ for the op amp itself, a $GBW$ for the noninverting stage, and a $GBW$ for the inverting stage. These figures may or may not be equal for each stage as shown in the following sections. For the following discussion it will be useful to emphasize these definitions:

$A$—The gain of the op amp as a function of frequency

$A_{MBoa}$—The midband gain of the op amp (no feedback)

$f_{2oa}$—The 3-dB bandwidth of the op amp (no feedback)

$GBW_{oa}$—The gain-bandwidth product of the op amp (no feedback)

$A_{ni}$—The gain of the noninverting stage as a function of frequency

$A_{MBni}$—The midband gain of the noninverting amplifier with feedback

$f_{2ni}$—The 3-dB bandwidth of the noninverting amplifier with feedback

$GBW_{ni}$—The gain-bandwidth product of the noninverting amplifier with feedback.

**Practical Considerations**

The output voltage swing of an op amp is limited by the dc power supplies used. A typical op amp may use +12 V and -12 V supplies. The output voltage swing is then limited to a range of values from approximately -11 V to approximately +11 V.

Because the op amp can exhibit a very high voltage gain, it is important to minimize dc power supply voltage fluctuations. These fluctuations might arise as the supply current value changes or if some type of noise is present. Generally, a small capacitor, perhaps 0.01 $\mu$F to 0.1 $\mu$F, is placed between ground and the IC pins that connect to each dc power supply voltage. This tends to hold the supply voltages applied to the chip at a constant voltage. The accompanying figure shows the usual configuration for the op amp and dc power supplies. This is referred to as *decoupling* the power supplies.

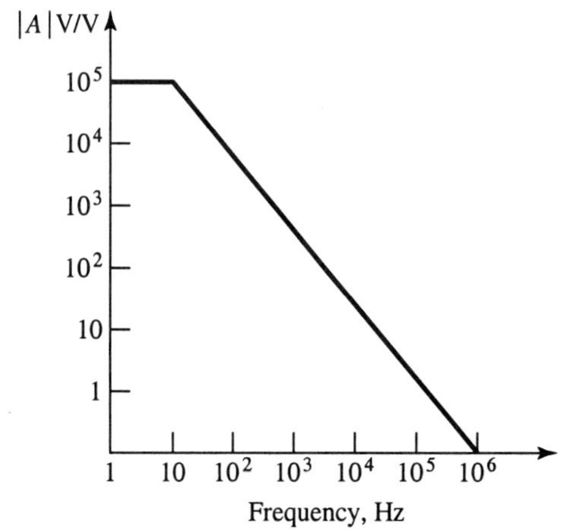


Figure 4.2: Voltage gain as a function of frequency.

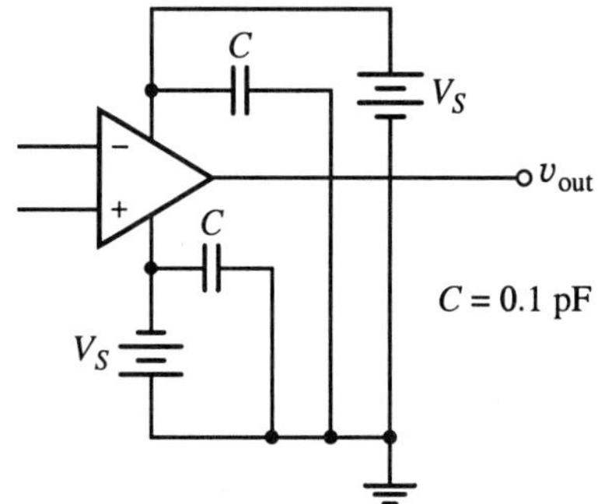


**Decoupling capacitors**

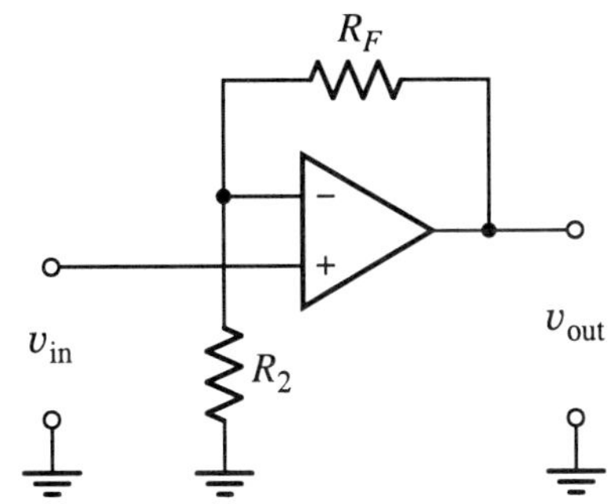


Figure 4.3: A noninverting amplifier.

### The Noninverting Stage

For the noninverting stage of Fig. 4.3, the voltage fed back from output to the inverting terminal of the noninverting stage is

$$v_1 = \frac{R_2}{R_2 + R_F} \times v_{\text{out}} = F v_{\text{out}}$$

where $F$ is the feedback factor or the transfer function of the feedback network. The general equation for gain of a noninverting stage is

$$A_{ni} = \frac{v_{\text{out}}}{v_{\text{in}}} = \frac{A}{1 + AF} \tag{4.1}$$

This is the well-known formula for closed-loop gain of a feedback amplifier.

The frequency-dependent expression for op amp gain, $A$, is

$$A = \frac{A_{MBoa}}{1 + jf/f_{2oa}} \tag{4.2}$$

If we now substitute Eq. (4.2) into Eq. (4.1), we find that the frequency-dependent expression for gain of the noninverting stage is

$$A_{ni} = \frac{A_{MBoa}}{1 + A_{MBoa}F} \frac{1}{1 + jf/(1 + A_{MBoa}F)f_{2oa}} \tag{4.3}$$

The overall midband gain of the amplifier is

$$A_{MBni} = \frac{A_{MBoa}}{1 + A_{MBoa}F} \tag{4.4}$$

whereas the upper 3-dB frequency is

$$f_{2ni} = f_{2oa}(1 + A_{MBoa}F) \tag{4.5}$$

The noninverting amplifier configuration results in a constant value of $GBW_{ni}$. That is, gain can be decreased by controlling the resistors $R_F$ and $R_2$, but the bandwidth increases by the same factor of gain decrease. This is demonstrated from Eqs. (4.4) and (4.5). The gain-bandwidth product of the noninverting stage is

$$\begin{aligned} GBW_{ni} &= A_{MBni}\, f_{2ni} = \frac{A_{MBoa}}{1 + A_{MBoa}F}\, f_{2oa}(1 + A_{MBoa}F) \\ &= A_{MBoa}\, f_{2oa} = GBW_{oa} \end{aligned} \tag{4.6}$$

Note that the op amp with feedback has the same $GBW$ as the op amp with no feedback. Gain and bandwidth are varied with $F$, but the $GBW$ does not change with feedback factor.

# Example 4.1

An amplifier is to be designed to couple a microphone to a resistive load. The microphone generates a peak output of 50 mV for a typical voice input level and has a 10-kΩ output impedance. The output voltage across the 2-kΩ load is to have a peak value of 10 V. The bandwidth of the voltage gain should be at least 40 kHz. If the $GBW$ of the op amp used is $3 \times 10^6$ Hz, calculate the overall bandwidth of the final design.

**Solution** From the specifications, the overall midband voltage gain must be

$$A_{MBo} = \frac{10}{0.05} = 200 \text{ V/V}$$

If a single noninverting stage were used to construct an amplifier with a midband gain of 200, the upper corner frequency would be

$$f_{2ni} = \frac{GBW_{ni}}{A_{MBni}} = \frac{GBW_{oa}}{A_{MBni}} = \frac{3 \times 10^6}{200} = 15 \text{ kHz}$$

This value of bandwidth does not satisfy the specification; thus, more than one stage must be used.

One possible solution might consist of two stages, the first having a gain of 10 and the second a gain of 20. Since the noninverting stages present negligible loading to source and load resistances in the kΩ range, the overall voltage gain would then be 200. Figure 4.4 shows this amplifier.

In order to set the proper gains, the following two equations must be satisfied:

$$1 + \frac{R_{F1}}{R_{21}} = 10$$

$$1 + \frac{R_{F2}}{R_{22}} = 20$$

There are four unknowns and only two equations here, a situation common in design problems. This allows us to select two of the resistances and then calculate the remaining two. If $R_{21}$ and $R_{22}$ are chosen (somewhat arbitrarily) to be 2 kΩ, then $R_{F1}$ and $R_{F2}$ can be found from the equations to be $R_{F1} = 18$ kΩ and $R_{F2} = 38$ kΩ.

The remaining calculation is that of the upper 3-dB frequency. This can be done by first finding the 3-dB frequency of each stage by dividing the respective gains into $GBW_{ni}$ or $GBW_{oa}$. The bandwidth of stage 1 is

$$f_{2ni1} = \frac{3 \times 10^6}{10} = 300 \text{ kHz}$$

The second-stage bandwidth is

$$f_{2ni2} = \frac{3 \times 10^6}{20} = 150 \text{ kHz}$$

The overall gain as a function of frequency can be written as

$$A_o(\omega) = \frac{10}{1 + jf/3 \times 10^5} \frac{20}{1 + jf/1.5 \times 10^5}$$

The magnitude of the denominator must equal $\sqrt{2}$ at the upper 3-dB point, $f_{2o}$. This results in the equation

$$2 = \left(1 + \frac{f_{2o}^2}{9 \times 10^{10}}\right)\left(1 + \frac{f_{2o}^2}{2.25 \times 10^{10}}\right)$$

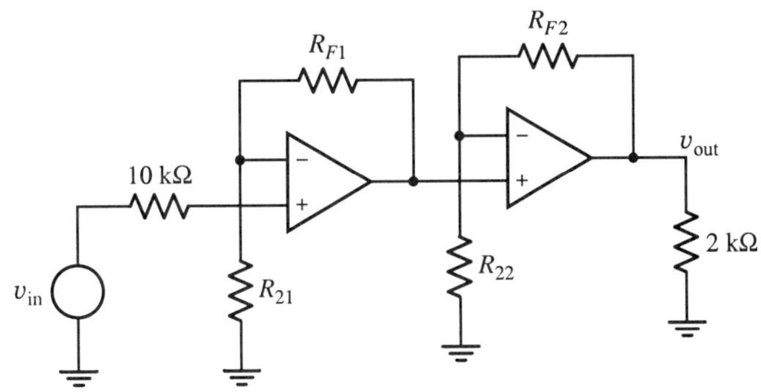


Figure 4.4: A two-stage amplifier.

Using trial-and-error methods shows the overall 3-dB frequency of the amplifier to be $f_{2o} = 126$ kHz.

---

### Practice Problems

**4.1** The midband gain of a noninverting op amp stage is 12 V/V and the upper corner frequency is $4 \times 10^5$ Hz. If the midband gain is changed to 5 V/V, what is the new bandwidth of the amplifier? *Ans:* 960 kHz.

**4.2** For the op amp of Practice Problem 4.1, what value of gain will result for a bandwidth of 1 MHz? *Ans:* 4.8 V/V.

**4.3** Two noninverting stages are cascaded. The first has a midband gain of 8 V/V and a bandwidth of 500 kHz. The second has a midband gain of 5 V/V and a bandwidth of 800 kHz. What are the overall midband gain and the overall bandwidth? *Ans:* $A_{MBo} = 40$ V/V, $f_{2o} = 392$ kHz.

**4.4** Change the individual stage midband gains of Practice Problem 4.3 to identical values to maintain an overall gain of 40 V/V. What is the new value of overall bandwidth? *Ans:* $A_{MB1} = A_{MB2} =$ 6.32 V/V, $f_{2o} = 405$ kHz.

### The Inverting Stage

For the inverting stage of Fig. 4.5, the frequency-dependent gain can be written as

$$A_i = \frac{-R_F}{R_1} \frac{1}{1 + j\frac{f}{f_{2oa}(1+A_{MBoa}R_1/[R_1+R_F])}} \tag{4.7}$$

The closed-loop, low-frequency gain is given by

$$A_{MBi} = \frac{-R_F}{R_1} \tag{4.8}$$

whereas the closed-loop bandwidth is

$$f_{2i} = f_{2oa}\,(1 + A_{MBoa}R_1/[R_1 + R_F]) \approx f_{2oa}A_{MBoa} \times \frac{R_1}{R_1 + R_F} \tag{4.9}$$

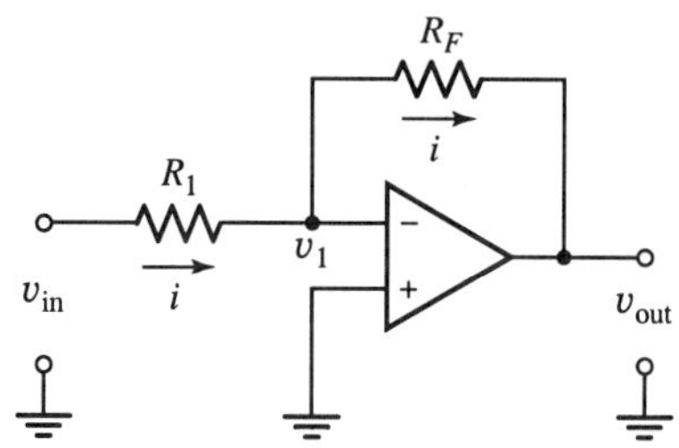


Figure 4.5: An inverting amplifier.

For closed-loop gains considerably greater than unity, which ensures that $R_F \gg R_1$, gain and bandwidth can again be directly exchanged.

If a signal generator resistance, $R_g$, is present, this value should be added to the series resistor $R_1$ in both Eqs. (4.8) and (4.9). This value will lower the midband gain and raise the bandwidth if $R_g$ is significant compared to $R_1$.

For an inverting stage, if $R_F/R_1 = 1000$, $A_{MBoa} = 2 \times 10^5$, and $f_{2oa} = 5$ Hz, the closed loop midband gain becomes

$$A_{MBi} = \frac{-R_F}{R_1} = -1000 \text{ V/V}$$

The gain decreases from the open-loop value by a factor of 200; hence, the bandwidth increases by this same factor to 1000 Hz.

If $R_F$ is not much larger than $R_1$, the $GBW$ of the inverting stage, designated $GBW_i$, can be calculated from Eqs. (4.8) and (4.9) to be

$$GBW_i = \mid A_{MBi} \mid \; f_{2i} \approx A_{MBoa} \, f_{2oa} \frac{R_F}{R_F + R_1} = GBW_{oa} \frac{R_F}{R_F + R_1} \tag{4.10}$$

For high-bandwidth, low-gain applications, the gain-bandwidth product of the inverting stage drops from the op-amp $GBW$. In such instances, the noninverting stage should be used to maximize the amplifier voltage gain.

A comparison of the $GBW$ as a function of midband voltage gain between the inverting and noninverting stages is shown in Fig. 4.6. At a voltage gain of unity, the noninverting stage will exhibit twice the bandwidth as that of the inverting stage.

Both the inverting and noninverting stages allow us to easily check the appropriateness of an op amp for a required amplifier design. If an amplifier with a gain of 100 and a bandwidth of 20 kHz is needed, an amplifying element with $GBW$ equal to or greater than $2 \times 10^6$ Hz must be used. The 741 op amp with a $GBW$ of $10^6$ would not be appropriate for this amplifier unless more than one stage were used.

**Practice Problems**

**4.5** An inverting stage has a midband gain of $-100$ V/V and a bandwidth of 40 kHz. Calculate the bandwidth when the midband gain is changed to $-2$ V/V. *Ans:* $f_{2o} = 1.33$ MHz.

**4.6** Repeat Practice Problem 4.5 if the midband gain is changed to $-10$ V/V. *Ans:* $f_{2o} = 364$ kHz.

### 4.1.2 Transistors

Although MOSFET usage in high-frequency communications is increasing, the BJT dominates this area from the low GHz range upward. The BJTs used in these applications are often Si-Ge heterojunction devices or gallium

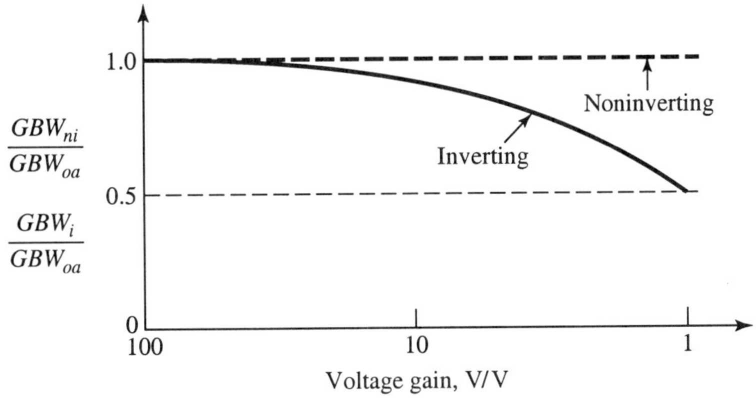


Figure 4.6: $GBW$ vs. voltage gain for inverting and noninverting stages.

arsenide, indium phosphide, or gallium nitride devices. Fortunately, the same model or equivalent circuit can often be used to analyze the BJT implemented in these various materials.

In this text, the MOSFET is discussed in connection with some circuits, and the BJT is considered in other circuits. A simulation program such as Spice is necessary to do useful circuit design. However, the concepts that determine circuit performance are often obscured by the use of a simulation program. Most of the circuits introduced in this text use equivalent circuits for the devices and discuss circuit operation in terms of these models. An understanding of these models is necessary to understand circuit operation. Although simulation will be used to fine-tune a circuit, design is almost always done using the equivalent circuits as the basis. Several single-stage MOSFET and BJT amplifiers are discussed in Chapter 1.

The $GBW$ of MOSFET or BJT stages may not be as meaningful as that for the op amp stage, however. We will consider these matters in the next section.

### The MOSFET Gain-Bandwidth Product

In theory, the simple common-source stage of Fig. 4.7 exhibits a near-constant $GBW$ as voltage gain is changed by changing $R_D$. If the voltage gain is low or the signal generator resistance is small, minimizing the effect of the Miller capacitance at the gate, the upper 3-dB frequency is determined by the output circuit. When this is the case, the upper corner frequency is

$$f_2 = \frac{1}{2\pi R_{\text{out}} C_{\text{out}}} \tag{4.11}$$

where

$$R_{\text{out}} = r_{ds} \mid\mid R_D$$

and

$$C_{\text{out}} = C_{gd} + C_{db} + C_{ds} + C_L \approx C_{gd} + C_{db} + C_L$$

In this expression, $C_L$ is any load capacitance or input capacitance from the following stage, if present.

The midband voltage gain is given by

$$A_{MB} = -g_m R_{\text{out}} \tag{4.12}$$

leading to a $GBW$ figure of

$$GBW = g_m R_{\text{out}} \times \frac{1}{2\pi R_{\text{out}} C_{\text{out}}} = \frac{g_m}{2\pi C_{\text{out}}} \tag{4.13}$$

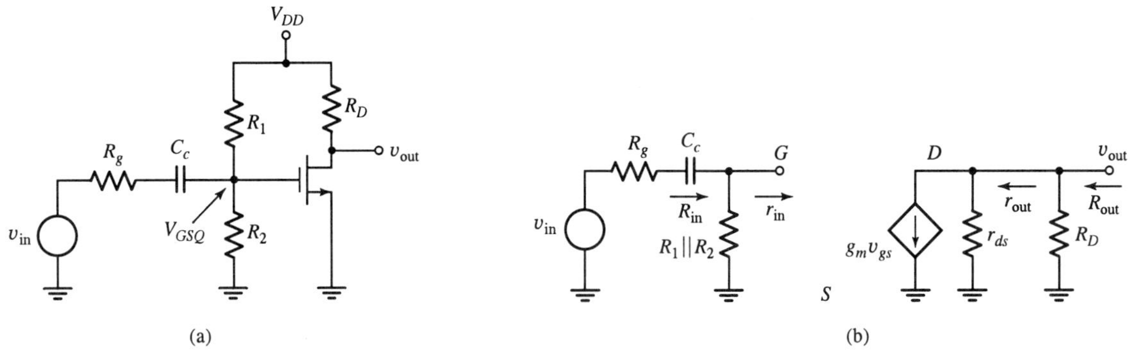


Figure 4.7: A common-source stage. (a) Actual circuit. (b) Equivalent circuit.

Since $R_{out}$ does not appear in the expression for $GBW$, but can be used to control the midband voltage gain, the $GBW$ is constant as gain and bandwidth are exchanged.

In practical IC design, a simple resistor is never used for the load. Rather, an active load of some sort is used to provide the drain with an incremental resistance. If a high-impedance current mirror provides the drain load, the midband voltage gain will be high, leading to a larger Miller capacitance. If signal generator resistance is significant as it might be if a stage with high output impedance drives the stage of Fig. 4.7, the input 3-dB frequency may influence the overall upper corner frequency. In such a case, the $GBW$ might vary as midband voltage gain is varied.

One configuration that can be useful in multistage design is the common-source stage with a diode-connected load shown in Fig. 4.8.

The diode-connected stage offers a relatively low load impedance to the amplifying stage and minimizes the Miller effect. The upper corner frequency is again determined by the output circuit and is given by Eq. (4.11). However, both $R_{out}$ and $C_{out}$ are different from the values used for the simple common-source stage. The output impedance is

$$R_{out} = \frac{1}{g_{m2} + g_{ds1} + g_{mb2} + g_{ds2}} \approx \frac{1}{g_{m2} + g_{mb2}} \tag{4.14}$$

where $g_{mb2}$ accounts for the body effect of device M2. The output capacitance is now influenced by device M2 and is

$$C_{out} = C_{gd1} + C_{db1} + C_{sb2} + C_{gs2} + C_L \tag{4.15}$$

The midband voltage gain is again expressed by Eq. (4.12) and the $GBW$ is given by Eq. (4.13).

Although the equations for $GBW$, $A_{MB}$, and $f_2$ apply for the circuits of Figs. 4.7 and 4.8, there are some subtle differences. In order to change the gain in the simple common-source stage, the load resistance $R_D$ is changed. For the diode-connected load stage, either the dc bias current can be changed to change the value of $R_{out}$ or the dimensions of $M2$ can be changed. In either case, the bias point changes, causing $g_{m1}$ and the output capacitance to change. In some cases, these changes are rather small and the $GBW$ remains constant. In others, the assumption of constant $GBW$ with gain change is invalid.

### Practice Problem

**4.7** A common-source device with $\mu C\text{ox}W/2L = 0.4\ \text{mA/V}^2$ is biased to a drain current of 100 $\mu$A. If $r_{ds} = 300\ \text{k}\Omega$, a drain resistance of 12 k$\Omega$ is used, and $C_{out} = 500$ fF, calculate the midband voltage gain, the upper corner frequency, and the $GBW$. Assume a zero-resistance signal generator drives the stage. *Ans:* $A_{MB} = 4.62$ V/V, $f_{2o} = 27.7$ MHz, $GBW = 12.8$ MHz.

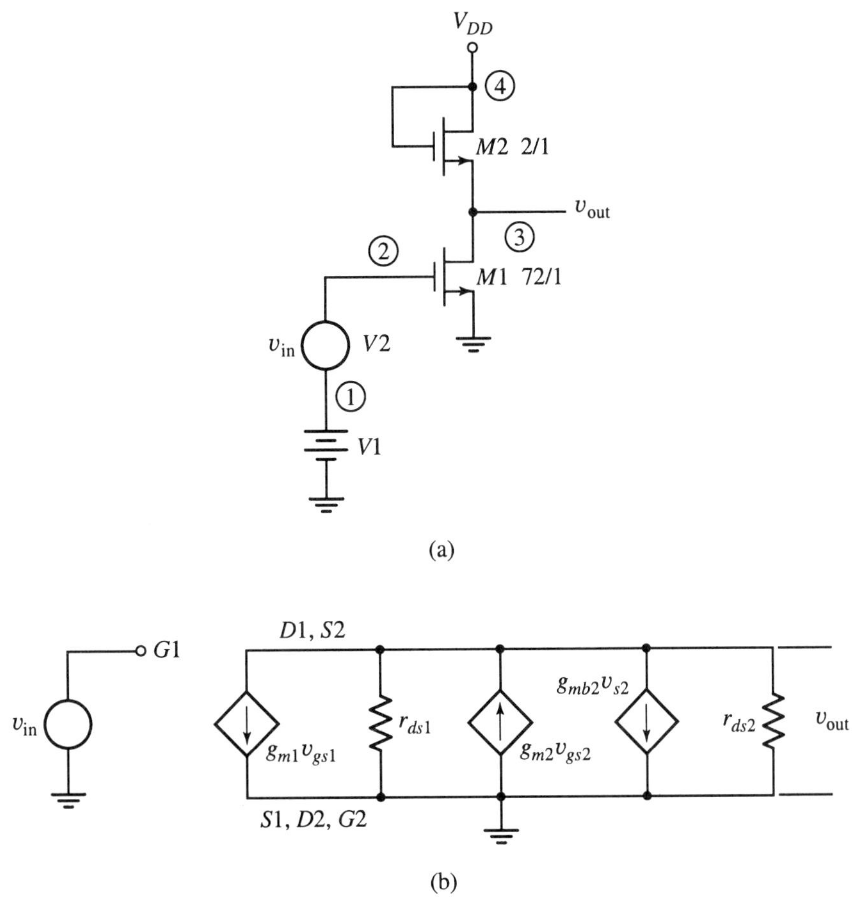


Figure 4.8: The diode-connected load stage.

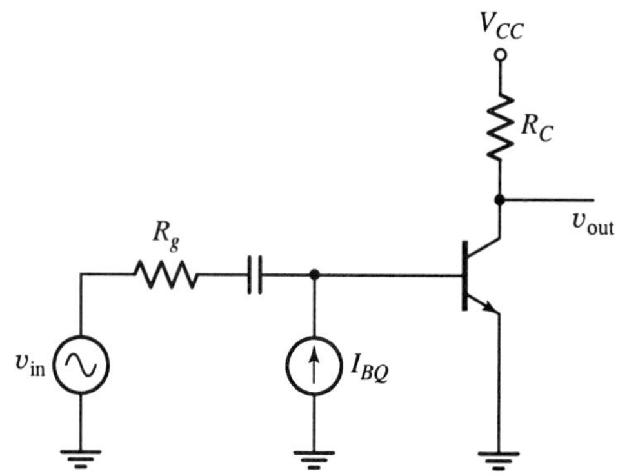


Figure 4.9: The common-emitter stage.

### The BJT Gain-Bandwidth Product

For the simple common-emitter stage of Fig. 4.9 the midband voltage gain is

$$A_{MB} = \frac{-\beta R_C}{r_\pi + r_x + R_g} \tag{4.16}$$

The bandwidth is

$$\omega_2 = \frac{B\omega_\beta}{D} = \frac{\omega_\beta}{D}\left(\frac{r_\pi + r_x + R_g}{r_x + R_g}\right) \tag{4.17}$$

where $B$ is the broadband factor given by

$$B = \frac{r_\pi + r_x + R_g}{r_x + R_g}$$

The $D$ factor accounts for the Miller effect on $C_\mu$ and is

$$D = 1 + R_C\omega_t C_\mu$$

The product of the midband gain and the bandwidth is

$$GBW = \frac{\omega_\beta}{D}\frac{\beta R_C}{r_x + R_g} = \frac{\omega_t}{D}\frac{R_C}{r_x + R_g} \tag{4.18}$$

Several variables can be controlled to extend the bandwidth of the common-emitter stage. In a given application, not all of these elements can be varied. For example, decreasing $R_g$ increases both the bandwidth and midband gain, resulting in a higher $GBW$ after broadbanding. There is a lower limit on this element determined by the signal generator resistance. Thus, a reasonable first step in any single-stage design procedure might consist of assuming the exact value of $R_g$ that must be used. If some choice of values is available, the lowest possible value should be selected.

If $C_\mu$ is decreased, the bandwidth will also increase with no corresponding loss of gain. This element can be lowered for a given transistor only by increasing the reverse bias of the collector-base junction. Here too, a limiting value is imposed by other circuit requirements. Perhaps the power supply voltage is specified or the collector voltage may be limited by the maximum rating, thus setting $C_\mu$ at its minimum value. Again, this element will be fixed and is not a convenient element to change for broadbanding purposes.

The gain-bandwidth product $\omega_t$ has been treated as a fixed quantity, but this figure drops off at lower values of emitter current. At very high values of emitter current $\omega_t$ also decreases. Figure 4.10 shows a typical variation of $\omega_t$ with emitter current. Although there is a range of emitter current over which $\omega_t$ is relatively constant, this

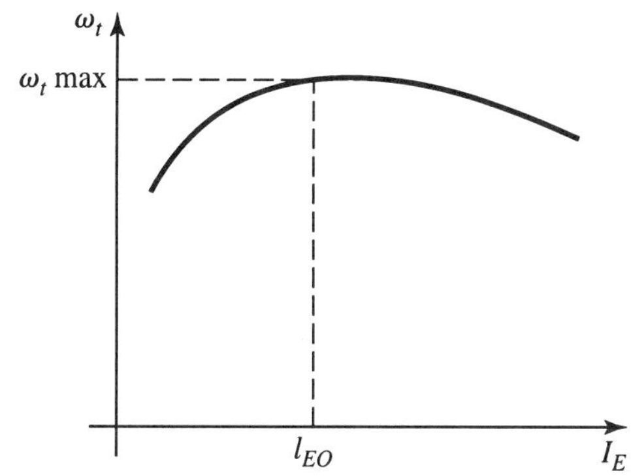


Figure 4.10: Variation of $\omega_t$ with emitter current.

figure decreases rapidly at low emitter currents. It is obviously advantageous to bias the stage to the current that leads to maximum $\omega_t$.

Returning to the equations for $A_{MB}$ and $\omega_2$, we see that still another parameter that can increase the bandwidth is $r_\pi$. If this element is increased the bandwidth increases while the gain decreases. The $GBW$ remains constant if we assume that $\omega_t$ is constant. An early proposal for multistage design suggested that emitter bias current be varied to control $r_\pi$, thus controlling the bandwidth. This method requires a low dc emitter current to achieve high bandwidths. Although Eq. (4.18) indicates that the $GBW$ remains constant, we must remember that this equation assumes a constant value for $\omega_t$. As indicated by Fig. 4.10, $\omega_t$ does not remain constant at low currents; hence, the $GBW$ falls off considerably as the bandwidth is increased. An effective broadbanding scheme should allow $I_E$ to be selected to optimize $\omega_t$, and should not require the bias to change from this value.

The last remaining variable over which we have control in many applications is the load resistance $R_C$. In certain applications a load to be driven is specified, but this value can be decreased by a parallel resistance. In other applications, the value might be allowed to vary over quite a wide range. It is perhaps more realistic to use $R_C$ to control bandwidth rather than any other element of the common-emitter stage. As the preceding paragraphs indicate, there are valid reasons for fixing these other element values.

As $R_C$ is made smaller, the $D$-factor decreases and bandwidth increases. The $D$-factor, given by

$$D = (1 + R_C\omega_t C_\mu) \tag{4.19}$$

does not vary directly with $R_C$ and in fact becomes a weak function of $R_C$ as $D$ approaches its minimum value of unity. Thus, as $D$ approaches unity the bandwidth increases much more slowly than $R_C$ decreases. On the other hand, the midband gain decreases directly with $R_C$ and the $GBW$ decreases as bandwidth is extended. This can be seen by expressing $D$ in terms of $R_C$ in Eq. (4.18) to give

$$GBW = \frac{\omega_t}{(1 + R_C\omega_t C_\mu)}\frac{R_C}{r_x + R_g} = \frac{\omega_t}{(1/R_C + C_\mu\omega_t)}\frac{1}{r_x + R_g} \tag{4.20}$$

As $R_C$ is lowered to extend the bandwidth, the denominator of the previous equation becomes larger, resulting in a lower $GBW$.

Although large values of $R_C$ result in very low bandwidths, perhaps we should consider the effects of increasing $R_C$ on the $GBW$. As $R_C$ becomes larger, the $D$-factor will approach a value of $R_C\omega_t C_\mu$. The $GBW$ will be given by

$$GBW = \frac{\omega_t}{\alpha R_C\omega_t C_\mu}\frac{R_C}{r_x + R_g} \approx \frac{1}{C_\mu(r_x + R_g)} \tag{4.21}$$

This equation gives the theoretical upper limit on $GBW$ for the common-emitter stage, but two items must be noted that prevent the attainment of this limit in practice: (1) large bandwidths cannot be obtained when

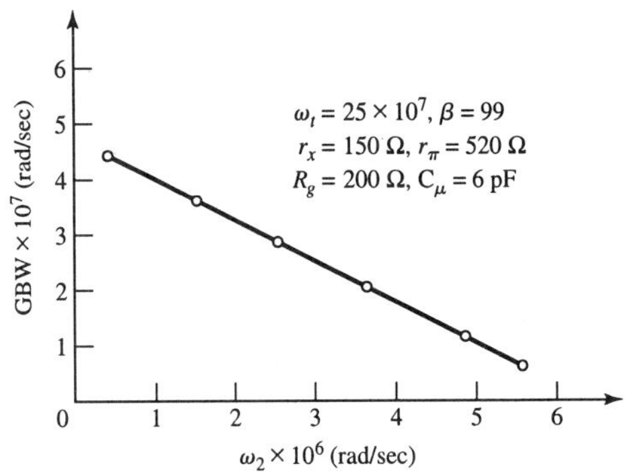


Figure 4.11: *GBW* as a function of bandwidth.

$R_C$ is large; and (2) there is always a certain amount of stray capacitance that will shunt the load. For small values of $R_C$ this capacitance may be negligible, but as $R_C$ takes on larger values, this capacitance can limit the actual bandwidth to a figure lower than that calculated by Eq. (4.17). Since bandwidth is now limited by these stray capacitance effects at large values of $R_C$, the *GBW* figure decreases as $R_C$ increases rather than remaining constant as indicated by Eq. (4.21).

We can plot the *GBW* figure as a function of bandwidth, noting that $R_C$ is varied to change the bandwidth. This information is displayed in Fig. 4.11. As this figure shows, the common-emitter stage is less efficient at higher bandwidths.

### Practice Problem

**4.8** In the common-emitter stage of Fig. 4.9, the generator resistance is 200 Ω and $R_C = 150\ \Omega$. If the transistor has values of $\beta = 120$, $r_x = 100\ \Omega$, $C_\mu = 3$ pF, and $f_t = 300$ MHz, choose a quiescent emitter current to create a gain of −6 V/V for the stage. Calculate the upper corner frequency and the *GBW* for the stage. *Ans:* $I_{EQ} = 1.17$ mA, $f_2 = 13.5$ MHz, $GBW = 81.1$ MHz.

### Base Compensation

One method of discrete stage broadbanding that allows the *GBW* to remain constant is that of base compensation, or the use of a parallel resistor-capacitor circuit in the base lead. This is referred to as a *compensating circuit*. In switching circuits, the capacitor $C_1$ is called a *speed-up capacitor*, due to the resulting improvement in switching speeds. This circuit is shown in Fig. 4.12.

When $R_g$ and $r_x$ are neglected, the input circuit reduces to that of an attenuator probe. Figure 4.13 shows the schematic of such a probe.

In the figure, $R_2$ and $C_2$ might represent the input resistance and capacitance to the vertical amplifier of an oscilloscope. Resistance $R_1$ is added to the circuit to attenuate the input signal or to give the circuit a higher input impedance. When $R_1$ is present, the output signal will be frequency-dependent. If $C_1$ is properly chosen, the output voltage will be frequency-dependent and will therefore have an infinite bandwidth. When $C_1 = 0$, the time constant of the circuit is

$$\tau = \frac{R_1 R_2}{r_1 + R_2} C_2 \tag{4.22}$$

The bandwidth in rad/sec equals the reciprocal of $\tau$:

$$\omega_2 = \frac{R_1 + R_2}{R_1 R_2 C_2} \tag{4.23}$$

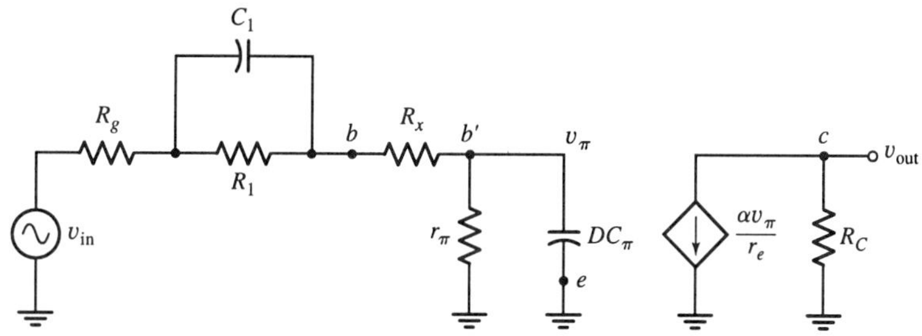


Figure 4.12: Equivalent circuit for a base-compensated stage.

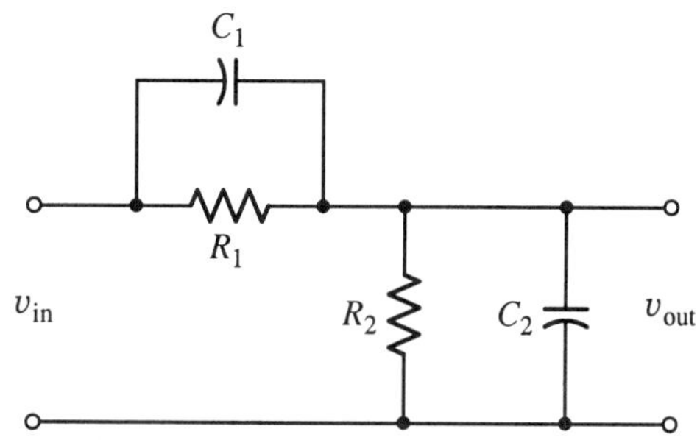


Figure 4.13: Attenuator probe circuit.

In an oscilloscope probe, the impedance added by using a large value of $R_1$ compared to $R_2$ adds a significant time constant to the actual input voltage being measured. It also reduces the loading of the probe, but does so at the expense of attenuation. The capacitor $C_1$ is used to remove the added time constant. When $C_1$ is included, the output voltage of Fig. 4.13 is

$$v_{out} = \frac{Z_2}{Z_1 + Z_2} v_{in} = \frac{\frac{R_2}{(1+j\omega R_2 C_2)}}{\frac{R_2}{(1+j\omega R_2 C_2)} + \frac{R_1}{(1+j\omega R_1 C_1)}} v_{in} \tag{4.24}$$

If $C_1$ is chosen such that $R_1C_1 = R_2C_2$, Eq. (4.24) reduces to

$$v_{out} = \frac{R_2}{R_1 + R_2} v_{in} \tag{4.25}$$

which is independent of frequency. The attenuator or probe circuit now has an infinite bandwidth and zero rise time. This same concept can be applied to discrete BJT design.

In the input loop of a transistor, $r_\pi$ and $DC_D$ correspond to $R_2$ and $C_2$ of the probe. The condition for compensation is

$$R_1C_1 = r_\pi DC_D = \frac{D}{\omega_\beta} \tag{4.26}$$

The capacitor $C_1$ is picked to be

$$C_1 = \frac{D}{R_1 \omega_\beta} \tag{4.27}$$

The selection of $R_1$ will come from gain considerations. Unfortunately, the transistor will not have an infinite bandwidth because $R_g$ and $r_x$ are not zero, as we assumed in this derivation. We have also neglected certain second-order effects that would become prominent as higher bandwidths are approached. The bandwidth can be found by analyzing the circuit of Fig. 4.12 and applying the condition for compensation. The output voltage is

$$v_{out} = \frac{-\alpha v_\pi}{r_e} R_C \tag{4.28}$$

and $v_\pi$ is found to be

$$v_\pi = \frac{\frac{r_\pi}{(1+j\omega r_\pi DC_D)}}{\frac{r_\pi}{(1+j\omega r_\pi DC_D)} + \frac{R_1}{(1+j\omega C_1 R_1)} + R_g + r_x} v_{\text{in}} \tag{4.29}$$

Combining these two equations and remembering that $R_1 C_1 = r_\pi DC_D$ gives

$$A_o = \frac{v_{\text{out}}}{v_{\text{in}}} = \frac{-\beta R_C}{r_\pi + R_1 + R_g + r_x}\left[1 + \frac{j\omega r_\pi DC_D(R_g + r_x)}{r_\pi + R_1 + R_g + r_x}\right] \tag{4.30}$$

The bandwidth is seen to be

$$\omega_2 = \frac{\omega_\beta}{D}\left(\frac{r_\pi + R_1 + R_g + r_x}{R_g + r_x}\right) \tag{4.31}$$

This can easily be compared to the case where $R_1$ and $C_1$ are not present. From Eq. (4.17), we have

$$\omega_2 = \frac{\omega_\beta}{D} B = \frac{\omega_\beta}{D}\left(\frac{r_\pi + R_g + r_x}{R_g + r_x}\right)$$

Since $R_1$ can be chosen to be several factors larger than $r_\pi + R_g + r_x$, the bandwidth from Eq. (4.31) can be much larger than that of the uncompensated case. The use of $R_1$ and $C_1$ to extend the bandwidth results in a corresponding decrease in gain. The midband gain is now

$$A_{MB} = \frac{-\beta R_C}{r_\pi + R_1 + R_g + r_x} \tag{4.32}$$

The gain decreases by the same factor that the bandwidth increases in this case. The $GBW$ is

$$GBW = \frac{\omega_\beta}{D}\frac{\beta R_C}{(R_g + r_x)} = \frac{\omega_t}{D}\frac{R_C}{(R_g + r_x)}$$

which is the same as the uncompensated case of Eq. (4.18). Note that since $R_1$ affects both gain and bandwidth, but does not appear in the $GBW$ equation, gain and bandwidth can be directly interchanged by this method. A high bandwidth can be achieved without decreasing the $GBW$ figure.

# Example 4.2

If the transistor in Fig. 4.12 has the following parameters: $f_t = 100$ MHz, $\beta = 100$, $r_x = 13\ \Omega$ ($I_E = 2$ mA), and $C_\mu = 5$ pF at the bias point, what are the midband gain and upper corner frequency for the cases where

**(a)** $R_1 = 2$ k$\Omega$, and $C_1$ has the correct value for compensation; and

**(b)** $R_1 = C_1 = 0$ (assume the same $I_E$ )?

**Solution (a)** The midband gain is found from Eq. (4.32) to be

$$A_{MB} = \frac{-100 \times 10^3}{1313 + 2000 + 600 + 100} = -25\text{V/V}$$

Converting Eq. (4.31) to Hz, we obtain

$$f_2 = \frac{f_\beta}{D}\left(\frac{r_\pi + R_g + R_1 + r_x}{r_x + R_g}\right)$$

Since

$$D = 1 + R_C \omega_t C_\mu = 1 + 1 \times 10^3 \times 2\pi \times 10^8 \times 5 \times 10^{-12} = 4.14$$

and

$$f_\beta = f_t/\beta = 1 \text{ MHz}$$

the bandwidth is

$$f_2 = \frac{1}{4.14} \times \frac{4013}{700} = 1.38 \text{ MHz}$$

The correct value of $C_1$ is

$$C_1 = \frac{D}{R_1 \omega_\beta} = \frac{4.14}{2 \times 10^3 \times 2\pi \times 10^6} = 329 \text{ pF}$$

**(b)** Without $R_1$ and $C_1$, the midband gain is

$$A_{MB} = \frac{-\beta R_C}{r_\pi + R_g + r_x} = \frac{-10^5}{2013} = -50 \text{ V/V}$$

The bandwidth is

$$f_2 = \frac{f_\beta}{D} B$$

where

$$B = 2013/700 = 2.88$$

so

$$f_2 = \frac{1}{4.14} \times 2.88 = 0.7 \text{ MHz}$$

---

A comparison of the efficiency of the base-compensated stage to the uncompensated stage is made in Fig. 4.14. As higher bandwidths are required, the *GBW* figure, and therefore the gain, are higher for the compensated stage than for the uncompensated stage. It should also be noted that the bandwidth can be varied without regard to the bias current. Consequently, emitter current can be set to maximize $\omega_t$ and broadbanding can take place without affecting this value.

Among other methods of broadbanding transistor stages are the emitter-compensated stage and shunt-peaked stage, both of which offer slightly better *GBW* figures than does the base-compensated stage. Both are somewhat more complex, however. The base-compensated stage offers an efficient stage that can form the basis of a practical design procedure, as will be shown in the following sections.

## 4.2 MULTISTAGE WIDEBAND AMPLIFIERS

**Important Concepts**

**1.** For stages with constant *GBW* figures as gain and bandwidth are exchanged, multistage theory can be used to optimize the design.

**2.** For a given required overall midband voltage gain, the overall bandwidth can be maximized with this theory.

**3.** If the required overall voltage gain is realized, but the bandwidth is greater than the specified value, the number of stages can be minimized.

**4.** The noninverting op amp stage has a constant *GBW* figure. The common-source MOSFET stage approximates a constant *GBW* stage and the base-compensated BJT stage has a constant *GBW* figure.

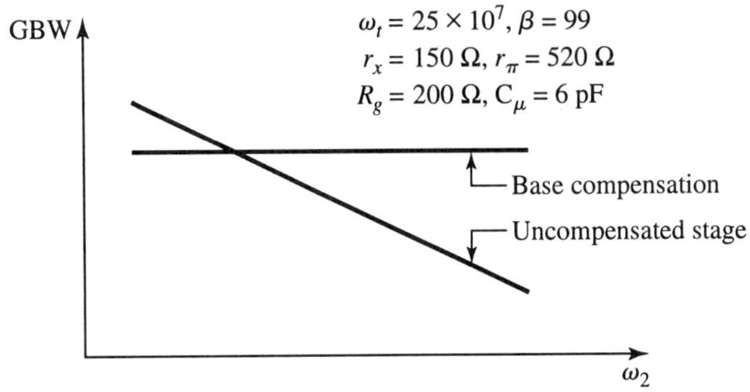


Figure 4.14: *GBW* as a function of bandwidth for base compensation.

### 4.2.1 Multistage Theory

When devices have constant $GBW$ figures as gain and bandwidth are exchanged, well-known theoretical methods can be applied to multistage amplifier design. When the $GBW$ figure is not constant, iterative or trial-and-error methods must often be applied.

A circuit consisting of $n$ identical stages is called an *iterative* stage amplifier. Iterative stages are often chosen to satisfy an amplifier design because equal bandwidths for each stage lead to the maximum overall bandwidth for a specified overall voltage gain. In wideband amplifiers, the lower 3-dB frequency is either zero or very small compared to the upper corner frequency. Consequently, the bandwidth is

$$BW = f_2 - f_1 \approx f_2$$

We will use the term *upper corner frequency* interchangeably with *bandwidth* for wideband stages, since the two parameters are almost equal. In practice the two major specifications for an amplifier are the overall voltage gain and the overall upper corner frequency.

#### Bandwidth Shrinkage in Iterative Stages

We will now consider the amplifier configuration of Fig. 4.15. In this $n$-stage amplifier, if each stage is down from its midband gain by 3 dB at a frequency of $f_2$, the overall amplifier will be down by $3n$ dB at $f_2$. For a five-stage circuit, the amplifier will be down from its midband gain by 15 dB at $f_2$. The frequency at which the overall midband voltage gain is down by 3 dB is called $f_{2o}$. This overall upper 3-dB point is always smaller than $f_2$. The decrease in $f_{2o}$ as more identical stages are added is referred to as *bandwidth shrinkage*. We will now derive the relationship between overall upper corner frequency, $f_{2o}$, and individual stage upper corner frequency, $f_2$.

The individual stage gains in the amplifier of Fig. 4.15 can be written as

$$A_1 = A_2 = A_3 = \cdots = A_n = \frac{A_{MB}}{1 + j\frac{f}{f_2}} \tag{4.33}$$

The overall voltage gain equals this expression raised to the $n$th power, or

$$A_o = \left[\frac{A_{MB}}{1 + j\frac{f}{f_2}}\right]^n = \frac{A_{MB}^n}{\left(1 + j\frac{f}{f_2}\right)^n} = \frac{A_{MBo}}{\left(1 + j\frac{f}{f_2}\right)^n} \tag{4.34}$$

where $A_{MBo}$ is the overall midband voltage gain of the amplifier. The magnitude of the overall gain is a function of frequency and can be expressed in dB as

$$A_{dB} = 20\ \log\ |A_o| = n\ \log \left|\frac{A_{MB}}{1 + j\frac{f}{f_2}}\right| \tag{4.35}$$

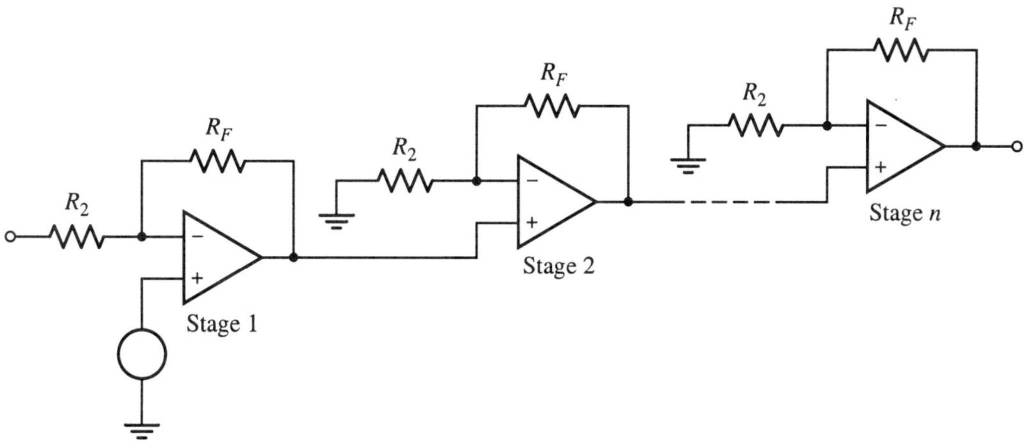


Figure 4.15: An iterative stage amplifier.

At low and midband frequencies, Eq. (4.34) becomes

$$A_o = A_{MBo} = A_{MB}^n \tag{4.36}$$

At higher frequencies the denominator increases, dropping the overall gain. The upper 3-dB frequency is that value of $f$ that causes the magnitude of the complex denominator of Eq. (4.34) to equal $\sqrt{2}$. When this occurs, the gain magnitude is

$$|A_o| = \frac{A_{MBo}}{\sqrt{2}}$$

and is 3 dB below the overall midband gain, $A_{MBo}$. We can equate the magnitude of the denominator to $\sqrt{2}$ and solve for the frequency that satisfies this relationship. This frequency will be $f_{2o}$, the overall upper 3-dB frequency. This results in

$$|(1 + jf_{2o}/f_2)^n| = |1 + jf_{2o}/f_2|^n = \sqrt{2}$$

Squaring both sides of this equation gives

$$|1 + jf_{2o}/f_2|^{2n} = 2$$

The magnitude of the complex term can be expressed as

$$|1 + jf_{2o}/f_2| = \sqrt{1 + \frac{f_{2o}^2}{f_2^2}}$$

Consequently, we can write

$$\left(\sqrt{1 + \frac{f_{2o}^2}{f_2^2}}\right)^{2n} = \left(1 + \frac{f_{2o}^2}{f_2^2}\right)^n = 2$$

This equation is solved for $f_{2o}$ to yield

$$f_{2o} = f_2\sqrt{2^{1/n} - 1} \tag{4.37}$$

This is the well-known bandwidth shrinkage equation.

To appreciate the magnitude of this effect, if two stages are used in the amplifier ($n = 2$) the overall bandwidth is 0.64 times the individual stage bandwidth. Three stages shrink the bandwidth to 51% of the individual value. Equation (4.37) can be approximated for $n \geq 3$ by the simpler expression

$$f_{2o} = \frac{f_2}{1.2\sqrt{n}} \tag{4.38}$$

# Example 4.3

Using the op amp stages of Example 4.1, realize the overall voltage gain of 200 V/V with equal gains and bandwidths of both stages. Calculate the overall upper 3-dB frequency of this two-stage amplifier.

**Solution** For identical stages, the gain of each stage must be set to

$$A_{MB1} = A_{MB2} = \sqrt{200} = 14.14 \text{ V/V}$$

For the noninverting stages of Fig. 4.3, these gains will be satisfied if $R_{F1}$ and $R_{F2}$ are chosen to be 26.3 kΩ and $R_{21}$ and $R_{22}$ are chosen to be 2 kΩ. The bandwidth of each stage will then be

$$f_{2ni1} = f_{2ni2} = f_{2ni} = \frac{GBW_{ni}}{A_{MBni}} = \frac{GBW_{oa}}{A_{MBni}} = \frac{3 \times 10^6}{14.14} = 212 \text{ kHz}$$

Using the bandwidth shrinkage equation we find that the overall bandwidth of the amplifier is

$$f_{2o} = f_{2ni}\sqrt{2^{1/2} - 1} = 0.64 f_{2ni} = 0.64 \times 212 = 136 \text{ kHz}$$

This value should be compared to that from Example 4.1 which is $f_{2o} = 126$ kHz. When bandwidth maximization is important, identical bandwidths are imperative.

**Practice Problems**

**4.9** Calculate the bandwidth shrinkage factor for eight identical stages using Eq. (4.37). Compare this to the value given by Eq. (4.38). *Ans:* 0.301; 0.294; -2.3% error.

**4.10** The $GBW$ of each stage of an amplifier with eight identical stages is 100 MHz. If the individual stage midband voltage gain is 1.9 V/V, what is the overall midband voltage gain and bandwidth? *Ans:* $A_{MBo} = 170$ V/V, $f_{2o} = 15.8$ MHz.

### Optimization of Overall Bandwidth

An interesting problem encountered in iterative stage design is the following:

*Given a specified overall midband voltage gain and overall bandwidth for an amplifier, what number of stages should be used to achieve this design?*

If a high value of individual stage gain is used, only a few stages are needed to realize the gain, thereby minimizing the bandwidth shrinkage. However, since $f_2 = GBW/A_{MB}$, a high value of $A_{MB}$ leads to a low value of $f_2$. If $A_{MB}$ is decreased to increase $f_2$, more stages are needed and the bandwidth shrinkage effect becomes more pronounced.

One would expect that there exists some value of gain or some number of stages that will optimize the overall bandwidth while realizing the specified value of overall gain. Such an optimizing value does exist and will be derived next.

We begin the derivation with Eq. (4.38):

$$f_{2o} = \frac{f_2}{1.2\sqrt{n}}$$

Since the gain-bandwidth product of the stage is

$$GBW = A_{MB} \times f_2$$

and

$$f_2 = \frac{GBW}{A_{MB}}$$

we can rewrite Eq. (4.38) as

$$f_{2o} = \frac{GBW}{1.2n^{1/2}A_{MB}} = \frac{GBW}{1.2n^{1/2}A_{MBo}^{1/n}} \tag{4.39}$$

Noting that $GBW$ and $A_{MBo}$ are constants, we could differentiate $f_{2o}$ with respect to $n$ and set this derivative equal to zero to find the value of $n$ that gives the maximum value of $f_{2o}$. The differentiation is simplified if we let $A_{MBo} = e^k$ where $k = \ln\ A_{MBo}$. This allows us to write the overall bandwidth as

$$f_{2o} = \frac{GBW}{1.2n^{1/2}e^{k/n}} \tag{4.40}$$

Differentiating gives

$$\frac{df_{2o}}{dn} = \frac{GBW}{1.2n^{3/2}e^{k/n}}\left[\frac{k}{n} - \frac{1}{2}\right] \tag{4.41}$$

This derivative is zero only if $n$ takes on a value of

$$n = 2k = 2\ \ln\ A_{MBo} \tag{4.42}$$

With this value of $n$, the individual stage gain becomes

$$A_{MB} = A_{MBo}^{1/n} = e^{k/n} = e^{1/2} = 1.65\ \text{V/V} \tag{4.43}$$

A quick check proves that this value of individual stage gain leads to a maximum rather than a minimum value of bandwidth.

In a given problem of amplifier design, we first assume an individual stage gain of 1.65, and then we calculate the required number of stages. The number of stages can be found from solving

$$1.65^n = A_{MBo}$$

giving

$$n = \frac{\ln\ A_{MBo}}{\ln\ 1.65} = 2\ \ln\ A_{MBo} \tag{4.44}$$

The individual stage bandwidth is found to be $f_2 = GBW/1.65$. The bandwidth shrinkage equation is then applied to find the maximum overall bandwidth. An example will demonstrate this method.

## Example 4.4

---

An amplifier requires a gain of 1000 V/V. If the op amps used have a $GBW$ of 6 MHz, calculate the maximum overall bandwidth that can be achieved and the number of stages that must be used.

**Solution** A single-stage gain of 1.65 V/V is required to lead to the maximum overall bandwidth. An overall gain of 1000 leads to a value of

$$n = 2 \ln 1000 = 13.8$$

Since $n$ must be an integer, 14 stages will be used. The individual stage bandwidth is

$$f_2 = \frac{GBW}{1.65} = \frac{6 \times 10^6}{1.65} = 3.636 \text{ MHz}$$

Applying the bandwidth shrinkage equation gives an overall bandwidth of

$$f_{2o} = \frac{f_2}{1.2n^{1/2}} = \frac{3.636 \times 10^6}{1.2 \times 14^{1/2}} = 810 \text{ kHz}$$

Let us suppose that the overall bandwidth resulting in Example 4.4 is not sufficient to satisfy the amplifier bandwidth. Using these op amps, we cannot realize this design. The only way to increase the overall bandwidth above the calculated maximum is to use op amps with higher $GBW$ values, if available.

On the other hand, if the bandwidth resulting from the design is higher than needed for the amplifier, the gain per stage can be increased, lowering the number of required stages. Often, the number of stages can be reduced considerably, by trial-and-error methods, to achieve the correct amplifier bandwidth. Thus, the major purpose of this theory is to verify that a given op amp or amplifying element can satisfy a design specification and suggest that a smaller number of stages than given by Eq. (4.44) can be used.

---

The next example demonstrates an approach to minimizing the number of stages required for a multistage, iterative amplifier.

## Example 4.5

---

A wideband amplifier is to be designed with a voltage gain of 1000 V/V and a bandwidth of 7 MHz. Using an op amp with $GBW = 8 \times 10^7$ Hz, minimize the number of stages required for this amplifier.

**Solution** As in the previous example, the overall bandwidth is maximized for an individual stage gain of 1.65 V/V. A total of 14 stages is needed to achieve a gain that is greater than or equal to 1000 V/V.

The individual stage bandwidth is then

$$f_2 = \frac{GBW}{1.65} = \frac{8 \times 10^7}{1.65} = 48.5 \text{ MHz}$$

Applying the bandwidth shrinkage equation gives an overall bandwidth of

$$f_{2o} = \frac{f_2}{1.2n^{1/2}} = \frac{48.5 \times 10^7}{1.2 \times 14^{1/2}} = 10.8\ MHz$$

Since this bandwidth exceeds the required value of 7 MHz, we will assume that a smaller number of stages with higher voltage gains might satisfy the design. Since we do not know how many stages might satisfy this design, we reduce the number of stages by a factor of two to seven stages. Each stage must now have a voltage gain of

$$A_{MB} = 1000^{1/7} = 2.683 \text{ V/V}$$

The individual stage bandwidth will now be

$$f_2 = \frac{GBW}{2.683} = \frac{8 \times 10^7}{2.683} = 29.8 \text{ MHz}$$

The overall bandwidth is found to be 9.39 MHz and still exceeds the specified value. Thus, we can lower the number of stages again.

With four stages, the midband gain must be $A_{MB} = 5.62$ V/V and the resulting overall bandwidth is calculated as $f_{2o} = 5.93$ MHz. This does not satisfy the bandwidth specification so the calculations are repeated for five stages and results in $A_{MB} = 3.98$ V/V and $f_{2o} = 7.49$ MHz. This multistage design could be realized with five stages rather than the 14 stages that maximize bandwidth. The savings resulting from five stages rather than 14 stages is obviously significant.

---

Although this theory applies directly to the noninverting op amp stage that has a constant *GBW* as gain and bandwidth are exchanged, it also has some value in BJT or MOSFET wideband amplifier design. The base-compensated BJT has been used for some years for iterative stages and the MOSFET diode-connected load stage can approximate the requirement of constant *GBW*. The most significant result of this theory is that iterative stages with low gains ranging from 1.65 V/V to about 3 V/V can lead to near-optimum overall bandwidth.

---

## SUMMARY

**1.** The *GBW* of a single amplifier stage is a significant figure of merit and can be used in multistage design if it remains constant as gain and bandwidth are exchanged.

**2.** The noninverting op amp stage has a constant *GBW* while the inverting stage *GBW* drops as the midband gain decreases toward a value of unity.

**3.** The MOSFET IC stage with a diode-connected load has a *GBW* that is approximately constant with gain.

**4.** The base-compensated BJT stage has a *GBW* that is constant with gain.

**5.** For a multistage amplifier based on identical, constant *GBW* stages, the maximum overall bandwidth results when the individual stage midband gain is 1.65 V/V.

---

## PROBLEMS

### Section 4.1.1 The Op Amp Single-Stage Amplifier

**4.1** Write the expression for gain $A$ of an op amp that has a dc gain of 400,000 V/V and a bandwidth of 10 Hz. Calculate the magnitude of the gain at a frequency of 5 kHz.

**4.2** Assuming $A_{MBoa} = 200{,}000$ V/V, find $R_{\text{in}}$, $R_{\text{out}}$, and $A_{MBi}$ for the circuit shown. If the op amp *GBW* is $4 \times 10^6$ Hz, what is the upper 3-dB frequency of the amplifier? *Hint:* $R_{\text{out}}$ can be approximated.

**4.3** Assuming $A_{MBoa} = 200{,}000$ V/V, find $R_{\text{in}}$, $R_{\text{out}}$, and $A_{MBi}$ for the circuit shown. If the op amp *GBW* is $4 \times 10^6$ Hz, what is the upper 3-dB frequency of the amplifier? *Hint:* $R_{\text{out}}$ can be approximated.

**4.4** Assuming $A_{MBoa} = 200{,}000$ V/V, find $R_{\text{in}}$, $R_{\text{out}}$, and $A_{MBni}$ for the circuit shown. If the op amp *GBW* is $4 \times 10^6$ Hz, what is the upper 3-dB frequency of the amplifier? *Hint:* $R_{\text{in}}$ and $R_{\text{out}}$ can be approximated.

**D4.5** Using an op amp with $GBW_{oa} = 4 \times 10^6$ Hz, design a noninverting amplifier with a midband voltage gain of 75 V/V. What is the bandwidth of the amplifier?

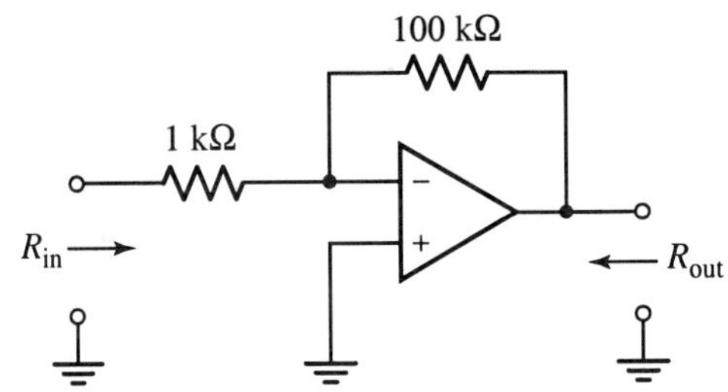


**Figure P4.2**

**D4.6** Using an op amp with $GBW_{oa} = 4 \times 10^6$ Hz, design an inverting amplifier with a midband voltage gain of -10 V/V. What is the bandwidth of the amplifier?

**4.7*** For the given circuit

**(a)** Calculate the midband value of voltage gain of the amplifier,

$$A_{MBi} = \frac{v_{out}}{v_{in}}$$

**(b)** Does the source resistance, $R_g$, affect the midband gain?

**(c)** Calculate the upper corner frequency of the overall voltage gain if the op amp $GBW$ is 3.2 MHz.

**(d)** Does the source resistance affect the upper corner frequency of the amplifier?

**4.8** Repeat Problem 4.7 for the circuit shown.

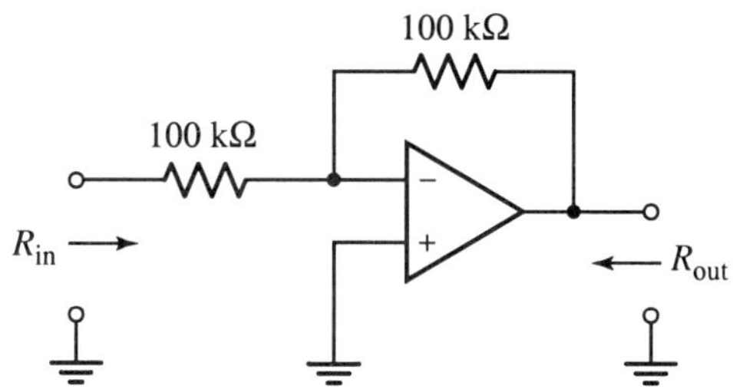


**Figure P4.3**

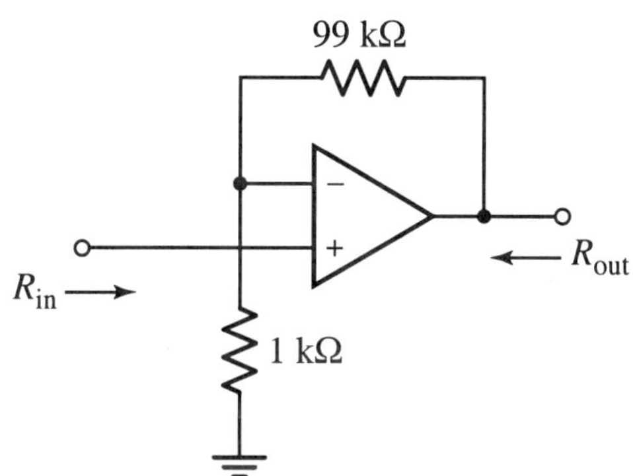


**Figure P4.4**

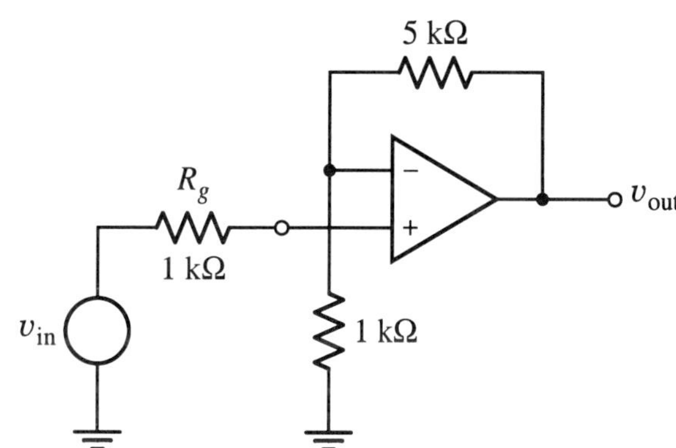


**Figure P4.8**

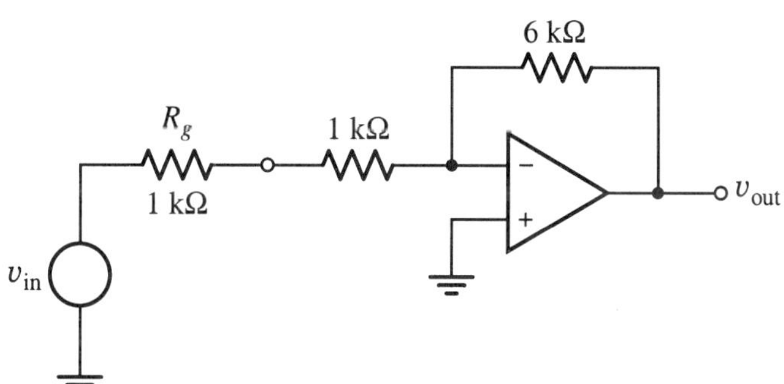


**Figure P4.7**

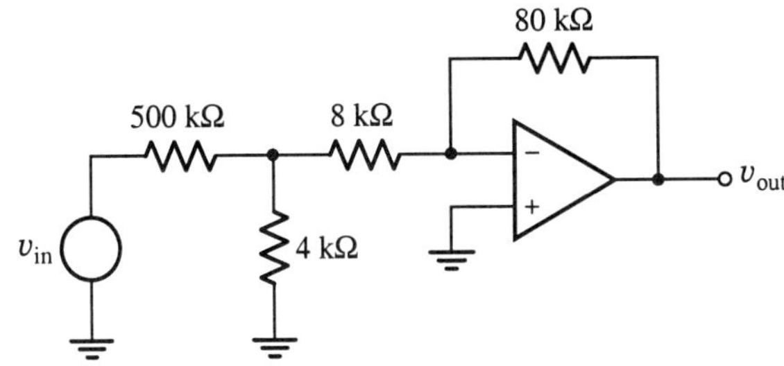


**Figure P4.9**

**4.9*** For the amplifier of the figure

**(a)** Calculate $A_{MB} = \frac{v_{out}}{v_{in}}$.
**(b)** If the $GBW$ of the op amp is 4 MHz, what is the upper corner frequency of the voltage gain?
**(c)** Write an expression for the voltage gain as a function of frequency.

**4.10** The upper corner frequency of the voltage gain for the circuit shown is found to be 150 kHz. What is the $GBW$ of the op amp?

**4.11** The upper corner frequency of the voltage gain for the circuit shown is found to be 150 kHz. What is the $GBW$ of the op amp?

**D4.12** Design an inverting stage to amplify a 0.2-V peak amplitude signal to a 3-V peak amplitude signal. The source has a 4-kΩ impedance and the load is 2 kΩ. Assuming the input source is referenced to ground and can be dc coupled to the op amp stage, use only one resistor with the op amp to complete this design. Calculate the bandwidth of your amplifier if the op amp $GBW$ is $6 \times 10^6$ Hz.

**D4.13** Repeat the previous problem using a noninverting stage and two resistors.

**4.14** Calculate the midband voltage gain of the noninverting circuit shown and the upper 3-dB frequency assuming that the op amp $GBW$ is $2 \times 10^6$ Hz.

**4.15** Calculate the midband voltage gain of the inverting circuit shown and the upper 3-dB frequency assuming that the op amp $GBW$ is $2 \times 10^6$ Hz.

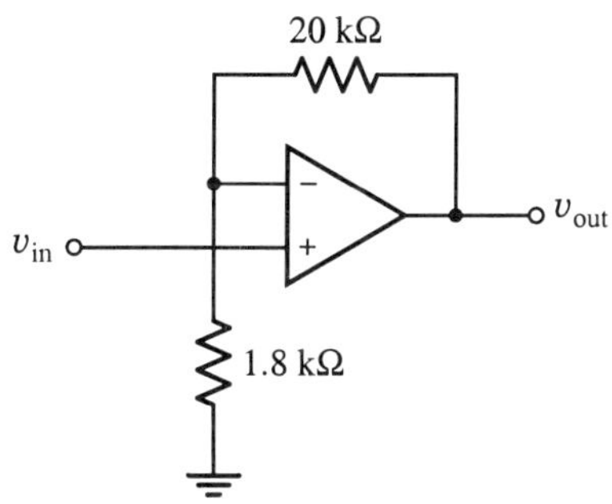


**Figure P4.10**

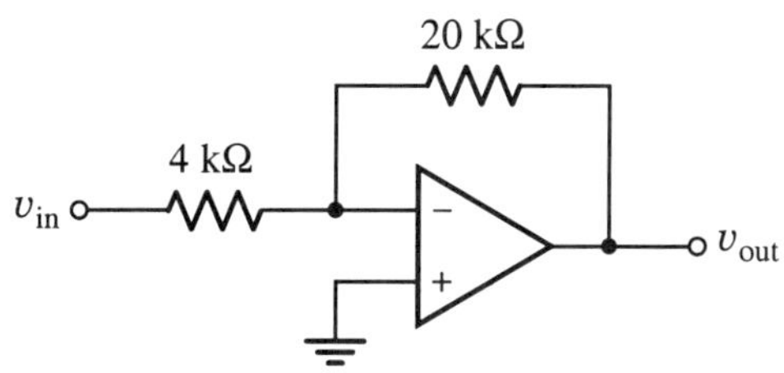


**Figure P4.11**

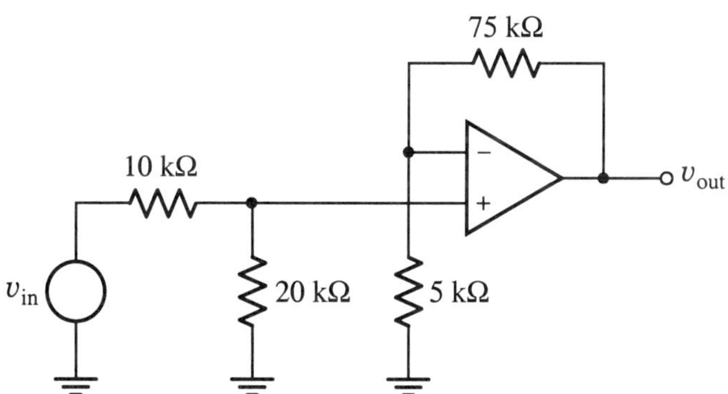


**Figure P4.14**

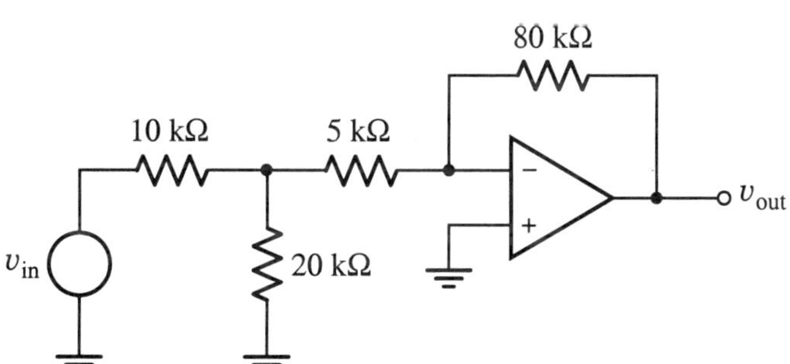


**Figure P4.15**

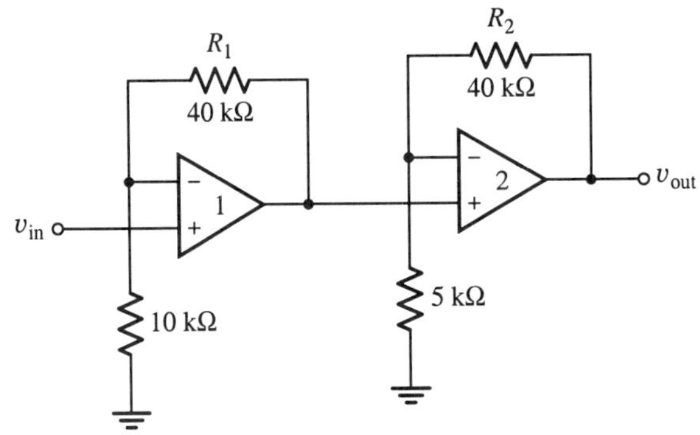


**Figure P4.16**

**4.16*** The op amps in the circuit shown are identical with $GBW_{oa} = 3.6$ MHz.

**(a)** Calculate the overall midband voltage gain of the amplifier.

**(b)** Calculate the upper 3-dB frequency of the amplifier voltage gain.

**(c)** Write an expression for voltage gain of the amplifier as a function of frequency.

**(d)** Modify the resistors $R_1$ and $R_2$ to keep the midband voltage gain constant while maximizing the upper corner frequency.

**(e)** Calculate this maximum upper corner frequency.

**4.17*** Repeat Problem 4.16 if $GBW_{oa1} = 3.6$ MHz and $GBW_{oa2} = 2.2$ MHz.

**Section 4.1.2 Transistors**

**4.18** A MOSFET has the following parameters: $g_m = 1.2$ mA/V, $r_{ds} = 10^5\ \Omega$, $C_{gs} = 35$ fF, $C_{db} = 5.0$ fF, and $C_{gd} = 3.0$ fF. If this device is used to couple a 100-Ω generator resistance to a 20-kΩ drain load, calculate the midband voltage gain and the upper corner frequency.

**4.19** Repeat Problem 4.18 if the load resistance is changed to 4 kΩ and the generator resistance is increased to 50 kΩ.

**D4.20*** If the gain of the stage of Problem 4.18 is to be increased to -40 V/V by increasing the load resistance, what is the new value of upper corner frequency?

**D4.21*** If the corner frequency of the stage of Problem 4.18 is to be increased to 1.5 GHz by decreasing the load resistance, what is the new value of midband gain?

**4.22*** A midband voltage gain of -20 V/V and a bandwidth of 560 kHz are required for an amplifier that couples a 200-Ω generator resistance to a variable load. A BJT is used with $r_x = 100\ \Omega$, $\beta = 100$, $C_\mu = 10$ pF and $f_t = 10$ MHz (at the bias point used). What is the smallest value of load resistance that will allow the transistor to satisfy this design? Using this value of $R_C$, calculate the required value of $r_\pi$.

**4.23** A common-emitter amplifier using the transistor of Problem 4.22 is constructed. If $R_g = 200\ \Omega$ and $R_C = 200\ \Omega$, what is the gain that will result when the bandwidth is 1 MHz?

**4.24** Given that a 3-kΩ resistor is inserted in the base lead of the circuit of Problem 4.23, calculate the required shunt capacitor for compensation. Evaluate the midband voltage gain and the upper corner frequency. Neglect the loading effect of bias network.

**4.25** Given that a 15-k$\Omega$ resistor is inserted in the base lead of the circuit of Problem 4.23, calculate the required shunt capacitor for compensation. Evaluate the midband voltage gain and the upper corner frequency. Calculate the lower corner frequency. Neglect the loading effect of bias network.

**D4.26** An amplifier is to be designed to couple a 700-$\Omega$ signal generator to a dc-coupled 2-k$\Omega$ load. The overall midband voltage gain is to be $-10$ V/V and the upper corner frequency is to exceed 80 kHz. Design this stage, using base compensation, if required, and the transistor parameters given in Problem 4.22. Explain why the uncompensated stage is inappropriate for this case.

**D4.27** Repeat Problem 4.26 given that the required voltage gain is -80 V/V and the upper corner frequency is to exceed 100 kHz.

**4.28*** Using base compensation, what is the maximum upper corner frequency that can be obtained with the transistor of Problem 4.22 when it couples a 3-k$\Omega$ load to a 500-$\Omega$ generator with an overall gain of $-15$ V/V? Neglect the loading effect of bias network. Assume that $r_e = 26\ \Omega$.

### Section 4.2 Multistage Wideband Amplifiers

**4.29** Given three nonidentical, noninteracting stages. The upper corner frequencies for the stages are $f_{21} =$ 1 MHz, $f_{22} = 2$ MHz, and $f_{23} = 10$ MHz. If these stages are cascaded, what is the overall upper corner frequency?

**4.30** Repeat Problem 4.29 given that $f_{21} = 1$ MHz, $f_{22} = 1$ MHz, and $f_{23} = 2$ MHz.

**4.31** An amplifier requiring an overall bandwidth of 3.1 MHz is constructed using identical stages with constant values of $GBW = 1 \times 10^8$ rad/sec. Calculate the maximum possible gain for this amplifier.

**4.32** Three stages are cascaded having the values $GBW_1 = 1 \times 10^8$ rad/sec, $GBW_2 = 2 \times 10^8$ rad/sec, and $GBW_3 = 8 \times 10^7$ rad/sec. The bandwidth of the stages are $\omega_{21} = 8 \times 10^6$ rad/sec, $\omega_{22} = 20 \times 10^6$ rad/sec, and $\omega_{23} = 40 \times 10^6$ rad/sec. Calculate the overall midband voltage gain and the overall bandwidth.

**4.33** Assuming the values of $GBW$ in Problem 4.32 remain constant as gain and bandwidth are exchanged, calculate the maximum possible overall midband voltage gain for an overall bandwidth of $\omega_{20} = 6.28 \times 10^6$ rad/sec.

**D4.34** Design an amplifier that has a midband gain of $\pm 600$ V/V and a bandwidth exceeding 1 MHz using transistors with the parameters of Problem 4.22. Use the minimum number of base-compensated stages. Assume that generator and load resistances must equal the interstage value.

**D4.35** Repeat Problem 4.34 if the required overall gain is 500 V/V and the overall bandwidth must exceed 800 kHz.

**D4.36** Repeat Problem 4.34 using op amp stages with overall open-loop gain of $100,000$ V/V and bandwidth of $1.1 \times 10^3$ rad/sec.

**D4.37** Identical op amp stages with $GBW_{oa} = 10^7$ Hz are available to construct an amplifier with an overall midband gain of 600 V/V and an overall bandwidth of 800 kHz or more.

- **(a)** What is the minimum number of stages required?
- **(b)** What is the midband voltage gain of each stage?

**D4.38** Repeat Problem 4.37 if the required midband gain is to be 1000 V/V.

**D4.39** An amplifier is to be designed with an overall midband gain of 800 V/V and an overall bandwidth of 300 kHz. If six identical op amp stages are used to implement this design, what is the minimum $GBW$ of the op amp?

**D4.40** Repeat Problem 4.39 if four stages are used.

# Chapter 5

# NARROWBAND AMPLIFIERS

## 5.1 NARROWBAND AMPLIFICATION

As indicated in Chapter 4, video or wideband stages are required to pass signals that contain a broad band of frequencies. A stage that has a lower corner frequency of 100 Hz and an upper corner frequency of 4 MHz is an example of a video amplifier, in which the ratio of the upper corner frequency to the lower is much larger than unity. There are other applications that require a narrowband amplifier. For example, the intermediate-frequency (IF) stages of conventional AM radio receivers are designed to amplify a narrow band of frequencies, approximately 10 kHz, centered about a frequency of 455 kHz. A signal containing only frequencies outside this band is not to be amplified. The upper corner frequency of an IF stage may be about 460 kHz while the lower corner frequency is near 450 kHz. The ratio of these frequencies is approximately unity. When this ratio is near unity, the amplifier is a narrowband circuit. A wideband amplifier may have a ratio of upper to lower corner frequency in the range of hundreds or thousands.

Narrowband amplifiers are very important in fields that depend on the simultaneous transmission of several channels of information. The general field of communication in its present form actually owes its existence to the narrowband amplifier. In radio broadcasting, several stations in a geographic area simultaneously transmit signals, with each station using a different carrier frequency. The ability of a receiver to select any one of these signals and amplify only the signal selected is the key factor in radio communication. The same general idea is applied in television receivers, two-way radio communications units, long-distance telephone facilities that carry several conversations simultaneously, cellular phones, and even in the field of satellite communications. Some of these fields use microwave equipment rather than transistor circuitry to achieve the required high frequencies, but tuned transistor amplifiers are involved in a large number of communication circuits. Solid-state radio and television receiving circuits account for a large portion of the total narrowband transistor amplifiers used.

The basis for the tuned amplifier is the resonant circuit containing a coil or inductor and a capacitor. A resistor is often added to the resonant circuit to control the bandwidth. We will discuss the basic operation of resonant circuits before proceeding to the tuned amplifier.

## 5.2 INDUCTORS AND RESONANT CIRCUITS

**Important Concepts**

**1.** The inductor is an imperfect energy storage element, exhibiting several energy loss mechanisms.

**2.** The quality factor or $Q$ of the inductor is a measure of the element's effectiveness in storing energy.

**3.** The losses of an actual inductor can be represented by a resistance placed in series with an ideal (lossless) inductor.

**4.** The $Q$ of a tuned circuit determines the selectivity of the circuit. This parameter can be related to the losses of the circuit.

### 5.2.1 Inductor Losses

A pure inductance has no loss of power when a current flows through the inductor. On the other hand, practical inductors do show a power loss due to several factors:

**1.** The coil wire has a finite resistance.

**2.** Eddy current and hysteresis effects in the core cause a power loss that is dependent on both frequency and the core material.

**3.** Skin effect causes the coil wire resistance to increase at high frequencies.

**4.** Proximity effects cause an increased resistance resulting from the interaction of magnetic flux of nearby windings or conductive paths on the coil.

**5.** Radiation of power to the surroundings may take place.

A coil generally consists of many windings of small-diameter wire. A larger diameter would cut down the ohmic resistance, but might make the coil too large to use. Some compromise between wire size and ohmic losses is made for practical coils. Small coils may also be fabricated on an integrated circuit substrate; however, these are very limited in size and value of inductance.

A metal core coil increases the inductance over an air core coil of the same dimensions. The metal increases the permeability of the core, which leads to a higher inductance. Unfortunately, the core exhibits losses such as eddy current and hysteresis losses that tend to decrease the quality of the coil. Ferrite cores having high resistance and high permeability minimize core losses due to eddy currents while offering large inductances. These types of cores have been used in radio and television receivers for many years. At very high frequencies approaching the GHz range, the air core coil is generally used, and the IC inductor is now finding more use in high-frequency applications.

*Skin effect* is a phenomenon that results in decreased current densities at the center of the conductor at higher frequencies. The current density increases near the surface of the conductor. The resistance of the conductor increases at higher frequencies compared to the dc ohmic resistance. The proximity effect also affects the losses as the flux produced from one particular winding interacts with the current in adjacent or nearby windings.

The coil often acts as an inefficient antenna due to the small dimensions, but some energy radiates to the surroundings. If certain items are placed near a coil, a significant amount of energy can be absorbed. In the extreme case, if another coil, loaded by a resistive load, were placed near the coil, excessive losses could result. Shielding of the coil with a metal cover will reduce these losses, but may introduce very small energy absorption losses due to the shield.

In recent years, small inductors of modest quality have been fabricated on monolithic chips, allowing inductors to be used in integrated circuits. Typical inductance values are 10 nH with quality factors of about 14, using copper conductors for the coil.

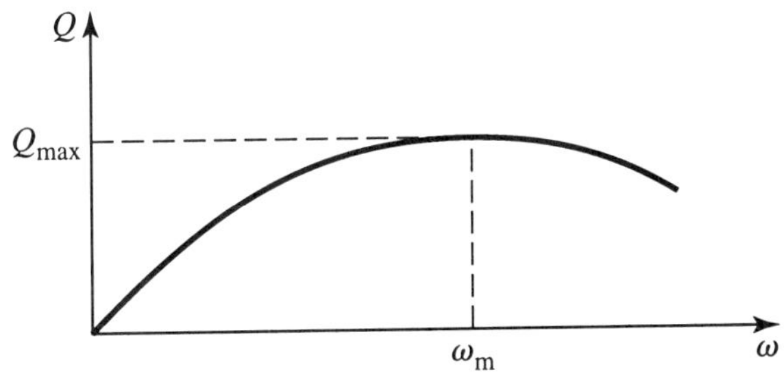


Figure 5.1: Variation of $Q$ with frequency.

The external source feels the effect of the many losses in a coil, but does not distinguish between the mechanisms responsible for the losses. As far as the external circuit is concerned, a resistance in series with a lossless inductor is equivalent to the practical inductor. The resistance, however, must be frequency-dependent to account for the frequency-dependent losses. An equation that expresses the variation of the resistance with frequency would be too complex to be of value to the circuit designer. Fortunately, such an equation is not needed for tuned amplifier design.

**Practical Considerations**

Since the losses of a coil vary with frequency, it is impossible to measure the equivalent resistive losses using an ohmmeter that measures resistance only at dc. A common method of evaluating losses is to measure the frequency response of a tuned circuit and relate the bandwidth and resonant frequency to the $Q$ of the coil. This method will be discussed in more detail after parallel resonance is considered.

The quality factor or $Q$ of a coil compares the energy stored in the coil to the energy dissipated by the resistance per cycle. For a lossless coil in series with a resistance, this value can be found to be

$$Q = \frac{\omega L}{R_s} \tag{5.1}$$

where $\omega$ is the operating frequency and $R_s$ is a series resistance representing the coil losses at the operating frequency. Coils approaching ideal behavior have very high values of $Q$. A lossless coil would have no losses, meaning that $R_s = 0$ and $Q \to \infty$.

Since the resistive losses increase at high frequencies, $Q$ is not a linear function of frequency. The variation might appear as shown in Figure 5.1.

There is quite a range of frequencies over which $Q$ is relatively constant, for a given coil, and this is the range in which $Q$ is specified by the coil manufacturer. When the coil is used in a resonant circuit, only a small range of frequency is of interest and, over this range, the resistive losses of the coil can be considered constant. Typical values of $Q$ for coils used in communication circuits range from about 20 to 150, depending on the type of coil used.

### 5.2.2 Parallel Resonance

A parallel LC-circuit, sometimes called a tank circuit, is popular in communication systems. This circuit is created by placing a capacitor in parallel with a coil as shown in Figure 5.2. Capacitors also have losses that can be represented by series resistances. However, capacitor losses are smaller than coil losses and can often be neglected. Capacitors may have values of $Q$ ranging from 50 for some paper capacitors to 300 for mylar or disk capacitors to 5000 for glass or quartz units.

It is conventional to neglect the losses associated with the capacitor and consider only those of the inductor. Although this assumption is invalid in some instances, sufficient accuracy is obtained in most circuits to justify its use.

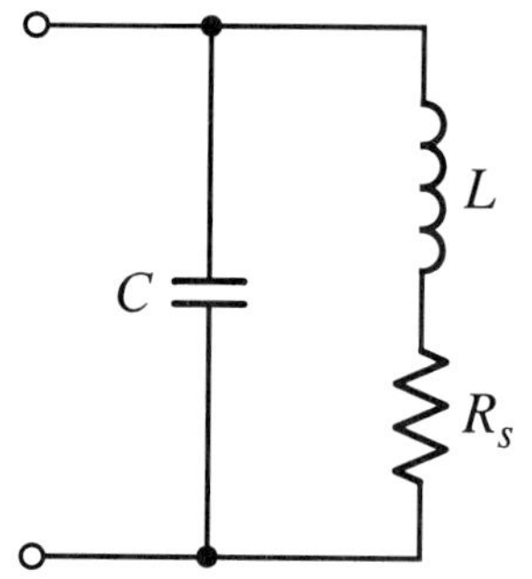


Figure 5.2: A parallel-tuned circuit (tank circuit).

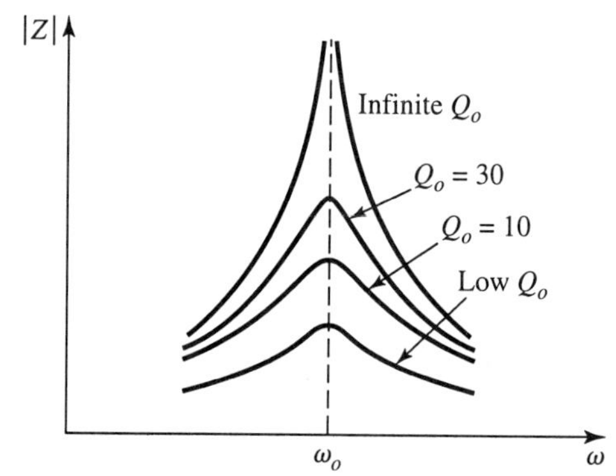


Figure 5.3: Magnitude of $Z$ as a function of frequency.

After adding the lossless capacitor, the $Q$ of the circuit remains equal to $\omega L/R_s$. Since we are interested in the resonant $Q$, we will use subscripts to indicate resonance. The $Q$ of the circuit at resonance is expressed as

$$Q_o = \omega_o L/R_{so} \tag{5.2}$$

where $\omega_o$ is the resonant frequency given by

$$\omega_o = \frac{1}{\sqrt{LC}} \tag{5.3}$$

and $R_{so}$ is the resonant value of resistance.

The impedance of the tank circuit is a function of $L$, $C$, and $R_s$ and is given by

$$Z = \frac{1}{j\omega C}||(R_s + j\omega L) = \frac{R_s + j\omega L}{-\omega^2 CL + j\omega R_s C + 1} \tag{5.4}$$

The magnitude of this function is plotted in Figure 5.3 for various values of $Q_o$. This plot demonstrates the selective properties of the tank circuit. If a current enters the tank at the resonant frequency, a large voltage will be developed across the circuit. If the frequency of the input current is far away from resonance, considerably less voltage will result.

The magnitude of the impedance can be calculated from Eq. (5.4) and becomes maximum when the real part of the denominator vanishes. This occurs when

$$1 - {\omega_o}^2 LC = 0$$

Solving this equation leads to Eq. (5.3).

Substituting this value of radian frequency into Eq. (5.4) results in the impedance of the tank circuit at resonance. For $\omega = \omega_o = 1/\sqrt{LC}$, we get a resonant impedance of

$$Z_o = \frac{R_{so}[1 + j(\omega_o L/R_{so})]}{j\omega_o R_{so} C} \tag{5.5}$$

The second term in the numerator of this equation equals $jQ_o$ and will be much larger than unity. Thus, we can neglect the first term of the numerator. This allows the resonant impedance to be written as

$$Z_o = \frac{L}{CR_{so}} \tag{5.6}$$

This equation takes on more significance if we eliminate $C$ from the expression by using Eq. (5.3). This gives $C = 1/\omega_o{}^2 L$ and, when substituted into the expression for resonant impedance, results in

$$Z_o = \frac{\omega_o{}^2 L^2}{R_{so}} = Q_o{}^2 R_{so} \tag{5.7}$$

**Practical Considerations**

The impedance at resonance is much greater than the resistance that represents the losses of the coil. If $R_{so} = 4\ \Omega$ and the resonant value of $Q_o$ is 50, the resonant resistance is

$$Z_o = 50^2 \times 4 = 10,000\ \Omega$$

The addition of the capacitor in parallel to the inductor establishes the resonant frequency and causes a large value of resistance to appear across the tank circuit at resonance.

It is possible to calculate the impedance of the tank circuit at frequencies near resonance by again using Eq. (5.4) and noting that $R_s \ll \omega L$ for values of frequency near $\omega_o$. Neglecting the first term in the numerator and assuming that $R_s$ remains constant at the resonant value, we can rewrite Eq. (5.4) as

$$Z = \frac{j\omega L}{-\omega^2 CL + j\omega R_{so} C + 1} \tag{5.8}$$

With a little more manipulation, this equation can be written as

$$Z = \frac{R_{so} Q_o{}^2}{1 + jQ_o \left[\frac{\omega}{\omega_o} - \frac{\omega_o}{\omega}\right]} \tag{5.9}$$

When a parallel resistance is placed across the tank circuit, additional losses are introduced. It is difficult to calculate the total losses from the circuit of Figure 5.2. It is more convenient to work with the equivalent circuit of Figure 5.4, which shows a lossless inductor, a lossless capacitor, and a parallel resistance $R_p$. This configuration allows $R_p$ to be easily combined with the added parallel resistor.

The resistor $R_p$ must be calculated to result in the same losses as the series resistor $R_s$ in order for both circuits of Figure 5.4 to be equivalent. The circuit involving $R_p$ allows the effect of an external parallel resistor to be easily calculated.

The two circuits of Figure 5.4 can be made approximately equivalent only near resonance. At zero frequency or dc, for example, the actual circuit presents a resistance of $R_s$ across the input terminals while the equivalent circuit presents zero dc resistance. Normally, we are interested in the operation of the tank circuit at frequencies near resonance. Thus, we can accept this inaccuracy at very low frequencies.

The key to making the two circuits approximately equivalent near resonance consists of making their impedances equal at resonance. By inspection of the equivalent circuit of Figure 5.4, the resonant impedance is seen to be $R_p$. This value is equated to the resonant impedance of the actual circuit given by Eq. (5.7), resulting in

$$R_p = Q_o{}^2 R_{so} \tag{5.10}$$

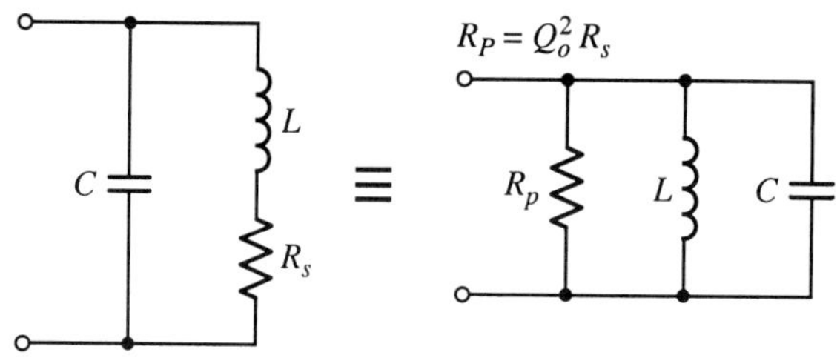


Figure 5.4: Equivalent circuits near resonance.

With this value of $R_p$ in the equivalent circuit, the two circuits are equivalent at resonance.

The question of accuracy of the equivalent circuit in representing the actual circuit at frequencies near resonance rather than only at resonance can be answered by calculating the impedance of the proposed equivalent circuit. The impedance of this circuit is

$$Z = \frac{j\omega L}{-\omega^2 CL + j\frac{\omega L}{R_p} + 1} \tag{5.11}$$

Equation (5.11) is very similar to Eq. (5.8) which gives the impedance of the actual circuit. The only difference is the imaginary terms in the denominators. In Eq. (5.8) this term is $j\omega CR_{so}$ and in Eq. (5.11) it is $j\omega L/R_p$. However, if $R_p$ is chosen according to Eq. (5.10), the two terms are equal and Eqs. (5.9) and (5.11) give exactly the same value of impedance.

The resonant $Q$, calculated from Eq. (5.2), must now be put in terms of $R_p$. To do this, we solve for $R_{so}$ from Eq. (5.10) and substitute this value into Eq. (5.2). This gives

$$Q_o = \frac{\omega_o L}{R_{so}} = \frac{\omega_o L}{R_p/Q_o^2} = \frac{Q_o^2 \omega_o L}{R_p} \tag{5.12}$$

Solving for $Q_o$ gives

$$Q_o = \frac{R_p}{\omega_o L} \tag{5.13}$$

Although we have shown that the equivalent circuit has the same impedance as the actual circuit near resonance, they are not equivalent at very low frequencies as previously mentioned. For low frequencies, the real term in the numerator of the accurate equation for $Z$, Eq. (5.4), cannot be neglected and Eq. (5.11) does not accurately represent the impedance of the circuit. However, the magnitudes of impedance for the two circuits are within 1% for frequencies different than $\omega_o$ by a factor of two. For tuned amplifier design, only near-resonant frequencies are important; consequently, the equivalent circuit is quite appropriate.

**Practice Problems**

**5.1** A coil of 80 $\mu H$ has a resonant $Q_o$ at $f_o = 920$ kHz of 22. Calculate the required resonating capacitance and the resistance $R_s$ that represents coil losses. *Ans:* $C = 374$ pF, $R_s = 21\ \Omega$.

**5.2** Repeat Practice Problem 5.1 if resonance is changed to 800 kHz with $Q_o = 18.5$. What is the value of $R_p$? *Ans:* $C = 494\ pF$, $R_s = 21.7\ \Omega$, $R_p = 7.44$ k$\Omega$.

### 5.2.3 Relation of Bandwidth to $Q$ in the Parallel Circuit

The upper and lower 3-dB points of the tuned-circuit impedance can be found from Eq. (5.9). Since the numerator of this equation is constant, the denominator will determine the 3-dB points. This equation can be expressed in terms of $R_p$ as

$$Z = \frac{R_p}{1 + jQ_o\left[\frac{\omega}{\omega_o} - \frac{\omega_o}{\omega}\right]} \tag{5.14}$$

The magnitude of this impedance is

$$|Z| = \frac{R_p}{\left(1 + {Q_o}^2\left[\frac{\omega}{\omega_o} - \frac{\omega_o}{\omega}\right]^2\right)^{1/2}} \tag{5.15}$$

At resonance, the denominator of Eq. (5.14) equals 1. The 3-dB points will be those frequencies that cause the imaginary term to also equal 1. This leads to two equations:

$$Q_o\left[\frac{\omega}{\omega_o} - \frac{\omega_o}{\omega}\right] = +1 \tag{5.16}$$

and

$$Q_o\left[\frac{\omega}{\omega_o} - \frac{\omega_o}{\omega}\right] = -1 \tag{5.17}$$

Each of these equations is a quadratic equation in $\omega$; consequently, two solutions for $\omega$ will result from solving each equation. Allowing only positive values of $\omega$ results in an upper 3-dB point of

$$\omega_2 = \omega_o + \frac{\omega_o}{2Q_o} \tag{5.18}$$

and a lower 3-dB point of

$$\omega_1 = \omega_o - \frac{\omega_o}{2Q_o} \tag{5.19}$$

The 3-dB bandwidth is found by subtracting $\omega_1$ from $\omega_2$ to give

$$\Delta\omega = \omega_2 - \omega_1 = \frac{\omega_o}{Q_o} \tag{5.20}$$

Converting this equation to Hertz leads to

$$\Delta f = f_2 - f_1 = \frac{f_o}{Q_o} \tag{5.21}$$

**Practical Considerations**

Manufacturers of inductors often specify a $Q$ for the coil rather than a value of $R_s$. This resistance varies with frequency and cannot be measured with an ohmmeter. Furthermore, it is easy to measure the coil $Q$ by adding a parallel capacitor to cause the desired resonant frequency. This tank circuit is then driven by a variable-frequency current source such as a signal generator and a series 20-MΩ resistor. The resonant frequency and the bandwidth are measured to allow $Q$ to be calculated as

$$Q_o = \frac{f_o}{\Delta f}$$

It follows that $R_p$ can then be calculated from

$$R_p = \omega_o L Q_o$$

Equation (5.15) can be used to find the magnitude of $Z$ at any frequency near resonance or to find the frequencies at which the magnitude equals a given percentage of the resonant value, $R_p$.

### Lowering the $Q$ of a Tank Circuit

When a parallel resistor $R_1$ is placed across the tank circuit, the overall $Q$ is lowered. The new value of $Q$ is called the loaded $Q$ and is designated as $Q_{\text{eff}}$. The total shunt resistance is called $R_{sh}$ and is simply the parallel combination of the two resistors or

$$R_{sh} = R_1 \parallel R_p \tag{5.22}$$

where $R_1$ is the added parallel resistance. The effective $Q$ is now expressed as

$$Q_{\text{eff}} = \frac{R_{sh}}{\omega_o L} = \frac{R_{sh}}{R_p} Q_o \tag{5.23}$$

Since the $Q$ of the circuit is lowered by this external resistor, the bandwidth increases to

$$\Delta f = \frac{f_o}{Q_{\text{eff}}} \tag{5.24}$$

The impedance of the circuit and the magnitude of the impedance near resonance are of the same form as Eqs. (5.14) and (5.15) with $R_p$ being replaced by $R_{sh}$ and $Q_o$ being replaced by $Q_{\text{eff}}$. This results in

$$Z = \frac{R_{sh}}{1 + jQ_{\text{eff}} \left[\frac{\omega}{\omega_o} - \frac{\omega_o}{\omega}\right]} \tag{5.25}$$

and

$$|Z| = \frac{R_{sh}}{\left(1 + Q_{\text{eff}}^2 \left[\frac{\omega}{\omega_o} - \frac{\omega_o}{\omega}\right]^2\right)^{1/2}} \tag{5.26}$$

**Practical Considerations**

The simplicity of equations such as Eq. (5.14) belie the exceptionally steep falloff of $|Z|$ with frequency. We recall that a lowpass $RC$ filter with a single capacitor can fall at a maximum rate of -6 dB/octave or -20 dB/decade. This rate is only approached as the frequency exceeds the corner frequency by at least an order of magnitude.

The tuned circuit falls most steeply at the 3-dB frequencies. The falloff in $|Z|$ at the upper 3-dB frequency is $6Q_{\text{eff}}$ dB/octave or $20Q_{\text{eff}}$ dB/octave. A $Q_{\text{eff}}$ of 30 results in a falloff at the upper 3-dB frequency of $30 \times 20 = 600$ dB/decade. It would require an $RC$ filter with 15 $RC$ sections or 30 elements to approach this incredible rate of falloff. Yet the tuned circuit does it with one capacitor, an inductor, and a resistance.

### Practice Problems

**5.3** Calculate the bandwidth of the resonant circuit of Practice Problem 5.1. *Ans:* 41.8 kHz.

**5.4** If a resistance of 40 k$\Omega$ is placed in parallel with the tank circuit of Practice Problem 5.2, calculate the effective $Q$. *Ans:* $Q_{\text{eff}} = 15.6$.

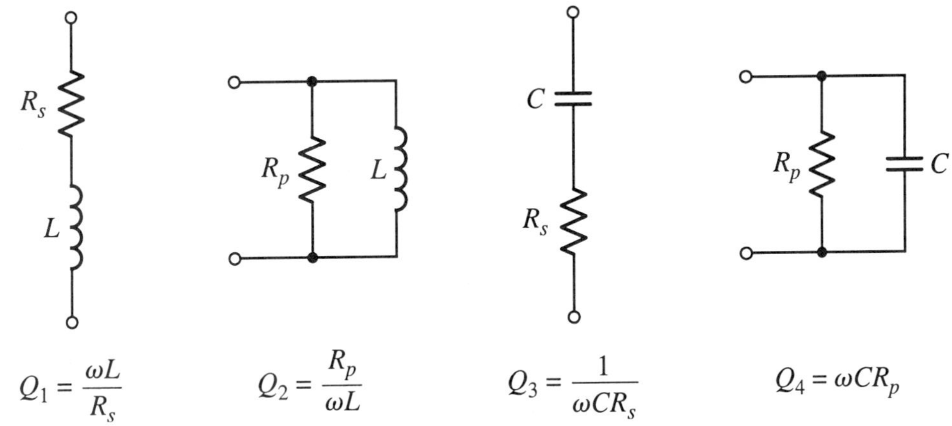


Figure 5.5: Four lossy circuits.

### 5.2.4 Review of Simple Tuned Circuit Relationships

The significance of the tuned circuit is such that it is appropriate to consider various forms of the circuit, as shown in Fig. 5.5, that may be encountered in practice.

**1.** The $Q$ of a circuit compares the ratio of energy stored to the energy dissipated per cycle. The basic definition is

$$Q = 2\pi \times \frac{\text{Energy stored}}{\text{Energy dissipated per cycle}}$$

A resistor has a theoretical $Q$ of 0 since it stores no energy. An ideal capacitor or inductor would have an infinite $Q$ since no energy would be dissipated. In practice, an inductor has several losses and may have $Q$ values in the frequency range for which it is designed that range from 10 to 200. Capacitor $Q$s depend on the construction process and may vary from 100 for so-called paper capacitors to several hundred for mylar or disk capacitors to 5000 or 10,000 for glass or quartz capacitors. Some piezoelectric crystals have effective $Q$s that are near 100,000. Often we neglect the losses of a capacitor in a tuned circuit because the inductive losses may exceed the capacitive losses by a large factor. Although we will neglect the capacitive losses in this text, many circuits must consider this source of loss.

**2.** The $Q$ has been derived elsewhere for four important cases and are tabulated in Figure 5.5. The equations for $Q$ given in the figure follow a pattern. For the series circuits, the $Q$ equals the ratio of reactance to resistance. For the parallel circuits, the $Q$ equals the ratio of resistance to reactance.

**3.** When a parallel tank circuit is constructed and all losses are attributed to the inductor, the configuration of Figure 5.6 pertains. The capacitor adds no losses and, perhaps surprisingly, no more energy is stored in the circuit per cycle. The inductor and capacitor continually exchange the sinusoidal energy.

**4.** The impedance of the parallel circuit is given by an expression that is not very useful in design situations, that is

$$Z = \frac{R_s + j\omega L}{-\omega^2 LC + j\omega R_s C + 1}$$

In order to control the circuit $Q$ and achieve the desired bandwidth, an external resistor is placed across the terminals of the tank circuit. Unfortunately, the expression for impedance becomes considerably more complex and does not immediately indicate the effect of the additional resistance.

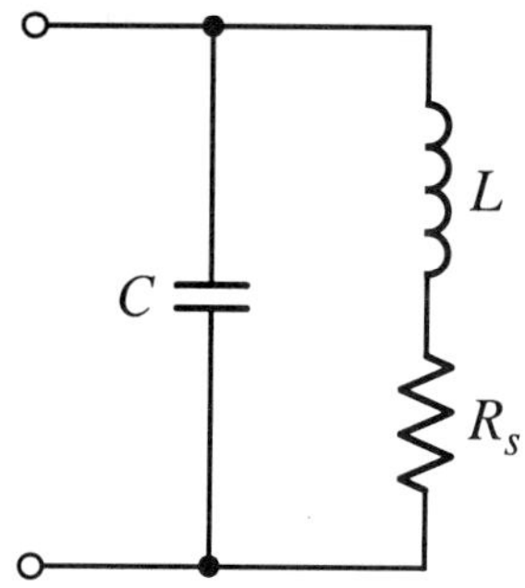

Figure 5.6: A tank circuit.

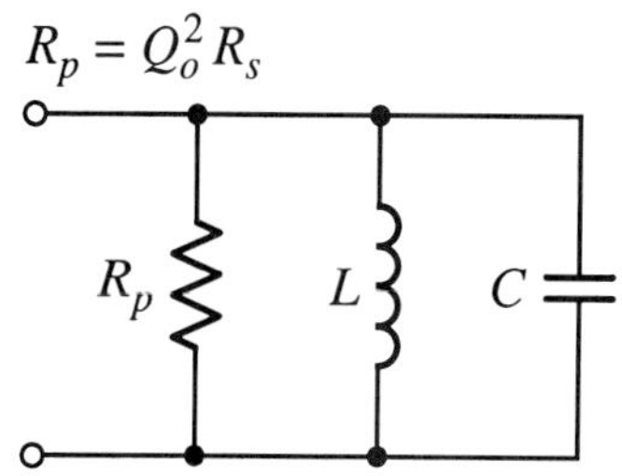

Figure 5.7: Equivalent circuit for the tank circuit.

**5.** In an effort to simplify the mathematics and relate the resistance to the circuit $Q$, the quasi-equivalent circuit of Figure 5.7 is used to model the original tank circuit. This circuit is a good model near resonance and becomes inaccurate only at frequencies far removed from the resonant frequency. A key point to remember is that typically we are only concerned with the frequency response near the resonant frequency. Thus, the quasi-equivalent circuit is very useful for our calculations.

The parallel resistance representing the losses of the tuned circuit has a value of $R_p = {Q_o}^2 R_s$. This value can easily be combined with any added parallel resistance, $R_1$, to give

$$R_{sh} = \frac{R_1 R_p}{R_1 + R_p}$$

and

$$Q_{\text{eff}} = \frac{R_{sh}}{R_p Q_o}$$

The expression for impedance as a function of frequency is now found from Eq. (5.25)

$$Z = \frac{R_{sh}}{1 + jQ_{\text{eff}}\left[\frac{\omega}{\omega_o} - \frac{\omega_o}{\omega}\right]}$$

## Example 5.1

A coil of value $L = 100\ \mu$H is to be used in a tank circuit that resonates at 150 kHz. A 10-k$\Omega$ resistance is placed in parallel with the tank circuit. If the coil $Q$ is 42 at the resonant frequency,

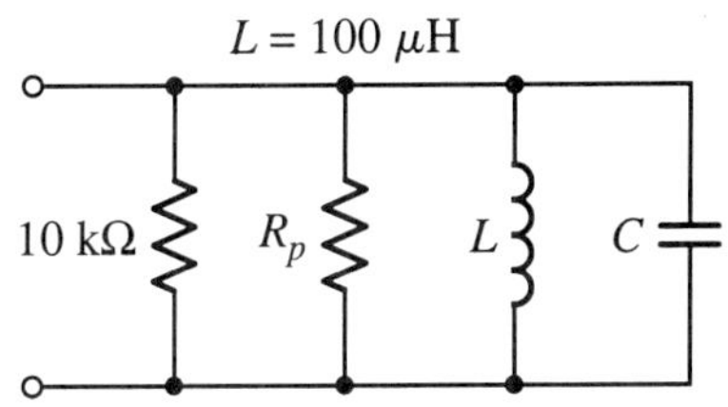


Figure 5.8: Tuned circuit for Example 5.1

**(a)** calculate the value of capacitance required;

**(b)** calculate the overall $Q$ of the circuit, $Q_{\text{eff}}$; and

**(c)** calculate the bandwidth of the circuit impedance.

**Solution** The circuit described is shown in Figure 5.8.

**(a)** The correct value of capacitance for $f_o = 150$ kHz is found from Eq. (5.3) to be

$$C = \frac{1}{\omega_o{}^2 L} = \frac{1}{(2\pi \times 1.5 \times 10^5)^2 \times 1 \times 10^{-4}} = 1.13 \times 10^{-2}\ \mu\text{F}$$

**(b)** To find the overall circuit $Q$, we must first find the value of $R_p$. This is done by using Eq. (5.13) to give

$$R_p = \omega_o L Q_o = 2\pi \times 1.5 \times 10^5 \times 10^{-4} \times 42 = 3.96\ \text{k}\Omega$$

When $R_p$ is combined with the 10-kΩ external resistance, the total resistance shunting the inductance is

$$R_{sh} = 3.96 \,||\, 10 = 2.84\ \text{k}\Omega$$

We now use Eq. (5.23) to calculate the effective $Q$ of the circuit. This results in

$$Q_{\text{eff}} = \frac{R_{sh}}{\omega_o L} = \frac{2.84 \times 10^3}{2\pi \times 1.5 \times 10^5 \times 1 \times 10^{-4}} = 30.1$$

**(c)** The bandwidth of the circuit impedance is found from Eq. (5.24) and is

$$\Delta f = \frac{f_o}{Q_{\text{eff}}} = \frac{150}{30.1} = 4.98$$

---

### 5.2.5 The Series Resonant Circuit

The tank circuit serves as a load for amplifying stages providing a higher voltage gain near resonance as the impedance increases. The series resonant circuit is used in filter applications that require a minimum impedance at resonance. While the tank circuit is used more often in communication circuits, the series resonant circuit is useful in certain applications. A simple series RLC circuit is shown in Figure 5.9.

The resistance $R_x$ is any added external resistance, whereas the resistance $R_s$ represents the losses of the coil and is a function of frequency. If $v_{in}$ is a sinusoidal signal with constant amplitude and varying frequency, there

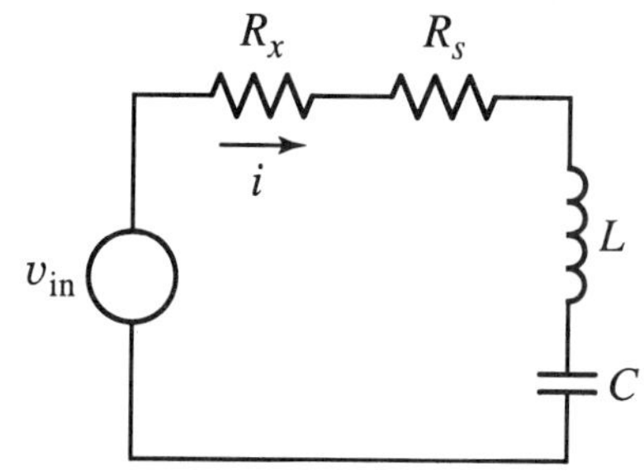


Figure 5.9: A series RLC circuit.

is one frequency at which the reactance of the inductor will cancel the reactance of the capacitor. This is the resonant frequency and the resulting current will be maximum at a value of

$$i = \frac{v_{in}}{R_x + R_{so}} \tag{5.27}$$

where $R_{so}$ is the resonant value of $R_s$.

The resonant frequency is again given by

$$\omega_o = \frac{1}{\sqrt{LC}} \tag{5.28}$$

It can be shown that the coil $Q$ at resonance is expressed by Eq. (5.1) as

$$Q_o = \frac{\omega_o L}{R_{so}} \tag{5.29}$$

The effective $Q$ at resonance, including losses from the external resistance, is

$$Q_{\text{eff}} = \frac{\omega_o L}{R_x + R_{so}} = \frac{\omega_o L}{R_{To}} \tag{5.30}$$

where $R_{To}$ is the total loop resistance at resonance.

Since $\omega_o L = 1/\omega_o C$, an alternate expression for $Q_{\text{eff}}$ is

$$Q_{\text{eff}} = \frac{1}{\omega_o C R_{To}} \tag{5.31}$$

The impedance of the series loop is

$$Z = R_T + j\left[\omega L - \frac{1}{\omega C}\right] \tag{5.32}$$

Near resonance, where $R_T \approx R_{To}$, this can be written as

$$Z = R_{To}\left(1 + j\left[\frac{\omega L}{R_{To}} - \frac{1}{\omega C R_{To}}\right]\right) \tag{5.33}$$

With a little manipulation, this equation becomes

$$Z = R_{To}\left(1 + jQ_{\text{eff}}\left[\frac{\omega}{\omega_o} - \frac{\omega_o}{\omega}\right]\right) \tag{5.34}$$

The magnitude of the impedance near resonance is

$$|Z| = R_{To}\left(1 + Q_{\text{eff}}^2\left[\frac{\omega}{\omega_o} - \frac{\omega_o}{\omega}\right]^2\right) \tag{5.35}$$

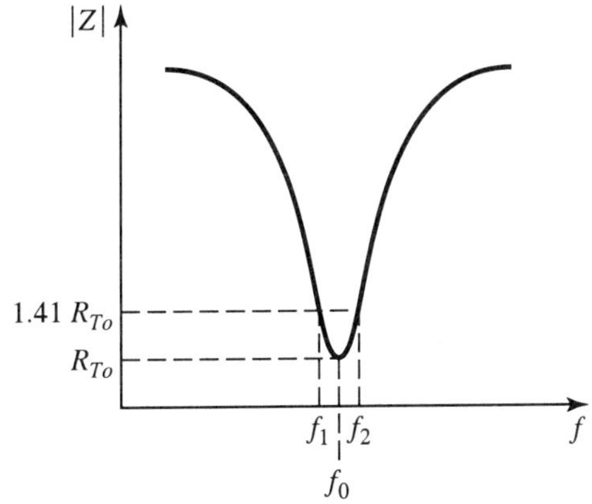


Figure 5.10: Impedance of a series RLC circuit.

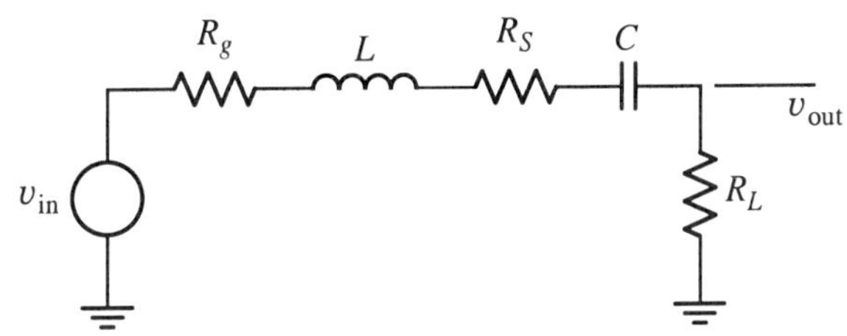


Figure 5.11: A series RLC bandpass filter.

This impedance magnitude is plotted as a function of frequency in Figure 5.10.

The minimum impedance of $R_{To}$ occurs at the resonant frequency. The frequencies at which the impedance has increased in magnitude by 3 dB are designated $f_2$ and $f_1$. These points can be found by equating the real and imaginary parts of the expression for $Z$ given by Eq. (5.34). The resulting bandwidth, $\Delta f = f_2 - f_1$, is

$$\Delta f = \frac{f_o}{Q_{\text{eff}}} \tag{5.36}$$

In theory, the series RLC circuit can be used as a bandpass filter when arranged as shown in Figure 5.11. The resistance $R_g$ is the signal generator resistance.

Unfortunately, for many practical communication applications, the sum of the source resistance and load resistance is much larger than $R_s$, and the effective $Q$ is too low to be useful. Some higher power applications that include lower impedances apply the series RLC circuit.

A more useful configuration is the band-reject filter of Figure 5.12.

At resonance, the attenuation of this filter is

$$\frac{v_{\text{out}}}{v_{\text{in}}} = \frac{R_{so}}{R_{so} + R_1} \tag{5.37}$$

If $R_1$ is chosen to be large, this attenuation factor is low. There is an upper limit imposed on $R_1$ by the 3-dB frequencies, $f_4$ and $f_3$. The rejection bandwidth, $\Delta f_{\text{rej}} = f_4 - f_3$, should be relatively small to have little effect on the signals that are to be passed to the output. The upper and lower 3-dB points can be found by assuming that $R_s$ will be negligible at these frequencies and solving the equation

$$\omega L - \frac{1}{\omega C} = \pm R_1$$

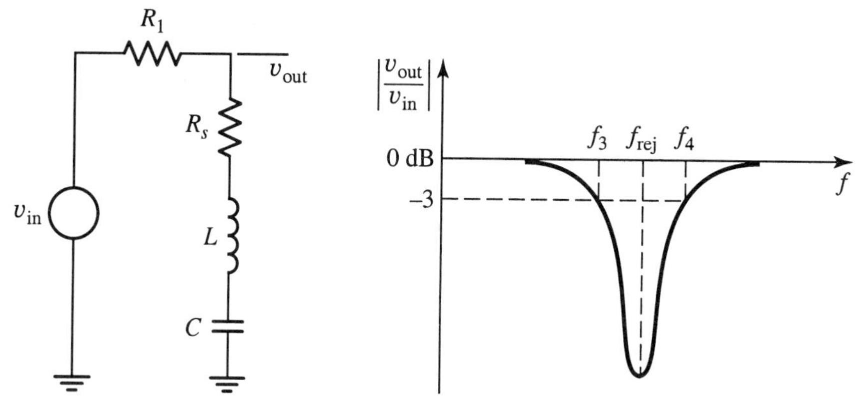


Figure 5.12: A band-reject filter.

for $\omega_3$ and $\omega_4$. The results lead to a rejection bandwidth of

$$\Delta f_{rej} = \frac{R_1}{2\pi L} \tag{5.38}$$

Selecting $R_1$ to be large compared to $R_{so}$ leads to good rejection or a deep notch, but if $R_1$ is too large, the rejection bandwidth will be broad and attenuate the frequencies to be passed. The starting point in selecting $R_1$ is to set the bandwidth to an acceptable value. For example, if the 3-dB points are to be 10% above and below the rejection frequency, the rejection bandwidth is $0.2f_o$. Using units of radians per second, we write

$$\Delta\omega_{rej} = \frac{R_1}{L} = 0.2\omega_o = \frac{0.2Q_oR_s}{L}$$

where Eq. (5.29) is applied to get the last term of the equation and it is assumed that $R_1 \gg R_s$. Thus, the value of $R_1$ should be

$$R_1 = 0.2Q_oR_s$$

This sets the desired bandwidth, but results in an attenuation at resonance of only

$$\frac{v_{out}}{v_{in}} = \frac{1}{1+0.2Q_o}$$

If the coil $Q$ is 100, the attenuation at resonance will be 1/21 in this case.

One example of the use of the band-reject filter is in conventional analog TV sets. The video signal is transmitted with a bandwidth of 3.75 MHz with an audio subcarrier superimposed on this signal at 4.5 MHz. When the signal is recovered, the video signal must be separated from the audio. A lowpass filter that passes the video signal does not attenuate the audio signal to the desired extent. The band-reject filter is used to attenuate the audio signal further.

## Example 5.2

(a) Select the capacitor in the filter of Figure 5.13 to cause a bandreject frequency of 4.5 MHz. Assume that $Q_o = 80$.

**(b)** Select $R$ to result in a rejection bandwidth of 1.2 MHz.

**(c)** Calculate the attenuation of the signal at 4.5 MHz.

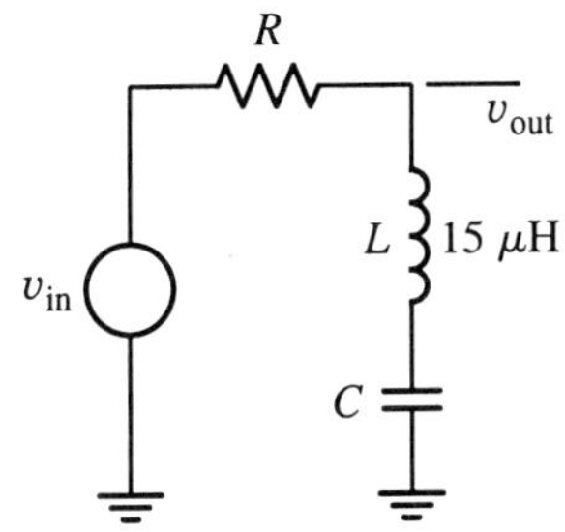


Figure 5.13: Filter for Example 5.2.

**(d)** Calculate the attenuation of the signal at 3.75 MHz.

**Solution (a)** The capacitor is found to be

$$C = \frac{1}{\omega_o^2 L} = \frac{1}{(2\pi \times 4.5 \times 10^6)^2 \times 15 \times 10^{-6}} = 83.4 \text{ pF}$$

**(b)** From Eq. (5.31), the resistance is calculated to be

$$R = 2\pi \times \Delta f_{\text{rej}} \times L = 2\pi \times 1.2 \times 10^6 \times 15 \times 10^{-6} = 113\ \Omega$$

**(c)** The attenuation at $f = 4.5$ MHz is found by first calculating $R_{so}$ as

$$R_{so} = \frac{\omega_o L}{Q_o} = \frac{2\pi \times 4.5 \times 10^6 \times 15 \times 10^{-6}}{80} = 5.3\ \Omega$$

This value is used to find that

$$\frac{v_{\text{out}}}{v_{\text{in}}}(4.5 \text{ MHz}) = \frac{R_{so}}{R_{so} + R} = \frac{5.3}{5.3 + 113} = 0.0448$$

The attenuation is 27 dB.

**(d)** The attenuation at $f = 3.75$ MHz can be found from a consideration of the transfer function. This gives

$$\left|\frac{v_{\text{out}}}{v_{\text{in}}}\right| = \left|\frac{R_{so} + j\left(\omega L - \frac{1}{\omega C}\right)}{R_{so} + R + \left(\omega L - \frac{1}{\omega C}\right)}\right|$$

An evaluation of this magnitude at $f = 3.75$ MHz results in an attenuation of 0.795 or 2 dB.

---

## 5.3 TAPPED CAPACITOR AND INDUCTOR CIRCUITS

**Important Concepts**

**1.** Connection of a load resistance across a tuned circuit may lower the $Q_{\text{eff}}$ to the point that bandwidth specifications cannot be met.

**2.** A tapped capacitor circuit or tapped inductor circuit can be used to maintain a higher value of $Q_{\text{eff}}$ when a small load resistance is connected.

**3.** Although the tapped circuits result in a high $Q_{\text{eff}}$, the output voltage is attenuated from the value across the tuned circuit.

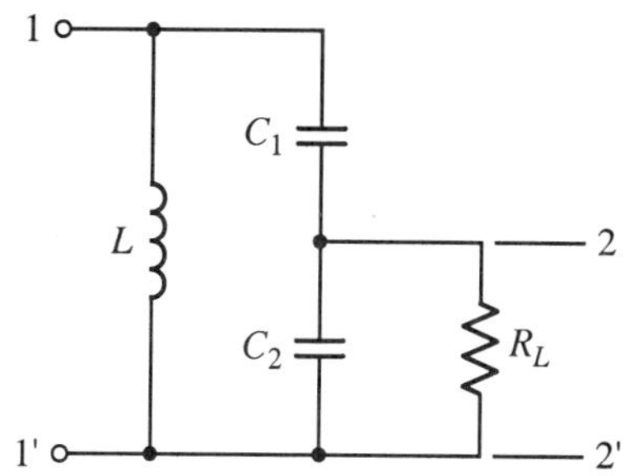


Figure 5.14: A tapped capacitor resonant circuit.

An earlier part of this chapter considers the design of parallel resonant circuits. In the simplest case, the resonant frequency and the required circuit $Q$ are specified, and the elements $L$, $C$, and $R$ are selected to result in the correct values of $f_o$ and $Q_{\text{eff}}$. Setting the value of $Q_{\text{eff}}$ properly leads to the desired value of bandwidth of the circuit.

In the practical situation, a resonant circuit of an amplifier stage will be loaded by the input impedance of the following amplifier stage. Since this input impedance may either have a specific value or fall within some practical limits, the freedom to choose any value of $R$ for the tuned circuit may not exist. The input impedance may be too small to achieve the required value of $Q_{\text{eff}}$, even with no additional parallel resistance in the tuned circuit. This section considers a few useful configurations that add flexibility to the design of the tuned circuit and are often used in communication circuits.

### 5.3.1 Tapped Capacitor Circuits

One of the simpler methods of preventing a resistive load from excessively decreasing the $Q$ of a tuned circuit is to use the tapped capacitor circuit of Figure 5.14. The two capacitors $C_1$ and $C_2$ are considered to be separate, lossless capacitors and the inductor is also considered to be lossless for the following discussion. It is assumed that the overall circuit $Q$ is at least 10.

In the configuration of Figure 5.14, the losses are due only to the load resistor $R_L$. It is not immediately obvious how this resistance affects the circuit $Q$. It is easy to calculate the circuit $Q$ if the resistor appears in parallel with the inductor, therefore, we want to develop an equivalent circuit that includes a parallel resistance. Such an equivalent is shown in Figure 5.15(a).

To calculate the values $C_{\text{eq}}$ and $R_t$, we use the sequence of equivalent circuits of Figure 5.15(b). The effect of the resistance $R_L$ is reflected by the series resistance $R_s$ with a value of

$$R_s = \frac{R_L}{Q_L^2} = \frac{1}{R_L \omega^2 C_2^2} \tag{5.39}$$

where $Q_L$ is the $Q$ corresponding to the parallel combination of $R_L$ and $C_2$. The two capacitors can now be combined in series to give $C_{\text{eq}}$. This capacitor has a value of

$$C_{\text{eq}} = \frac{C_1 C_2}{C_1 + C_2} \tag{5.40}$$

The series form of $C_{\text{eq}}$ and $R_s$ can now be transformed to the parallel form. In this form, the capacitor value is unchanged while the resistor is transformed to a value of

$$R_t = Q_s^2 R_s = \frac{1}{R_s \omega^2 C_{\text{eq}}^2}$$

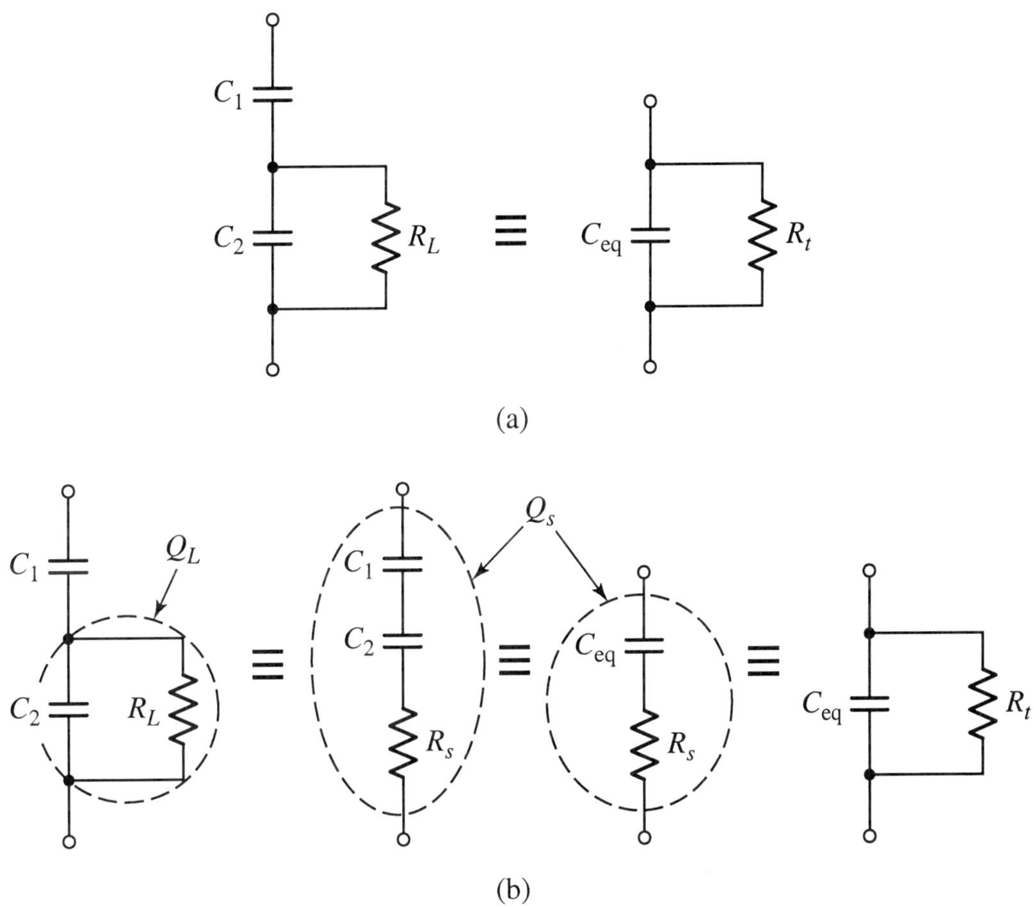


Figure 5.15: (a) An equivalent circuit. (b) Steps required in deriving equivalent circuit.

If we now replace $R_s$ in this equation with the value from Eq. (5.39), we can write

$$R_t = \frac{1}{\frac{1}{R_L \omega^2 C_2^2} \omega^2 C_{\text{eq}}^2} = R_L \left( \frac{C_2^2}{C_{\text{eq}}^2} \right) = R_L \left( \frac{C_1 + C_2}{C_1} \right)^2 \tag{5.41}$$

The inductor can now be assumed to be in parallel with the capacitance $C_{\text{eq}}$ and the resistor $R_t$. This leads to the following equations,

$$f_o = \frac{1}{2\pi\sqrt{LC_{\text{eq}}}} \tag{5.42}$$

and

$$Q_{\text{eff}} = \frac{R_t}{\omega L} \tag{5.43}$$

This equivalence can be used to replace a simple tuned circuit and maintain the required $Q$ when a low value of $R_L$ is present.

The voltage across the terminals 2-2′ in Figure 5.14 can be expressed in terms of the voltage across terminals 1-1′ as

$$v_{2-2'} = v_{1-1'} \frac{C_1}{C_1 + C_2} \tag{5.44}$$

The capacitors $C_1$ and $C_2$ can be selected to result in a value of $R_t$ that is much higher than $R_L$ to maintain a high value of circuit $Q$. The price paid for this, however, is a decreased output voltage compared to the voltage across the entire circuit.

**Practical Considerations**

The derivation of circuit $Q$ resulting in Eq. (5.43) was based on a lossless coil. In the practical situation, the circuit $Q$ will be determined by $R_{sh}$ rather than $R_t$. The shunt resistance is

$$R_{sh} = R_t \parallel R_p \parallel R_1$$

where $R_p$ represents the coil losses at resonance and $R_1$ is the external resistance added in parallel with the coil.

The following example will demonstrate the use of the tapped capacitor circuit. Although we have not yet studied the tuned amplifier stage yet, this example introduces a simple configuration of such a stage.

## Example 5.3

The collector output impedance of the circuit of Figure 5.16 is 140 kΩ. The inductor is 60 μH and has a $Q_o = 80$ at the resonant frequency of 800 kHz. The desired bandwidth of the tuned circuit is 14 kHz. If the input impedance to the amplifier is 10 kΩ,

**(a)** calculate the bandwidth of the loaded circuit.

**(b)** Design a tapped capacitor circuit to result in a bandwidth of 14 kHz.

**Solution (a)** The losses of the coil can be represented by a resistance of value

$$R_p = \omega L Q_{\text{coil}} = 2\pi \times 8 \times 10^5 \times 60 \times 10^{-6} \times 80 = 24.1 \text{ k}\Omega$$

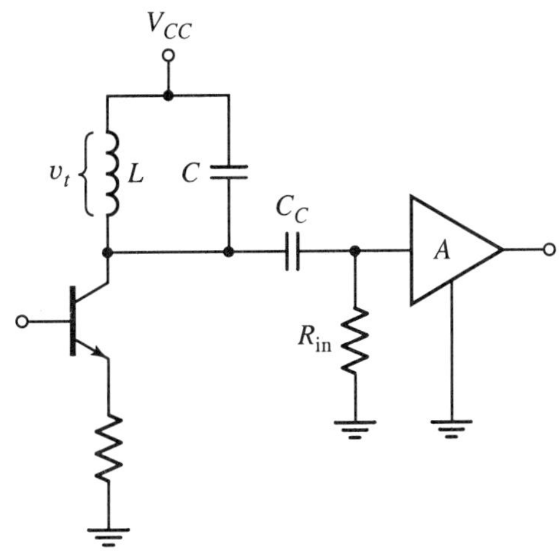


Figure 5.16: A tuned stage followed by an amplifier.

The total impedance of the tuned circuit, including the input impedance to the next stage, is

$$R_{sh} = \frac{1}{\frac{1}{R_p} + \frac{1}{R_{\text{out}}} + \frac{1}{R_{\text{in}}}} = \frac{1}{\frac{1}{24.1} + \frac{1}{140} + \frac{1}{10}} = 6.73 \text{ k}\Omega$$

The overall circuit $Q$ is now

$$Q_{\text{eff}} = \frac{R_{sh}}{R_p} Q_o = \frac{6.73}{24.1} \times 80 = 22.3$$

The bandwidth of the circuit is

$$BW = \frac{f_o}{Q_{\text{eff}}} = \frac{800}{22.3} = 35.8 \text{ kHz}$$

Obviously, the circuit $Q$ is too low to achieve the 14-kHz bandwidth required.

**(b)** With the tapped capacitor circuit, the total impedance of the tuned circuit will be the parallel combination of $R_{\text{out}}$, $R_p$, and $R_t$. The circuit $Q$ required for a 14 kHz bandwidth is

$$Q_{\text{eff}} = \frac{f_o}{BW} = \frac{800}{14} = 57.1$$

This leads to a resonant impedance of

$$R_{sh} = \frac{Q_{\text{eff}}}{Q_o} R_p = \frac{57.1}{80} \times 24.1 \text{ k}\Omega = 17.2 \text{ k}\Omega$$

The impedance across the resonant circuit due to $R_{\text{out}}$ and $R_p$ is found to be 20.6 kΩ. We must now solve for the required value of $R_t$ to place in parallel with this impedance to give a total resistance of 17.2 kΩ. A value of $R_t = 104.2$ kΩ will result in the desired value. In order to achieve this value of $R_t$ with a load resistance equal to the 10-kΩ input impedance of the following amplifier, the capacitors must satisfy Eq. (5.41), or

$$\left(\frac{C_1 + C_2}{C_1}\right)^2 = \frac{R_t}{R_L} = \frac{104.2}{10} = 10.42$$

In addition to this condition, the equivalent capacitance must cause a resonant frequency of 800 kHz, giving

$$\frac{C_1 C_2}{C_1 + C_2} = C_{eq} = \frac{1}{\omega_o^2 L} = 660 \text{ pF}$$

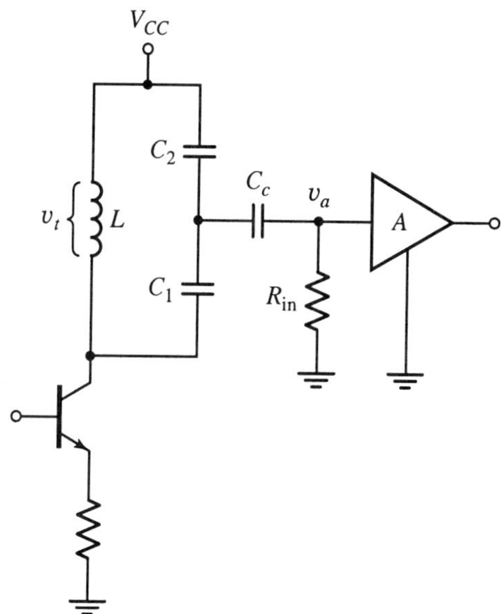


Figure 5.17: A tapped capacitor design.

Combining these two conditions on the capacitors leads to values of

$$C_1 = 956 \text{ pF}$$

and

$$C_2 = 2130 \text{ pF}$$

The tapped capacitor circuit is shown in Figure 5.17.

Whereas the circuit $Q$ is now high enough to achieve the desired bandwidth, the overall circuit gain is reduced. The voltage reaching the amplifier terminals will be only a fraction of the voltage appearing across the tuned circuit at resonance. This attenuation factor is

$$\frac{v_a}{v_t} = \frac{C_1}{C_1 + C_2} = 0.310$$

This attenuation factor can be offset by increasing the gain of the amplifier that follows the tuned stage.

---

**Practice Problem**

**5.5** Rework Example 5.3 if the desired bandwidth is 15 kHz. *Ans:* $C_1 = 1048$ pF, $C_2 = 1784$ pF.

### 5.3.2 Tapped Inductors and Transformers

Tapped inductor circuits and transformers can also be used to increase the impedance of the tuned circuit to the proper level. The derivation of the appropriate equations for the tapped inductor circuit will be left to the student. The equations corresponding to the circuits of Figure 5.18 are

$$L_{eq} = L_1 + L_2 \tag{5.45}$$

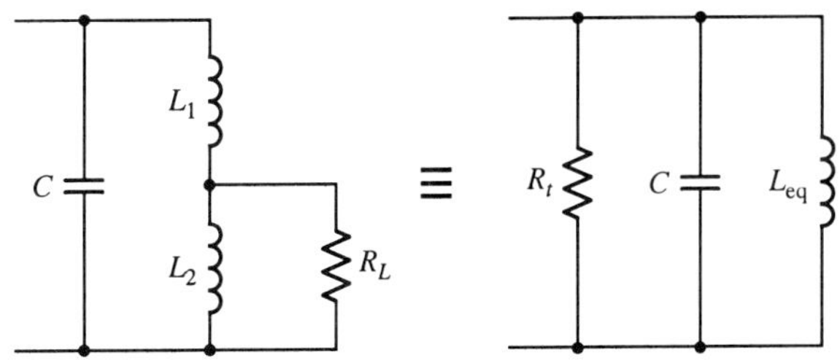


Figure 5.18: A tapped inductor circuit equivalent.

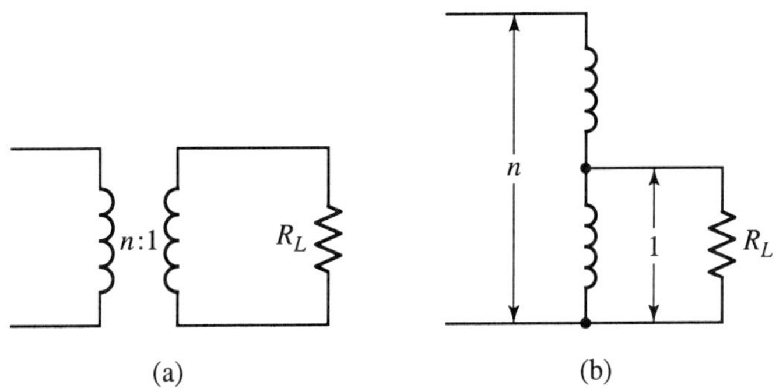


Figure 5.19: Transformers. (a) Conventional. (b) Auto-transformer.

and

$$R_t = R_L \left( \frac{L_1 + L_2}{L_2} \right)^2 \tag{5.46}$$

These equations again assume a high-$Q$ situation and pertain to lossless inductors with no mutual coupling.

Two types of transformer circuits are used in high-frequency circuit design. The first is a conventional transformer with separate primary and secondary windings. The second is the auto-transformer that shares common windings between primary and secondary. Both types of transformers are depicted in Figure 5.19.

The ratio of turns between primary and secondary is designated by $n$. If this number is greater than unity, the transformer steps the voltage down from primary to secondary by this factor. For example, if $n = 4$, then the secondary voltage is 1/4 the primary voltage. The current steps up by the same factor that the voltage decreases. If a load is placed across the secondary windings, the current in the secondary will be $n$ times the current in the primary. The turns ratio, $n$, can also be less than unity.

Figure 5.20 shows a common transformer configuration with a resistive load on the secondary. The impedance seen looking into the primary, as a result of the load resistance, is found by noting that

$$v_{\text{out}} = \frac{1}{n} v_{\text{in}}$$

and

$$i_{\text{out}} = n i_{\text{in}}$$

where $i_{\text{in}}$ is the primary current and $i_{\text{out}}$ is the secondary current. Taking the ratio of $v_{\text{in}}$ to $i_{\text{in}}$ yields the input resistance and is

$$R_{\text{in}} = \frac{v_{\text{in}}}{i_{\text{in}}} = n \frac{v_{\text{out}}}{\frac{1}{n} i_{\text{out}}} = n^2 \frac{v_{\text{out}}}{i_{\text{out}}} = n^2 R_L \tag{5.47}$$

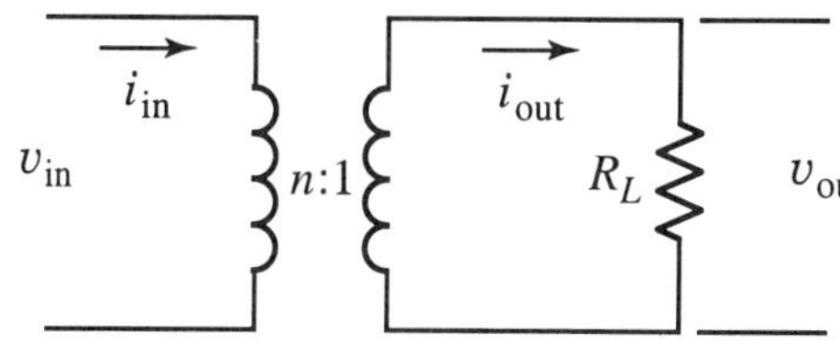


Figure 5.20: A transformer with a resistive load.

Equation (5.47) indicates the reflected effects on the primary of the secondary load, but the primary also has an inductance associated with the windings. This inductance appears in parallel with the reflected load impedance.

## 5.4 STAGGER-TUNED STAGES

**Important Concepts**

1. Amplifiers that fall very sharply above and below a resonant frequency must have high $Q$s. As a result, the 3-dB bandwidth is often too small to meet specifications.

2. Two amplifiers in cascade, with proper tuning, can offer sharp falloff while providing a broader bandwidth.

A common use of the tuned amplifier in communication circuits is to amplify a small band of frequencies while discriminating against a nearby band of frequencies. For example, an amplifier may be designed to amplify frequencies in the band 450 kHz to 460 kHz with a high gain. Signals of frequency 440 kHz and lower or 405 kHz and higher must have a much lower gain. This example is a problem faced in the design of AM radio receivers when two incoming signals differ in carrier frequency by 20 kHz. The desired carrier frequency is translated to the intermediate frequency of 455 kHz with a 10-kHz bandwidth, while the undesired carrier frequency may be translated to 435 kHz or 475 kHz. The frequency range of these unwanted signals varies from 430 to 440 kHz and 470 to 480 kHz.

If we attempt to solve this problem with a single tuned amplifier stage we might choose a resonant gain at 455 kHz of 100. This resonant gain is proportional to $R_{sh}$ of the tuned circuit. In practice, we might want only 1 dB of attenuation at 450 kHz and 460 kHz. For ease of comparison of several circuits, we will allow 3 dB of attenuation at these frequencies. This requires a $Q_{\text{eff}}$ of

$$Q_{\text{eff}} = \frac{f_o}{\Delta f} = \frac{455}{10} = 45.5$$

Although this value of $Q_{\text{eff}}$ amplifies the desired frequency range by a factor that varies from 70.7 V/V (at 450 kHz and 460 kHz) to 100 V/V (at 455 kHz), it does not discriminate against the other frequency bands to the extent desired. Using Eq. (5.26) we calculate the magnitude of the impedance at 440 kHz and 460 kHz to be approximately 44% of the resonant value (-7 dB). The gain at these frequencies would then be about 44. This difference in gain from the resonant value does not lead to much discrimination against these unwanted frequencies.

The first response to a need for higher attenuation of off-resonant frequencies might be to use a tuned circuit with a higher $Q$. After all, a $Q$ higher than 45.5 may not be difficult to obtain for this example. If a higher $Q$ is used, however, the attenuation of the desired components at the frequencies of 450 kHz and 460 kHz may become excessive because the bandwidth shrinks with increasing $Q$. We want to keep a reasonable bandwidth to avoid attenuation of signals within the desired frequency range, but the circuit should attenuate the undesired signals more heavily.

One method that might be applied to achieve these conflicting goals is to use multiple tuned stages. In doing so, we could tune the stages to the same resonant frequency or we could use stagger-tuning. In practical radio or

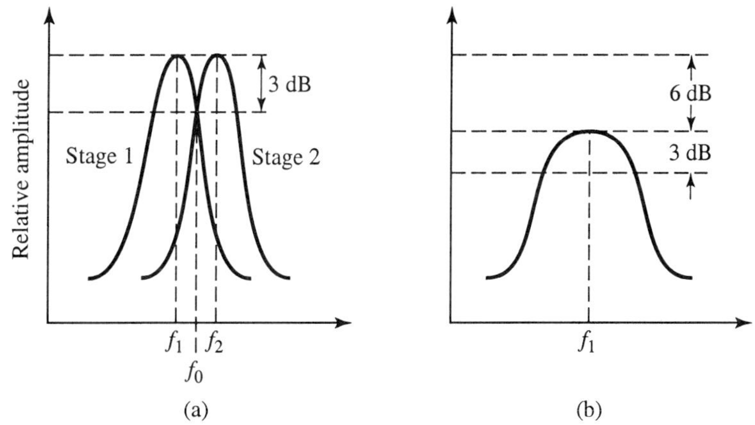


Figure 5.21: (a) Response of individual stagger-tuned circuits. (b) Response of cascaded stagger-tuned stages.

television receiver circuits, stagger-tuned methods are generally employed. In this approach a pair of stages are detuned from the resonant frequency, one of the stages has the lower 3-dB point set to the resonant frequency, and the other stage has its higher 3-dB point tuned to the resonant frequency. The resulting overall response is flatter than a single stage in the passband and has a sharper falloff as the frequencies depart from the passband.

In order to demonstrate the two approaches, we will first consider two identical stages with buffering between stages and an overall bandwidth defined by the same 3-dB points as the previous single tuned amplifier. For two identical stages, each stage should be down by 1.5 dB at the frequencies 450 kHz and 460 kHz. We use Eq. (5.26) to find that the required $Q$ for each stage is then 29.1. These two stages would cause the resonant gain of 100 V/V to be down 13.6 dB at 440 and 460 kHz to a value of approximately 21 V/V. This is a slight improvement over the single tuned stage that had a gain of 44 V/V at these frequencies.

It might be appropriate to use $n$ tuned stages to improve the situation even more. If we consider the 3-dB bandwidth of $n$ identical tuned stages, it can be shown that the same bandwidth shrinkage factor used for wideband stages applies to multiple tuned stages with equal $Q$s. For $n$ such stages the bandwidth can be expressed as

$$\Delta f_n = \Delta f \sqrt{2^{1/n} - 1} \tag{5.48}$$

where $\Delta f_n$ is the overall bandwidth resulting from cascading $n$ identical stages. With this new bandwidth, the overall $Q$ of the $n$ stages can be written as

$$Q_n = \frac{f_o}{\Delta f_n} = \frac{f_o}{\Delta f \sqrt{2^{1/n} - 1}} = \frac{Q_{\text{eff}}}{\sqrt{2^{1/n} - 1}} \tag{5.49}$$

Generally, when more than one tuned stage is cascaded, it is more effective to use stagger-tuning. The simplest form of this method uses two stages, each detuned in opposite directions from the resonant frequency by one-half the bandwidth. Figure 5.21 shows the response of two stagger-tuned stages. Although the detuning can be less than one-half the bandwidth, maximal flatness in the passband occurs when the detuning equals one-half the bandwidth.

A single tuned circuit would be tuned to a resonant frequency of $f_o$ with upper and lower 3-dB frequencies of $f_2$ and $f_1$. As Figure 5.21 demonstrates, one of the stagger-tuned circuits is tuned to resonance at $f_2$ and the other is resonant at $f_1$. The response of two cascaded stagger-tuned amplifier stages with equal $Q$s would be

$$G = \frac{K}{\left[\left(1 + Q_{\text{eff}}^2 \left[\frac{f}{f_1} - \frac{f_1}{f}\right]^2\right)\left(1 + Q_{\text{eff}}^2 \left[\frac{f}{f_2} - \frac{f_2}{f}\right]^2\right)\right]^{1/2}} \tag{5.50}$$

where $K$ is the resonant gain. Since each stage is down by 3 dB at the old resonant frequency, $f_o$, the overall response of the stagger-tuned stages is one-half the value of the single tuned stage at resonance. In other words, Eq. (5.50) results in a resonant value of

$$G = \frac{K}{2}$$

The 3-dB bandwidth of the stagger-tuned amplifier is given by

$$\Delta f = \frac{\sqrt{2} f_o}{Q_{\text{eff}}} \tag{5.51}$$

Typically, two tuned stages with no loading between stages are used to create the stagger-tuned circuit.

This response can be compared to the single tuned stage and to two stages tuned to the same resonant frequency. Of course, the gain must be restored to a resonant value of 100 V/V since stagger-tuning drops the resonant gain by a factor of two. Equation (5.51) is applied to set the $Q_{\text{eff}}$ of the individual stages to have a resonant frequency of 455 kHz with a 3-dB bandwidth of 10 kHz. This requires a $Q_{\text{eff}}$ of 64.3. Then from Eq. (5.50), we find that the response of the stagger-tuned circuit is down from the resonant value by 19.4 dB at 440 kHz and 470 kHz. This is about 6 dB more attenuation than the two identical tuned stages. For most radio receiver circuits, an attenuation of 19.4 dB is not sufficient, requiring multiple stagger-tuned sections to be used. Cascading two stagger-tuned amplifiers results in very sharp falloffs, but requires accurate tuning to create the desired response. This situation requires four tuned circuits to construct. Most radio or television receivers use several stagger-tuned stages in cascade. These stages are carefully tuned by the manufacturer.

**Practice Problem**

**5.6** A stagger-tuned stage is designed with each individual stage having a $Q_{\text{eff}}$ of 76. If the resonant frequency of the circuit is 860 kHz, what is the overall $Q$ of the circuit? *Ans:* 53.7.

## 5.5 TUNED AMPLIFIERS

**Important Concepts**

**1.** The tuned stage places a tank circuit in the collector of a transistor. The gain is proportional to the magnitude of the impedance of the tuned circuit.

**2.** The high resonant impedance of a tuned circuit requires that the output impedance of the transistor be considered.

**3.** At higher resonant frequencies, the collector-base transition capacitance must be considered.

**4.** Instability of tuned circuits may result unless neutralization or mismatching is used.

Tuned amplifiers intended for relatively low-frequency amplification are rather simple to design. In such cases, the low-frequency transistor equivalent circuit can be applied. For high-frequency tuned stages the situation becomes more complex due to the parasitics of the transistor.

In low-frequency design, the major difference from wideband amplifiers is the large magnitude of transistor load impedance at resonance compared to the wideband amplifier load. It is not uncommon to see a 100-$\Omega$ load in a discrete wideband stage, whereas the resonant resistance of a tuned circuit might be 30 $k\Omega$ or more. The output resistance of the transistor becomes more prominent in the gain and bandwidth calculations for the resonant stage.

At high frequencies, both output impedance and feedback parasitics such as the collector-base depletion-layer capacitance affect the design. The transition between low-frequency and high-frequency design depends largely

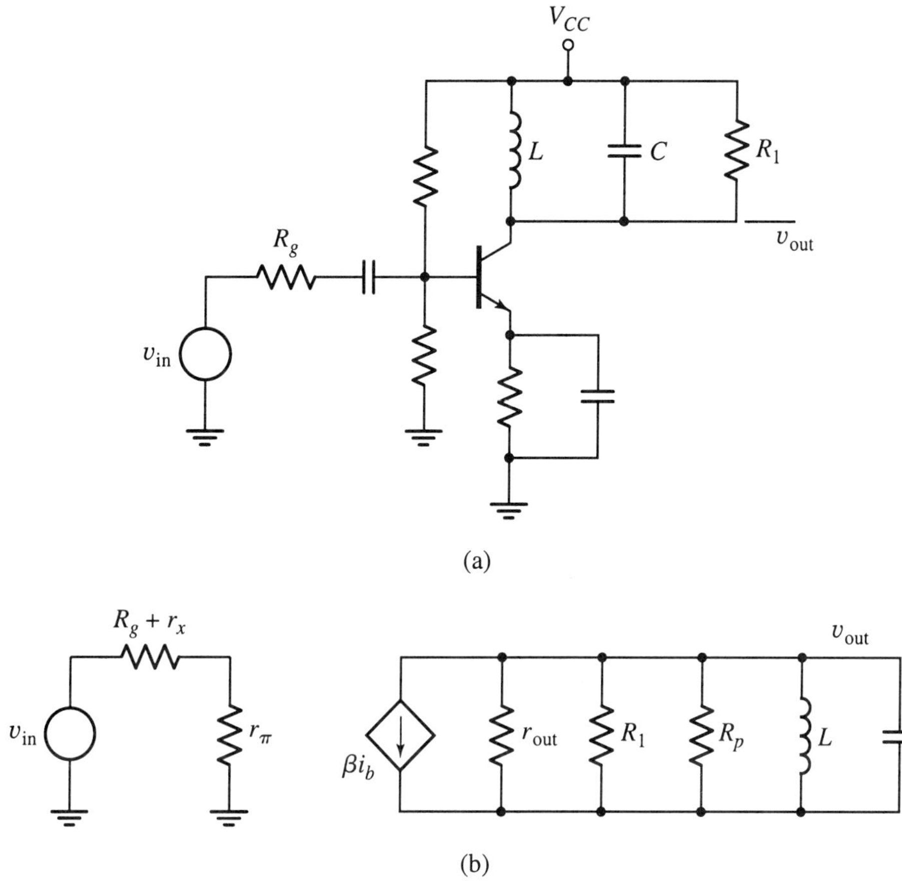


Figure 5.22: (a) Tuned amplifier. (b) Simple equivalent circuit.

on the transistors used. Generally, if the resonant frequency is below $f_\beta = f_t/\beta$ of the transistor, low-frequency design methods apply. In fact, it is often possible to use low-frequency techniques up to a resonant frequency of $10f_\beta$ with reasonable accuracy. Above this value, other considerations must be made. Both situations will be treated in succeeding paragraphs.

### 5.5.1 The Transistor Tuned Amplifier at Low Frequencies

Figure 5.22 shows a tuned transistor amplifier with a resistance $R_1$ shunting the tank circuit.

When the tank circuit is added to the collector circuit, the output resistance of the transistor and $R_1$ appear in parallel with $R_p$. The equivalent resistance shunting the tank at resonance is

$$R_{sh} = R_p \,||\, R_1 \,||\, r_{\text{out}}$$

The effective $Q$ is lower than the coil $Q$ as a result of these additional losses. The magnitude of the load impedance will exhibit a resonant peak at $\omega_o$, but the bandwidth will be broader than that for the LC circuit by itself due to the lower effective $Q$.

The voltage gain can be expressed as

$$A = \frac{-\beta Z_L}{r_\pi + r_x + R_g}$$

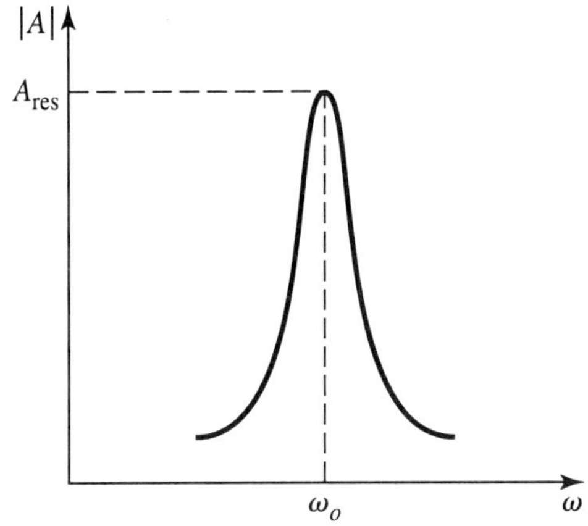


Figure 5.23: Gain of a tuned stage as a function of frequency.

where $R_g$ is the signal generator impedance and $Z_L$ is the total collector load impedance that is a function of frequency. The magnitude of the gain will be

$$|A| = \frac{\beta |Z_L|}{r_\pi + r_x + R_g} \tag{5.52}$$

The gain of the tuned stage has a frequency response that is proportional to the response of the tank circuit as shown in Figure 5.23.

Since the gain of the circuit is proportional to the magnitude of the loaded tuned circuit impedance, the gain at resonance is expressible as

$$A_{\text{res}} = \frac{-\beta \omega_o L Q_{\text{eff}}}{r_x + r_\pi + R_g} = \frac{-\beta R_{sh}}{r_x + r_\pi + R_g} \tag{5.53}$$

A tuned amplifier can have a very high resonant gain if $Q_{\text{eff}}$ is high. For example, if the sum of $r_\pi + r_x + R_g$ is 3 kΩ, $\beta = 100$, $\omega_o = 6.28 \times 10^5$ rad/sec, $L = 1$ mH, and $Q_{\text{eff}} = 50$, a resonant gain of $-1047$ V/V results.

Equation (5.24) for bandwidth applies to the tuned collector stage. The bandwidth in this case would be $f_o/Q_{\text{eff}} = 2$ kHz. The resonant collector impedance for this circuit is $R_{sh} = 31$ kΩ.

# Example 5.4

A transistor with $r_x = 100\ \Omega$, $V_A = 52$ V (Early voltage), $f_t = 300$ MHz, and $\beta = 100$ is used to construct a tuned stage with a bandwidth of 16 kHz and a resonant frequency of 600 kHz. An output voltage of at least 3-V peak is required. A coil with $Q_o = 64$ and $L = 300\ \mu$H is used. Design this stage for a power supply of $V_{CC} = 10$ V. Assume that $V_{CEQ} = 8$ V. Calculate the resonant gain of the stage if the signal generator resistance is 1 kΩ.

**Solution** The required capacitor is easily calculated as

$$C = \frac{1}{\omega_o^2 L} = \frac{1}{(2\pi \times 600{,}000)^2 \times 3 \times 10^{-4}} = 235 \text{ pF}$$

The value of $R_p$ is found from

$$R_p = \omega_o L Q_o = 2\pi \times 600{,}000 \times 10^{-4} \times 64 = 72.4 \text{ k}\Omega$$

The required value of $Q_{\text{eff}}$ is calculated from

$$Q_{\text{eff}} = \frac{f_o}{\Delta f} = \frac{600}{16} = 37.5$$

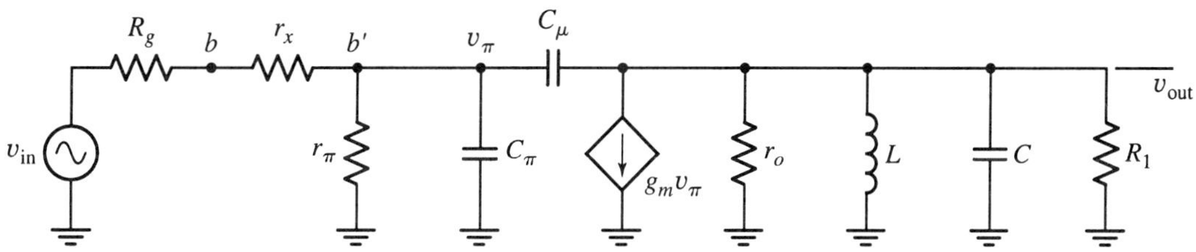


Figure 5.24: A tuned stage at high frequencies.

The shunt resistance to create this value of $Q_{\text{eff}}$ is

$$R_{sh} = \omega_o L Q_{\text{eff}} = 2\pi \times 600{,}000 \times 3 \times 10^{-4} \times 37.5 = 42.4 \text{ k}\Omega$$

A quiescent collector current of 0.4 mA will easily provide for the required output swing. (Why is this statement valid?) With this value of collector current, the output impedance of the transistor is

$$r_{\text{out}} \approx \frac{V_A + V_{CEQ}}{I_C Q} = \frac{60}{0.4} = 150 \text{ k}\Omega$$

In order to achieve a value of 42.4 k$\Omega$ for $R_{sh}$, the following must be true

$$R_{sh} = 42.4 = R_p \parallel R_1 \parallel r_{\text{out}} = 72.4 \parallel R_1 \parallel 150$$

From this equation the value of $R_1$ is found to be 323 k$\Omega$. The resonant gain is then

$$A_{res} = \frac{-\beta R_{sh}}{r_\pi + r_x + R_g} = \frac{-100 \times 42.4}{7.67} = -553 \text{ V/V}$$

In this equation, $r_\pi$ was calculated as

$$r_\pi = (\beta + 1) r_e = 101 \times \frac{26}{0.4} = 6.57 \text{ k}\Omega$$

The remaining task in this design is that of selecting a bias circuit to produce 0.4 mA for the collector current. This task will be left to the student.

---

## 5.5.2 The Tuned Stage at High Frequencies

We have assumed that the output impedance of the transistor is real and that the simple transistor equivalent circuit applies to the tuned amplifier stage. For narrowband applications at higher frequencies, the transistor can prove to be very troublesome. The inherently low impedances associated with the transistor and a tendency toward instability under certain conditions are the two major problems encountered for tuned stages. The tendency for unstable operation is due primarily to the presence of the collector-base capacitance and resistance. The circuit of Figure 5.24 shows a tuned stage with the hybrid-$\pi$ equivalent circuit used for the transistor.

The value of $r_{\text{out}}$ can be neglected for small load resistances, however, the resonant circuit of the narrowband stage does not always satisfy this requirement. At resonance, the impedance of the tuned circuit might range from several kilohms up to a few megohms, depending on the components used. Since the output impedance of the transistor can be less than 50 $k\Omega$, it can lower the effective $Q$ of the circuit and cause less selectivity.

If no external resistance is added to the tank the shunt resistance is $R_{sh} = R_p \parallel r_{\text{out}}$. It is not unusual for $r_{\text{out}}$ to be much smaller than $R_p$, meaning that $Q_{\text{eff}}$ is much smaller than $Q_o$ and that the selectivity is much poorer than that of the unloaded circuit. It can be shown that the low output impedance of the transistor at

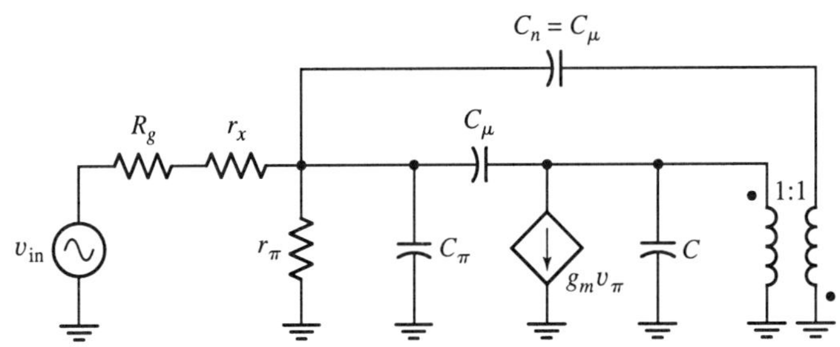


Figure 5.25: Theoretical neutralization.

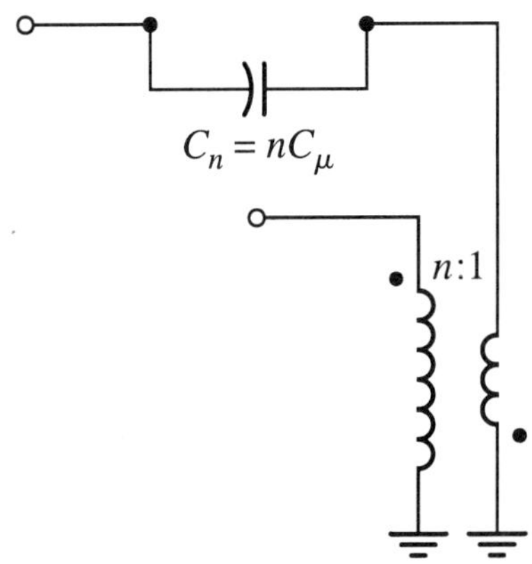


Figure 5.26: Practical neutralizing network.

high frequencies is due mainly to the presence of $C_\mu$. For good selectivity, the loading of the tank circuit by the transistor output impedance must be minimized.

The instability problem is also due to the presence of $C_\mu$. At frequencies just below resonance, the load impedance is quite high and slightly inductive. A portion of the output voltage is fed back through $C_\mu$ to the input. Under these conditions, the feedback voltage can be greater than the initial applied voltage. As will be seen in Chapter 6, this will cause oscillations to occur; that is, an output voltage can be present even when there is no input signal. Without going into further detail at present, it is sufficient to note that the presence of $C_\mu$ is responsible for most of the difficulties that occur in BJT tuned amplifiers.

### Neutralization of the Effects of $C_\mu$

The feedback current through $C_\mu$ can be canceled by adding an equal and opposite amount of current. Consider the circuit of Figure 5.25.

The transformer inverts the phase of the output voltage, and since the neutralizing impedance equals the feedback impedance, the neutralizing current is equal in magnitude, but opposite in direction to the feedback current. The net current from output to input is now approximately zero. A scheme allowing more practical element values uses a stepdown transformer with a turns ratio of $n$:1, and a neutralizing capacitance of $nC_\mu$ as shown in Figure 5.26.

The above theory has assumed that the circuit designer has access to the $b'$-terminal, which of course is a false assumption. In practice, the neutralizing network must be connected to terminal $b$ rather than $b'$ since $b'$ is inaccessible. The method of calculating $C_n$ remains the same, but the neutralization is now a function of frequency and some adjustment in $C_n$ may be required.

The impedance of the neutralizing network will be reflected to the primary side of the transformer. The

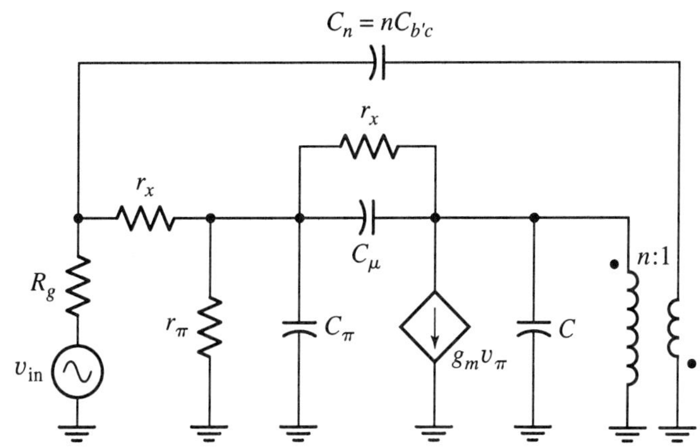


Figure 5.27: Practical tuned circuit.

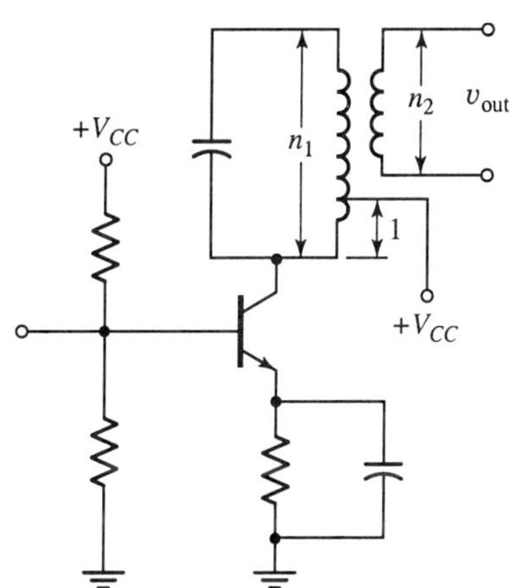


Figure 5.28: A mismatched stage.

capacitance of the resonant circuit is $C + C_\mu/n$. Figure 5.27 shows the practical neutralized circuit.

### Mismatching of Stages

If the load impedance presented to the transistor is small, more of the output current will flow through this impedance and less will be fed back to the input through $C_\mu$. This results in a more stable circuit. On the other hand, a high impedance must be presented to the tank circuit to obtain a high $Q$ circuit. Both of these impedance conditions can be met by using a step-up transformer between the transistor output and the tuned circuit. In practice, a tapped transformer is used, which might appear as shown in Figure 5.28.

The load presented to the tuned transformer by the transistor is $n_1^2 r_{\text{out}}$, which can be quite high. The following stage will also cause an impedance of $(n_1/n_2)^2 r_{\text{in}2}$ to appear across the tuned circuit. The impedance presented to the transistor by the tuned circuit is $(1/n_1)^2 R_p$ and the secondary reflects an impedance of $(1/n_2)^2 r_{\text{in}2}$ back to the collector of the transistor. The tuned circuit has a shunt impedance of

$$R_{sh} = R_p \; || \; n_1^2 r_{\text{out}} \; || \; \left(\frac{n_1}{n_2}\right)^2 r_{\text{in}2}$$

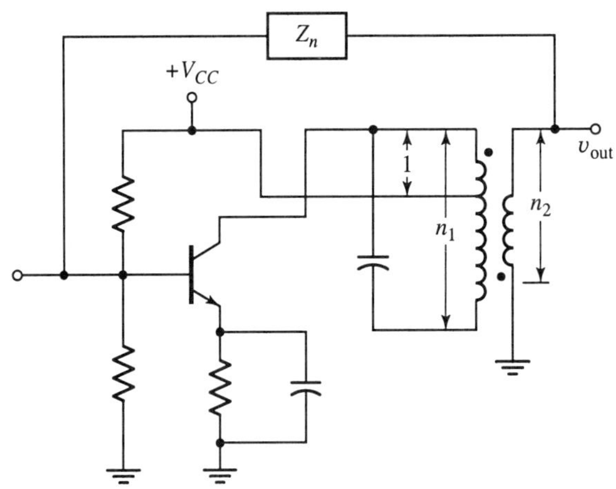


Figure 5.29: Mismatching and neutralization.

whereas the transistor sees an impedance of

$$r = r_{\text{out}} \,||\, \left(\frac{1}{n_1}\right)^2 R_p \,||\, \left(\frac{1}{n_2}\right)^2 r_{\text{in2}}$$

Obviously these impedances will allow a high-circuit $Q$, while causing the circuit to be more stable. For rf-circuits the necessary turns ratio generally ranges from 3:1 to 10:1.

The big advantage of mismatching is that it leads to stability for all frequencies, unlike neutralization which guarantees stability only for a certain band of frequencies. Mismatching also reduces the alignment problem when several tuned stages are used. Each tank circuit can be tuned with little or no effect on adjacent stages. When neutralization is used, the tuning often affects the neutralization of adjacent stages.

### Combination of Mismatching and Neutralization

A combination of the two techniques can be used for critical applications. The circuit of Figure 5.29 demonstrates one scheme of applying both methods.

The output impedance of the transistor will be high due to neutralization. This allows a higher $Q$ to be obtained by using both techniques rather than either one alone.

## 5.6 TUNED STAGES WITH NONSINUSOIDAL INPUTS

**Important Concepts**

**1.** Nonsinusoidal periodic waveforms can be applied to the tuned stage, but the output voltage can be sinusoidal if the resonant circuit is tuned properly.

**2.** The output sinusoid will have a frequency equal to that of the fundamental or some harmonic of this frequency, depending on the resonant frequency of the tuned circuit.

**3.** If the transistor reaches saturation at the peak of the output signal, an amplitude-limiting effect occurs.

There are several important communication applications wherein a nonsinusoidal periodic signal is applied to a tuned stage. One example is that of the class-C amplifier in which the collector current is sinusoidal for less than one-half of the period and equal to zero for the remainder of the period. A second example is in a base

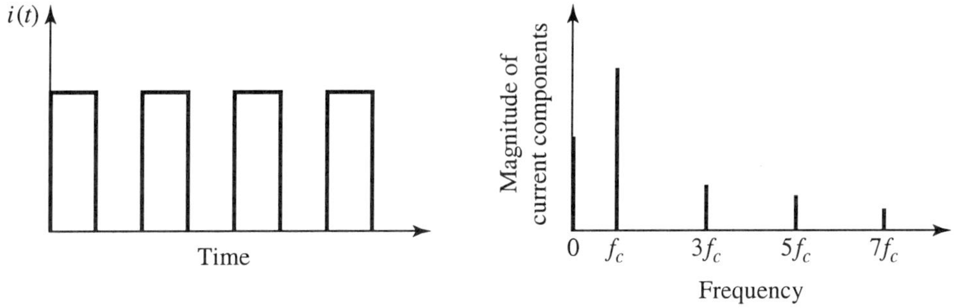


Figure 5.30: Components of a repetitive, nonsinusoidal waveform.

modulation stage when a square wave is applied to the input of the stage to achieve very linear modulation. These stages will be discussed in detail in Chapter 8.

A periodic signal can be expressed in terms of the sum of a fundamental frequency and several harmonic components plus a dc component if present. This type of waveform is often analyzed by Fourier series methods to evaluate the amplitude and phase of each component. For example, the Fourier series for a square wave of current, varying from 0 to $I_{\text{peak}}$ with a 50% duty cycle can be written as

$$i = \frac{I_{\text{peak}}}{2} + \frac{2I_{\text{peak}}}{\pi}\left(\sin\,\omega_c t - \frac{1}{3}\,\sin 3\omega_c t + \frac{1}{5}\,\sin 5\omega_c t - \cdots\right) \tag{5.54}$$

The waveform as a function of time, along with the magnitude of the components as a function of frequency, are shown in Figure 5.30. This waveform contains only odd harmonics.

If this current passes through an impedance, each component will develop a voltage that depends on the amplitude of the component and the magnitude of the impedance at the frequency of that component. If the impedance is a resistance, presenting the same impedance to all frequencies, a square-wave voltage will result. When this current flows into a high-$Q$ tank circuit, tuned to one of the ac components of the signal, only this component will develop a voltage across the tank circuit. Since the tuned circuit will offer negligible impedance to dc and all other components, negligible voltage develops at other than the tuned frequency. Application of this square wave of current to a parallel tuned circuit produces a voltage at the resonant frequency, assuming this frequency equals that of one of the ac components that make up the square wave.

### 5.6.1 Nonsaturating Tuned Stages

The results of the preceding paragraphs can be applied to a nonsaturating tuned stage such as that shown in Figure 5.31. The equivalent circuit shows a tank circuit being driven by the collector source. This model neglects the output impedance of the transistor stage. The input signal is a square wave with an amplitude of $V_1$ having symmetry about the $V = 0$ level. The frequency of the square wave is 50 kHz.

The transistor will cut off when the input voltage goes negative. At the positive peak of the input, the collector current will have a peak value of

$$I_{C\text{peak}} = \frac{\beta[V_1 - V_{BE(on)}]}{R_B} \tag{5.55}$$

As an example, we will assume that $V_1 = 3$ V, $\beta = 80$, and $V_{BE(on)} = 0.7$ V. If $R_B$ is 100 k$\Omega$ and $V_{CC} = 16$ V, the resulting peak collector current is

$$I_{C\text{peak}} = \frac{80 \times [3 - 0.7]}{100} = 1.84\text{ mA}$$

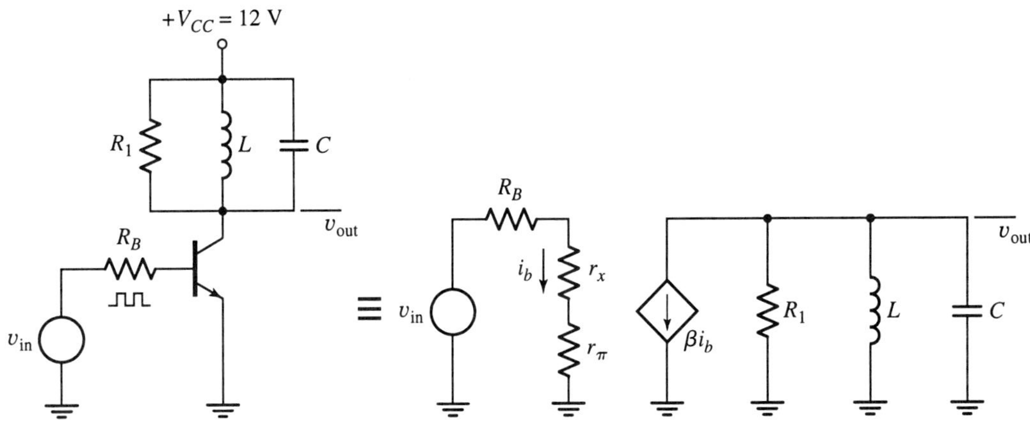


Figure 5.31: A tuned amplifier driven by a square wave.

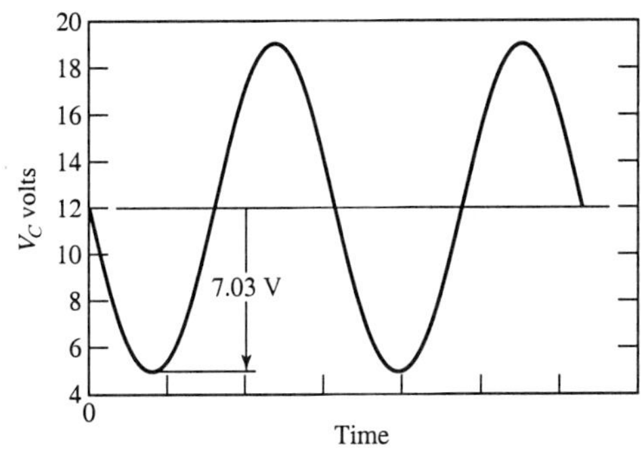


Figure 5.32: Output waveform for amplifier of Figure 5.31.

If the tank circuit is tuned to a resonant frequency of 50 kHz and has a 6 kΩ resonant resistance, the sinusoidal output voltage at a frequency of 50 kHz will have a magnitude of

$$|v_{\text{out}}| = R_{sh} \times I_{C\text{fund}} = R_{sh} \times \frac{2I_{C\text{peak}}}{\pi} = \frac{6 \times 2 \times 1.84}{\pi} = 7.03$$

The total output voltage including the dc component is shown in Figure 5.32.

It is possible to tune the tank circuit to the third harmonic. Assuming this is done and the resonant resistance of the tank is 8 $k\Omega$, the magnitude of $v_{\text{out}}$ will be 3.12 V.

### Frequency Multiplication

A frequency multiplication scheme may be based on this circuit. A rectangular wave or other nonsinusoidal periodic waveform of a particular fundamental frequency can be applied to a tuned stage, resulting in an output at some multiple of the fundamental. Since all common waveforms used in electrical engineering tend to lead to decreasing components with frequency, multipliers that double or triple the input frequency are common. Generating higher order harmonic outputs becomes more difficult due to the small amplitudes involved. The class-C stage considered in Chapter 2 and Chapter 8 can perform this function of frequency multiplication. This has been a very popular method in the production of frequency modulation waveforms using the Armstrong method. This method will also be considered in Chapter 9.

# Example 5.5

A 100 kHz square wave with a 50% duty cycle, varying from 0 to 6 V, is applied to the circuit of Figure 5.31. The tuned circuit has a $Q_{\text{eff}}$ at 300 kHz of 21 using a 400-$\mu$H coil. Calculate the value of $R_B$ to lead to a 4.5-V peak output value at 300 kHz. Assume that $\beta = 110$ and $V_{BE(\text{on})} = 0.7$ V.

**Solution.** The peak collector current in this case is given by Eq. (5.55) as

$$I_{C\text{peak}} = \frac{110[6-0.7]}{R_B} = \frac{583}{R_B}$$

Since the tuned circuit is tuned to the third harmonic of the square wave, the peak value of this component can be found from Eq. (5.54) as

$$\frac{2I_{C\text{peak}}}{3\pi} = \frac{2}{3\pi}\frac{583}{R_B} = \frac{124}{R_B}$$

The shunt impedance of the tuned circuit at 300 kHz is given by

$$R_{sh} = \omega L Q_{\text{eff}} = 2\pi \times 300,000 \times 400 \times 10^{-6} \times 21 = 15.8 \text{ k}\Omega$$

The magnitude or peak value of the output voltage is then

$$v_{\text{out}} = \frac{124}{R_B} \times 15.8 = 4.5 \text{ V}$$

Solving for $R_B$ leads to

$$R_B = 435 \text{ k}\Omega$$

In this case, a 100 kHz signal is applied to the input, whereas a 300 kHz signal is developed at the output.

**Practice Problem**

**5.7** Rework Example 5.5 if the input square wave varies from 0 to 3 V. *Ans:* 189 kΩ¿

## 5.6.2 Saturating Tuned Stages

If the periodic input signal to a tuned stage is increased to the point that the sinusoidal collector voltage reaches the edge of the saturation region at the peaks of the signal, an interesting result occurs. In this situation, the amplitude of the ac output is equal to the power supply voltage $V_{CC}$, neglecting the saturation voltage of the transistor. If the input signal is small and the input source has low source impedance, the collector will be constrained on these peaks at the saturation voltage. If the input drive is increased beyond this point, the peak value of the output voltage remains at $V_{CC}$. Even though the collector current may be very distorted as the stage is driven into saturation, the output voltage remains sinusoidal for a high-$Q$ tank circuit load.

In practice, the tuned circuit does not exhibit zero impedance at off-resonant frequencies. Furthermore, when the transistor saturates, the collector does not resemble the current source of the small-signal transistor model. At saturation, the diode between base and collector becomes forward biased. If the base voltage is relatively large, the collector voltage of an *npn* transistor can be distorted at the negative peaks of the output voltage by feedthrough from the input source via the base-collector diode. The output voltage may then appear as in Figure 5.33. The distortion is a function of the amplitude of the input signal and the resistance in series with this signal. The figure indicates the general change that may occur as input series resistance is increased.

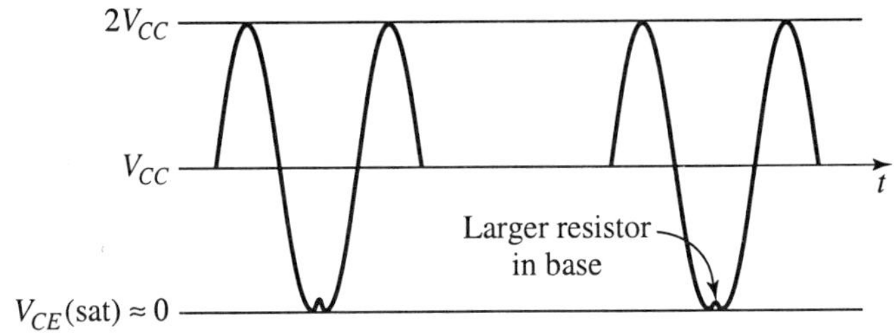


Figure 5.33: Output voltage of a transistor when driven into saturation.

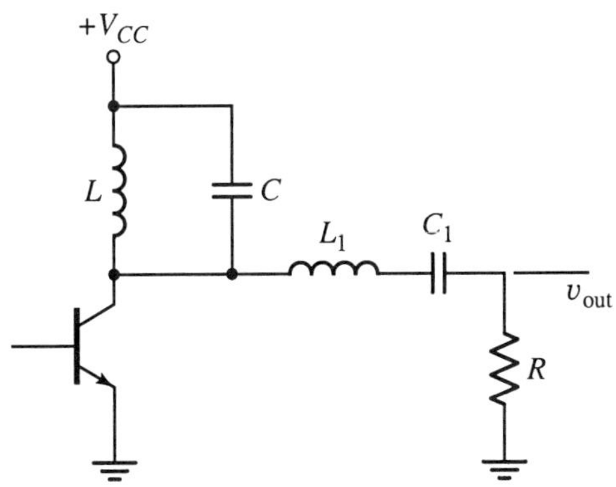


Figure 5.34: Additional filtering to achieve a low-distortion sinusoid.

The amplitude of the output voltage will still be limited to approximately $V_{CC}$ and still approximates a sinusoid. Obviously, higher harmonics exist in the output signal since the distortion is periodic. A series resonant circuit can be added to the circuit to filter or remove the components introduced by the distortion. A tuned stage with this filter is shown in Figure 5.34. The series resonant circuit is tuned to the same frequency as the parallel resonant circuit. At series resonance, the impedance of the filter is negligible and the output signal amplitude is

$$|v_{\text{out}}| = \frac{R}{R + R_{sh}} V_{CC}$$

where $R_{sh}$ is the resonant impedance of the tank circuit. At second, third, and higher harmonic frequencies, the series circuit has a significant impedance, thereby greatly attenuating these components. This type of arrangement is often used in practical communication circuits such as amplitude modulators that utilize the tuned stage.

A significant increase in the amplitude of the input signal will negligibly affect the amplitude of the output signal. This characteristic leads to the use of this circuit configuration in amplitude-limiting applications as well as class-C modulation circuits. This saturating tuned stage forms the basis of the popular collector modulator to be discussed in Chapter 8.

The saturating stage introduces additional losses into the tuned circuit since the output impedance of the transistor drops drastically during the time that the device is saturated. The circuit $Q$ is lowered considerably making the filtering less effective. A diode between ground and the collector that becomes forward biased as the collector voltage drops below ground improves the limiting effect on the output signal.

## SUMMARY

**1.** Narrowband amplification is very important in communication applications. Tuned circuits using inductors and capacitors are used to construct tuned circuits.

**2.** The inductor is an energy storage element that also exhibits some power loss. The parameter $Q$ is a measure of the effectiveness of energy storage.

**3.** The $Q$ of a circuit determines the bandwidth of a tuned circuit and equals the ratio of resonant frequency to 3-dB bandwidth.

**4.** Tapped capacitor or inductor circuits can minimize the reduction in circuit $Q$ when a load resistance is added to the circuit.

**5.** Stagger-tuned stages are used to provide a broader bandwidth while providing more attenuation at frequencies outside the bandwidth.

**6.** Tuned amplifier stages use tuned circuits for loads and provide narrowband amplification. At high frequencies, mismatch or neutralization must be used to avoid instability in the amplifier.

**7.** An input waveform to a tuned amplifier, containing several frequency components, can develop a sinusoidal output if the tuned circuit load is resonant at a frequency that equals one of the input components.

## PROBLEMS

### Sections 5.2.1 - 5.2.3 Inductors and Resonant Circuits

**5.1** A coil has a $Q_o$ of 75 and an inductance of 40 $\mu$H. If the coil is to be used at a resonant frequency of 10 MHz, what is the required capacitance? Given that a resistance of 200 $k\Omega$ is placed in parallel with the tank circuit, calculate the $Q$ of the circuit and the bandwidth.

**5.2** Repeat Problem 5.1, assuming that $Q_o = 50$ and $f_o = 1$ MHz.

**5.3** A parallel tank circuit includes a capacitor of 100 pF and a resistor of 100 kΩ. Given that the resonant frequency is 8 MHz and the bandwidth is 80 kHz, find the circuit $Q$, the inductance, and the effective parallel resistance of the coil $R_p$.

**5.4** The resonant frequency of the circuit shown is 860 kHz, and the 3-dB bandwidth is 25 kHz. Calculate $Q_o$, the $Q$ of the coil.

**5.5** If $R_B$ is changed to 10 MΩ in the circuit of Problem 5.4, what will the bandwidth now be?

Explain how this circuit could be used to measure the $Q$ of a coil.

### Section 5.2.4 Review of Tuned Circuit Relationships

**5.6** Repeat Problem 5.4 if the capacitor $Q$ is 180.

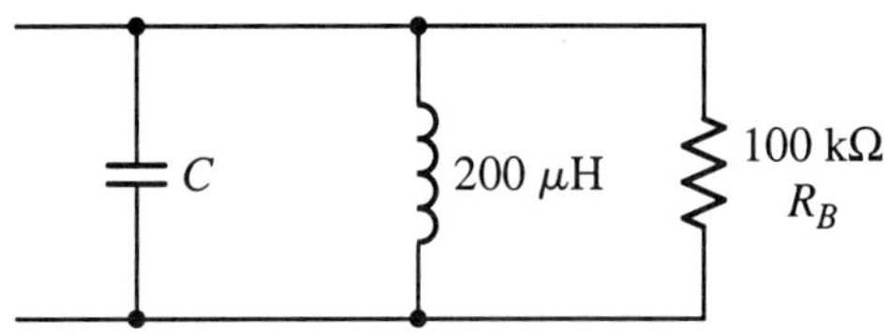


**Figure P5.4**

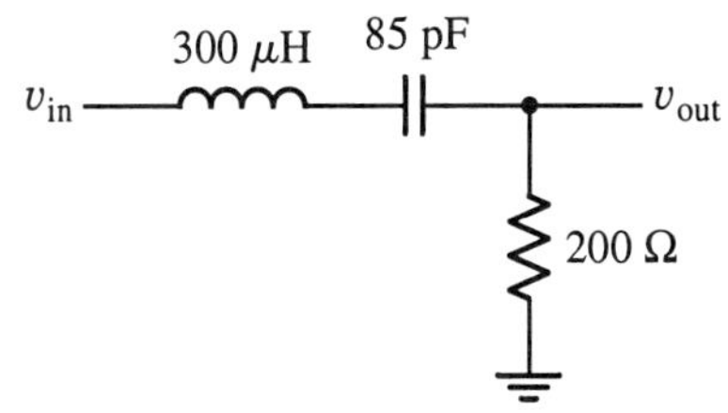


**Figure P5.8**

**5.7** A capacitor of value 200 pF has a $Q$ of 200. Calculate the parallel resistance representing the capacitive losses at 660 kHz. Calculate the series resistance that represents the capacitive losses at 660 kHz.

### Section 5.2.5 The Series Resonant Circuit

**5.8** The coil $Q$ is 80 while the $Q$ of the capacitor is 200 at the resonant frequency. Plot the frequency response of the circuit shown.

### Section 5.3 Tapped Capacitor and Inductor Circuits

**5.9** The circuit shown resonates at 600 kHz and has a bandwidth of 15 kHz. Assuming ideal capacitors, calculate the inductance of the coil and the coil $Q$.

**5.10** The coil $Q$ is 50 at the resonant frequency. The peak magnitude of $v_{\text{in}}$ is 2 V.

**(a)** Select $C_1$ and $C_2$ to result in $f_o = 300$ kHz and a bandwidth of $\Delta f = 15$ kHz.

**(b)** Calculate $|v_{\text{out}}|$ at resonance.

**(c)** Calculate $|v_{\text{out}}|$ at $f = 280$ kHz.

**5.11** Assume all inductors and capacitors are lossless.

**(a)** Calculate the resonant resistance $R_o$.

**(b)** Calculate $v_{\text{out}}$ at resonance.

**5.12** Calculate the impedance seen looking into the primary windings.

### Section 5.4 Stagger-Tuned Stages

**D5.13** Using coils with $L = 160\ \mu$H and $Q_o = 68$, design a stagger-tuned stage with a resonant frequency of 800 kHz and a bandwidth of 20 kHz. Assume a buffer stage between each tuned circuit to eliminate loading between

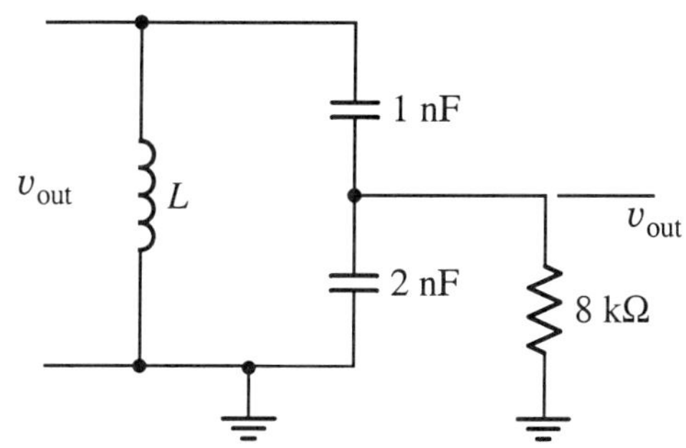


**Figure P5.9**

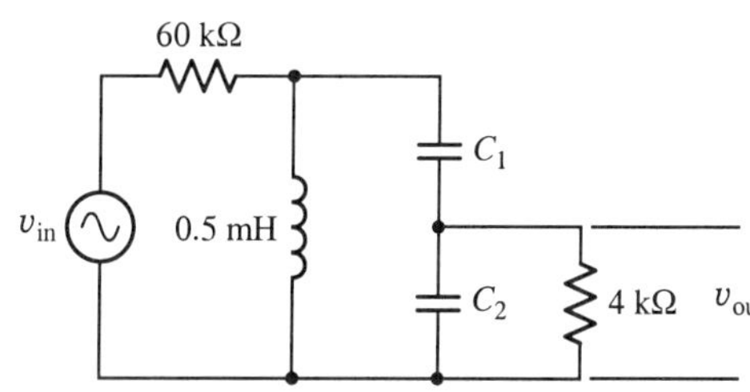


**Figure P5.10**

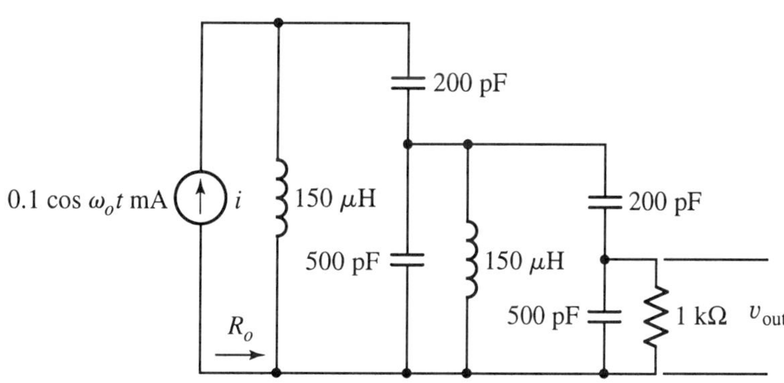


**Figure P5.11**

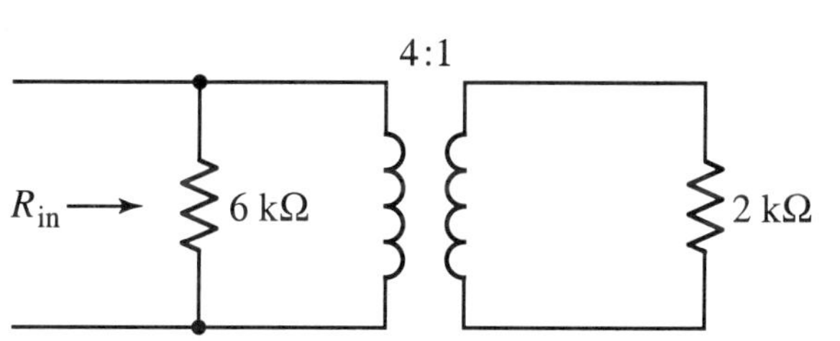


**Figure P5.12**

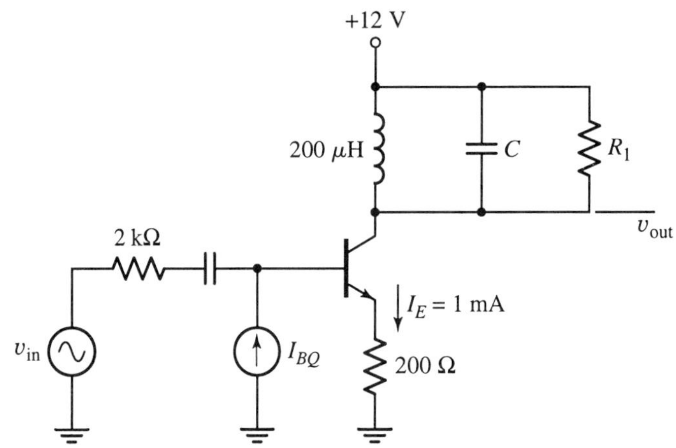


**Figure P5.14**

stages.

### Section 5.5.1 The Transistor Tuned Amplifier at Low Frequencies

**D5.14** The transistor $\beta$ is 100 and $r_{\text{out}} = 200$ kΩ. The coil $Q$ is 40 at the desired resonant frequency of 1 MHz. The bandwidth of $v_{\text{out}}$ is to be 50 kHz.

**(a)** Calculate the capacitance needed.
**(b)** Select the value of $R_1$ required.
**(c)** Calculate the resonant voltage gain, $\frac{v_{\text{out}}}{v_{\text{in}}}$.

**D5.15** The tuned circuit of Problem 5.14 is modified as shown.

**(a)** Calculate $C_1$ and $C_2$ to maintain a resonant frequency of 1 MHz and a bandwidth of 50 kHz.
**(b)** Calculate the resonant voltage gain, $\frac{v_{\text{out}}}{v_{\text{in}}}$.

**D5.16** The coil has an inductance of 200 μH and a $Q$ of 40 at a resonant frequency of 1 MHz. Assume that both transistors have values of $\beta = 100$ and $V_{BE(\text{on})} = 0.7$ V.

**(a)** Select $C_1$, $C_2$, and $R$ to result in a bandwidth of 60 kHz at $f_o = 1$ MHz.
**(b)** Calculate $v_{\text{out}}$ at resonance.

### Section 5.5.2 The Tuned Stage at High Frequencies

**D5.17** In the circuit of Figure 5.27, $C_\mu = 5$ pF and the output impedance of the transistor after neutralization is 50 kΩ. If the inductance appearing at the transformer primary is 40 μH with a $Q_o = 80$, and a resonant frequency of 1.6 MHz along with a bandwidth of 40 kHz are required, what are suitable values for $C$, $C_n$, and the transformer turns ratio $n$? If $r_x = 100$ Ω, $r_\pi = 2$ kΩ, $R_g = 4$ kΩ, and $\beta = 80$, what value of resistance must be shunted across the tank circuit? Neglect the $f_t$ of the transistor.

**5.18** Repeat Problem 5.17 given that the secondary of the transformer is to drive a stage with 8-kΩ input impedance in addition to the neutralizing network.

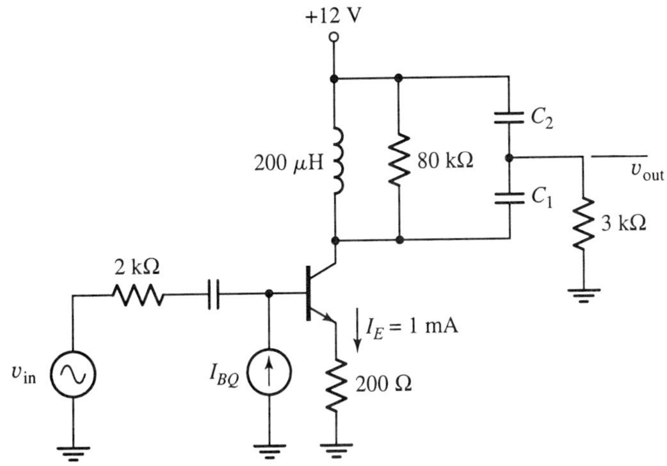


**Figure P5.15**

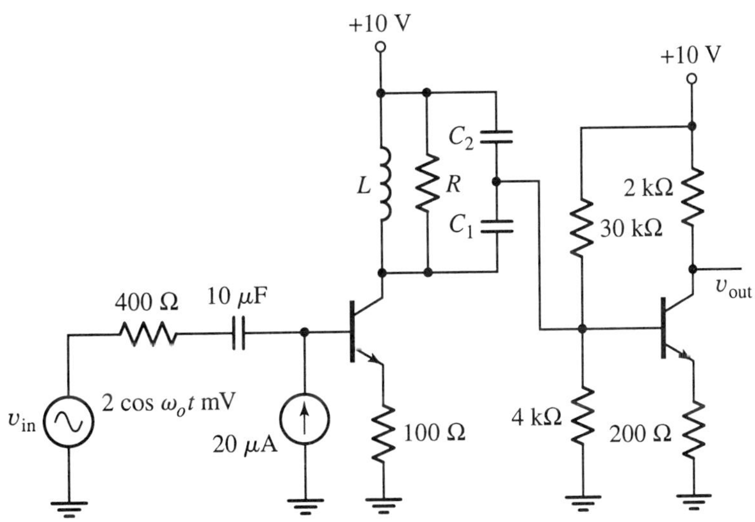


**Figure P5.16**

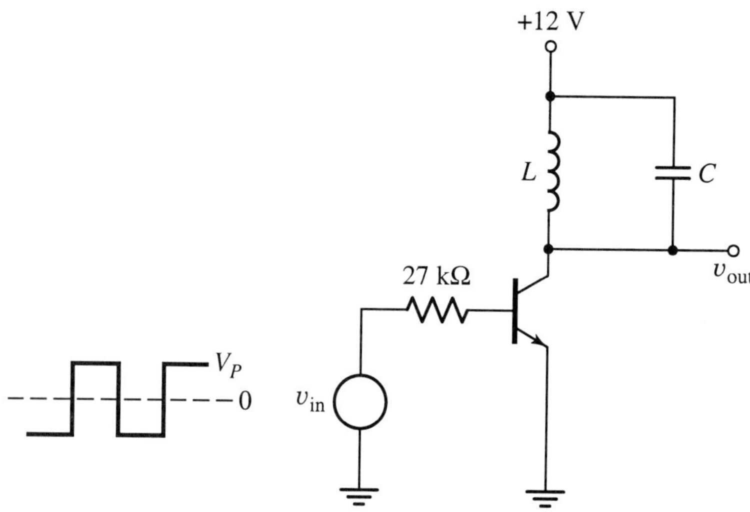


**Figure P5.20**

**5.19** In the circuit of Figure 5.29, $n_1 = 10$, $n_2 = 1/2$, $C_\mu = 5$ pF, and the output impedance of the transistor after neutralization is 20 kΩ. The tank circuit has an unloaded $Q$ of 100 and an inductance of 60 $\mu$H. Given that the input impedance to the following stage is 1 kΩ, find the proper values of $C_n$ and $C$ to give a resonant frequency of 2 MHz. Calculate the bandwidth of the circuit.

### Section 5.6 Tuned Stages with Nonsinusoidal Inputs

**5.20** A square wave of current can be represented by

$$i = \frac{I_{\text{peak}}}{2} + \frac{2I_{\text{peak}}}{\pi}\left[\sin\omega_1 t - \frac{1}{3}\sin 3\omega_1 t + \frac{1}{5}\sin 5\omega_1 t - \cdots\right]$$

where $\omega_1 = 2\pi \times 10^4$ rad/s. The tuned circuit in the figure is resonant at $3 \times 10^4$ Hz with a shunt impedance of 3 kΩ. The transistor has values of $\beta = 120$, $V_{BE(\text{on})} = 0.7$ V, and $V_{CE(\text{sat})} = 0.2$ V.

Sketch the output signal referenced to the input signal for

(a) $V_P = 2$ V.
(b) $V_P = 6$ V.

# Chapter 6

# SINUSOIDAL OSCILLATORS

Power is supplied to electronic circuits by either a dc voltage source or a fixed frequency ac source. If the power is supplied by an ac waveform it is converted to a dc signal by means of a power supply. This dc power is then transformed into the waveform required by various wave generation or wave-shaping techniques. Examples of important signal types are sinusoids, ramps, rectangular waveforms, and pulses.

One of the most important signal generators is the sinusoidal oscillator. This chapter will first consider the design of the oscillator and then proceed to a consideration of various practical circuits.

## 6.1 DESIGN CONSIDERATIONS

**Important Concepts**

**1.** Positive feedback is necessary to create a sinusoidal oscillator.

**2.** Sinusoidal oscillation is possible when the gain around the feedback loop is unity and the phase shift is 0°.

**3.** Amplitude-stabilizing circuits are used to maintain a constant-amplitude output signal as circuit parameters drift.

### 6.1.1 General Design Considerations

When several stages of amplification are used in a negative feedback amplifier, reactive effects often cause an additional 180° phase shift around the loop at some frequency, changing negative feedback to positive feedback. Either nonsinusoidal or sinusoidal oscillations may result at the output of the amplifier when positive feedback occurs. It is not unusual that compensating networks must be added to avoid oscillation problems. One might then conclude that oscillator design is very simple inasmuch as that instability and oscillations occur so often in amplifier design. It is, however, quite a different matter to construct a circuit that oscillates at a prescribed frequency with a prescribed stable amplitude than to simply achieve oscillations. A useful design procedure must lead to a circuit that meets specific requirements.

Figure 6.1 shows a basic configuration for a positive feedback circuit. Under certain conditions, this circuit

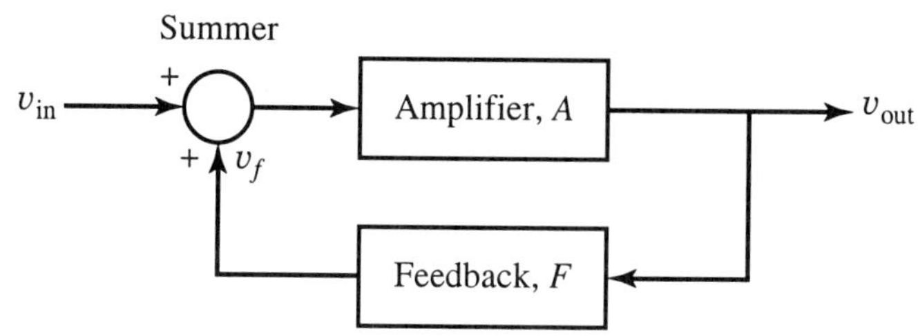


Figure 6.1: Block diagram of a positive feedback circuit.

will become a sinusoidal oscillator.

The overall gain for the system in Figure 6.1 can be derived as

$$G = \frac{v_{out}}{v_{in}} = \frac{A}{1 - AF} \tag{6.1}$$

As $AF$ approaches unity, the overall gain approaches infinity and a finite output voltage can result when no input signal is present. This results in an oscillating output signal. Other factors determine whether this is a sinusoidal or nonsinusoidal signal.

It is not quite as easy to apply Eq. (6.1) as it first appears. Other factors tend to complicate the direct application of this equation. For example, both $A$ and $F$ are generally frequency-dependent and $A$ varies with the operating point of the active devices used in the amplifier. Temperature or aging effects can also modify values of $A$ and $F$, to further complicate reliable design.

In sinusoidal oscillators, the term $AF$ is called the *return ratio* and must have a magnitude of unity with a phase shift of 0° or some multiple of 360° at a single frequency. When the return ratio, $AF$, has unity magnitude and the correct phase angle, the output signal feeds back the correct input signal to exactly reproduce the output. Random noise or transients can cause the oscillations to begin, but they are sustained by recirculating the appropriate signal. It may appear that the circuit is manufacturing its own energy since the input signal is amplified to produce an output signal which, in turn, produces the input signal. In the process of amplification, dc power enters the oscillator and is converted to ac power and thus, there is a net power flow into the system.

If the return ratio is increased drastically the output signal will tend to follow this increase. In general, at least one device will be driven out of its active region and the return ratio will then drop to unity or lower. In relaxation oscillators such as the astable multivibrator, the return ratio tends to a large value, forcing the active devices to switch rapidly across the operating region. The gain around the loop is then very low until a charging capacitor forces a device into the active region once more. At this point the astable switches rapidly to an alternate state, again dropping the value of the loop gain or return ratio.

The basic design goals for the sinusoidal oscillator are:

**1.** to create a return ratio magnitude of unity or slightly larger at the desired oscillation frequency;

**2.** to create a phase shift in the return ratio of 0° or 360° at this same frequency;

**3.** to ensure that the first two conditions are not satisfied at other frequencies; and

**4.** to ensure that the first two conditions will continue to be satisfied as parameter values change due to aging, temperature change, or device replacement.

The first three conditions can be met in a variety of ways, but the fourth condition often requires special consideration. An example will demonstrate the basic concept of sinusoidal oscillator design.

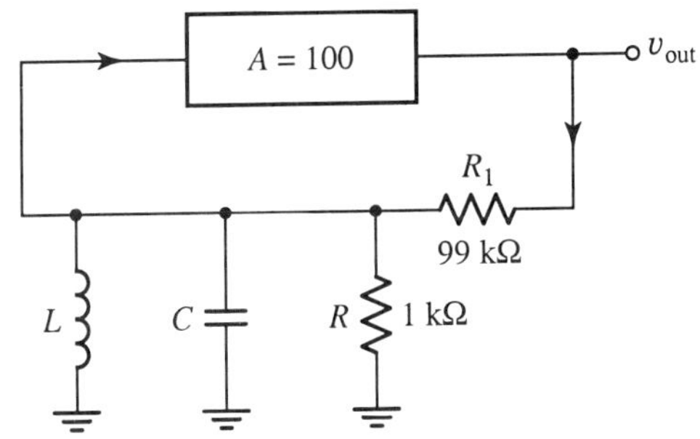


Figure 6.2: Ideal oscillator circuit.

## Example 6.1.

Given that the gain of the ideal amplifier in Figure 6.2 is 100 V/V, find a suitable feedback network to convert this amplifier to a sinusoidal oscillator. Assume the amplifier has infinite input and zero output impedances.

**Solution** The circuit of Figure 6.2 will lead to a magnitude for $AF$ of unity and a zero-degree phase shift when the tank circuit is resonant. We can express the feedback factor as

$$F = \frac{Z_t}{Z_t + R_1}$$

where $Z_t$ is the tank circuit impedance. This quantity is

$$Z_t = \frac{j\omega L}{1 - \omega^2 CL + j\omega L/R}$$

giving a return ratio of

$$AF = \frac{j\omega LA}{R_1(1 - \omega^2 CL) + j\omega L(1 + R_1/R)}$$

In examining this equation, we see that a zero-degree phase shift will occur only when the real part of the denominator goes to zero. This leaves an expression for $AF$ that is real, implying a zero-degree phase shift. This happens to be the resonant frequency of the tuned circuit and is given by

$$\omega_o = \frac{1}{\sqrt{LC}}$$

At this frequency the magnitude of $AF$ becomes

$$|AF(\omega_o)| = \frac{AR}{R_1 + R} = \frac{100R}{R_1 + R}$$

Although an infinite number of values can cause oscillations, the values $R = 1$ kΩ and $R_1 = 99$ kΩ lead to a return ratio of unity at the resonant frequency. An examination of the expression for $AF$ shows that there is only one possible oscillation frequency since neither the phase shift nor magnitude conditions can be satisfied at any frequency other than resonance.

**Practical Considerations**

In practice, the value of $R$ in Example 6.1 might be chosen slightly larger than 1 kΩ, resulting in an open-loop value for $|AF|$ that is greater than unity. When the loop is closed, the gain $A$ must readjust to a value that causes $|AF| = 1$ to sustain sinusoidal oscillations. Otherwise, the output would continue to grow until the extremes of the active region are reached. We next examine the mechanics of this readjustment of gain since it is a very important mechanism in the operation of a sinusoidal oscillator.

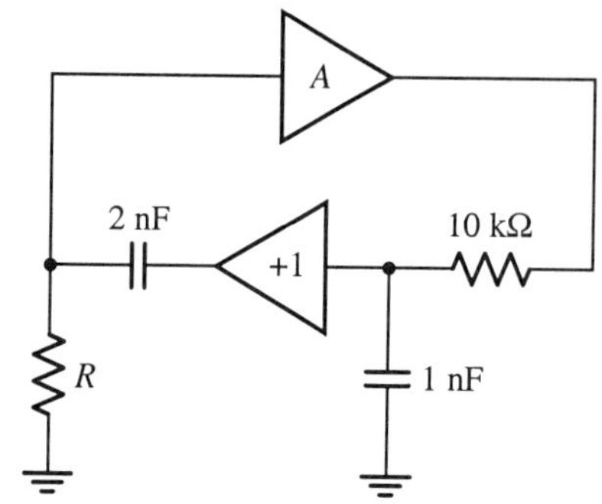


**Figure PP6.3**

**Practice Problems**

**6.1** Derive Eq. (6.1).

**6.2** If $A$ is changed to 50 V/V in the circuit of Figure 6.2, choose $R_1$ and $C$ to cause sinusoidal oscillation at $f_o = 120$ kHz. Assume that $L = 100\ \mu$H. *Ans:* $R_1 = 49$ kΩ, $C = 17.6$ nF.

**6.3** Assume that the amplifiers in the circuit have infinite input and zero output impedances. Find $R$ and $A$ to cause sinusoidal oscillation at $f = 16$ kHz. *Ans:* $R = 4.94$ kΩ, $A = 2.01$ V/V.

### 6.1.2 Amplitude Stability

If the magnitude of the return ratio dropped below unity, the circuit would cease oscillations. Each time the signal circulated around the loop, the magnitude would become smaller and the signal would not sustain itself. Since temperature changes, age, and operating-point changes can affect gain, the magnitude of $AF$ might drop below unity as gain changes, even if it were initially set to unity. It is standard procedure to design the circuit to lead to a *nominal gain* value of $|AF|$ that is slightly greater than unity at the oscillation frequency. The nominal value of $|AF|$ is the return ratio before the loop is closed. The greater the nominal value of $|AF|$, the greater will be the amplitude of the output signal and the amount of distortion. In BJT or MOSFET amplifiers, the distortion increases as output signal magnitude increases. This distortion will affect the gain $A$ and, in many cases, will automatically lower the gain to the value that leads to $|AF| = 1$. Because the variation of $A$ with operating point may be a complex nonlinear relationship, it is difficult to derive quantitative design equations. In order to demonstrate the nature of amplitude limiting, however, the following simplified example is used.

## Example 6.2.

In the circuit of Figure 6.2, the gain of the amplifier is given by

$$A = \frac{B}{r_e}$$

where $B$ is a constant and $r_e$ is the small-signal emitter-base diode resistance of the last transistor stage. We will also assume that the dc distortion present in the final-stage collector current is directly proportional to the amplitude of the output voltage swing; that is

$$I_C = I_{CQ} - K|v_{\text{out}}|$$

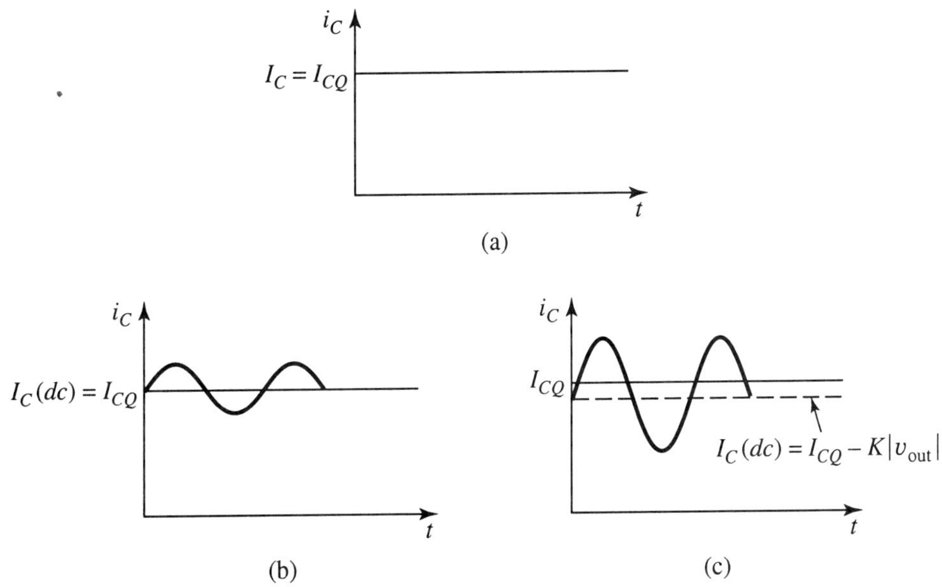


Figure 6.3: Collector current. (a) No output signal. (b) Undistorted output signal. (c) Distorted signal.

where $I_{CQ}$ is the no-signal quiescent current, $K$ is a constant, and $I_C$ is the dc collector current that flows in the presence of the distorted signal. Find the amplitude of the output voltage.

**Solution** The waveforms of Figure 6.3 show the collector current before and after distortion.

As the distortion increases, $I_C$ decreases which in turn increases $r_e$. The net result is a decreased gain. If $A$ is designed to have a nominal value of 105 and $F = 1/100$, at the oscillation frequency, the nominal value of $|AF|$ will be 1.05 V/V. When the loop is closed, the output will tend to become infinite for this value of return ratio. The output voltage could then reach a very large value if not for the distortion in the output signal. Since the distortion in $I_C$ increases with output signal, the gain will automatically limit at a value of $A$ that is approximately 100 V/V. In other words, $v_{out}$ will stabilize at the value that limits the in-circuit value of $|AF|$ to approximately unity. If we know the value of $K$ and $I_{CQ}$, the amplitude of $v_{out}$ can be found.

Since

$$A = \frac{B}{r_e} = \frac{BI_E}{0.026} = \frac{BI_C}{0.026\alpha}$$

and $A = 105$ when $I_C = I_{CQ}$, then the gain can be written as

$$A = \frac{B}{0.026\alpha}(I_{CQ} - K|v_{out}|) = 105 - \frac{BK|v_{out}|}{0.026\alpha}$$

Because $A$ must equal 100 V/V to limit $|AF|$ to unity, the amplitude of the output voltage can be found as

$$|v_{out}| = \frac{0.026 \times \alpha \times 5}{BK} = \frac{0.130\alpha}{BK}$$

---

### Practical Considerations

We speak of a sinusoidal oscillator as being *unstable* or *marginally stable*. This simply means that no input is required to generate the output signal. On the other hand, if the magnitude of the sinusoidal output signal remains constant, we refer to this as a stable magnitude. Thus, a marginally stable (unstable) circuit can have a stable output voltage magnitude.

If the value of $K$ is small in Example 6.2, meaning that the gain variation with $I_C$ is weak, the output voltage will be large. If $K$ is large, the output voltage will be small. The amplitude stability is greater for a larger value of $K$. If a temperature change causes $I_{CQ}$ to increase, $K|v_{\text{out}}|$ must increase to keep $A$ constant at a value of 100 V/V. If $K$ is large, then the change in $|v_{\text{out}}|$ can be small, but if $K$ is small then $|v_{\text{out}}|$ must change a great deal to offset the change in $I_{CQ}$.

For good stability the change in gain with output voltage amplitude should be large and an increase in amplitude must result in decreased gain. Expressed mathematically, the derivative $dA/dv_{\text{out}}$ must be a large negative number for good amplitude stability.

In a practical amplifier the gain $A$ depends on many parameters that are dependent on operating point or quiescent current. BJT voltage gain depends on $r_e$, and MOSFET gain depends on transconductance, $g_m$. Both $r_e$ and $g_m$ are functions of operating point or quiescent current.

Input and output impedances may vary with bias points and the feedback factor $F$ may be affected by impedance changes. In some cases, the gain may tend to increase rather than decrease, with output signal. In this case, the operating point can be affected drastically as the amplifier either leaves the operating region or moves to a region that causes the gain to decrease with further distortion.

For oscillators requiring stable, predictable operation, a dc signal proportional to output amplitude is generated and used to control the amplifier gain. In this way the gain $A$ can be made to vary appropriately with $v_{\text{out}}$ to create an amplitude-stabilized oscillator. Such an oscillator will be discussed in a later section.

**Practice Problem**

**6.4** For the oscillator described in Example 6.2, $B = 2600$ and $K = 2 \times 10^{-4}$. Calculate $|v_{\text{out}}|$. *Hint:* Assume $\alpha \approx 1$. *Ans:* 0.25 V.

### 6.1.3 Frequency Stability

The oscillation frequency of the ideal oscillator of Figure 6.2 is a function only of $L$ and $C$. These elements can change with temperature and the frequency at which zero-degree phase shift occurs will then change. In a practical oscillator, parasitic capacitances will exist that may have a slight influence on the phase shift of the circuit. For example, the collector depletion-layer capacitance of a bipolar transistor or the gate capacitance of a MOSFET may affect the oscillation frequency to some degree. As operating point changes, temperature changes, or device replacement modify these parasitic elements, the frequency of oscillation changes.

The key to good frequency stability is to make the phase shift a strong function of frequency near the frequency of zero-degree phase shift. If this is the case, a change in value of a capacitance will lead to a phase shift, $\theta$, resulting in a shift of frequency to restore the return ratio to zero-degree phase shift. If $|d\theta/d\omega|$ is large, only a slight change in $\omega$ is required to correct the phase shift.

Let us now return to the idealized oscillator of Figure 6.2. The phase shift of $AF$ can be expressed as

$$\theta = 90° - \tan^{-1} \frac{\omega L(R + R_1)}{(1 - \omega^2 CL)RR_1}$$

Since $R_1$ is much larger than $R$, we can write

$$\theta = 90° - \tan^{-1} \frac{1}{Q(1 - \omega^2/\omega_o^2)}$$

where $Q = R/\omega L$ and $\omega_o^2 = 1/LC$. Differentiating this phase angle with respect to radian frequency, assuming $Q$ is approximately constant near resonance, gives

$$\frac{d\theta}{d\omega} = \frac{-2\omega Q/\omega_o^2}{1 + Q^2\left(1 - \frac{\omega^2}{\omega_o^2}\right)^2}$$

At the resonant frequency $\omega = \omega_o$, and this derivative becomes

$$\frac{d\theta}{d\omega} = -\frac{2Q}{\omega_o} \quad \text{at} \quad \omega = \omega_o$$

The phase shift will change rapidly with frequency if the resonant circuit has a large value of $Q$. We conclude that the frequency stability is better for high-$Q$ circuits because the variation of phase shift with frequency near resonance is greater than that for a low-$Q$ circuit.

For extremely good frequency stability, crystal-controlled oscillators are used. Very high values of $Q$ apply to the equivalent electrical circuit of the crystal leading to correspondingly high values of $d\theta/d\omega$ for these oscillators. A sometimes serious problem of the quartz crystal is the variation of electrical parameters with temperature. If temperature stability is important, the crystal can be placed in a heat oven that maintains a constant crystal temperature.

## 6.2 PRACTICAL OSCILLATOR CONFIGURATIONS

**Important Concepts**

**1.** Several circuit configurations can be used to create sinusoidal oscillators. The $RC$ oscillator, the Wien bridge oscillator, and the tuned-circuit oscillator are examples of popular circuits.

**2.** Crystal-controlled oscillators can be designed to have very good frequency stability due to the high $Q$ of the crystal.

Many configurations are used in oscillator design. This section will cover a few of the popular configurations available at the present time.

### 6.2.1 The RC Oscillator

A phase shift or $RC$ oscillator is shown in Figure 6.4. The circuit can be constructed without using an emitter follower, but there will be a much narrower range of element values that can satisfy the loop-gain requirements. A MOSFET version can also be used and need not follow the gain stage with a source follower due to a less severe loading problem.

Some approximations can be made in the circuit of Figure 6.4 which allow several simplifications without involving a significant loss of accuracy. For example, the load of stage 1 is $R_L$ in parallel with the input impedance of the emitter follower, but since the input impedance to the emitter follower is very high, this is approximately $R_L$. This simplifies the calculation of voltage gain of the first stage. Another simplification results from assuming that the voltage gain of the emitter follower is unity and its output impedance is zero.

If only two $RC$ stages were used, the phase shift around the loop could not equal 0°. The amplifier stage shifts the phase 180° so that each RC stage would then be required to shift the phase by 90° . Since a single $RC$ stage cannot pass a finite signal with a 90° phase shift, three stages are required as a minimum to produce oscillations.

In order to simplify the design, the capacitors are chosen to be equal, and both $R_1$ and $R_2$ are chosen to equal the resistance $R$, which is defined as

$$R = (R_B \mid\mid R_3) \mid\mid (R_F + r_x + r_\pi) \tag{6.2}$$

This definition allows all three resistors of the $RC$ network to be equal, resulting in a simpler analysis. The equivalent circuit then appears as shown in Figure 6.5.

To analyze the circuit, the forward gain from node $a$ to the collector of the common-emitter stage, $A = v_{C1}/v_a$, is first found. Since the emitter follower has unity gain, this value also equals the gain from node $a$ to the output

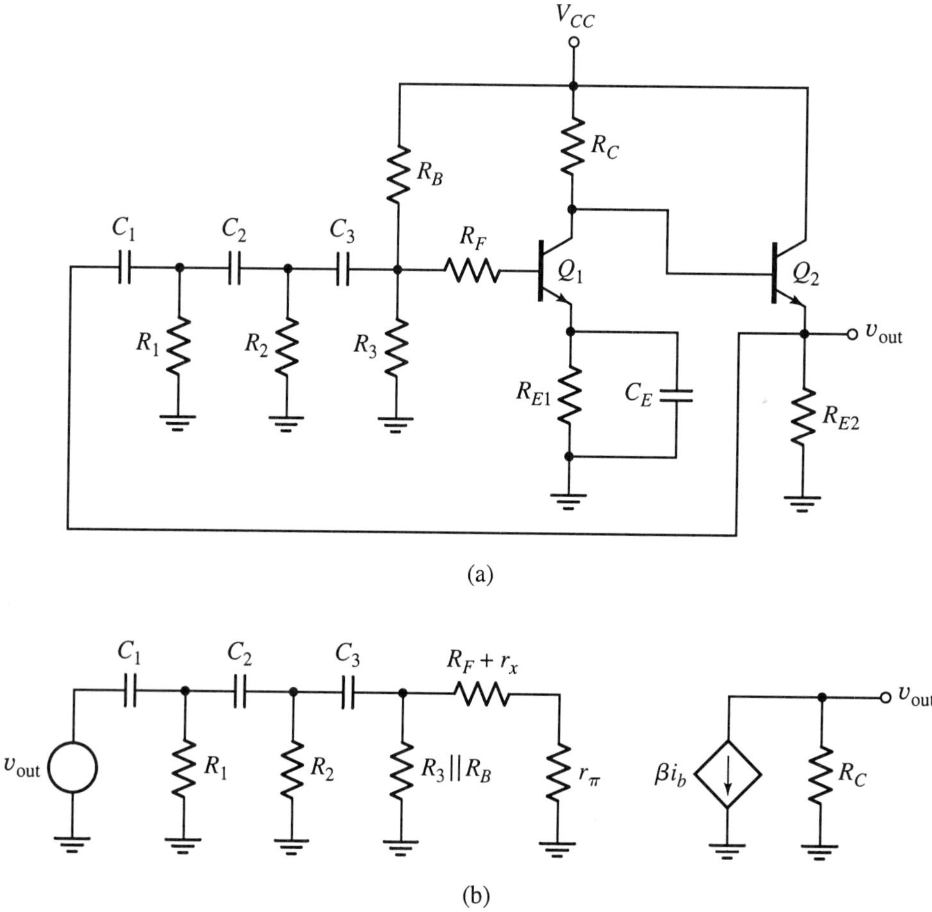


Figure 6.4: *RC* oscillator. (a) Actual circuit. (b) AC equivalent circuit.

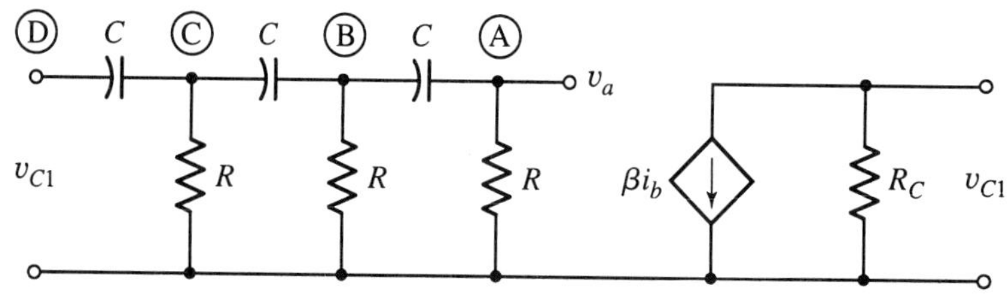


Figure 6.5: Equivalent circuit of simplified *RC* oscillator.

of the emitter follower. The feedback factor $F$ is a measure of the fraction of the emitter-follower output voltage that reaches node $a$. This value is the transfer function, $F = v_a/v_{C1}$, around the feedback loop.

The gain is

$$A = \frac{v_{C1}}{v_a} = \frac{-\beta R_L}{R_F + r_x + r_\pi} \tag{6.3}$$

The feedback transfer function is found by assuming that a voltage $v_{C1}$ is applied to the feedback loop (node $d$). The voltage $v_a$ is then found by solving the appropriate loop equations. The result of this step is

$$F = \frac{v_a}{v_{C1}} = \frac{1}{1 - \frac{5}{\omega^2 C^2 R^2} + j\frac{1}{\omega^3 C^3 R^3} - j\frac{6}{\omega CR}} \tag{6.4}$$

The product $AF$ is then

$$AF = \frac{\frac{\beta R_L}{R_F + r_x + r_\pi}}{-1 + \frac{5}{\omega^2 C^2 R^2} - j\frac{1}{\omega^3 C^3 R^3} + j\frac{6}{\omega CR}} \tag{6.5}$$

In order for $AF$ to have zero phase shift, the sum of imaginary terms in the denominator must equal zero at the frequency of oscillation. Hence

$$\frac{1}{\omega_o^3 C^3 R^3} = \frac{6}{\omega_o CR}$$

giving

$$\omega_o = \frac{1}{\sqrt{6}\, CR} \tag{6.6}$$

Thus the zero phase-shift condition has been imposed on $AF$. The magnitude condition must now be satisfied; that is,

$$|AF| = 1 = \frac{\beta R_L/(R_F + r_x + r_\pi)}{5/\omega_o^2 C^2 R^2 - 1} = \frac{\beta R_L}{29(R_F + r_x + r_\pi)} \tag{6.7}$$

At the oscillation frequency, $F$ equals $-1/29$. We then select $R_L$ and $(R_F + r_x + r_\pi)$ to satisfy Eq. (6.7). However, we must impose the previous constraint on the term $(R_F + r_x + r_\pi)$ as a result of a simplifying step. In Eq. (6.2), for convenience, we set

$$R = (R_B \,||\, R_3) \,||\, (R_F + r_x + r_\pi)$$

Since our analysis was based on this assumption, we require that $(R_F + r_x + r_\pi)$ be greater than $R$ to satisfy Eq. (6.2) The three conditions associated with the $RC$ oscillator (including one condition imposed to simplify the analysis) are

$$\omega_o = \frac{1}{\sqrt{6}\, CR}$$

$$|AF| = 1 = \frac{\beta R_L/(R_F + r_x + r_\pi)}{5/\omega_o^2 C^2 R^2 - 1} = \frac{\beta R_L}{29(R_F + r_x + r_\pi)}$$

$$R = (R_B \,||\, R_3) \,||\, (R_F + r_x + r_\pi)$$

An example will demonstrate the application of these equations.

## Example 6.3

Assume that the transistors in Figure 6.4(a) have current gains of $\beta = 100$, $r_x = 200\ \Omega$, and $Q_1$ is to be biased with an emitter current of 2 mA, giving $r_\pi = 1313\ \Omega$. Select $R$, $C$, $R_C$, $R_3$, $R_B$, $R_{E1}$, $C_E$, $R_{E2}$, and $R_F$ to give an oscillation frequency of $f_o = 100$ kHz. Use a power supply of $V_{CC} = 12$ V.

**Solution** Many combinations of values will satisfy this design. One starting point is to select $R$ as a reasonable value, say 2 k$\Omega$. Then $C$ is calculated from Eq.(6.6) as

$$C = \frac{1}{2\pi \times 10^5 \times \sqrt{6} \times 2 \times 10^3} = 325 \text{ pF}$$

We then choose $R_{E1}$ to be 0.5 k$\Omega$ and $C_E$ to be 1 $\mu$F to ensure complete bypassing of $R_{E1}$ at the frequency of oscillation. If we choose $R_3$ to be 10 k$\Omega$, the calculation of $R_B$ can be based on the bias requirement that $I_E = 2$ mA. The appropriate equations are (neglecting the small drop across $R_F$):

$$\frac{V_B}{10} + I_B = \frac{V_{CC} - V_B}{R_B}$$

and

$$V_B = I_E R_{E1} + V_{BE} = 1.0 + 0.7 = 1.7 \text{ V}$$

Solving these equations gives a value of 54 k$\Omega$ for $R_B$.

We can select $R_{E2}$ as 5 k$\Omega$ in order to obtain a high input impedance to $Q_2$. Next $R_F$ is selected from Eq. (6.2). This value is found to be $R_F = 1.1$ k$\Omega$. The minimum value of load resistance is found from Eq. (6.7) to be

$$R_C = \frac{29(R_F + r_x + r_\pi)}{\beta} = \frac{29 \times 2613}{100} = 758\ \Omega$$

This value would be chosen slightly larger to cause the open-loop value of $|AF|$ to be larger than unity.

---

The amplitude of the $RC$ oscillator can be stabilized by deriving a signal that is proportional to the output amplitude and using this signal to control the gain. The arrangement shown in Figure 6.6 would provide this stabilization.

The positive portion of the output waveform is amplified, inverted, and rectified by the combination of $Q_3$ and $D_1$. The capacitor $C_F$ charges to the peak value of this amplified signal and is selected so that it will hold this charge for several cycles of the output waveform. The dc signal on the capacitor is proportional to the amplitude of the output signal. This voltage is applied to the bias network in such a way as to lower the base current when the signal on $C_F$ becomes larger. Hence, $r_\pi$ will increase and cause the voltage gain of $Q_1$ to decrease. This circuit can be analyzed by the method used in Example 6.2.

Figure 6.7 shows an $RC$ oscillator using an n-channel MOSFET as the amplifying element. If $r_{ds} \parallel R_D$ is much smaller than $R$, Eq. (6.6) gives the oscillation frequency. The voltage gain from gate to drain must again equal $-29$; thus,

$$g_m \times (r_{ds} \parallel R_D) = 29$$

where $g_m$ is the transconductance of the MOSFET. The bias resistors must establish a proper value of $V_{GSQ}$ and satisfy the relation

$$R_1 \parallel R_2 = R$$

The transconductance is a function of the bias voltage, $V_{DSQ}$. Therefore, care must be taken to establish the proper bias when designing this circuit.

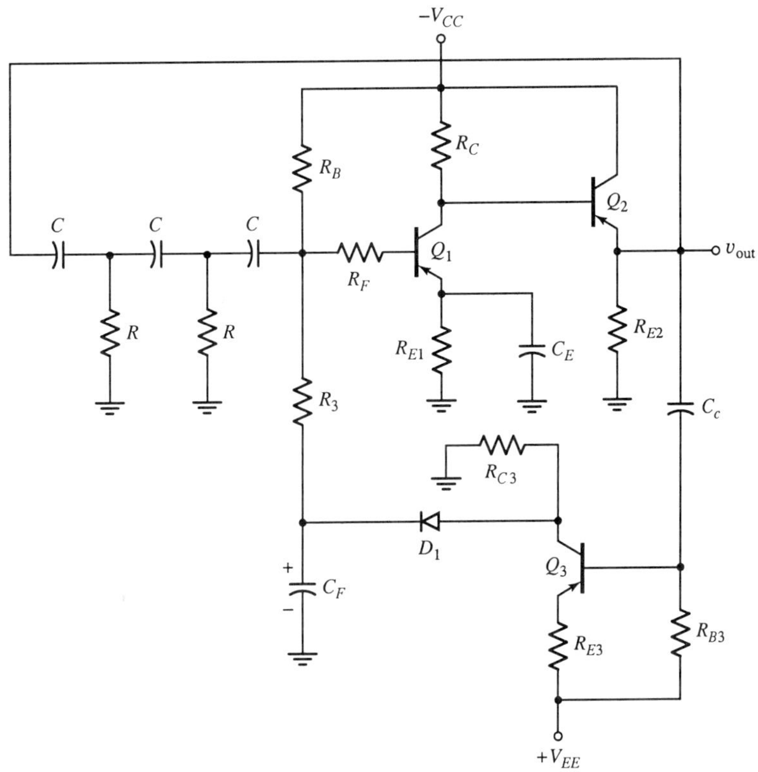


Figure 6.6: An amplitude-stabilized $RC$ oscillator.

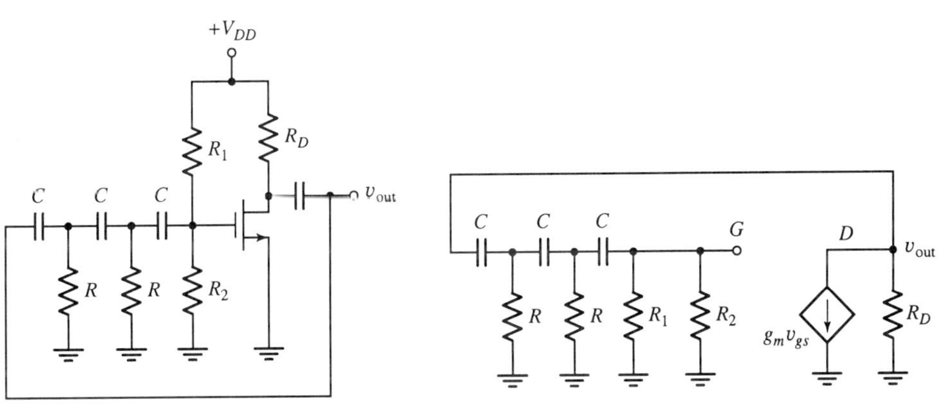


(a) (b)

Figure 6.7: (a) $RC$ oscillator using a MOSFET. (b) Equivalent circuit.

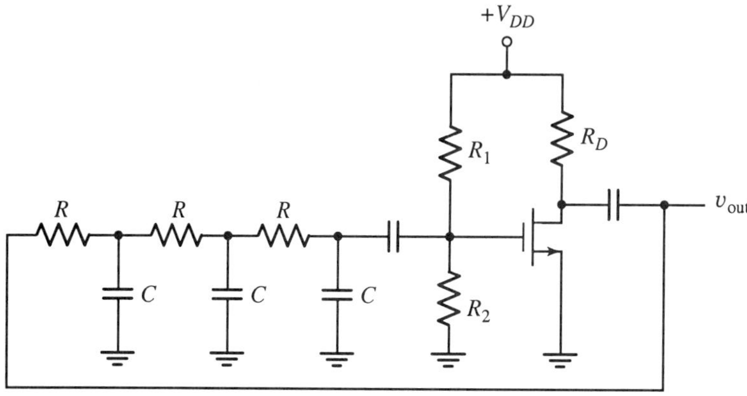


Figure 6.8: Noise-insensitive $RC$ oscillator.

The range of element values generally used with the phase-shift oscillator dictates relatively low-frequency applications up to a few hundred kilohertz. The frequency can be varied without changing the amplitude of oscillations by tuning all three capacitors simultaneously. When this type of tuning is used, the resonant value of $F$ remains at $1/29$.

An alternate version of the $RC$ oscillator is shown in Figure 6.8.

This version typically requires a few more element values, but it is less sensitive to noise transients than the oscillators of Figures 6.4 and 6.7. This is because the $RC$ network forms a lowpass circuit in Figure 6.8, whereas a highpass circuit results from the configurations of Figs. 6.4 and 6.7. Noise signals, which are typically higher in frequency, are filtered out of the loop gain when a lowpass filter is included.

**Practice Problems**

**6.5** If $C$ is selected to be 250 pF in Example 6.3, calculate the required value of $R$. *Ans:* 2.6 kΩ.

**6.6** In Example 6.3, the bias is changed to 1.5 mA and $R_F$ is changed to 1.2 kΩ. Calculate the required value of $R_C$ to achieve sinusoidal oscillation. *Ans:* 914 Ω.

### 6.2.2 Active Phase-Shift Oscillator

The $RC$ oscillator is sometimes called a phase-shift oscillator since the $RC$ network causes 180° of phase shift at the oscillation frequency. It is possible to use active phase shifters to achieve the required phase relation for the return ratio. Consider the circuits of Figure 6.9.

It can be shown that the transfer function of the op amp circuit is

$$A = \frac{A_{oa}}{A_{oa}+2}\,\frac{-j\omega CR+1}{j\omega CR+1} \approx \frac{-j\omega CR+1}{j\omega CR+1} \tag{6.8}$$

This transfer function has a unity magnitude at all frequencies while the phase shift is

$$\phi = -2\,\tan^{-1}\omega CR \tag{6.9}$$

A similar transfer function results from the MOSFET circuit, provided that $R_1 << R$ and $g_m r_{ds} >> 2$. These circuits are called allpass networks since all signals pass through the circuits with no magnitude change, but have

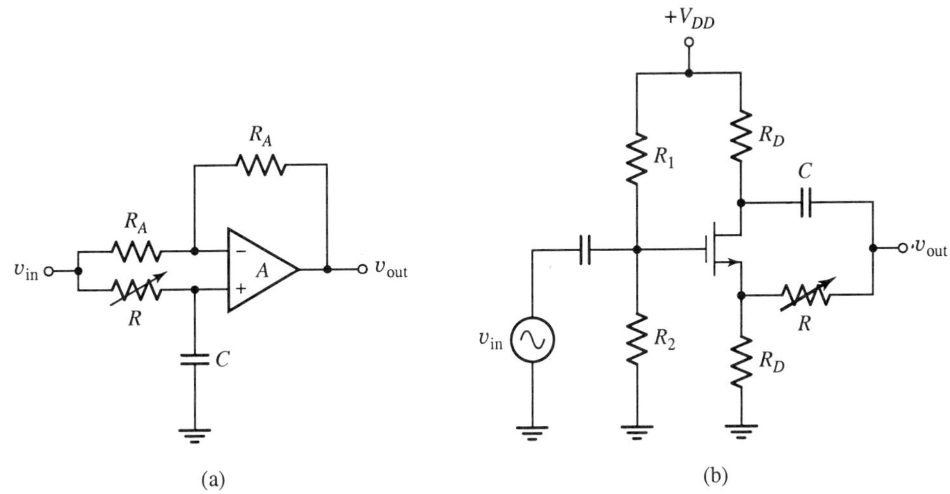


Figure 6.9: Active phase shifters. (a) Op amp circuit; (b) MOSFET circuit.

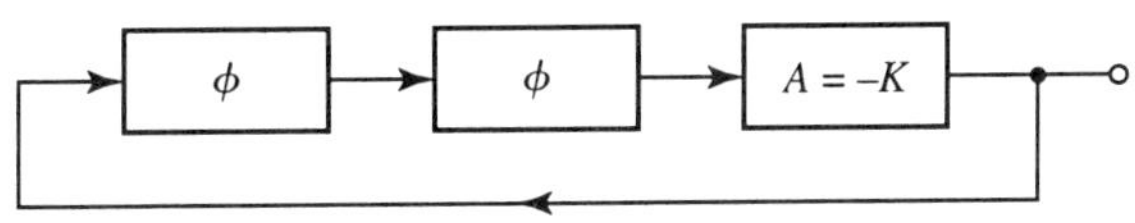


Figure 6.10: Active phase-shift oscillator.

a change in phase. They are also called active phase shifters. Figure 6.10 shows a block diagram of an active phase-shift oscillator.

With two active phase shifters and an amplifier of gain $-K$ having high input and low output impedances, the return ratio is

$$AF = -K\left(\frac{1 - j\omega CR}{1 + j\omega CR}\right)^2 \tag{6.10}$$

Rationalizing the denominator of this function gives

$$AF = -K\frac{(1 - j\omega CR)^4}{(1 + \omega^2 C^2 R^2)^2} \tag{6.11}$$

The zero-degree phase-shift condition on $AF$ will occur when $\omega CR = 1$ or at a frequency of

$$\omega_o = \frac{1}{CR} \tag{6.12}$$

The inverting amplifier should have a unity gain, or $K = 1$, to satisfy the magnitude condition for oscillations. This circuit can be tuned by varying both of the phase-shift resistors. Although tuning can be accomplished by varying only one resistor, Eqs. (6.11) and (6.12) would not apply. Note that variation of neither $R$ nor $C$ affects the magnitude of $AF$; thus, the output signal remains constant as frequency changes.

When used with op amps, the frequency range of the active phase-shift oscillator is limited to the range of a few hundred kilohertz. If implemented with MOSFET or BJT phase shifters, the frequency of oscillation is extended into the hundred MHz area.

### 6.2.3 The Tuned-Circuit Oscillator

The BJT or MOSFET can be used as the active device in the tuned-circuit oscillator. These oscillators can operate in the class-A or class-C modes and can achieve very high-frequency operation. We will consider the

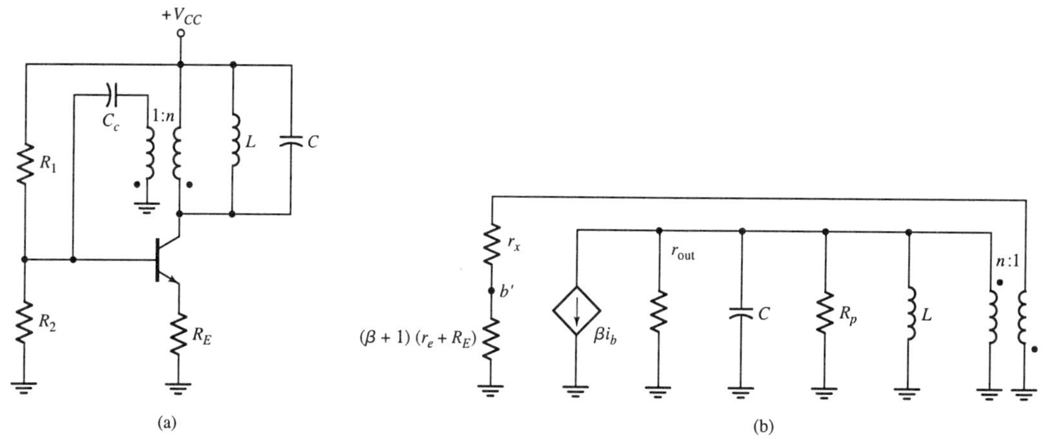


Figure 6.11: (a) Actual tuned-collector oscillator. (b) Equivalent circuit.

tuned-collector BJT oscillator in the class-A mode as shown in Figure 6.11.

The tuned-collector oscillator is analyzed quite simply when it is assumed that the transformer is ideal. The equivalent circuit for this stage is shown in Figure 6.11(b). In this circuit, $r_{\text{out}}$ is the output impedance of the transistor and $R_p$ is the parallel resistance that represents the losses of the tuned circuit. The turns ratio of the transformer is $n$, and the signal from primary to secondary must be inverted through the transformer. The equivalent circuit does not include $C_\mu$, but for high-frequency oscillators, this element should be included. The resonant frequency of the oscillator is

$$\omega_o = \frac{1}{\sqrt{LC}} \tag{6.13}$$

At this frequency, the collector load is real and consists of the shunt combination of $r_{\text{out}}$, $R_p$, and the impedance reflected through the transformer. This load is

$$R_{sh} = r_{\text{out}} \,||\, R_p \,||\, n^2(r_x + r_\pi + (\beta+1)R_E) \tag{6.14}$$

The voltage gain from base to collector is

$$A = \frac{-\beta R_{sh}}{r_x + r_\pi + (\beta+1)R_E} \tag{6.15}$$

and the voltage fed back from collector to base is

$$F = -\frac{1}{n} \tag{6.16}$$

The value of $AF$ is then

$$AF = \frac{\beta R_{sh}}{n(r_x + r_\pi + (\beta+1)R_E)} \tag{6.17}$$

Equating the return ratio at resonance to unity, the value of $n$ required for oscillations is

$$n = \frac{\beta R_{sh}}{r_x + r_\pi + (\beta+1)R_E} \tag{6.18}$$

The term $R_{sh}$ appears to be a function of $n$ from Eq. (6.14), but generally both $r_{\text{out}}$ and $n^2(r_x + r_\pi + (\beta+1)R_E)$ are much larger than $R_p$. This allows $R_{sh}$ to be approximated by $R_p$, and the return ratio can be expressed as

$$AF = \frac{\beta R_p}{n(r_x + r_\pi + (\beta+1)R_E)} \tag{6.19}$$

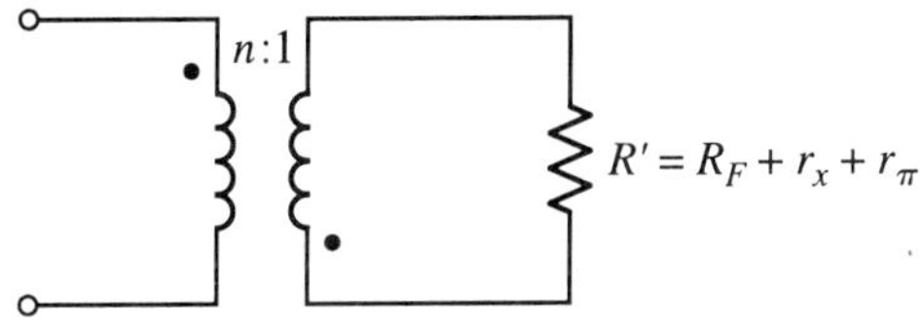


Figure 6.12: Transformer measurement.

The value of $n$ is then given by

$$n = \frac{\beta R_p}{(r_x + r_\pi + (\beta + 1)R_E)} \tag{6.20}$$

In order that the oscillator possess good frequency stability the effective $Q$ of the tank circuit must be quite high. Inclusion of $R_E$ in the emitter circuit also improves the frequency stability of the tuned-collector oscillator by leading to higher values of $r_{\text{out}}$ and large values of impedance reflected through the transformer. These higher impedances lead to a higher value of effective $Q$.

In practice, an ideal transformer will not be used. The inductance of the transformer can serve as the resonating inductance, eliminating the need for a separate inductor. The equation for oscillation frequency, Eq. (6.13), and Eqs. (6.19) and (6.20) can be put into terms of primary, secondary, and mutual inductances of the transformer. This is not a very practical method of design because these inductances must be measured. If the elements of the transformer must be measured, it is simpler to measure the input inductance of the transformer when the secondary is loaded in the same way that it will be loaded in the circuit. The circuit of Figure 6.12 demonstrates this. The input inductance and the $Q$ of the transformer can now easily be measured, and the previously developed theory can be applied directly.

**Practice Problems**

**6.7** In the tuned-collector circuit of Figure 6.11, $R_p$ = 24 kΩ, $r_x + r_\pi$ = 3.2 $k\Omega$, $R_E$ = 1.5 $k\Omega$, and $\beta = 124$. Calculate the required transformer turns ratio for sinusoidal oscillation. *Ans:* $n = 15.6$.

**6.8** Calculate the value of $R_{sh}$ for Practice Problem 6.7 by including the impedance reflected through the transformer to the output. *Ans:* $R$ = 24 kΩ.

### 6.2.4 The Wien Bridge Oscillator

The circuit of Figure 6.13 depicts the popular Wien bridge oscillator.

This oscillator can operate without the resistors $R_1$ and $R_2$, but these legs of the bridge form the basis for the amplitude-stabilized circuit to be considered shortly and will therefore be included in the present discussion. The $RC$ network exhibits 0° of phase shift at one specific frequency and, if return ratio magnitude conditions are satisfied, the circuit will oscillate at this frequency.

This circuit is slightly different from previous oscillator circuits discussed because there are two feedback paths rather than one. The path to the noninverting terminal causes positive feedback at the oscillation frequency, but the gain around this loop is less than unity. The feedback to the inverting terminal represents negative feedback.

The return ratio or gain around the loop can be found from the gain expression

$$v_{\text{out}} = A(v_a - v_b)$$

and the two feedback equations

$$v_a = \frac{Z_2}{Z_1 + Z_2} v_{\text{out}}$$

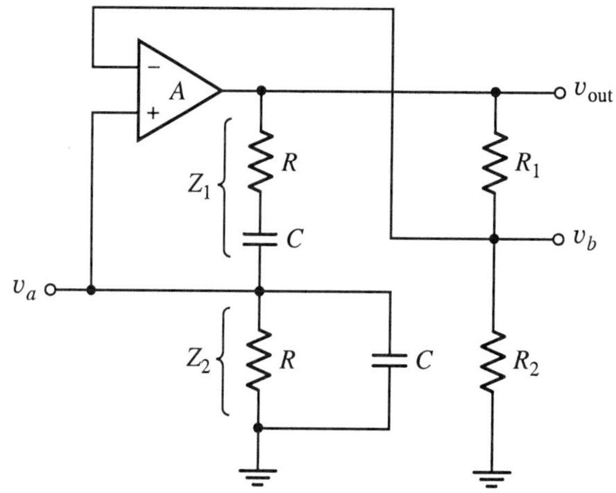


Figure 6.13: Wien bridge oscillator.

and

$$v_b = \frac{R_2}{R_1 + R_2} v_{out}$$

Combining these equations gives a loop-gain value of

$$AF = A\left[\frac{j\omega CR}{1 - \omega^2 C^2 R^2 + 3j\omega CR} - \frac{R_2}{R_1 + R_2}\right] \tag{6.21}$$

The feedback factor is the bracketed term in Eq. (6.21). This return ratio will exhibit zero-degree phase shift only at the frequency

$$\omega_o = \frac{1}{CR} \tag{6.22}$$

This is the only possible oscillation frequency for the oscillator. The magnitude of $AF$ at this frequency becomes

$$|AF| = A\left[\frac{1}{3} - \frac{R_2}{R_1 + R_2}\right] \tag{6.23}$$

It is obvious that $R_2/(R_1 + R_2)$ must be smaller than 1/3 or the phase shift will be 180° rather than 0°. Since $A$ is a very large number, $R_2/(R_1 + R_2)$ is selected to be only slightly smaller than 1/3. In fact, the theoretical value of this ratio to cause unity magnitude of $AF$ is found from Eq. (6.23) to be

$$\frac{R_2}{R_1 + R_2} = \frac{1}{3} - \frac{1}{A} \tag{6.24}$$

A small change in the ratio $R_2/(R_1 + R_2)$ can lead to a large change in $|AF|$ at the oscillation frequency. This could lead to serious distortion problems unless $A$ changes very sharply with dc distortion.

Tuning can be accomplished by varying the two capacitors or resistors simultaneously. Commercial oscillators often use capacitor tuning for continuous frequency variation over a given range and resistive tuning to switch to different frequency ranges. With an op amp as the amplifying element, the frequency range of the Wien bridge circuit is again limited to the hundred kilohertz range. Transistor or MOSFET oscillators may be designed to reach the megahertz frequency range.

A depletion FET can be used as a variable resistance to stabilize the amplitude of oscillations, as shown in Figure 6.14. The variable resistance of the FET is controlled by the gate bias voltage. A potentiometer in the comparator circuit establishes the nominal amplitude of oscillations along with the FET gate voltage. If parameter changes now occur causing an increase in amplitude, the peak detection and comparator circuit generates a

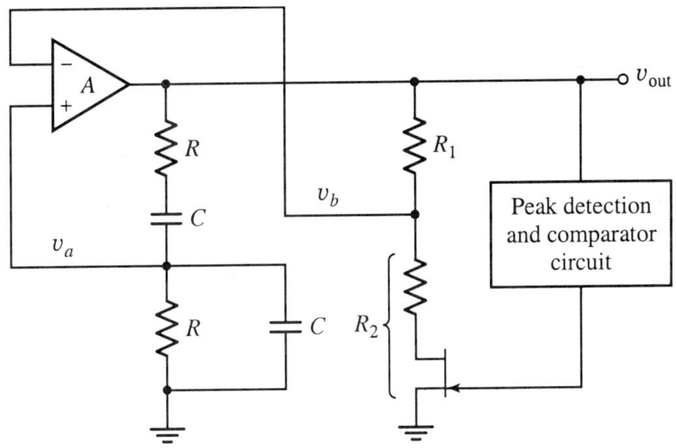


Figure 6.14: Amplitude-stabilized Wien bridge oscillator.

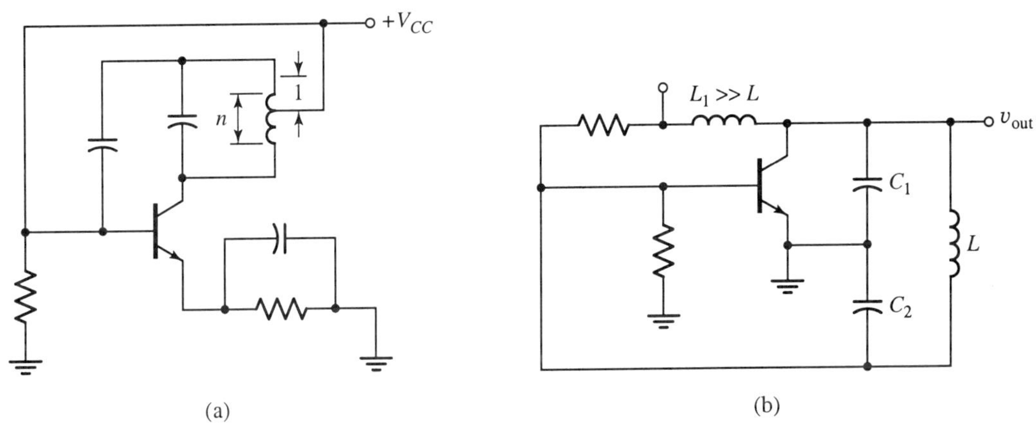


Figure 6.15: (a) A Hartley oscillator. (b) A Colpitts oscillator.

more negative gate voltage which increases the total resistance $R_2$. This causes a larger voltage $v_b$, which tends to offset the increase in $v_{out}$.

**Practice Problem**

**6.9** In the circuit of Figure 6.13, $A = 100,000$ V/V, $R = 12$ kΩ, and $R_2 = 10$ kΩ. Select $R_1$ and $C$ to cause sinusoidal oscillation at 30 kHz. *Ans:* $R_1 = 20$ kΩ, $C = 442$ pF.

### 6.2.5 Hartley and Colpitts Oscillators

The Hartley oscillator is very similar to the tuned-collector oscillator, replacing the transformer with a tapped coil, as shown in Figure 6.15(a). This arrangement does not require mutual coupling between inductors. Two separate inductors can be used in this circuit.

The Colpitts oscillator uses only one inductor and two capacitors to obtain the required phase shift through the feedback loop. Figure 6.15(b) shows one version of a Colpitts oscillator. Both of these oscillators can operate at very high frequencies if high-frequency transistors are used as the amplifying element. The circuits will not be analyzed here, although a Colpitts crystal oscillator will be analyzed in the next section.

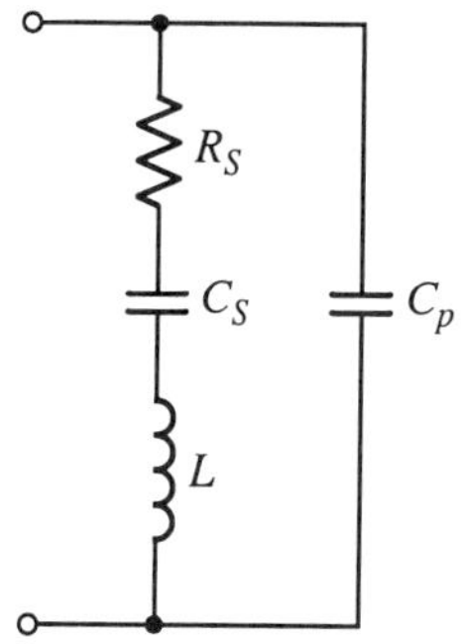


Figure 6.16: Equivalent circuit for the crystal.

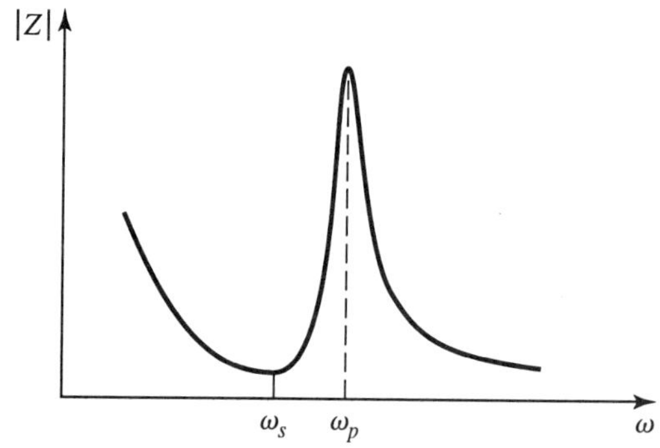


Figure 6.17: Magnitude of impedance vs. frequency for the crystal.

Both feedback networks of these two oscillators must cause a phase shift of 180° so that the angle of $AF$ is 0°.

### 6.2.6 Crystal-Controlled Oscillators

It has been mentioned previously that a high-$Q$ circuit leads to good frequency stability. A piezoelectric quartz crystal represents an element that behaves as a high-$Q$ resonant circuit. The equivalent circuit of the quartz crystal appears in Figure 6.16.

For typical crystals, the $Q$ might fall somewhere in the range of 20,000 to 100,000. The parallel capacitance $C_p$ is much larger than the series capacitance $C_s$. For a 500 kHz crystal, typical values might be $C_p = 4$ pF and $C_s = 8$ fF. The inductance for this crystal could be 12.7 H while the series resistance is 2 kΩ with a $Q$ of 20,000.

The magnitude of the impedance of the crystal is plotted in Figure 6.17 with an exaggerated gap between series and parallel resonance. These frequencies are generally separated by less than 1%.

The equation for series resonance is

$$\omega_s = \frac{1}{\sqrt{LC_s}} \tag{6.25}$$

The equation for parallel resonance can be found by equating the positive reactance of the series branch to the negative reactance of the parallel capacitance. This gives

$$\omega_p L - \frac{1}{\omega_p C_s} = \frac{1}{\omega_p C_p}$$

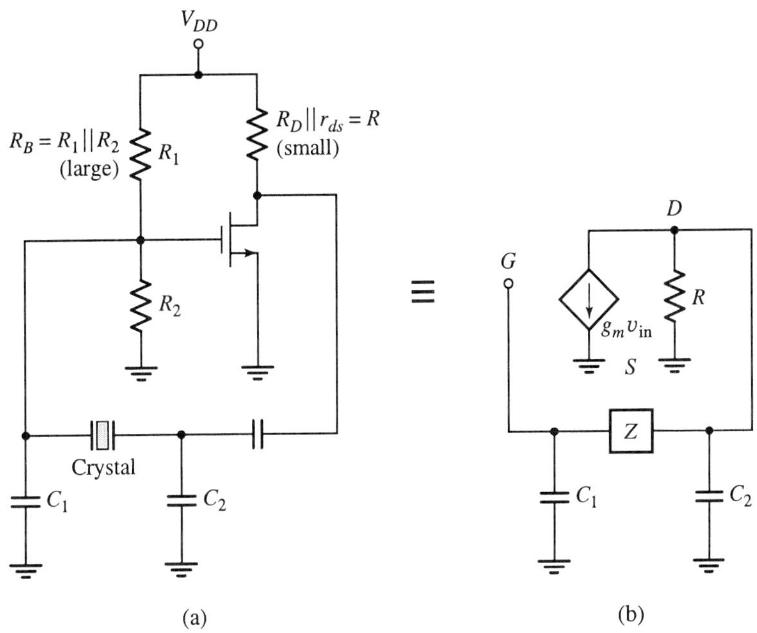


Figure 6.18: (a) A crystal-controlled Colpitts oscillator. (b) Equivalent circuit.

Solving for $\omega_p$ results in

$$\omega_p = \frac{1}{\sqrt{LC_sC_p/(C_s + C_p)}} \tag{6.26}$$

Since $C_s$ is hundreds of times smaller than $C_p$, this equation leads to an $\omega_p$ that is very slightly higher than $\omega_s$.

Between the two resonant frequencies, the impedance of the crystal appears to be inductive and changes value from zero reactance at $f_s$ to several hundred $k\Omega$ near $f_p$. This behavior is similar to a very high-$Q$ tank circuit operating just slightly below the resonant frequency. The crystal can then be used as an inductor in a conventional oscillator circuit and will operate somewhere between $f_s$ and $f_p$ because this is the range over which the crystal looks inductive.

The impedance of the crystal over its inductive range can be found to be approximately

$$Z = \frac{j(\omega^2 - \omega_s^2)}{\omega C_p(\omega_p^2 - \omega^2)} \tag{6.27}$$

Note that this impedance has a zero at $\omega_s$ and a pole at $\omega_p$.

Figure 6.18 shows a Colpitts crystal oscillator using a MOSFET as the amplifying element. The crystal replaces the inductor in the standard Colpitts circuit. The bias resistors, $R_1$ and $R_2$, can be selected to be quite large. It can often be assumed that the parallel resistance of the bias resistors is much larger than the reactance of the capacitor $C_1$. This allows loading effects of the input resistance of the amplifier to be neglected.

For convenience of calculation, we will assume that the load resistance in parallel with the output impedance of the MOSFET is given by $R$, that is,

$$R = R_D \,||\, r_{ds}$$

We will further assume that the output capacitance, $C_{gd} + C_{db}$ (gate-to-drain capacitance plus drain-to-bulk capacitance) can be lumped into the capacitance $C_2$ whereas the value of $C_1$ accounts for the external capacitance added plus the gate to source capacitance $C_{gs}$.

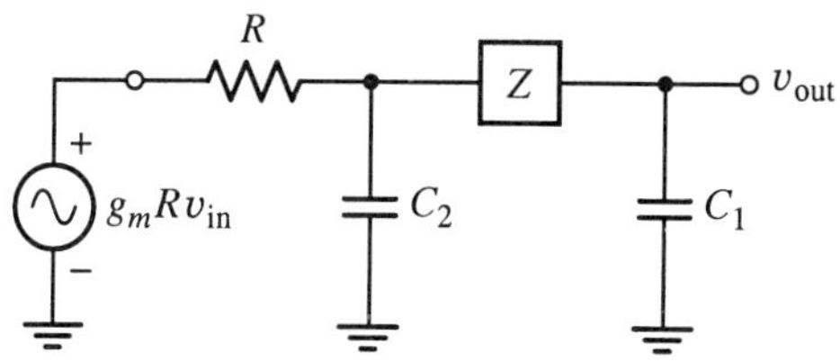


Figure 6.19: An equivalent circuit of the crystal oscillator.

With these assumptions, and after taking a Thévenin equivalent circuit of the drain-to-source elements, the circuit reduces to that of Figure 6.19.

The return ratio can be evaluated from the two-loop circuit of Figure 2.19 to be

$$\frac{v_{\text{out}}}{v_{\text{in}}} = \frac{-g_m R}{j\omega R(C_1 + C_2) - j\omega^2 C_1 C_2 |Z| R - j\omega C_1 |Z| + 1} \tag{6.28}$$

The first two terms in the denominator are imaginary, since $Z$ is imaginary, while the third and fourth terms are real. In order that the return ration have a phase shift of 0°, the imaginary part of the denominator must go to zero, giving possible oscillations when

$$\omega R(C_1 + C_2) = \omega^2 C_1 C_2 |Z| R \tag{6.29}$$

If we set the oscillation frequency to the geometric mean of $\omega_s$ and $\omega_p$ for convenience, we can show from Eq. (6.29) that this requires

$$\frac{C_1 C_2}{C_1 + C_2} = C_p \tag{6.30}$$

Of course, the circuit will oscillate only if the magnitude of the return ratio is unity at the oscillation frequency. Returning to Eq. (6.28) using the condition of Eq. (6.30) results in the requirement that

$$\frac{C_1}{C_2} = g_m R \tag{6.31}$$

With the conditions of Eqs. (6.30) and (6.31) satisfied, the crystal-controlled Colpitts circuit will oscillate somewhere between $f_s$ and $f_p$. Since these two frequencies are less than 1% different, the oscillation frequency is rather accurately established.

The popular Pierce oscillator is a crystal-controlled Colpitts oscillator with no external capacitances added. The parasitic capacitances determine the values of $C_1$ and $C_2$ while the resistive load is adjusted to achieve the condition of Eq. (6.31).

# Example 6.4.

Assume that a crystal is available with the following parameters: $R_s = 2$ kΩ, $L = 12.7$ H, $C_s = 8$ fF, and $C_p = 4$ pF. The MOSFET used is biased to a point where $g_m = 3$ mA/V and $R$ is selected to be 10 kΩ. Select the capacitors $C_1$ and $C_2$ to cause sinusoidal oscillations near the crystal frequency.

**Solution** It is an easy matter to calculate the parallel resonant frequency of the crystal as 499.8 kHz. In order to cause oscillations at approximately this frequency, we must satisfy Eqs. (6.30) and (6.31). This leads to

$$\frac{C_1 C_2}{C_1 + C_2} = C_p = 4 \text{ pF}$$

and

$$g_m R = 30 = \frac{C_1}{C_2}$$

Solving these two equations simultaneously gives $C_1 = 4.133$ pF and $C_2 = 124$ pF.

### 6.2.7 General Comments on Oscillators

The oscillator circuits considered previously have been analyzed for low and midband frequencies. At higher frequencies, the wiring and parasitic capacitances of the device must be included in the analysis. This complicates the design procedure, but does not change the basic steps. Equations for gain and feedback factor or for loop gain are written to include these additional effects and the same conditions for oscillation apply. In both the tuned-circuit and the Hartley oscillators, neutralization of the collector-base feedback path can be done to remove some high-frequency problems. The techniques of Section 5.4.2 can be applied directly to these high-frequency stages.

## SUMMARY

**1.** A sinusoidal oscillator is created by applying positive feedback at the desired oscillation frequency with a return ratio of unity.

**2.** Amplitude-stabilizing circuits can be used to maintain a constant output signal as element values drift.

**3.** Several configurations have been developed over the years to create sinusoidal oscillators.

**4.** The crystal-controlled oscillator can have very good frequency stability.

## PROBLEMS

### Section 6.1.1 General Design Considerations

**6.1*** The open-loop gain of an amplifier is

$$A = \frac{-1000}{1 + j\omega/10^7}$$

and the feedback factor is

$$F = \frac{10^{-2}}{(1 + j\omega/4 \times 10^7)^2}$$

**(a)** Will the circuit oscillate? If so, at what frequency will oscillation occur?
**(b)** If the low-frequency gain of the amplifier is changed to $-1250$, at what frequency will oscillation occur?

**6.2*** Repeat Problem 6.1 if the two corner frequencies of $F$ are changed to $8 \times 10^7$ rad/sec.

**D6.3** Assume that the amplifier of the circuit has infinite input and zero output impedances and each inductor has a $Q$ of 30. A sinusoidal oscillation is desired.

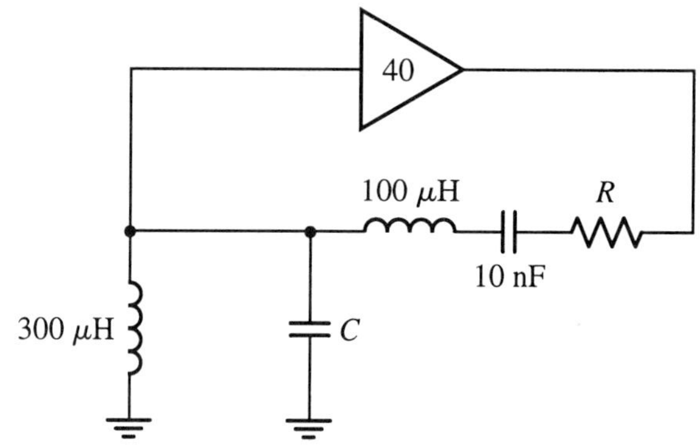


**Figure P6.3**

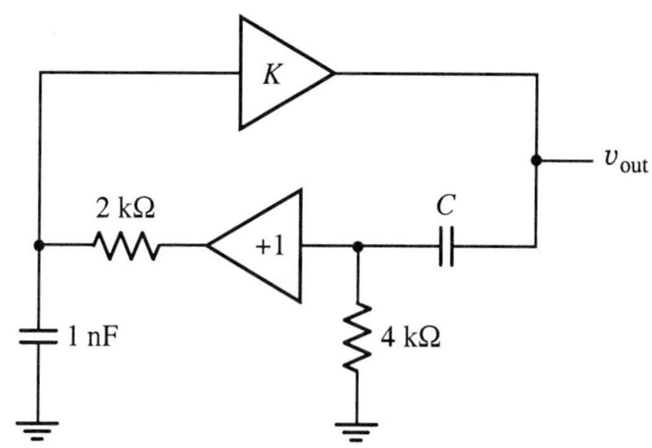


**Figure P6.4**

**(a)** Select $C$ to cause the circuit to oscillate at a single frequency.
**(b)** Select $R$ to cause the circuit to oscillate at a single frequency.
**(c)** At what frequency will oscillation occur?
**(d)** If the amplifier input impedance drops from infinity to 20 $k\Omega$, what value of $R$ should be used?

**6.4** Assume that the amplifier has infinite input and zero output impedances.

**(a)** Find the value of $C$ in the circuit to cause an oscillation frequency of $f_o = 100$ kHz.
**(b)** Find the minimum value of $K$ to result in sinusoidal oscillation.

**6.5** For the circuit shown

**(a)** Calculate the only possible frequency of sinusoidal oscillation in Hz.
**(b)** Calculate the required value of $A$ for sinusoidal oscillation.

**6.6** Assume infinite input and zero output impedances for both active stages in the circuit shown.

**(a)** Calculate the value of $C$ to lead to oscillation at $f = 10$ kHz.
**(b)** Find the minimum value of $A$ to cause sinusoidal oscillation at $f = 10$ kHz.

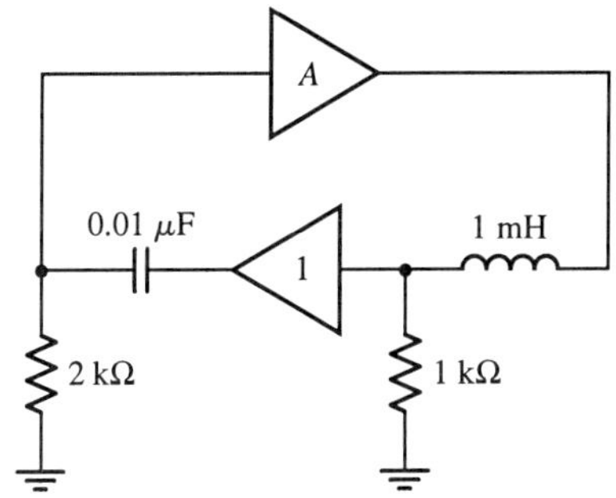


**Figure P6.5**

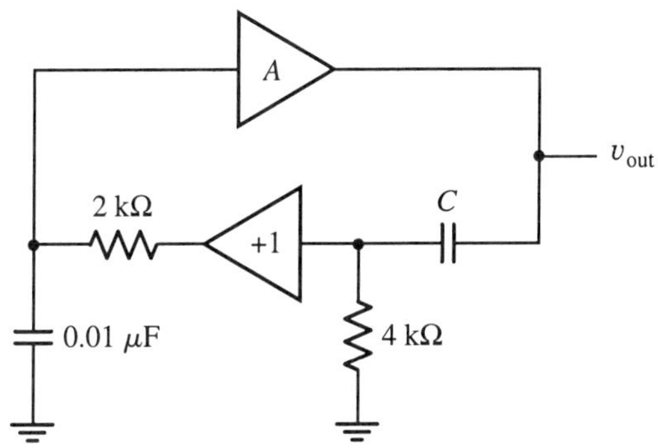


**Figure P6.6**

**6.7** In the circuit of the figure

**(a)** Should the gain of the amplifier be positive or negative to create the possibility of oscillation in the circuit?
**(b)** Explain your choice in part (a).
**(c)** What value should $|A|$ have to cause sinusoidal oscillation?
**(d)** Select $C$ to lead to an oscillation frequency of 8 kHz.

### Section 6.1.2 Amplitude Stability

**6.8** In the circuit of Figure 6.2, the gain is given by $A = 3000/r_e$. If $r_e = 26/I_E$ where $I_E$ is in mA and

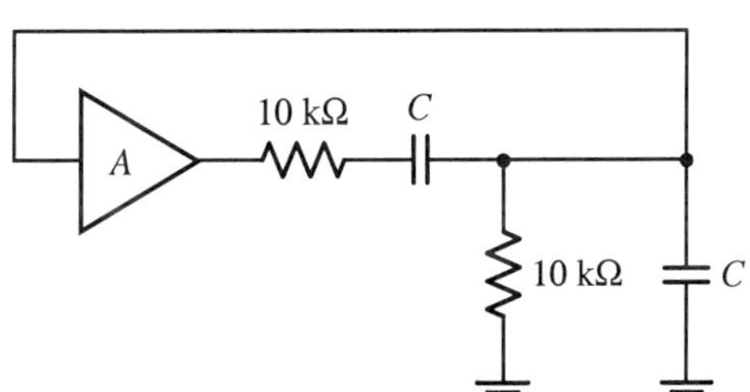


**Figure P6.7**

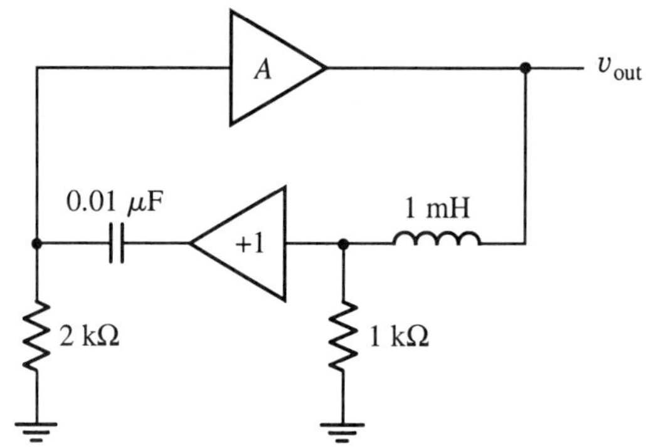


**Figure P6.10**

$I_E = 2 - 0.1|v_{out}|^2$, find the peak value of $v_{out}$.

**6.9*** The forward gain of an ideal amplifier is given by $A = -(8 - 6|v_{out}|)$ where $|v_{out}|$ is the peak value of the output signal. The feedback factor is

$$F = \frac{1}{4}\left(\frac{j\omega RC + 1}{j\omega RC - 1}\right)^2$$

If $R = 9.2$ kΩ and $C = 0.0012$ μF,

**(a)** at what frequency will the circuit oscillate?
**(b)** What is the peak value of $v_{out}$?

**6.10** Assume an ideal amplifier and buffer stage with infinite input and zero output impedances in the figure.

**(a)** Calculate the only possible oscillation frequency.
**(b)** Calculate the value of required gain, $A$, for oscillations to occur.
**(c)** If $A = 1.5 - 0.1|v_{out}|$, find the peak value of $v_{out}$ for the circuit.

**6.11** Assume ideal buffers and amplifier with infinite input and zero output impedances in the circuit shown.

**(a)** At what frequency can the circuit oscillate?
**(b)** What is the maximum value of $K$ that will lead to sinusoidal oscillation?
**(c)** If $K = 25 - 0.1|v_{out}|$, what is the peak value of the output voltage?

**6.12** Assume infinite input and zero output impedances for both amplifiers of the figure. The gain $A$ is given by

$$A = 0.6 - \frac{|v_{out}|}{40}$$

**(a)** Find the frequency of oscillation in Hz.
**(b)** Find the peak value of $v_{out}$.

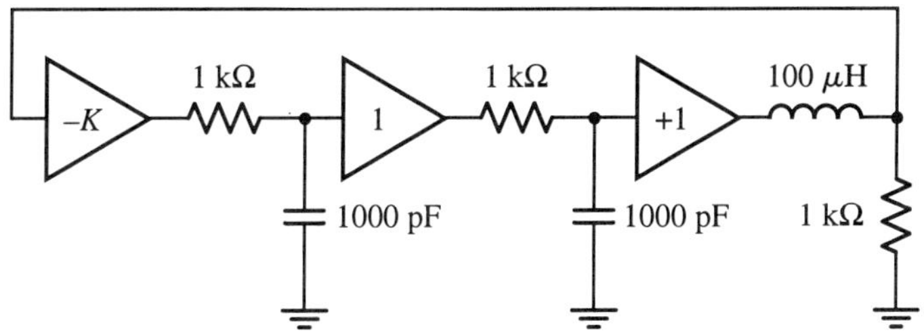


**Figure P6.11**

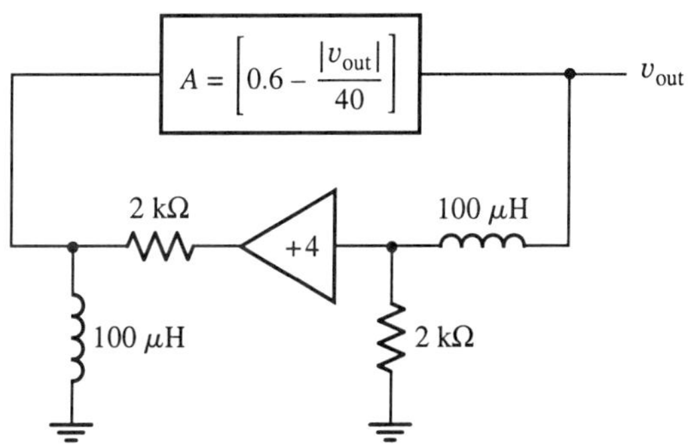


**Figure P6.12**

**D6.13** Assume that $\beta = 100$ and $r_x = 0$ for the BJT and that the amplifiers of unity gain both have infinite input and zero output impedances.

**(a)** Calculate $R$ to lead to a sinusoidal oscillation at $f = 50$ kHz.
**(b)** Find the minimum value of $R_C$ for oscillation.
**(c)** If emitter current is $I_E = (1.1 - 0.02|v_{out}|)\ mA$, how large is $|v_{out}|$?

### Section 6.1.3 Frequency Stability

**6.14** The circuit of Figure 6.2 oscillates at 10 kHz when $L = 0.1$ mH. If $L$ changes to 0.096 mH, find the new oscillation frequency.

### Section 6.2.1 The RC Oscillator

**6.15** The circuit shown oscillates at 10 kHz with a value of $R = 2$ kΩ. Assume all amplifier stages are ideal.

**(a)** Find the necessary value of $C$.

**(b)** Calculate the minimum value of $A$.

**6.16** Find the frequency of oscillation and the gain requirement for the $RC$ oscillator of Figure 6.4 if $C_3 = 2C_2 = 2C_1$ and $R_1 = R_2 = R_B \parallel R_3 \parallel (R_F + r_x + r_\pi)$.

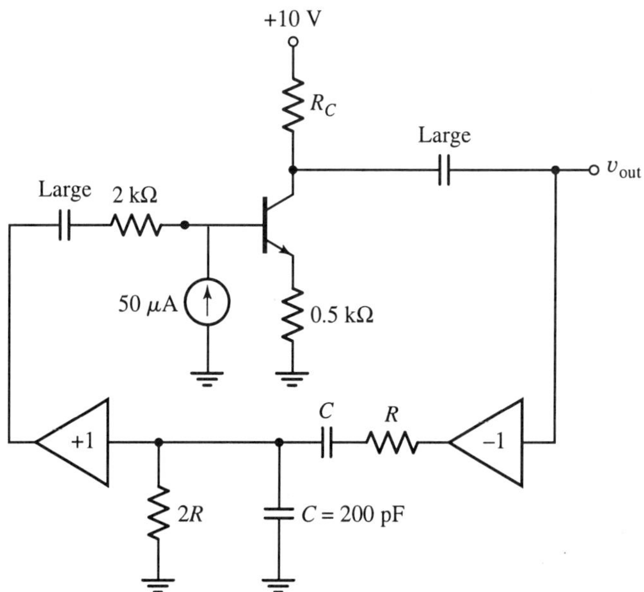


**Figure P6.13**

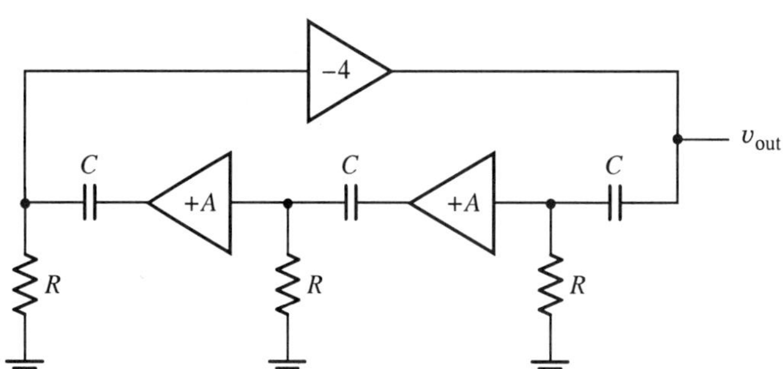


**Figure P6.15**

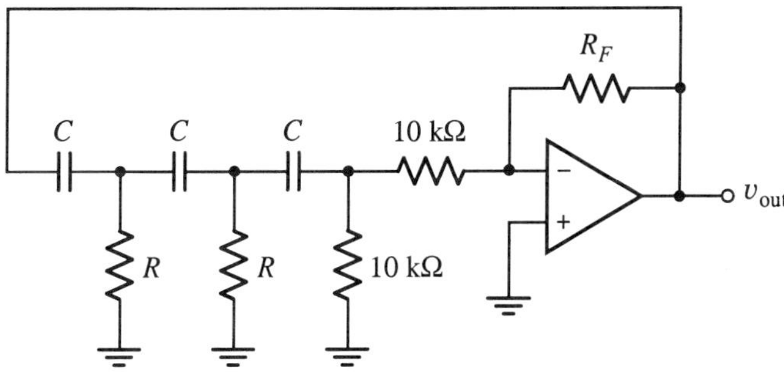


**Figure P6.19**

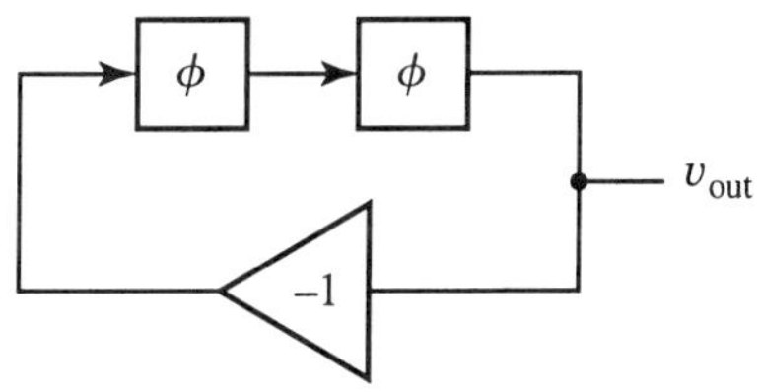


**Figure P6.20**

**D6.17** The MOSFET of Figure 6.7 has a $g_m$ of 4 mA/V and an $r_{ds}$ of 30 kΩ when biased at $V_{GSQ} = 2.6$ V. Select values for $R_1$, $R_2$, $R$, and $C$ to cause oscillation at a frequency of 24 kHz. Assume that $V_{DD} = 10$ V.

**6.18** In the circuit of Figure 6.8, assume that the input impedance of the amplifier is much larger than the reactance of the capacitance in the $RC$ network at the oscillation frequency.

**(a)** What is the possible oscillation frequency?
**(b)** What is the requirement on gain for the MOSFET to cause sinusoidal oscillations?

**D6.19** Select $R$, $C$, and $R_F$ to lead to sinusoidal oscillation at $f = 20$ kHz in the circuit.

**Section 6.2.2 Active Phase-Shift Oscillator**

**6.20** Both phase-shift circuits shown have a transfer function of

$$G = \frac{-j2\omega + 1/RC}{j2\omega + 1/RC}$$

If the circuit is to oscillate at 5 kHz, find suitable values of $C$ and $R$.

**D6.21** Design an active phase-shift oscillator using op amp stages with $A = 300,000$ V/V. The oscillation frequency should be adjustable from 4 kHz to 40 kHz using capacitive tuning. State all assumptions made and specify minimum and maximum capacitor values.

**D6.22** Repeat Problem 6.21 using MOSFET stages with $g_m = 6$ mA/V and $r_{ds} = 24$ kΩ.

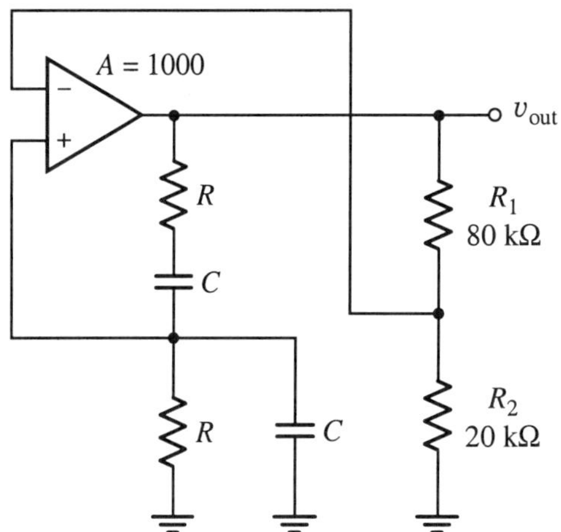


**Figure P6.27**

**Section 6.2.3 The Tuned-Circuit Oscillator**

**D6.23** Select appropriate values for $n$, $L$, $C$, $R_1$, $R_2$, $C_c$, and $R_E$ in the tuned-circuit oscillator of Figure 6.11(a) to result in a 220-kHz oscillation frequency. Assume that the coil $Q$ is 50, $\beta = 100$, and $V_{CC} = 12$ V.

**D6.24** Repeat the design of Problem 6.23 if a MOSFET biased at $V_{GSQ} = 2.5$ V is used. Assume that $g_m = 6$ mA/V, $r_{ds} = 60$ kΩ, and $V_{DD} = 12$ V.

**Section 6.2.4 The Wien Bridge Oscillator**

**D6.25** In the Wien bridge oscillator of Figure 6.13, $R = 10$ kΩ, $C = 0.01\ \mu$F, and $A = 100$ V/V. Select $R_1$ and $R_2$ to cause sinusoidal oscillation. At what frequency will the circuit oscillate?

**D6.26** Using an op amp with $A = 200,000$ V/V, design a Wien bridge oscillator to operate at $f = 50$ kHz.

**D6.27** With the values of $R_1$ and $R_2$ shown in the circuit, would you expect the Wien bridge circuit to (a) not oscillate; (b) oscillate with a sinusoidal output; or (c) oscillate with a nonsinusoidal output. Explain your answer.

Without changing $A$ or $R_2$, select values of $R$, $C$, and $R_1$ to result in sinusoidal oscillation.

**Section 6.2.6 Crystal-Controlled Oscillators**

**D6.28*** Assume that a crystal is available with the following parameters: $R_s = 1.2$ kΩ, $L = 12.7$ H, $C_s = 10$ fF, and $C_p = 6$ pF. The MOSFET used is biased to $V_{GSQ} = 2.1$ V where $g_m = 4.4$ mA/V. Choose capacitances that allow parasitics to be ignored.

**(a)** Design a Colpitts crystal-controlled sinusoidal oscillator.
**(b)** At what frequency will your circuit oscillate?

**D6.29** If the parasitic capacitances for the MOSFET in Problem 6.28 are $C_{gs} = 4$ pF while $C_{gd} + C_{db} = 1.2$ pF, design a Pierce oscillator. What is the oscillation frequency?

# Chapter 7

# BASIC CONCEPTS IN COMMUNICATIONS

## 7.1 BRIEF HISTORY OF ELECTRONIC COMMUNICATIONS

### 7.1.1 The Vacuum Tube and Radio

The general-purpose vacuum tube is no longer used in the electronics field although a few special purpose, high-power tubes are used in the communications area. The invention of this device is credited with ushering in the era of electronics. The development of the vacuum tube triode [7] in 1906 is attributed to Lee de Forest (1873–1961). Doctor de Forest received the patent on the triode, then called the audion, in 1908. This device could amplify small signals such as those produced by a microphone. By using several triode stages, the signal could be amplified sufficiently to drive the speaker of a public address system.

The triode amplifier was also used in amplitude modulation (AM) receivers to amplify the small output signal of the antenna, but the gain was often insufficient to result in clear reception of the transmitted signal.

The next significant advancement was the development of the regenerative receiver, an important radio component. Edwin H. Armstrong had received the patent in 1914 for this circuit which used a triode. In 1924, after a long legal battle, the Supreme Court ruled in favor of de Forest although most knowledgeable observers and the lower courts felt that Armstrong deserved credit for this development [7]. AM radio flourished for decades using the basic idea of the regenerative receiver.

What may have been the first audio wireless transmission from a vacuum-tube transmitter occurred in 1912 in the lab of Frederick Lowenstein of New York City. Using an antenna, he was able to transmit a voice signal for several blocks. The construction of crystal sets became the hobby of thousands of people that same year [7].

On April 14, 1912, a coded distress message was received by a telegraph operator of the Marconi Radio Station in New York from the S. S. *Olympic*, a ship that was about 1400 miles away from New York in the Atlantic Ocean. The message stated, "S. S. *Titanic* ran into iceberg–sinking fast." The *Titanic* sank, with a resulting loss of 1517 lives. Mainly as a result of the wireless message, rescuers saved 705 lives. It was immediately obvious that ships equipped with wireless systems could reduce loss of life in such tragic accidents. Congress then passed an act requiring all ships with 50 or more passengers to include wireless systems on board.

The next few years saw a great development of interest in radio. The U. S. Navy, private firms, and the government constructed several stations. In 1914 the Federal Telegraph Company used 608-foot antenna towers in San Francisco and Heeia, Hawaii, to transmit sound over a 2400-mile distance.

The necessity of good communications during World War I served as the impetus for rapid development of vacuum tubes and radio circuits. The tetrode (1919) and pentode (1926) improved on the capability of the triode. These devices were indispensable to the tremendous growth of the field from the 1920s to the early 1960s.

Much of the fundamental development in the commercial radio and communications areas took place between 1920 and 1940, although John Carson [5] of AT&T applied for a patent on single-sideband transmission in 1915. The patent was granted in 1923. This basic idea was used for many years to simultaneously transmit several long-distance phone conversations over a single channel. Not until the 1960s did pulse code modulation methods begin to replace single-sideband methods of telephone transmission.

During World War I, President Woodrow Wilson ordered the dismantling of all amateur radio equipment in the United States. In April 1917, station 9XM at the University of Wisconsin began transmitting voice and music, but was shut down because of the war. Later that year, President Wilson placed all radio facilities in the United States under government control. The war ended on November 11, 1918, but it was not until October 1, 1919 that the secretary of the navy removed the restrictions on amateur radio.

Several stations returned to the air in 1919. Regular radio broadcasts using amplitude modulation (AM) began from Detroit and Pittsburgh in 1920 [7]. For many years, the AM broadcast ruled the airways. Although frequency modulation (FM) was conceived around 1912, many well-known engineers such as John Carson felt this form of transmission was impractical. Carson concluded in 1922 that FM would experience low signal-to-noise ratios compared to AM signals [5]. His conclusion was based on the use of a very small deviation in frequency from the carrier frequency value.

Edwin Armstrong (1890-1954) began pursuing FM in 1925 and received a patent for it in 1933. In a paper he presented in 1935 [2], Armstrong indicated that he had demonstrated FM transmission using larger frequency deviations with very successful results. Commercially however, FM transmission did not become popular until the 1950s. Armstrong demonstrated the advantages of FM over AM by setting up a receiver of each type in a side-by-side arrangement. The same information was then transmitted over FM and AM transmitters. A static noise source was added, and the AM signal could hardly be discerned. The FM signal remained almost crystal clear.

Although Armstrong demonstrated to RCA that FM was far less sensitive to interference than was AM, RCA apparently did not want to convert to the better system. It was the nation's dominant radio company and had sold millions of AM sets to the American public [13]. In fact, it opposed further development of FM, and from this point, Armstrong's relationship with RCA deteriorated rapidly.

After Armstrong failed to interest RCA in his FM system in the mid-1930s, he began promoting this transmission method on his own. He was quite successful in this promotion, and in 1941 there were over a dozen licensed FM transmitters, 500 applications before the Federal Communications Commission, and 25 FM receiver manufacturers [10].

At about this time, Armstrong rejected an offer of over $1 million for the rights to the system from RCA which was now quite interested in FM [4]. RCA developed its own version of FM, but apparently used many of Armstrong's concepts. After World War II, Armstrong accused RCA of infringement on his FM patents. The case began in 1949, but dragged on in the courts for years. In February 1954, Armstrong wrote a note to his wife and then jumped from the thirteenth floor of his apartment house. While there were other problems in his life, the legal battles he had fought had certainly contributed to his state of mind at the time of his death.

Before the end of that same year, a settlement was reached between RCA and Armstrong's estate. RCA was to pay just over $1 million to the estate, with the first payment of $125,000 dollars to be made on Christmas Eve 1954. This settlement was perhaps smaller than Armstrong would have exacted had he remained in the picture, but the court costs to RCA probably reached two to three times that amount.

In addition, Armstrong's wife, Marion, used this settlement money to pursue damages from many other radio companies including Philco, Motorola, Bendix, Admiral, DuMont, Sylvania, Packard Bell, and Raytheon. Most settled by 1957, but Emerson Radio held out until 1959 while Motorola fought the settlement until 1967 when the Supreme Court refused to rehear the case. Marion Armstrong lived until 1979 in quiet luxury from the various settlements with these radio companies [13]. Perhaps her greatest joy was not the money, but the establishment of her husband as the unquestioned developer of the FM system.

The remainder of the FM saga is known to anyone who listens to the radio. This type of modulation grew during the 1950s and 1960s, although certain FCC (Federal Communications Commission) decisions slowed its rate of growth. In 1977, the number of commercial FM transmitters finally surpassed the number of commercial AM transmitters. Although the carrier frequencies assigned to FM limit the distance of transmission compared to AM distances, the quality of the sound is much greater. FM is now used for music, while traditional AM is used primarily for news and informational programs. Only in the late 1980s did some improvement in AM quality occur as the FCC allowed a broader bandwidth to be used in AM transmission. With the improved AM systems that are used only by FCC-approved stations, the quality of transmission approaches that of FM signals.

### 7.1.2 Television

Television was conceived as early as the 1870s [14]. Although a mechanical/electrical system was patented in 1885 by Paul Nipkow, [14], this system was not implemented. Vladimir K. Zworykin applied for patents on the iconoscope in 1923 and the kinescope in 1927 while working for the Westinghouse Electric Company [4]. The iconoscope was never constructed and in 1934, a U.S. Patent Board gave priority to Philo T. Farnsworth on the basis that Farnsworth had developed a working model of the image dissector, the first electrical scanning TV camera.

As a 15-year-old farm boy in Idaho, Farnsworth had conceived the idea of using electrons to create a means for transmitting pictures as a 15-year old farm boy in Idaho. He later discussed these ideas with his high school chemistry teacher, who in 1934, would give testimony that was valuable in assigning Farnsworth's work priority over Zworykin's.

Farnsworth was largely self-educated, having spent just one year at Brigham Young University before embarking on his technical career. He produced the first working model of an all-electric system in 1928 [1]. This system was observed by employees of the Pacific Telephone Company and was reported in the *San Francisco Chronicle.*

Farnsworth's pioneering efforts have evidently been forgotten by more recent historians who instead credit RCA with the dominant developments in TV. In fact, this then very powerful company did more to hinder the development of TV than to foster it. Even the well-publicized image orthicon tube was unpatentable by RCA owing to a prior patent held by Farnsworth [9]. After years of attempting to develop its own system, RCA finally agreed to pay a fee to use many of Farnsworth's basic patents in 1939. Farnsworth held Several patents considered imperative to the operation of TV systems [1]. The significance of his pioneering work is reflected in the fact that the agreement with RCA represented the first time this powerful firm had agreed to pay continuing royalties for the use of a patent. Farnsworth had already signed licensing agreements with Philco, Baird (England), Fernseh (Germany), and AT&T [1], thus his work was now recognized by the leading electronics firms of the day. In addition to his pioneering efforts in American television, Farnsworth's systems were used in England and other parts of Europe in the mid-1930s.

Farnsworth never received the recognition or the wealth commensurate with his contributions to this field primarily because of the government's refusal to set TV standards and then the outbreak of World War II. After the war, many of Farnsworth's patents had expired or were near the end of the 17-year expiration time for U.S. patents.

Throughout the 1930s, RCA had lobbied to keep the government from developing broadcast standards for TV since the RCA system was inferior to Farnsworth's and those of others. When RCA finally purchased Farnsworth's patent rights and was ready to burst onto the TV scene, World War II was imminent, further delaying the development of TV standards. While six companies produced TV receivers for the New York market prior to 1940 (American Television Corp., Andrea Radio Corp., Allen B. DuMont Labs, General Electric, Philco, and RCA), the field did not gain significance until the late 1940s.

The transmission of radio and TV signals, along with the reception of these signals, were based on electronic circuits using vacuum tubes as the device of primary importance. The conversion of TV to transistors took place in the 1960s. Up to this time, many stores, including supermarkets, had vacuum-tube testers available to check the tubes of a TV set. These stores also sold vacuum tubes to replace defective devices.

Color television was first demonstrated to the U.S. public in 1951 [8]. Private industries had been working on color schemes since the mid-1940s, but different companies used different approaches to the production of colored images.

In the United States, changes or additions to public communication procedures must be approved by the Federal Communications Commission (FCC). CBS was one of the first to petition the FCC for acceptance of its wideband scheme for color transmission. This scheme required a 15 MHz bandwidth rather than the 6 MHz required by the black-and-white scheme. In March 1947, the FCC denied this petition. In July 1949, CBS demonstrated a second system to the American Medical Association convention in Atlantic City, New Jersey. This system could transmit color on a 6 MHz bandwidth. Although the resolution was not as good as existing black-and-white sets, the FCC adopted the CBS standard in 1950. The methods used by this color TV were not compatible with the black-and-white receivers in use at the time. All other TV firms were now forced by the FCC/CBS standard to adopt this method of transmission or attempt to get the standard changed.

RCA which had been involved in its own development of a color scheme, opposed the FCC's acceptance of the CBS standard, taking the matter to court. In May 1951, the U.S. Supreme Court upheld the FCC order. On June 25, 1951, regular color telecasts to the public began in New York using the CBS system.

RCA continued work on its scheme which was compatible with black-and-white receivers. Ultimately, the National Television Standards Committee (NTSC) recommended a standard that was closely related to the RCA system. In December 1953, the FCC reversed itself and adopted the NTSC recommendation as the U.S. standard, rescinding the CBS standard. This reversal by the FCC was probably a wise decision, but cost those manufacturing companies using the CBS system dearly, forcing some to file for bankruptcy.

The NTSC standard has been used with little change for five decades. This standard is not that observed by the entire world, however. Only Japan and South America adopted a similar standard while many other countries developed their own standards.

A similar problem to that experienced with color TV arose in conjunction with the high-definition television (HDTV) schemes initially proposed. This system uses digital image processing methods and greatly improves the resolution and capabilities of the older analog TV system. The FCC adopted its necessary "wait-and-see" posture to determine if one method will ultimately prove to be the best before setting a standard in the United States. A step in the right direction with HDTV was taken in 1993 when the FCC encouraged the competing companies to form a group referred to as "the Grand Alliance." This group set several standards that now define the field. Ultimately the companies that meet the standards with the highest quality and reasonable complexity will prosper.

Transmission by selected TV stations in the HDTV format began taking place in 1998. Although TV receivers for HDTV were introduced for a cost of approximately $10,000, the price has dropped considerably and is expected to continue to drop as HDTV becomes more popular. The FCC had mandated that all transmissions must be in the HDTV format by the year 2006, but as a result of continuing pressure from TV broadcasters, it appears that, as this date is approached, the deadline will not be met.

### 7.1.3 The Telephone

Only a few brief milestones in the development of the telephone will be considered here, beginning with Alexander Graham Bell's development and patent of the first telephone in 1876. Originally, the voice signal transmitted over a phone line was the direct audio signal. In order to serve more subscribers, the party line was used to minimize the number of lines necessary. Perhaps this scheme represented the first "time-division multiplexed" system as users simply waited until the line was free before making a call.

In the early 1920s, the concept of transmitting multiple conversations over a pair of wires simultaneously was developed. This system used single-sideband transmission with an accompanying pilot tone that made recovery of each separate conversation easier at the receiving phone company office. The basis for this scheme was a mathematical development by AT&T pioneer, John R. Carson. For years, this single-sideband system has been used to carry multiple conversations over a single cable. A version placed in service in 1974 handled 108,000 conversations simultaneously over 10 pairs of coaxials in the same cable [11].

In the early 1960s, the pulse-code modulation scheme began to replace the single-sideband method of transmitting multiple conversations over phone lines. This system allows more conversations over a pair of lines and is more immune to noise. Although pulse-code modulation is the predominant system in use, outlying areas still use the single-sideband method in the year 2000.

Four significant developments in telephony are

1. The development of microwave links for long-distance transmission
2. The development of fiber optic channels
3. The development of satellite communications
4. The development of cellular phones

Microwave links can replace phone lines to transmit multiple long-distance phone conversations through the earth's atmosphere via microwaves. These links use frequencies in the GHz range and became prominent in the 1950s. Directional signals are transmitted to a repeater which, receives the signal, amplifies it, and transmits it to another repeater. These microwave links allow very rough terrain to be spanned and lower the cost of an equivalent cable link.

Fiber optic cables or "light pipes" carry information via light rather than the lower frequency electromagnetic waves. Many more conversations can be carried over a single fiber optic strand than can be carried over a microwave link or a pair of wires.

Communication satellites, carrying phone conversations, television signals, and other data became available in the early 1960s. The first communication satellite, Telstar I, relayed television signals between the United States and Europe in 1962. Later, geosynchronous satellites were launched that allow communication signals to be transmitted and received from all countries of the Earth.

Cellular phones were demonstrated to the FCC by Motorola in 1972 [6]. Within 15 to 20 years, these devices became quite significant in the lives of a majority of citizens in the United States and other countries.

### 7.1.4 Satellite Communications

The first dependable communication satellites were deployed in the mid-1960s. These satellites are now called geosynchronous Earth-orbit or GEO satellites. At an altitude of almost 36,000 km above the equator, each satellite illuminates about one-third of the Earth with its electrical signals. Three such satellites, equally spaced around the equator and rotating at the same angular velocity as the Earth, cover the entire globe except for the polar regions. These satellites maintain the same relative position to the Earth. At present, there are about 180 commercial GEO satellites handling TV distribution to broadcasting stations and cable companies, distribution of TV to homes, land mobile telecommunications, and long-haul telephone communications. These satellite systems have been enormously successful in linking the world, especially in allowing TV broadcasting from one part of the world to be received in homes on the other side of the globe.

One problem with the GEO satellites arises from the long delays associated with transmitting a signal over a path length of 80,000 km. For TV signals, this delay is acceptable, but for phone conversations, the delay is very annoying. Total time delay for GEO satellites can be kept to about 400 ms, including processing and propagation time. Another problem that prohibits direct transmission of signals to a hand-held telephone receiver is the limited power that reaches the Earth. A large dish antenna is required to accurately receive this signal.

For these reasons, a new satellite system was developed to allow almost unlimited coverage of the Earth by a mobile phone system. These satellites are put into low Earth-orbit (LEO), typically 500 to 1500 km above the Earth's surface. Although these satellites are not synchronous relative to the Earth's rotation, a large number are deployed to cover the entire surface. The power transmitted to the Earth from these satellites is sufficient for reception by a hand-held receiver and the total delay is approximately 250 ms less than that of a GEO satellite. Companies' initial efforts to market these systems met with little success due to the high cost of the handsets.

**Practical considerations**

The first experiments with TV transmission from an airborne transmitter took place in 1948. Charles E. Nobles, a Westinghouse engineer, conceived the idea of transmitting TV from an airplane in the stratosphere to increase the area of coverage [12]. Westinghouse engineers had a workable design by September 1946, but difficulty with financing delayed tests until 1948. Flights began over western Pennsylvania, rebroadcasting signals from Baltimore and Washington. The transmissions were identified as "Airborne Station X10A." Viewers from remote areas in Ohio, Virginia, and Michigan reported reception of the signals.
Westinghouse petitioned the FCC on August 13, 1948 for a permit to build a TV station in Pittsburgh that would broadcast from a transmitter flying 30 miles west of the city at an altitude of 25,000 feet. It was hoped that this transmitter would cover 127,000 square miles and reach more than 12 million people. The FCC denied the request and put a freeze on new permits in order to study the signal interference problem between stations. During the four-year period of the freeze, AT&T developed a coaxial cable system to connect the Atlantic seaboard to the Midwest. Further expansion of AT&T's system to broadcast to large national audiences eliminated the need for stratospheric broadcasting.

## 7.2 ELECTRONIC COMMUNICATION SYSTEMS

**Important Concepts**

**1.** One important communication scheme uses a carrier signal combined with the information to be conveyed to form a signal that is transmitted to a receiver.

**2.** The carrier signal determines the channel frequency, while the information causes only small variations in frequency about the carrier frequency.

**3.** A transmission channel can consist of the atmosphere, conductive cables, the space above the Earth's atmosphere, or light pipes. In some cases, a channel may be made up of more than one transmission medium.

**4.** The frequency of the carrier signal determines how well the signal can be propagated over the channel to be used. Lower frequency signals propagate in the Earth's atmosphere and follow the Earth's curvature better than high-frequency signals.

**5.** The Federal Communications Commission (FCC) determines the transmission frequencies and power levels as well as other important transmission parameters for signals to be transmitted in the United States.

### 7.2.1 Communication Systems

Figure 7.1 shows a basic communication system.The information to be transmitted is generally an electrical signal that may represent voice, music, data, or video signals from a TV camera. The frequency bandwidth occupied by the information signal is called the *baseband*, and the signal is often referred to as the *baseband signal.* The information or baseband signal interacts with a *carrier* signal, and the resulting signal is transmitted over a *channel* to a receiver. The carrier frequency is typically much higher than the highest baseband frequency, perhaps by a factor of several hundred or thousand times.

The channel is the medium by which the transmitted signal is propagated. A very common channel is the atmosphere, or different layers of the atmosphere, over which radio and TV signals are propagated. A cable or pair of wires serves as the channel in other applications such as long-distance telephone transmission, telephone data links, or cable TV systems. A few applications have a transmitter that converts the electrical signal to a

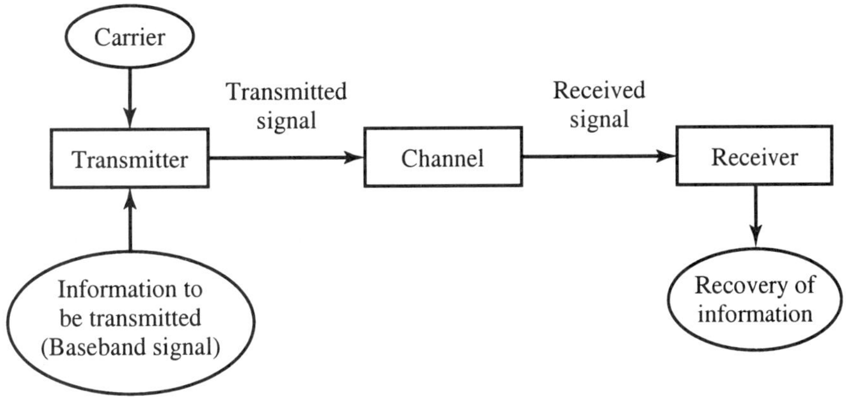


Figure 7.1: A communication system.

corresponding light signal that is propagated over a fiber optic channel. In many cases, the channel might consist of two or more different media or *links*. For example, a phone conversation may be transmitted over a cable link for a portion of its path and then be transmitted through the atmosphere via a terrestrial microwave link to reach the receiving station. For overseas transmission, phone conversations or TV signals travel by a cable link to a special transmitter that beams the signal up to a satellite. The satellite receives this signal and transmits it to another satellite that can direct the signal down to the receiving country. Still another example of a multilink channel is the cable TV system that receives signals via a satellite link and then distributes these signals to users over a cable.

The interaction of the information signal and the carrier signal is generally called *modulation*. The process of modulation is central to communication systems and will be discussed in depth in this and following chapters.

### 7.2.2 A Communication System Example

The block diagram of Figure 7.2 represents a radio communication system. All communication systems implement several concepts included in this diagram. The information source could be a microphone or the output of a tape player. This signal is amplified and applied as one input to a modulator which has a carrier frequency applied to the other input. The carrier signal has a much higher frequency than any of the frequency components of the information signal. The modulated signal derived from the modulator output consists of a signal with possible frequency components that are very close to the frequency of the carrier. This is called a radio frequency or RF signal. The RF signal is applied to the transmitting antenna that converts the pure electrical signal to an electromagnetic wave that propagates through the atmosphere.

If a communication system uses a cable for the channel, the modulated signal is transmitted to the receiver in the form of an electrical signal rather than an electromagnetic signal. The cable conducts the electrical signal over low-resistance conductors to the receiver although repeating amplifiers may be distributed over the cable to boost the signal strength if the signal attenuation becomes excessive. A disadvantage of this channel is the necessity of a cable. An advantage is that unwanted signals cannot easily penetrate the cable to cause deterioration of the desired signal. Another advantage is that antennas are not required to convert the signal into an electromagnetic wave.

The conductivity of the atmosphere is very low compared to a conducting cable. Electrical currents cannot be conducted through space using the methods used for cable transmission. Electromagnetic waves can, however, be propagated through space or the atmosphere for relatively long distances.

The atmosphere is a particularly noisy channel. Electromagnetic waves generated by other radio and TV stations are also being transmitted over this same channel. Furthermore, many other sources of electromagnetic waves may add to the total unwanted signal power of the channel. Large abrupt changes of current can release electromagnetic energy to the atmosphere, depending on the physical characteristics of the conductors through

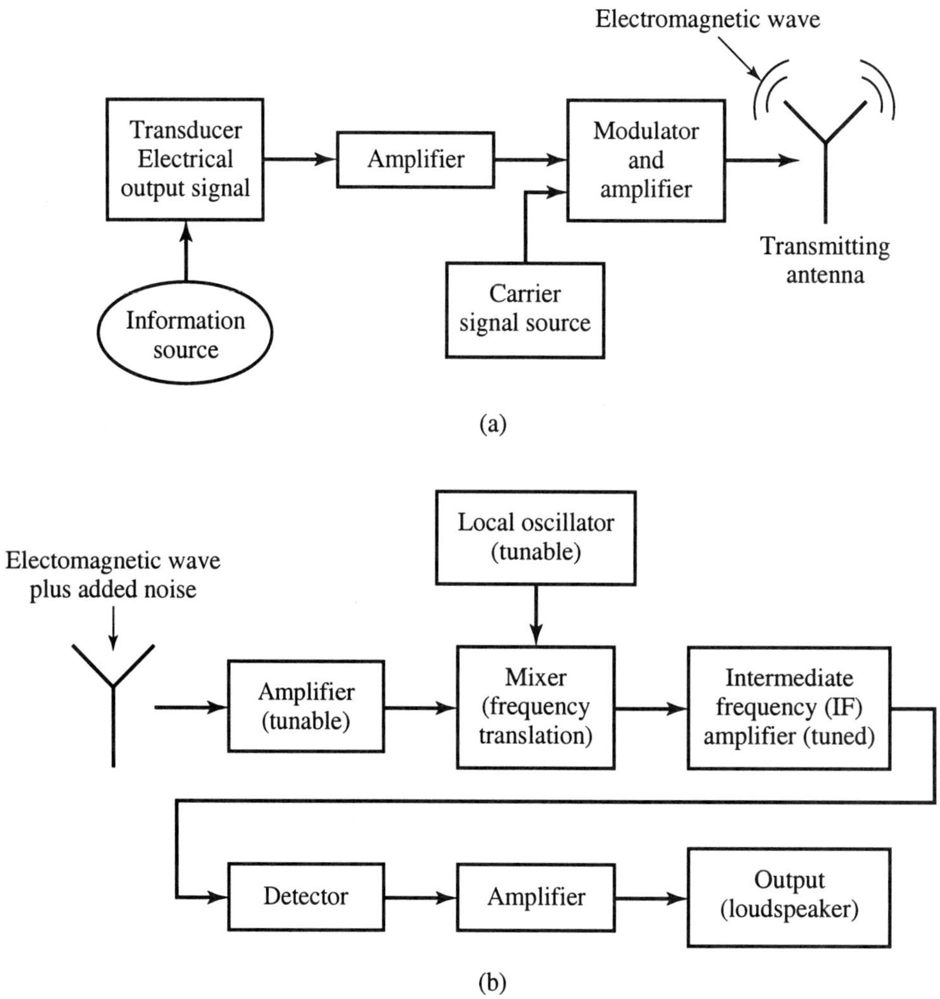


Figure 7.2: Radio communication system. (a) Transmitter; (b) Receiver.

which the current flows. An arc welder or a heavy duty solenoid that interrupts some current flow can emit noise in the channel. The currents used to create the spark for the ignition of an internal combustion engine radiate electromagnetic waves that can interfere with the desired signal to be received. The high-resistance carbon spark plug wires on automobiles minimize the transmission of electromagnetic energy into the atmosphere. Although these wires deteriorate more rapidly than the old stranded conductive wires of the 1950s, they reduce the noise effects on the vehicle's radio and nearby TV sets. Amateur radio transmission or mobile services such as police and ambulance also transmit signals via this channel. All signals except the desired one represent noise to the communications system.

The receiving antenna of Figure 7.2(b) picks up many electromagnetic signals including the desired one and converts the energy of the electromagnetic wave to electrical energy. This signal is many orders of magnitude smaller than the signal that was used to drive the transmitting antenna. The antenna is somewhat selective in that it will reject noise signals having frequencies far removed from the frequency range for which the antenna is designed. Noise components with frequencies near that of the desired signal will be passed to the input amplifier. This stage amplifies the antenna signal and discriminates against noise signals of other frequencies. A tuned stage with an adjustable resonant frequency is used for the input amplifier. This stage is tuned to the carrier frequency of the desired transmitted signal.

The desired signal along with the remaining noise is then translated or changed in frequency. The magnitude of this translation depends on the frequency of the desired carrier. The mixer changes the carrier frequency to some predetermined value called the *intermediate frequency.* This signal is then selectively amplified by several stages tuned to the intermediate frequency. Noise signals with frequencies that differ very little from the carrier frequency are eliminated by the IF amplifier. The information contained in the transmitted signal is then extracted by a *detector circuit*, amplified, and applied to a speaker.

Upon initial exposure to a communication system, one might question the use of such a complex scheme to transmit information from one point to another. Would it be possible to transmit the information directly through the atmosphere without involving a carrier signal? The answer to this question is "No!" We will understand why this is true as we consider three important advantages of the carrier method of transmission.

For transmission of electromagnetic waves through the atmosphere antennas must be used to radiate the desired signal. Antennas tend to operate efficiently only when the antenna dimensions are of the same order of magnitude as the wavelength of the transmitted signal. An electromagnetic wave traveling in free space has a wavelength of

$$\lambda = c/f$$

where $c$ is the velocity of light and $f$ is the frequency of the signal. A very low-frequency signal wavelength may be several kilometers long while a carrier signal wavelength may be only a few meters. The transmission of low frequencies requires very large antenna systems sometimes occupying acres of land, whereas high-frequency antennas are far more compact. The lower limit on antenna dimensions for relatively efficient transmission is $\lambda/10$, although quarter-wavelength antennas are often used in significant applications [6].

A second advantage of the carrier method relates to the fact that the baseband signal is often a wideband signal. For example, an audio signal may contain frequencies ranging from 100 Hz to 5 kHz. The ratio of highest to lowest frequency is 50 to 1 in this case. The corresponding wavelength ratio is 1 to 50. An antenna could not efficiently radiate all components of such a signal. When a carrier is used, the transmitted signal contains frequencies that are all very close to that of the carrier. Thus, the antenna dimension for efficient radiation is constant and of reasonable size.

A third advantage of the carrier method is that many different signals can be transmitted simultaneously, each using a different carrier. The receiver can then be tuned to the desired signal. This allows the transmission of many AM radio stations, FM radio stations, and TV stations simultaneously over the same channel or propagation medium (the atmosphere). It also allows the transmission of many TV signals over the same cable. Even if direct transmission of the video and sound information by many stations simultaneously could be accomplished efficiently, it would not allow the separation and detection of a single station.

An example of an AM radio station transmitting antenna will show the dimensions that must be used. One AM radio station broadcasts with a carrier frequency of 1160 kHz. This station has a "clear" rating from the FCC, which means that up to 50,000 W of average power can be transmitted on this frequency. The wavelength of this signal is

$$\lambda = \frac{c}{f} = \frac{3 \times 10^8}{1.16 \times 10^6} = 258.6 \text{ m}$$

The wavelength is about 848 feet. A quarter wavelength antenna would then be 212 feet in length. The particular antenna used for this station is a half-dipole antenna that places one portion 212 feet from the ground while using the ground as the other part of the antenna.

A radio receiver can function with a signal strength as low as 1 $\mu$V/m. This signal strength may result in a 50-$\mu$V (rms) signal across a 300-$\Omega$ receiving antenna. If the transmitted power is 50,000 W and the received power is

$$P_r = \frac{v^2}{R} = \frac{(50 \times 10^{-6})^2}{300} = 8.33 \times 10^{-12} \text{ W}$$

then the ratio of transmitted power to received power is a factor of $6 \times 10^{15}$! The emphasis on efficient antenna design is directed toward the transmitting antenna while very little emphasis is placed on the receiving antenna for commercial radio.

**Practice Problems**

**7.1** An FM station transmits on a frequency of 101.1 MHz. Calculate the wavelength of this signal in feet. *Ans:* 9.74.

**7.2** A quarter wavelength antenna is 169.7 feet in length. What AM frequency will this antenna efficiently transmit? *Ans:* 1450 kHz.

### 7.2.3 The Electromagnetic Spectrum

Table 7.1 lists the various frequency ranges of interest to the communications field [6]. Although modulation has not yet been discussed, it is useful to consider the transmission characteristics and usage of each frequency band.

The release of electromagnetic energy into the atmosphere of the United States is controlled by the FCC. This agency allocates carrier frequencies, bandwidths, and energy levels of transmitted waves for all nonmilitary applications. Lower frequency electromagnetic waves with frequencies extending into the low-frequency (LF) range can be propagated effectively by *ground waves.* These waves travel parallel to the Earth's surface and suffer little attenuation by moisture, smog, or other particles in the lower atmosphere. As these waves leave the curved surface of the Earth, they are reflected by the ionosphere back to the surface and are therefore capable of following the curvature of the Earth. This mode of propagation is also called *surface ducting* as the wave is constrained to travel through a duct between the Earth's surface and an atmospheric layer. Large antennas are required for transmission of these frequencies, and most applications fall in the military category. Long-range navigation, telegraphy, submarine communication, and radio beacons represent some of the uses of these lower frequencies. A very limited range of frequency transmission results from antenna size restrictions.

**Table 7.1 Classification of the Frequency Spectrum**

| Frequency Range | Name | Approximate Transmission Distance of Electromagnetic Wave through the Atmosphere |
|---|---|---|
| 30 to 300 Hz | Extremely low frequency (ELF) | Over 1600 km<br>Ground wave |
| 0.3 to 3 kHz | Voice frequency (VF) | Over 1600 km<br>Ground wave |
| 3 to 30 kHz | Very low frequency (VLF) | Over 1600 km<br>Ground wave |
| 30 to 300 kHz | Low frequency (LF) | 1600 km<br>Ground wave |
| 300 to 3000 kHz | Medium frequency (MF) | 0-1600 km<br>Ground wave and night sky wave |
| 3 to 30 MHz | High frequency (HF) | *Long distance or horizon<br>Sky wave or line-of-sight |
| 30 to 300 MHz | Very high frequency (VHF) | Horizon<br>Line-of-sight |
| 300 to 3000 MHz | Ultrahigh frequency (UHF) | Horizon<br>Line-of-sight |
| 3 to 30 GHz | Superhigh frequency (SHF) | Horizon<br>Line-of-sight |
| 30 to 300 GHz | Extrahigh frequency (EHF) | Horizon<br>Line-of-sight |

*Long distance if used in sky-wave mode.

If the communication channel is cable or wires, the antenna problem is eliminated. A major application of such a system is the tone transmission of digital data between computer terminals over 3-kHz telephone lines.

The LF band is less affected by ionospheric storms than are higher bands. This band is occasionally used for backup communications in conjunction with systems in higher bands that may be adversely affected by weather conditions. It is also used for marine communications and radio beacons.

The medium-frequency (MF) band contains the familiar commercial radio band of 535 kHz to 1705 kHz. In the United States and most parts of the world, this band is reserved for transmission of amplitude-modulated information to general audiences. This band uses antennas that achieve reasonable radiation efficiency with reasonable size. The commercial AM band depends on ground-wave propagation in the daytime because these frequencies are absorbed by the atmosphere rather than reflected. Consequently, these signals can only be transmitted to the horizon in the daytime. Under more moist atmospheric conditions the wave is reflected back to the earth's surface and can follow the curvature. Nighttime transmission results in longer distances traveled, allowing reception from distances of 750 to 1000 miles under optimum weather conditions.

Several smaller frequency ranges within the MF band are allocated to aeronautical services. The range from 1705 kHz to 2850 kHz is used primarily for short-distance fixed circuits such as police, fire, and other emergency calls. Aeronautical frequencies extend from the MF band into the HF band with short distance point-to-point and ground-air-ground communications falling in the 2850-kHz to 4063-kHz range. Aircraft flying scheduled routes are allocated specific channels within this band while nonscheduled routes are allocated separate channels. In the United States, the band from 3.5 MHz to 4 MHz is reserved for amateur radio operation, although most other countries use this band for fixed and mobile services.

The high-frequency (HF) band is also used for long-distance point-to-point and transoceanic ground-air-ground communications. Maritime ship-to-coast frequencies use several channels in this band for telephone and telegraph communications. Land mobile frequencies extend from the MF band through the HF band and on

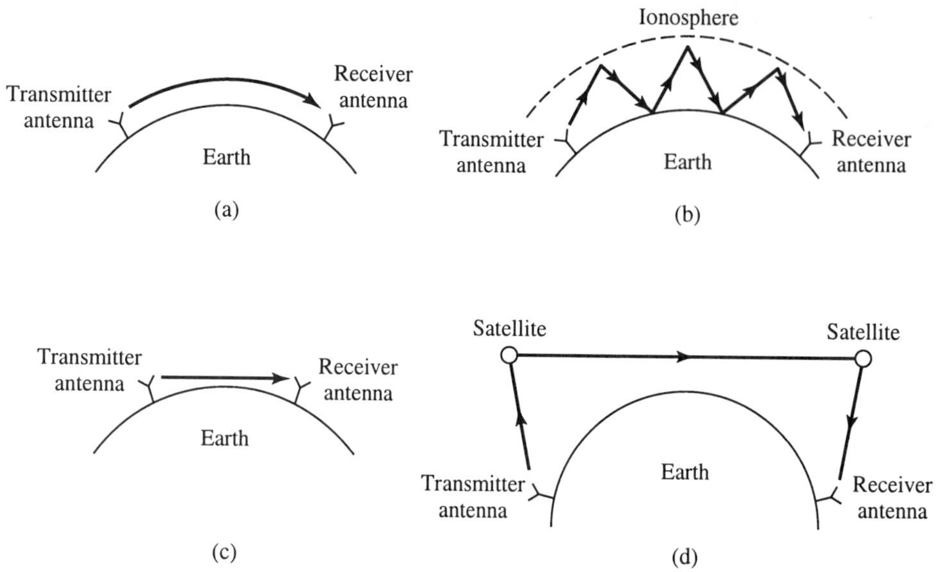


Figure 7.3: Modes of electromagnetic wave propagation. (a) Ground-wave propagation (below≈2 MHz). (b) Sky-wave (2 to 30 MHz). (c) Line-of-sight (above≈30 MHz). (d) Satellite transmission (above ≈ 3 MHz).

up through the higher bands. In the United States specific frequency allocations for several services have been assigned by the FCC. For example, the petroleum industry is assigned the frequency ranges 25.02 to 25.3 MHz, 30.66 to 30.82 MHz, and several others. Assignments have been made to industries related to business, forest products, manufacturing, motion pictures, power, telephone maintenance, and others.

The dominant mode of propagation in the HF band is sky-wave propagation. Ground waves travel from the transmitting station to the horizon, but a portion of the wave energy is reflected by the ionosphere back to the Earth's surface. Figure 7.3 shows four different modes of transmission. Note in Figure 7.3(b) that the sky-wave propagation mode can travel around the Earth's curvature as a result of the reflections from the ionosphere and the Earth's surface. Large areas between the areas of coverage do not receive significant signal strength to receive the transmitted signal. The frequency range from 1600 kHz to 30 MHz used to be referred to as short waves. The altitude of reflection of the sky-wave may vary from about 50 km to 400 km, consequently the distance of transmission can vary. With directional antennas and the proper angle of transmission, one-hop transmission of 4000 km is obtainable. Two- or three-hop transmission increases this distance considerably, but is less dependable and predictable.

An important user of the very high-frequency (VHF) band is the television industry. Television channel assignments for analog transmission range from 54 MHz to 88 MHz (channels 2–6) and from 174 MHz to 216 MHz (channels 7–13) in the VHF band and from 470 MHz to 890 MHz in the ultrahigh-frequency (UHF) band. Commercial FM radio is inserted into the spectrum between television channel 6 and channel 7, from 88 MHz to 108 MHz. FM radio stations transmit carrier frequencies ranging from 88.1 MHz to 107.9 MHz. Signals in the VHF and UHF bands are not reflected back to the Earth's surface and are line-of-sight systems, as shown in Figure 7.3(c). A VHF TV signal may be received up to distances of about 160 km by a typical TV receiver. In the early days of cable TV, a sensitive receiving antenna was placed on the highest mountain near the outlying area to be served. The signals from all available channels were amplified and transmitted over cable to the subscribers' homes. This system was referred to as community antenna television or CATV.

Commercial AM, FM, and television, except for HDTV, presently transmit signals using analog modulation methods. With the exception of AM stereo transmission and an increased AM bandwidth approved for some stations, little has changed in these modulation schemes since the 1960s. Of course, HDTV is changing TV transmission to digital methods.

Since the mid-1960s, when synchronous satellites became available for communications, many TV signals have been transmitted by this means. Three satellites, revolving at the same angular velocity as that of the Earth, are placed in orbit. Since the satellites maintain the same position relative to the Earth, they are said to be placed in geostationary orbits. Typically, these orbits are about 36,000 km in altitude. Each of the satellites contains a communication system that can receive signals from the Earth or from another satellite and then transmit the received signal back to the Earth or to another satellite. Although a signal of frequency above the HF band cannot be propagated for long distances along the surface of the Earth, it is easy to propagate such signals through free space. Once the various atmospheric layers are penetrated by the signal, little or no attenuation takes place in space. Furthermore, the transmission of a satellite signal is directional, whereas a TV station must use an omnidirectional antenna. Rather low levels of energy are then required to transmit from Earth to a communications satellite.

The combination of a transmitter and receiver is called a *transponder.* A typical communications satellite launched in the 1980s has 12 to 24 transponders [6]. The electronics for the satellite is powered by a solar cell/battery system that delivers a power of somewhat less than 1 kW. A popular frequency for the transmission from Earth to the satellite (uplink) is 6 GHz while the transmission from the satellite back to Earth (downlink) often uses a 4-GHz carrier. With these two carrier frequencies, a satellite is allocated a 500-MHz bandwidth. Each transponder might then use a bandwidth of 36 MHz. Many systems simply transmit to a properly positioned satellite which returns the signal to the desired section of the earth rather than transmitting to another satellite as depicted in Figure 7.3.

The communication satellite has replaced or minimized the use of earlier communication systems. For example, the bulk of transoceanic telephone traffic is now carried over satellite channels rather than by the older submarine cables. Satellite communication systems carry data, telephone, and television signals to and from almost all parts of the world [6].

Phase and frequency modulation is generally used in satellite communications for transmission of TV signals. Data and voice frequency phone signals may use a combination of digital and analog modulation methods when transmitted over a satellite link.

**Practical Considerations**

An important addition to TV signal transmission over satellite has been the introduction of digital transmission. The familiar satellite frequencies near 4 to 6 GHz are also used in microwave telephone ground links. As a result, the power transmitted back to Earth from a satellite is limited by the FCC to avoid interference with these ground links. Consequently, large dish antennas must be used to receive the satellite signal properly. The digital systems use frequencies in the vicinity of 10 to 14 GHz and can transmit higher powers to Earth without interfering with telephone ground links. The antennas required for this reception have a diameter of approximately 18 inches compared to diameters of perhaps 100 inches for the lower frequency satellites. The digital satellites typically use phase-shift keying as the modulation method.

---

### 7.2.4 Selected FCC Specifications

Table 7.2 presents a comparison of a few standard communication systems.

**Table 7.2 Some FCC Specifications**

| **Class D Citizens' Band** | |
|---|---|
| Carrier frequencies | 26.965 to 27.405 MHz |
| Average power input to antenna | 5 W |
| Estimated transmission distance | 2 to 4 miles in urban areas<br>5 to 15 miles in open terrain<br>20 to 30 miles over water |
| SSB transmission<br>peak envelope power input to antenna | 12 W |
| Estimated transmission distance | Double that of AM |
| **Commercial AM Radio**<br>(Old standards) | |
| Carrier frequencies | 540 to 1600 kHz |
| Modulation bandwidth | 5 kHz |
| Local stations average power | 0.1 to 1.0 kW |
| Regional stations average power | 0.5 to 5 kW |
| Clear stations average power | 0.25 to 50 kW |
| **Commercial AM Radio**<br>Voluntary standard - 1987 | |
| Carrier frequencies | 540 to 1700 kHz |
| Modulation response | 15 dB down at 10 kHz<br>30 dB down at 10.5 kHz<br>40 dB down at 11 kHz<br>50 dB down at 15 kHz |
| **Commercial FM Radio** | |
| Carrier frequencies | 88 to 108 kHz |
| Modulation bandwidth | 15 kHz |
| Average power input to antenna | 0.25 to 100 kW |
| **Television** | |
| Channels 2 to 6 (VHF) | 54 to 88 MHz |
| Average power to antenna | 100 kW |
| Channels 7 to 13 (VHF) | 174 to 216 MHz |
| Average power to antenna | 316 kW |
| Channels 14 to 83 (UHF) | 470 to 890 MHz |
| Average power to antenna | 5 MW (6700 horsepower!) |

## 7.3 MULTIPLEXING

**Important Concepts**

**1.** When multiple information streams are sent simultaneously over the same channel, either time-division multiplexing or frequency-division multiplexing is used.

**2.** Frequency-division multiplexing modulates each baseband signal to be transmitted with a different carrier frequency. At the receiver, these different frequencies are separated to recover each individual baseband signal.

**3.** Time-division multiplexing generally encodes the signal to be transmitted into some type of digital code. Each coded signal to be transmitted has bits sent onto the channel only during preselected timing intervals.

Information such as digital data can be transmitted from a terminal to a computer over a pair of wires in its original form, that is voltage levels that represent binary ones and zeros. Local telephone conversations may also be transmitted to nearby locations via wires that carry the voltage signal produced by the microphone. This signal is amplified and routed to the proper phone which then converts these signals back to sound pressure waves by means of an earpiece or speaker. Whereas this direct transmission of information is common in several applications, there are important limitations that lead to the use of *frequency-division multiplexing* (FDM) or *time-division multiplexing* (TDM) methods.

A very severe limitation of direct transmission is that one pair of wires must be connected between each pair of communication points and each pair of wires can convey only one source of information at any given time. For example, if 10,000 people in San Francisco simultaneously called 10,000 people in New York by telephone, this direct method would require 10,000 pairs of wires or channels.

Even if the bandwidth of a channel is several factors greater than that of the source of information, a second source cannot be directly combined with the first and transmitted simultaneously. The atmosphere surrounding the Earth is a channel with a tremendous bandwidth, but FDM methods are used to transmit more than one source at a time.

### 7.3.1 Frequency-Division Multiplexing

One of the first methods used to maximize the number of conversations carried over a single pair of telephone lines was frequency-division multiplexing. This method uses each conversation or baseband signal to modulate a different carrier frequency. The resulting output signal of each modulator consists of frequencies that are near the corresponding carrier frequency applied to that modulator. As long as the frequency separation between all carrier signals is sufficient, none of the modulated signals overlaps another in frequency. These signals are then added and transmitted over the pair of lines. At the receiving end of the lines, the signals can be separated into individual signals using frequency-selective networks. Each individual signal is detected or demodulated to recover the original conversation. This method of transmission has been used for telephone signal transmission for over 50 years. Since the mid 1960s, TDM circuits have been replacing the older FDM circuits in telephone applications, but FDM circuits have not been completely eliminated.

Microwave relay stations connecting toll offices have used FM modulation for several years, but in 1980, a single-sideband system using less bandwidth was introduced [6].

Another important example of FDM is that of the commercial broadcast industry. Commercial AM, FM, and TV stations each transmit over the same range of frequencies. All of the AM stations in a geographic area radiate electromagnetic energy into the atmosphere simultaneously within the AM frequency band. The FCC assigns different carrier frequencies to each transmitter. An AM radio receiver can then tune to the desired carrier frequency, and recover the voice or music carried by this station. All other signals, with different carrier frequencies are ignored.

Cellular phones have used FM modulation for voice transmission, moving from initial wideband FM systems to narrowband systems to accommodate more subscribers.

Satellite communications apply both digital and analog methods to transmission and reception. While phase-shift keying is popular, so also is FM. Many systems are complex enough that both digital and analog are used in the same system [6].

### 7.3.2 Time-Division Multiplexing

The speed with which electronic circuits can operate is many factors greater than that of the information signals to be transmitted. For example, the output of a microphone or telephone mouthpiece produces a voice signal that contains almost all energy in the 100 Hz to 3000 Hz range. The period of the highest frequency is 333 $\mu$s. This signal will change almost negligibly in 125 $\mu$s. Consequently, the value of the waveform at any time could be approximated quite accurately if the amplitude were measured at 125-$\mu$s intervals. Electronic circuits such as analog-to-digital converters (ADC) can measure the value of an analog voltage in a fraction of this time. This process of measuring an analog voltage and converting it to an equivalent digital code is called sampling.

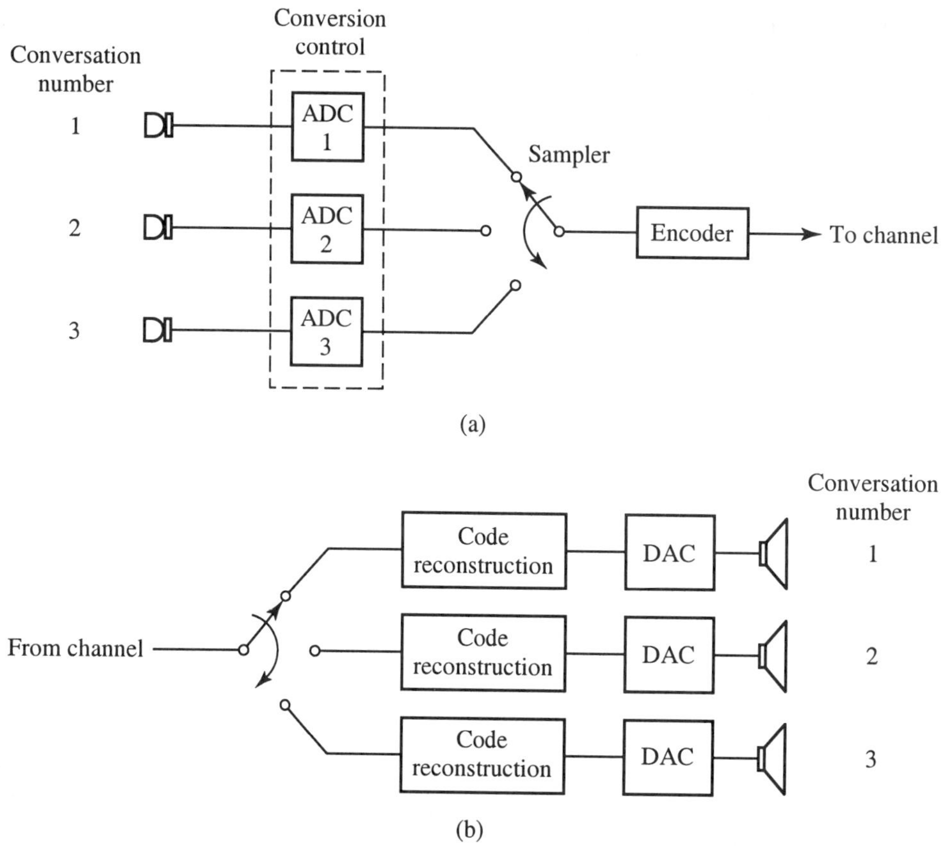


Figure 7.4: A TDM model. (a) Transmitter. (b) Receiver.

The signals from several telephone conversations can be repetitively sampled within the 125-$\mu$s interval, and each sampled value can be transmitted to a receiver during different time slots within this same interval. The receiver must recover the samples and associate each with its corresponding information source. Although the model depicted in Figure 7.4 is not the actual time-division multiplexing (TDM) system used, it demonstrates the basic concept involved in this method.

Conversation number 1 is sampled and the ADC converts the voltage level to a digital code. For telephone systems, the code is typically an 8-bit code and the sample rate is 8 kHz. The same process takes place for the other conversations. The codes for all three conversations now reside in the ADC output register until the next samples are taken. Rather than send the entire 8-bit code for the first conversation followed by the next 8-bit code, the system interleaves the bits. The sampler first sends bit 1 of ADC 1 to the encoder followed by bit 1 of ADC 2 and bit 1 of ADC 3. Then the sampler sends bit 2 of each ADC, then bit 3, and so on until the eighth bit of all ADCs has been sent. All eight bits of all ADCs are encoded and transmitted over the channel before the next sample of each conversation is taken.

At the receiving end of the channel, the signal is routed to the proper code reconstruction circuitry at the proper times. The first bit from ADC 1 fills the corresponding bit position of a register. The first bit from ADC 2 fills the corresponding bit position of a second register, and the first bit from ADC 3 fills the corresponding bit position of a third register. This process is repeated for all eight bits from each ADC. Each of the code reconstruction registers now holds the corresponding ADC values for one sample time. The DACs convert each of these values to an analog voltage that may then be filtered and amplified before applying to the proper speaker.

The TDM method has proven to be less susceptible to noise problems, and it also allows more conversations to be transmitted over a given pair of lines than the FDM method. As a result, the telephone system has been evolving toward the TDM method over the last two or three decades. The area of digital data transmission, whether using TDM or single-source transmission, is an important one.

## 7.4 AMPLITUDE MODULATION

**Important Concepts**

**1.** Amplitude modulation allows the baseband signal to vary the amplitude of a carrier signal in accordance with the instantaneous voltage of the baseband signal.

**2.** The carrier frequency may be 200 to 300 times the highest frequency of the baseband signal or more.

**3.** The index of modulation describes the maximum change in carrier signal amplitude compared to the unmodulated value.

**4.** The process of modulation leads to new frequency components in the output signal that consist of sums and differences of the carrier frequency and the baseband or modulating signal components.

**5.** The output signal of an amplitude modulator can be modified to create more efficient transmission. The double-sideband suppressed-carrier signal and the single-sideband signal are used in some applications.

Three types of modulation are of interest to the communication industry: amplitude modulation, phase modulation, and frequency modulation. Amplitude and frequency modulation are the methods used by commercial broadcasters. Phase modulation is important as a result of its usage in producing frequency-modulated signals as well as its usage in digital transmission systems such as digital TV.

### 7.4.1 The Amplitude-Modulated Waveform

As previously mentioned, a carrier frequency scheme is always used to transmit audio frequency information. The required broad range of wavelengths of an audio signal cannot be effectively transmitted directly due primarily to antenna limitations. The carrier frequency used is always much greater than that of the highest audio frequency to be transmitted. For example, the commercial AM broadcast band extends from 535 kHz to 1705 kHz, while the bandwidth of the information to be transmitted is typically 5 kHz. The carrier signal is combined in such a way with the lower frequency audio signal to produce a high-frequency signal. The transmitted signal frequency varies less than 1% from the carrier frequency. The wavelength of this modulated signal is nearly constant and much shorter than that of a frequency in the audio range; thus, the antenna design problem is considerably simplified. The advantage of allowing several different sources of information to be transmitted simultaneously using different carriers is also very significant.

Not only are audio signals transmitted by amplitude modulation, so also is the video information for television. Again this video information is transmitted by AM waveforms, with each station transmitting a different carrier frequency.

In amplitude modulation, the amplitude of the carrier signal is caused to vary in accordance with the instantaneous value of the signal to be transmitted. Figure 7.5 demonstrates the concept of amplitude modulation, using a sine wave as the information signal. The information to be transmitted is called the modulating signal or baseband signal and is applied to one input of an amplitude modulator circuit. This baseband signal, shown in Figure 7.5(a), determines the amplitude of the output carrier. The modulator has a second input to which the carrier signal is applied. The carrier waveform appears in Figure 7.5(b).

The output of the modulator is the amplitude-modulated signal of Figure 7.5(c), which can easily be transmitted since it is a high-frequency signal. In Figure 7.5, the modulating signal is shown with a frequency of about one-sixteenth the value of the carrier frequency. In actual practice, the carrier frequency might be 100 to

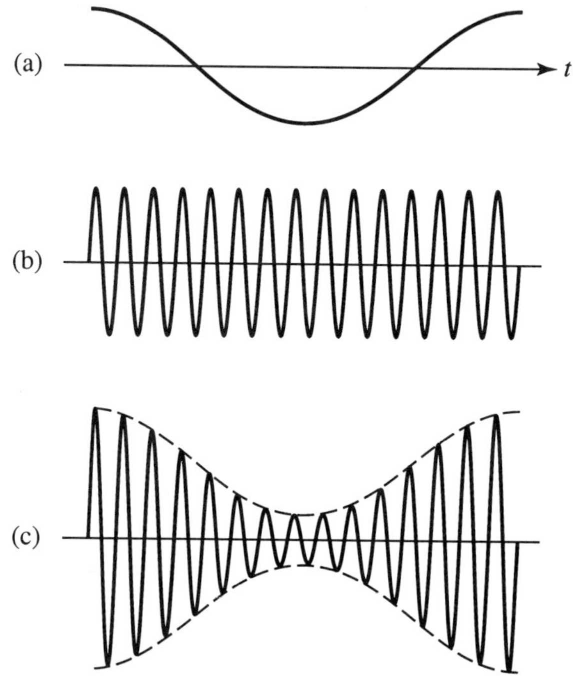


Figure 7.5: Amplitude modulation. (a) Information to be transmitted (baseband or modulating signal). (b) Carrier signal. (c) Carrier signal modulated by baseband signal.

10,000 times the frequency of the modulating signal, but this difference would be impossible to represent in a figure.

The AM signal is received and demodulated by the receiver to obtain the desired baseband signal information. That is, a signal is generated by the demodulator that varies in accordance with the peaks of the AM waveform, giving the original modulating signal.

It is convenient to base the theoretical discussion of amplitude modulation on the assumption of a sinusoidal modulating signal. This assumption is useful in practice because most audio sounds to be transmitted are made up of harmonic components. The extension of the single component input to an audio input modulating signal is thus rather straightforward.

### 7.4.2 Index of Modulation

When the applied modulating signal is zero, the output of the modulator is called the unmodulated carrier. As the amplitude of the modulating signal is increased, the change in amplitude of the output signal increases. The index of modulation relates the change in amplitude of the output to the unmodulated output amplitude. Figure 7.6 depicts the modulator waveform used to define the index of modulation.

The index of modulation, designated $m$, is defined as the ratio of the maximum deviation of amplitude from the unmodulated value of the AM waveform to the unmodulated amplitude. The equation for $m$ can be written

$$m = \frac{\Delta}{V_c} \tag{7.1}$$

where $\Delta$ is the maximum change of amplitude from the unmodulated value and $V_c$ is the amplitude of the unmodulated carrier.

We note in Figure 7.6 that there is a positive deviation in amplitude of the carrier when the modulating voltage goes positive and a negative deviation when the modulating signal goes negative. The positive deviation is designated $\Delta^+$, while the negative value is designated $\Delta^-$. For practical systems, $m$ is a function of the

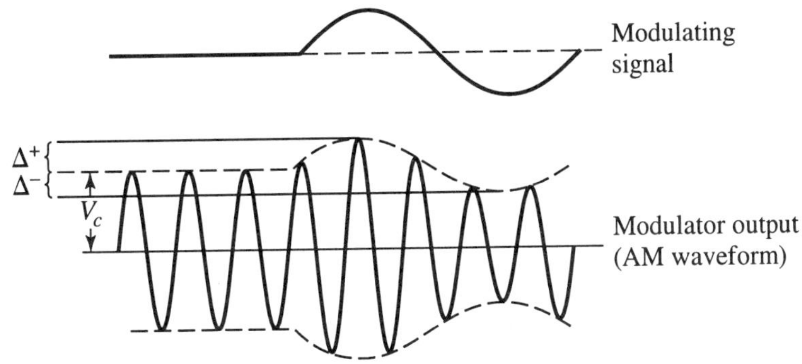


Figure 7.6: Waveform used to define $m$.

amplitude of the modulating signal and certain parameters of the modulator. For nonsinusoidal signals or nonlinear modulators, the deviation in carrier amplitude may be greater in the positive direction than in the negative direction. For this case, two different values of $m$ can be defined; however, we will generally limit our discussion to symmetrical modulation with a single applicable value of $m$. In other words, we will assume that

$$\Delta^+ = \Delta^- = \Delta$$

for the tone modulators to be discussed in this chapter.

The percent modulation of an AM waveform is obtained by multiplying $m$ by 100. When $\Delta = V_c$, 100% modulation is present. For $\Delta < V_c$ as shown in Figure 7.5 or 7.6, something less than 100% modulation is present. If $\Delta > V_c$, the waveform is said to be overmodulated, and the envelope of the AM waveform may not follow the value of the modulating signal. Figure 7.7 shows a 100% modulated waveform and an overmodulated signal. Many physical modulators will remain in the cutoff state or the saturated state when the amplitude of the carrier tends to be negative as shown in Figure 7.7(b).

**Practice Problems**

**7.3** A carrier signal with a 10-V amplitude is modulated to cause a maximum amplitude of 17.8 V. What is the index of modulation for this signal in percent? *Ans:* 78%.

**7.4** A modulated signal has a maximum amplitude of 24 V. If the index of modulation is 90%, what is the amplitude of the unmodulated carrier? *Ans:* 12.6 V.

### 7.4.3 Mathematical Description of the AM Waveform

If the modulating signal controls the amplitude of the carrier to produce the AM waveform, the expression for the modulator output can be written as

$$v_{\text{out}} = [V_c + \Delta \cos \omega_m t] \cos \omega_c t \tag{7.2}$$

This equation assumes that a sinusoidal modulating signal is applied. The expression included in brackets represents a time-varying amplitude of the carrier signal. Compared to the period of the carrier signal, the period of this time-varying term is very long.

Manipulating Eq. (7.2) and applying Eq. (7.1) allows us to write

$$v_{\text{out}} = V_c\,(1 + m \cos \omega_m t) \cos \omega_c t \tag{7.3}$$

From a practical standpoint, Eq. (7.3) is valid only for values of $m$ less than or equal to unity. While larger values of $m$ would lead to a negative amplitude for the carrier signal, practical modulators generally produce zero output for this situation, as noted earlier.

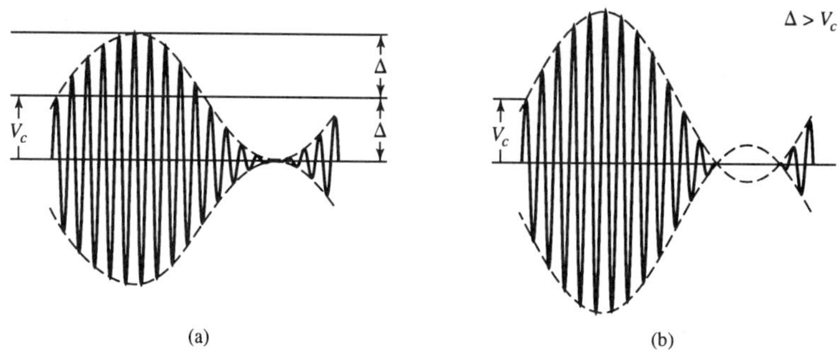


Figure 7.7: Amplitude modulated waveforms. (a)100% modulation. (b) overmodulation.

The AM waveform is not a pure sinusoid as is the unmodulated carrier signal. For sinusoidal modulation the AM waveform is repetitive, however, and from Fourier series considerations, we know that it must be composed of sinusoidal components. Returning to Eq. (7.3) and applying the trigonometric identity

$$\cos\,\omega_c t \cos\,\omega_m t = \frac{1}{2}\left[\cos\,(\omega_c - \omega_m)t + \cos\,(\omega_c + \omega_m)t\right]$$

we can write

$$v_{\text{out}} = V_c\,\cos\,\omega_c t + \frac{mV_c}{2}\,\cos\,(\omega_c - \omega_m)t + \frac{mV_c}{2}\,\cos\,(\omega_c + \omega_m)t \tag{7.4}$$

The AM waveform or the modulator output voltage may be expressed as three sinusoidal components: one of radian frequency $\omega_c$, one of frequency $\omega_c - \omega_m$, and one of frequency $\omega_c + \omega_m$. Since $\omega_m$ is generally much smaller than $\omega_c$, the frequency content of the AM waveform is limited to a small range of frequency centered about $\omega_c$. Figure 7.8 shows the components of a 100% modulated waveform. For this case $m = 1$, and the signal becomes

$$v_{\text{out}} = V_c\,\cos\,\omega_c t + \frac{V_c}{2}\,\cos\,(\omega_c - \omega_m)t + \frac{V_c}{2}\,\cos\,(\omega_c + \omega_m)t \tag{7.5}$$

The component of frequency $\omega_c - \omega_m$ is called the lower sideband frequency whereas the component of frequency $\omega_c + \omega_m$ is known as the upper sideband frequency. The three constant-amplitude signals add together to give the AM waveform.

Audio signals contain several components rather than a single component. Furthermore, in the dynamic speech process, the components vary with time. A typical speech sound might appear as shown in Figure 7.9(a). If this sound serves as the modulating signal, the AM waveform of Figure 7.9(b) results. The spectral representation of part (c) is valid only while the speech sound generates the repetitive waveform shown in part (a). The upper sideband contains all components of frequency greater than $\omega_c$, while the lower sideband contains all components of frequency less than $\omega_c$.

The AM modulated waveform is used in commercial AM broadcasting on the frequency range from 535 kHz to 1705 kHz, with carrier frequencies ranging from 540 kHz to 1700 kHz. In addition to limiting the total power of the transmitted wave, the FCC places a restriction on the index of modulation also. For commercial AM transmission, the average value of $m$ should fall in the range of 0.85 to 0.95. Amateur radio transmitters use the AM waveform as well as single-sideband (SSB) transmission.

**Practice Problems**

**7.5** A 700-kHz carrier signal with a 10-V amplitude is tone modulated by a 3500-Hz signal to cause a maximum amplitude of 17.8 V. What is the amplitude of the sideband components of this signal?

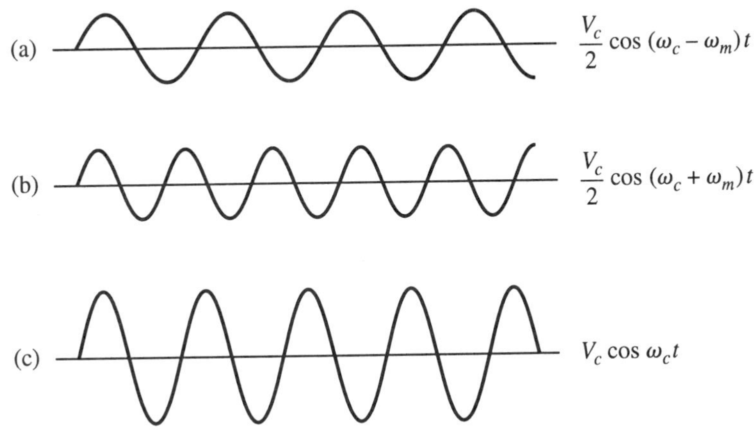


Figure 7.8: The three components of a 100% sinusoidally modulated waveform. (a) Lower sideband signal; (b) upper sideband signal; (c) carrier.

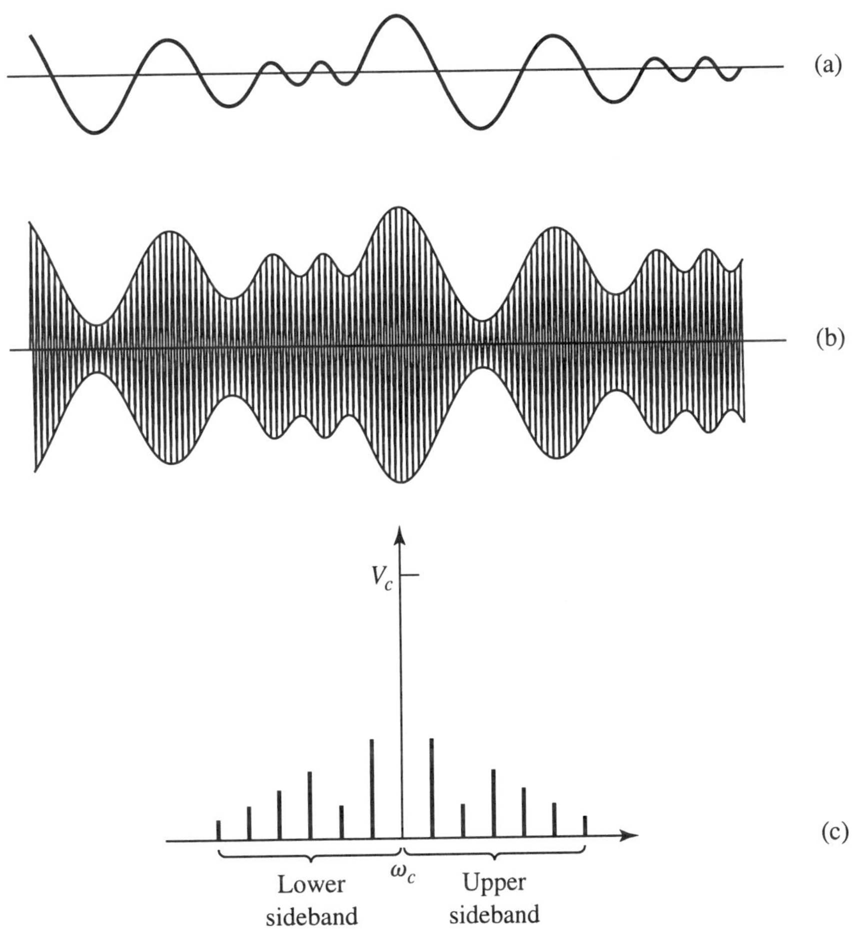


Figure 7.9: (a) Modulating signal using speech. (b) Resulting AM waveform. (c) Spectral representation at a given instant.

What are the sideband frequencies? *Ans:* 3.9 V, 696.5 kHz, 703.5 kHz.

**7.6** A tone-modulated signal has a maximum amplitude of 24 V. If the index of modulation is 90%, what is the amplitude of the sidebands? *Ans:* 5.67 V.

### 7.4.4 Power Relations in AM Waveforms

One form of expressing an AM waveform modulated by a sinusoid is

$$v_{\text{out}} = V_c \cos \omega_c t + \frac{mV_c}{2} \cos (\omega_c - \omega_m)t + \frac{mV_c}{2} \cos (\omega_c + \omega_m)t \tag{7.6}$$

Since the impedance across which these components appear is approximately constant for the three frequencies, the power in each component is proportional to the square of the voltage amplitudes. The average carrier power is

$$P_C = \frac{V_c^2}{2R} \tag{7.7}$$

where $R$ is the resistive load of the modulator. The lower sideband power, $P_{LS}$, is equal to the upper sideband power, $P_{US}$, and is

$$P_{LS} = P_{US} = \frac{m^2 V_c^2}{8R} \tag{7.8}$$

Each sideband carries a power of

$$P_{LS} = P_{US} = \frac{m^2}{4} P_C \tag{7.9}$$

The carrier component does not depend on the modulating signal and does not carry any of the desired information from the modulating signal. The sidebands depend on the index of modulation, which is a linear function of the modulating voltage. Each of these sidebands carry information relative to the frequency and magnitude of the modulating signal.

For 100% modulation, the value of $m$ is unity, and each sideband amplitude equals one-half the carrier amplitude. Each of the sidebands then contains one-fourth the power of the carrier, giving a total power of

$$P_T = P_C + P_{LS} + P_{US} = P_C \left(1 + \frac{1}{4} + \frac{1}{4}\right) = \frac{3}{2} P_C \tag{7.10}$$

If no modulating signal is being applied to the modulator, no information is being transmitted. Yet the carrier power continues to be transmitted. Even with 100% modulation, the carrier power equals two-thirds the total transmitted power. Certain classes of commercial radio stations are allowed to transmit a maximum average power of 50 kW. Assuming such a station could use an index of modulation of 100%, we can calculate the distribution of power in the transmitted signal. At peak output, 33.33 kW is contained in the carrier signal, which basically carries no information, while 16.67 kW is contained in the two information-carrying sidebands.

This highly inefficient method of transmission has been used since 1920 when station WWJ of Detroit and station KDKA of Pittsburgh began regular AM broadcasts. Use of the AM waveform, however, is influenced by other factors in addition to efficiency. One overriding factor in AM transmission is the ease with which the received signal can be demodulated. The simpler demodulators of the AM radio allowed the cost of these receivers to be minimal. For the commercial AM band, it is more important to maximize the number of listeners by minimizing receiver cost than to minimize the transmitted power.

**Practice Problems**

**7.7** An unmodulated carrier signal delivers 5 W to an antenna. If the index of modulation for a sinusoidal modulating signal is 0.8, what is the power delivered by each sideband? *Ans:* 0.8 W.

**7.8** A tone-modulated signal has a maximum amplitude of 24 V and delivers 13 W to a load. If the unmodulated carrier power is 9 W, calculate the index of modulation. *Ans:* 0.943.

### 7.4.5 Carrier Suppression

If receiver cost minimization is not a major consideration, it is possible to develop more efficient methods of transmitting information using amplitude-modulation methods. The carrier signal conveys little or no information and yet requires the bulk of the power used in conventional AM waveform transmission. For a specified limit on total power transmission, it is possible to suppress the carrier and transmit only the two sidebands at an increased power level. It is further possible to eliminate the carrier and one sideband, concentrating the total available transmission power into a single sideband which theoretically conveys all necessary information. These possibilities result in suppressed-carrier transmission and single-sideband (SSB) transmission, respectively. Suppressed-carrier transmission is also referred to as double-sideband (DSB) or as double-sideband suppressed-carrier (DSB-SC) transmission. The SSB method of transmission requires only half the bandwidth of the AM system. Figure 7.10 shows the output waveforms for sinusoidally modulated DSB and SSB modulators.

In general, the transmitter is somewhat more complex for these systems, but the receiver is far more complex than the simple AM waveform receiver. Furthermore, the quality of the demodulated signal is generally poorer. A carrier-suppressed signal can be created by a balanced modulator or by a perfect multiplier, as we shall see later.

### 7.4.6 Single-Sideband Systems

Before TDM circuits became popular in telephone transmission of speech, single-sideband (SSB) transmission was the only method used to carry multiple conversations over a pair of lines. This method is still used, but is being phased out as the overall system evolves to TDM circuits. In telephone transmission of the human voice, quality is not of prime importance and the single-sideband method is applicable. In this instance, the voice frequencies transmitted range from 300 to 3400 Hz. The spacing in such a system is typically 4 kHz for each voice channel. Up to 12 channels, for example, are assembled into a group band with a frequency range of 60 to 108 kHz. As many as five group bands can be used to allow a 60-channel system to transmit simultaneous conversations over a single transmission line. This scheme reduces the power and bandwidth requirements per conversation compared to the corresponding values that would be required for the AM method.

Amateur radio operators often transmit an SSB signal to increase the range of transmission. All transmission power allowed by the FCC can be used for the single band of frequencies in one sideband. This results in a higher-energy transmitted signal that can be received over a greater range of distances than can the conventional AM signal. As we shall see, the accompanying distortion in the demodulated signal must be minimized in order to make this system usable.

Another application of SSB is in land mobile services such as emergency services. In 1985, the FCC approved SSB for use in the $150-170$ MHz range in 5-kHz bandwidth channels [6]. The bandwidth of an SSB signal equals the bandwidth of the baseband signal, although a pilot tone may also be transmitted to simplify detection or recovery of the baseband signal. Before 1985, these frequencies were approved for narrowband FM transmission, which requires much more bandwidth than SSB transmission.

Until 1980, microwave links for public telephone networks used FM transmission. Through advances in klystron power amplifiers, SSB became a feasible transmission method around this time. Again, because of the reduction in bandwidth needed for each signal, more conversations can be relayed over a fixed total bandwidth when using SSB instead of FM.

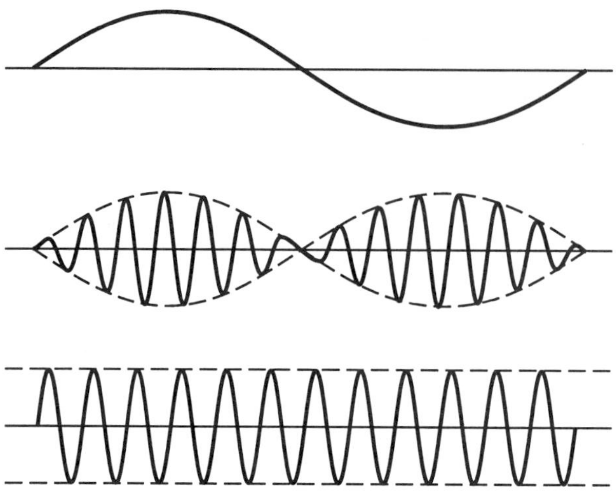

Figure 7.10: (a) Modulating voltage. (b) Carrier-suppressed waveform. (c) Single-sideband waveform.

### 7.4.7 Vestigial-Sideband (VSB)

This system transmits the carrier and one sideband with little or no attenuation, along with the other sideband that has been heavily attenuated. This method is used to transmit the video information for analog television. Some distortion can be tolerated, but if the lower sideband is completely eliminated, the distortion is unacceptable. Only a portion of lower sideband is transmitted to reduce distortion while conserving bandwidth.

### 7.4.8 On-Off Keying (OOK)

Amplitude-shift keying refers to changing the amplitude of the signal to some fixed amplitude, depending on a discrete-level modulating signal. In one very simple case, the input modulation or information is a straight binary or two-level non-return-to-zero (NRZ) signal. Each level produces a fixed-amplitude carrier at the output. Often one amplitude level is finite and the other is zero. When this is the case, it is referred to as on-off keying or OOK. Figure 7.11 demonstrates the waveforms for an OOK system. If $T_b$ is the duration of the sinusoidal signal and $T$ is the total period between bits, the duty cycle is defined as $d = T_b/T$.

The fundamental frequency for this waveform is $f_o = 1/T$ and the carrier frequency is $f_c$. Generally $f_c$ is much larger than $f_o$. When the modulating signal equals 1, the carrier signal is expressed as $V_1 \cos \omega_c t$.

The higher order frequency components will be larger when the input continues to repeat a 1-0 code rather than other codes that consist of strings of ones or strings of zeros. With a duty cycle of $d$, a Fourier analysis of the repetitive 1-0 code indicates that the coefficients of the harmonics are

$$a_n = V_1 d \left[ \frac{\sin\left(n - \frac{\omega_c}{\omega_o}\right) d\pi}{\left(n - \frac{\omega_c}{\omega_o}\right) d\pi} + \frac{\sin\left(n + \frac{\omega_c}{\omega_o}\right) d\pi}{\left(n + \frac{\omega_c}{\omega_o}\right) d\pi} \right] \tag{7.11}$$

The number $n$ is the number of the harmonic component.

Since the frequency $\omega_c$ is much larger than $\omega_o$, the denominator of the second fraction in Eq. (7.11) is very large, making this term much smaller than the first term and thus, the second fraction can be neglected.

The coefficients of the components can then be closely approximated by

$$a_n = V_1 d \left[ \frac{\sin\left(n - \frac{\omega_c}{\omega_o}\right) d\pi}{\left(n - \frac{\omega_c}{\omega_o}\right) d\pi} \right] \tag{7.12}$$

The most common value for the duty cycle is 0.5, although some special applications may use a smaller value.

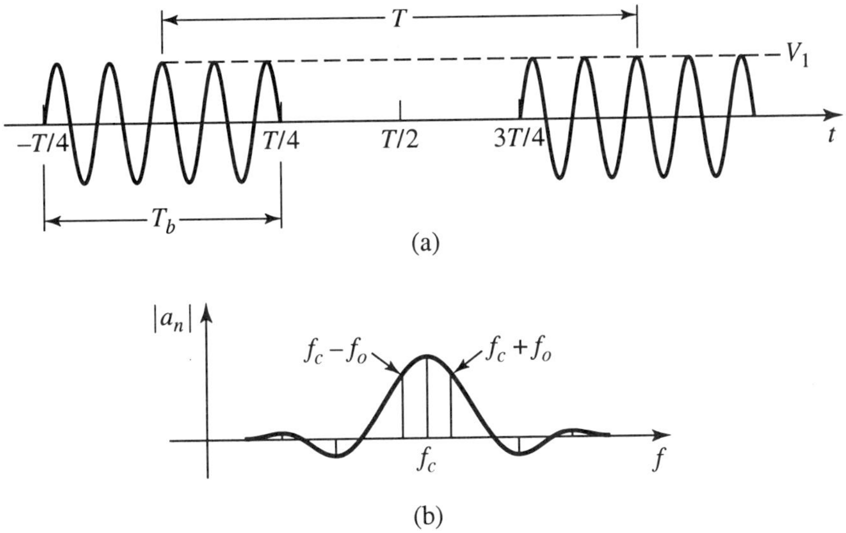


Figure 7.11: (a) An OOK signal. (b) The spectrum for a repetitive 10 code.

For small values of $n$, the coefficients are approximately zero. This means that the fundamental frequency component at $\omega_o$ and the lower harmonics carry infinitesimal power. For a duty cycle of 0.5, the maximum value of the coefficients occurs when $n = \omega_c/\omega_o$ (or the nearest integer value). This component has a frequency of $n\omega_o = \omega_c$, and the value of the coefficient is

$$a_n = \frac{V_1}{2}$$

The falloff of components around this harmonic is symmetrical and falls off as

$$\frac{V_1}{2}\frac{\sin x}{x}$$

would fall around the point $x = 0$, as shown in Figure 7.11(b). Note that the two zeros of the $(\sin\ x)/x$ function nearest to $\omega_c$ are $\omega_c + \omega_o$ and $\omega_c - \omega_o$.

The three components within this bandwidth of $2\omega_o$ carry over 90% of the total energy of the OOK signal. This can be shown by calculating the power carried by the waveform of Figure 7.11(a) and then comparing this value to the power carried by the three components in question. The average power of the waveform of repeated 1-0 bits is found as

$$P_{\text{total}} = \frac{1}{2}\left[\frac{V_1^2}{2R}\right] = \frac{V_1^2}{4R}$$

where $R$ is the resistance across which this signal is developed. The values of the three coefficients included between the frequencies $\omega_c - \omega_o$ and $\omega_c + \omega_o$ can be calculated from Eq. (7.12). This gives values of $V_1/\pi$ at frequencies of $\omega = \omega_c - \omega_o$ and $\omega = \omega_c + \omega_o$. At a frequency of $\omega_c$, the coefficient is $V_1/2$. The power carried by these three components is

$$P_3 = \frac{1}{2R}\left[\frac{V_1}{2}\right]^2 + \frac{1}{2R}\left[\frac{V_1}{\pi}\right]^2 + \frac{1}{2R}\left[\frac{V_1}{\pi}\right]^2 = 0.226\left[\frac{V_1^2}{R}\right]$$

This power represents slightly over 90% of the total power of the waveform. The transmitting and receiving circuitry need only have a bandwidth of slightly over $2\omega_o$ to receive 90% of the power in such a system.

Fiber optic systems in the United States and transoceanic fiber optics apply OOK modulation in TDM signaling.

**Practice Problem**

**7.9** If a 10-MHz carrier is modulated with a digital stream of alternating ones and zeros with a bit time of 0.01 ms, calculate the frequency of the largest component and the frequency of the two sidebands nearest this component. Assume a 50% duty cycle. *Ans:* 10 MHz, 9.9 MHz, 10.1 MHz.

### 7.4.9 Quadrature Amplitude Modulation (QAM)

A quadrature modulation system is capable of transmitting two separate sources of information without increasing the required transmission bandwidth. This method has found use in systems that must increase the information transmitted, yet maintain the previously allocated bandwidth. Two notable examples are color TV and stereo AM systems. The bandwidths allocated for black and white TV transmission were not increased when color TV was introduced. In addition, a color broadcast must be received to be displayed in black and white with the older black and white receivers. If a new transmission method is designed so that it can be received by the older receivers, it is referred to as being compatible. Color TV transmission is compatible with black and white receivers.

Since the FCC's approval of FM stereo transmission in 1961, this type of radio broadcasting has become very popular. The quality of musical programs broadcast by stereo FM is much higher than AM programs. The FM stations can transmit an audio or baseband bandwidth of 15 kHz while the AM stations are limited to a 5-kHz bandwidth. Although the bandwidth of the AM signal is not sufficient for high-quality music, approval for stereo AM broadcasting was granted by the FCC in 1982. In the late 1980s, however, the FCC approved a wider bandwidth for AM transmission by selected stations. The actual audio bandwidth transmitted by these stations approaches 10 kHz and improves the quality of AM music considerably. More will be said about this newer system in a later section.

In order to develop a compatible AM stereo system, the following conditions must be satisfied:

1. Two baseband modulating signals, a left and a right audio signal, must be transmitted with the same channel allocation used to transmit a single baseband signal (monaural).

2. The stereo signal must be received by an older monaural or monophonic receiver. This receiver should drive the speaker with a single signal that is proportional to the sum of the right and left audio signals.

The basic quadrature system uses one AM modulator and one balanced modulator. The AM modulator is driven by a modulating signal and the carrier signal. The balanced modulator is driven by the second modulating signal and by the same carrier signal shifted in phase by 90°. The block diagram for the quadrature system is shown in Figure 7.12.

The outputs of both modulators are added to produce the transmitted signal. This voltage can be written for tone modulation as

$$v_{\text{out}} = V_c \cos\, \omega_c t + m_1 V_c \cos\, \omega_1 t\; \cos\, \omega_c t \quad \text{(AM signal)}$$

$$+ m_2 V_c \cos\, \omega_2 t\; \cos\,(\omega_c t - 90°) \quad \text{(DSB signal)} \tag{7.13}$$

At the receiver, the two signals $v_{m1}$ and $v_{m2}$ can be detected by multiplying the incoming signal by cos $\omega_c t$ or by sin $\omega_c t$. Neglecting the attenuation of the channel and the gain of the receiver, we find that the output of

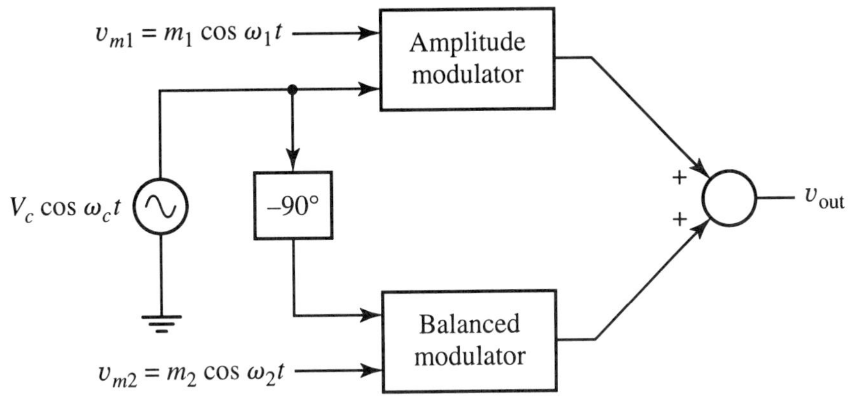


Figure 7.12: Quadrature amplitude modulator.

the first multiplier ($\cos\ \omega_c t$) is

$$\begin{aligned} v_1 &= V_c \cos^2\ \omega_c t + m_1 V_c \cos\ \omega_1 t\ cos^2\ \omega_c t + m_2 V_c \cos\ \omega_2 t\ \cos\ \omega_c t\ \sin\ \omega_c t \\ &= \frac{V_c}{2} + \frac{m_1 V_c}{2} \cos\ \omega_1 t + \frac{V_c}{2} \cos\ 2\omega_c t + \frac{m_1 V_c}{2} \cos\ \omega_1 t\ \cos\ 2\omega_c t \\ &+ \frac{m_2 V_c}{2} \cos\ \omega_2 t\ \sin\ 2\omega_c t \end{aligned} \tag{7.14}$$

The second term of this equation represents the first information or modulating signal. The first term is a dc term and the third, fourth, and fifth terms are high-frequency terms that can easily be filtered out.

The output of the second multiplier ($\sin\ \omega_c t$) is

$$\begin{aligned} v_2 &= V_c \cos\ \omega_c t\ \sin\ \omega_c t + m_1 V_c \cos\ \omega_1 t\ \cos\ \omega_c t\ \sin\ \omega_c t + m_2 V_c \cos\ \omega_2 t\ \sin^2\ \omega_c t \\ &= \frac{m_2 V_c}{2} \cos\ \omega_2 t - \frac{m_2 V_c}{2} \cos\ \omega_2 t\ \cos\ 2\omega_c t \\ &+ \frac{V_c}{2} \sin\ 2\omega_c t + \frac{m_1 V_c}{2} \cos\ \omega_1 t\ \sin\ 2\omega_c t \end{aligned} \tag{7.15}$$

The first term of this equation represents the second information or modulating signal. The other terms are higher frequency terms that can again be filtered. Thus, two channels can be transmitted over a single-channel bandwidth and can be separated at the receiver. The carrier signal at the receiver can be generated with a phase-locked loop.

A stereo AM signal could be produced by allowing the two modulation sources of a QAM system to be the right and left audio channels. Although this would allow transmission of the two independent stereo signals, it would not be compatible with a monaural receiver. For stereo AM transmission, the first modulating signal consists of the sum of the right and left audio channels (L + R), and the second modulating signal consists of the difference of these audio signals (L - R). A stereo receiver detects the two signals separately, then adds to get a 2L signal and subtracts to get a 2R signal. These signals are used to drive the two speakers.

One of the five stereo AM systems approved by the FCC is called C-QUAM. In this system, both modulator outputs are attenuated or multiplied by a factor of cos Θ where Θ is the angle resulting from the phasor addition of the two modulator outputs. This is indicated in the phasor diagram of Figure 7.13.

A monaural receiver detects the phasor sum of the two modulated signals, which in this case, will be multiplied by cos Θ. The signal that results is a signal that is proportional to the AM waveform modulated by L + R. A conventional AM detector is used to demodulate this signal. This demodulated signal, proportional to L + R, then drives the single speaker. The stereo receiver applies a 1/cos Θ correction term to the incoming signal to allow accurate stereo demodulation.

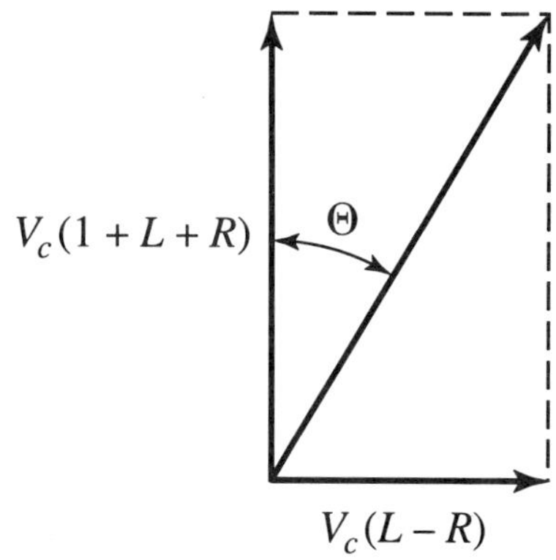


Figure 7.13: Phasor representation of AM stereo modulator outputs.

In addition to stereo AM transmission, QAM is used in microwave telephone relays operating in the 6- to 11-GHz range. As mentioned, analog color television applies a scheme similar to quadrature modulation in order to transmit color information over the same bandwidth used by black-and-white transmission.

### 7.4.10 Frequency Requirements for AM Broadcasting

An examination of Eq. (7.4), that is

$$v_{\text{out}} = V_c \cos\, \omega_c t + \frac{mV_c}{2} \cos\,(\omega_c - \omega_m)t + \frac{mV_c}{2} \cos\,(\omega_c + \omega_m)t$$

shows that a tone-modulated AM signal has three frequency components: the carrier frequency, $f_c$, and the two sideband frequencies, $f_c - f_m$ and $f_c + f_m$. The bandwidth necessary to accurately transmit or receive this signal is

$$\Delta_f(\text{AM}) = (f_c + f_m) - (f_c - f_m) = 2f_m$$

or twice the modulation frequency.

In transmitting an audio or musical baseband signal, the maximum frequency transmitted determines the required bandwidth of the AM signal. The FCC limits the maximum baseband signal frequency to 5 kHz for a commercial AM signal. While this frequency limit is quite acceptable for pure voice transmission, it offers rather low-quality transmission of musical programs that contain harmonics and overtones above 5 kHz.

The channel bandwidth for commercial AM signals is then twice this maximum baseband signal or 10 kHz. The transmitting circuitry must eliminate all baseband frequencies above 5 kHz before applying this signal to the modulator. Guidelines based on older FCC specifications [6] specify the 1-kHz baseband signal at 0 dB for the reference while the band from 100 Hz to 5 kHz must be within 2 dB of the reference output.

The FCC allows a 15-kHz baseband signal to be used in commercial FM broadcasting. The poor musical quality of commercial AM transmission compared to commercial FM transmission has nothing to do with the modulation processes, but is a result of the bandwidth limitation imposed by the FCC on the AM signal. This statement is true only for noiseless signals since one advantage of FM over AM is its noise immunity.

The 15-kHz limit on FM transmission is arguably overkill because not much energy of a musical signal falls between 10 kHz and 15 kHz. Furthermore, as people age, the ear's sensitivity to higher frequency signals decreases. However, the quality of FM transmission of music since the 1960s relegated AM transmissions to programs that were primarily voice, such as news or talk shows.

In the late 1970s stereo AM was proposed to the FCC. Several different methods were proposed to accomplish stereo transmission on the existing 10-kHz bandwidth allocated for AM transmission. AM broadcasters felt that this improvement was needed to allow AM to be more competitive with FM transmission. On the other hand, the baseband frequency bandwidth limit would still limit the tonal quality of the AM signal, even for stereo systems. In recent years, the FCC has allowed some stations to increase the transmission bandwidth, as will be discussed later.

When stereo AM transmission was first proposed, the FCC required that it be compatible with existing monophonic receivers. This requirement is designed to protect the consumer so that an older monophonic receiver can still receive a stereo transmission. Several different schemes of transmitting two separate signals over the same transmission bandwidth were developed by different companies. In 1980, the FCC approved a scheme developed by Magnavox. When other companies threatened legal action against the FCC, this agency rescinded this decision in 1982 and announced it would *not* settle on a specific system [3].

There were then five different methods of stereo AM transmission, each requiring its own method of reception. The Motorola C-QUAM system discussed in the previous section has emerged as the clear leader in the AM stereo race, having many more transmitters in the United States as well as more receivers [3]. The only other system that now competes with C-QUAM is the Hahn system that transmits the left stereo signal over the lower sideband and the right signal over the upper sideband. Some AM receivers include circuitry for both systems, with a switch that selects the system used according to the type of stereo transmission being received.

In 1987, the FCC approved a voluntary standard that allowed the increase of the maximum baseband frequency to something approaching 10 kHz. This greatly improves the tonal quality of music transmission. Not all stations transmit under this voluntary standard and a special receiver is required to recover the baseband signal. Figure 7.14 shows the filter requirements on the baseband signal. Note that the actual filter response can and normally will be less abrupt, falling on or below the ideal response shown in the figure.

For a well-designed filter, the 3-dB bandwidth can be extended above 8 kHz and the frequencies at 10 kHz may be down by a factor of 10. The quality of music programs is improved greatly when transmitted with this voluntary standard and becomes comparable to that of FM.

**Practical Considerations**

In areas with lower population density, the FCC may assign carrier frequencies to commercial AM stations that differ by 20 kHz. This minimizes interference when the voluntary standard is used. In highly congested areas such as Los Angeles, carrier frequencies may differ by only 10 kHz. Some receivers that apply the voluntary standard have adjustable filters that further limit the frequency response to achieve less interference from adjacent stations. Of course, tonal quality suffers in this compromise.

In spite of the fact that AM transmission has been around for over 80 years, the field remains dynamic as we enter the twenty-first century.

## 7.5 ANGLE MODULATION

**Important Concepts**

**1.** The amplitude of the carrier remains constant with modulation in angle modulation schemes. The angle and frequency of the signal are modified in accordance with the voltage of the baseband signal.

**2.** Phase modulation is used today in transmission of digital signals such as digital TV and has been traditionally important in creating the frequency-modulated signal for FM radio.

**3.** Frequency modulation (FM) is used for commercial radio and has also been used for the audio in analog TV receivers.

**4.** FM signals are less affected by noise than are AM signals, but require more bandwidth to transmit an identical baseband signal.

Amplitude modulation was the first type of modulation used to transmit radio broadcasts, beginning in the 1920s. The concept of frequency modulation was proposed around 1930, but did not become popular until the

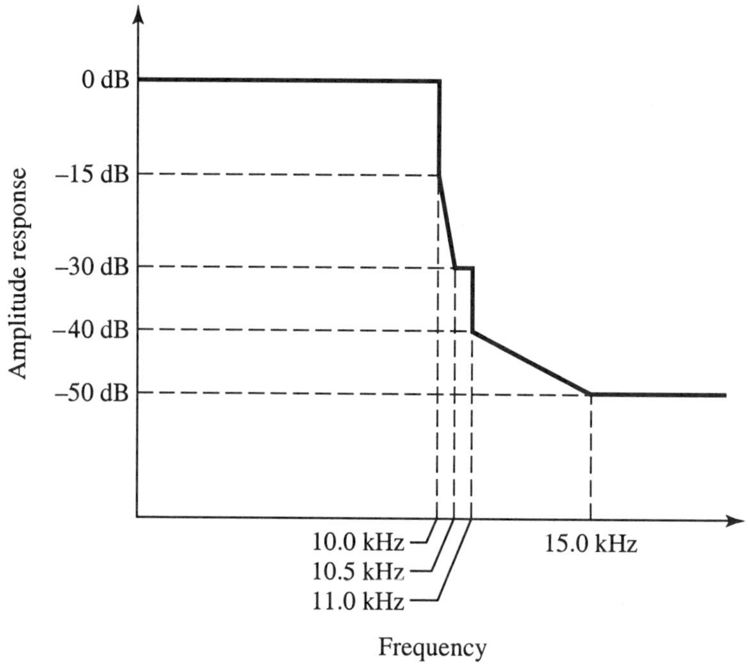


Figure 7.14: Ideal filter response for the voluntary standard.

1950s. At the present time FM broadcasting is the preferred method of music transmission because of its higher fidelity and better signal-to-noise ratio. Television uses FM to convey audio information while using VSB for the video. Mobile radio services and microwave telephone relays also use FM transmission.

Phase modulation is closely related to frequency modulation and is used in the production of the FM waveform in some modulation schemes. We will first become familiar with phase modulation before proceeding to frequency modulation. Phase-shift keying modulation methods are prevalent in satellite communications, especially when used for digital television signal transmission.

### 7.5.1 Phase Modulation

Phase modulation occurs when the phase of a carrier signal is caused to change from its quiescent value by the application of a modulating signal. Generally, the phase deviation varies directly with the amplitude of the modulating signal. If a sinusoidal modulating signal, $M \sin \omega_m t$, is applied to a phase modulator, the resulting phase-modulated waveform can be expressed as

$$v_{\text{out}} = V_c \cos\left(\omega_c t + \phi_0 + k_p M \sin \omega_m t\right) \tag{7.16}$$

The amplitude of the carrier signal is unchanged, but the phase shift of the waveform now changes with time in proportion to the modulating signal. The constant $k_p$ is determined by certain circuit parameters to be discussed in the following paragraphs. When the maximum value of $M$ is applied to the modulator, the largest phase deviations occur. This maximum possible phase deviation is called the phase modulation index and is

$$m_p = k_p M_{\max} \tag{7.17}$$

Figure 7.15 shows the comparison of a waveform with zero phase shift to a second of advancing phase. Just as in the case of the AM waveform, the difference in carrier and modulating frequencies is much larger than implied from the waveforms of Figure 7.15. The change in phase for typical values of modulating signal requires several cycles of carrier signal before a significant phase change is noted.

Comparison of the two waveforms shows that the frequency of the advancing phase waveform is higher than the zero phase signal. If the zero phase signal is the unmodulated carrier while the advancing phase signal

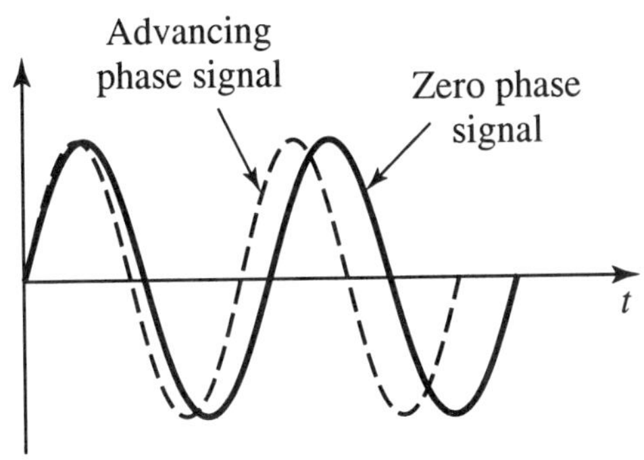


Figure 7.15: Effect of changing phase.

corresponds to the phase modulated waveform, we see that phase modulation results in both a phase change and a frequency change.

### Frequency Deviation of a Phase Modulator

Examination of Figure 7.15 reveals that a change in phase of a signal also leads to a change in frequency. Given the phase-modulated waveform represented by Eq. (7.16),

$$v_{\text{out}} = V_c \cos\left(\omega_c t + \phi_0 + k_p M \sin\ \omega_m t\right)$$

we can find the instantaneous frequency by differentiating the argument or instantaneous phase with respect to time. The argument of the waveform is given by

$$\theta = \left(\omega_c t + \phi_0 + k_p M \sin\ \omega_m t\right)$$

The instantaneous radian frequency is found from

$$\omega = \frac{d\theta}{dt} = \omega_c + k_p M \omega_m \ \cos\ \omega_m t \tag{7.18}$$

The unmodulated frequency ($M = 0$) is $\omega_c$, whereas the modulated frequency deviates from this value in both the positive and negative directions by a maximum value of

$$\Delta\omega = k_p M \omega_m \quad \text{or} \quad \Delta f = k_p M f_m \tag{7.19}$$

The frequency deviation of the phase-modulated waveform varies with modulating frequency. A constant value of modulating voltage leads to a constant phase deviation, regardless of the modulating frequency, but the frequency deviation increases directly with modulating frequency. This result is not surprising when the process of phase modulation is considered. During the time of one modulating voltage period, the phase is changed sinusoidally with an amplitude equal to the phase deviation. This phase change results in a frequency change. If the frequency of the modulating voltage is increased while the phase deviation remains the same, it follows that a higher frequency deviation results.

### Practice Problem

**7.10** A phase modulator has a value of $m_p = 0.12$. What is the maximum frequency deviation for a 2-kHz modulating signal? Repeat for an 8.6-kHz signal? *Ans:* 240 Hz, 1032 Hz.

### 7.5.2 Frequency Modulation

Frequency modulation occurs when the frequency of a carrier signal changes in accordance with a modulating signal. For most communication systems, the frequency deviation is related linearly to the amplitude of the modulating signal. The expression for frequency of an FM signal with sinusoidal modulation is

$$f = f_c + k_f M \cos \omega_m t \tag{7.20}$$

where $k_f$ is a constant and $M$ is the magnitude of the modulating signal applied to the frequency modulator. The frequency deviation of the FM signal is the maximum change in frequency and from Eq. (7.20) is seen to be

$$\Delta f = k_f M$$

The index of modulation for a tone-modulated FM signal is defined as the maximum frequency deviation of the waveform divided by the modulation frequency, thus,

$$m_f = \frac{\Delta f}{f_m} = \frac{k_f M}{f_m} \tag{7.21}$$

The FCC allows a maximum frequency deviation of 75 kHz in commercial FM broadcasting. This deviation is typically imposed on a carrier frequency of approximately 100 MHz. The maximum frequency deviation is only 0.075%, requiring a very stable carrier frequency. A small drift in frequency of the carrier appears as a frequency deviation and will lead to a false signal being detected at the receiver.

The frequency deviation of the FM signal is constant with the modulating frequency. In comparing this situation to that of a phase modulator, we see from Eq. (7.19) that the frequency deviation varies with modulating voltage for the phase modulator. If the modulating signal applied to a phase modulator were made to vary inversely with modulation frequency, the phase modulator would produce an FM signal. This fact is used to create the Armstrong frequency modulator that has been used for commercial FM transmission for decades.

Rather than produce a phase deviation that is constant at all frequencies, the Armstrong system causes the phase deviation to vary inversely with modulation frequency. At higher modulation frequencies, the phase deviation is less and the frequency deviation is constant. Instead of applying a modulation signal of $M \sin \omega_m t$ to the phase modulator, a signal of

$$k_I \frac{M}{\omega_m} \sin \omega_m t$$

is applied, where $k_I$ is a constant. The output of the modulator is given by Eq. (7.16) with $\phi_0$ taken as zero. This gives

$$v_{\text{out}} = V_c \cos \left( \omega_c t + \frac{k_I k_p M}{\omega_m} \sin \omega_m t \right)$$

When the argument of this waveform is differentiated with respect to time to find the instantaneous frequency, the result is

$$\omega = \omega_c + \frac{k_I k_p M}{\omega_m} \omega_m \cos \omega_m t = \omega_c + k_I k_p M \cos \omega_m t$$

The radian frequency deviation is now constant with modulation frequency which is the desired result. For this case, the deviation in Hertz is

$$\Delta f = \frac{k_I k_p M}{2\pi} \tag{7.22}$$

Figure 7.16 shows an FM waveform with a highly exaggerated frequency deviation.

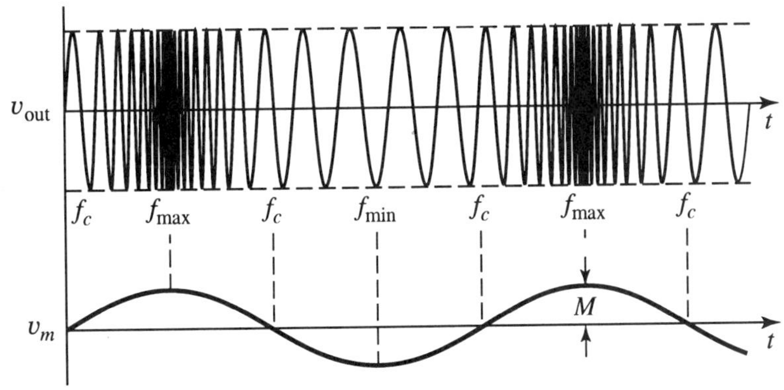


Figure 7.16: An FM waveform.

In summary,

1. During each cycle of the modulating signal, $v_m$, the frequency of $v_{out}$ moves from $f_c$ to $f_{max}$, then back to $f_c$, then to $f_{min}$, then back to $f_c$. The magnitude of $v_{out}$ remains constant.

2. The frequency deviation is

$$\Delta f = f_{max} - f_c = f_c - f_{min}$$

3. For FM, a change in modulating frequency, $f_m$, does not change $\Delta f$ as it would in PM.

4. An increase in modulating signal amplitude, $M$, increases $f_{max}$, decreases $f_{min}$, and thereby increases $\Delta f$.

5. For a typical $f_c$ of 100.1 MHz in commercial FM, the maximum frequency deviation, limited by the FCC, is $\Delta f_{max} = 75$ kHz. Thus, $f_{max} = 100,175$ kHz and $f_{min} = 100,025$ kHz. The maximum frequency change is less than 0.1%.

### Bandwidth Requirements of FM Waveforms

The instantaneous frequency of an FM waveform varies from a low value of $f_c - \Delta f$ to a high value of $f_c + \Delta f$. It might appear that a bandwidth of $2\Delta f$ is sufficient to transmit the FM signal. This conclusion is incorrect inasmuch as additional sidebands are generated as the instantaneous frequency varies. Carson's rule [15] states that the bandwidth required to transmit the entire FM waveform is twice the sum of the maximum frequency deviation and the modulating frequency; that is,

$$BW = 2(\Delta f + f_m) \tag{7.23}$$

We will not derive this result here, but will refer the interested reader to the reference given where a complete derivation of the frequency spectrum of an FM signal is given. Any frequency-selective network through which the FM signal must pass with negligible distortion must have a bandwidth given by Eq. (7.23). For example, a commercial FM station with an allowed maximum frequency deviation of 75 kHz and a maximum modulation frequency of 15 kHz would apparently require a bandwidth of

$$BW = 2(75 + 15) \text{ kHz} = 180 \text{ kHz}$$

In fact, the minimum separation between FM radio stations is 200 kHz.

**Practical Considerations**

The preceding calculation of bandwidth is based on a modulating signal of 15 kHz that is large enough to drive the output signal to a frequency deviation of 75 kHz. If an FM station is transmitting music, any component that has a frequency of 15 kHz will have a very small amplitude compared to the components below 1 kHz. Thus, Carson's rule predicts a much broader bandwidth than is actually required. This excess bandwidth is used to advantage in stereo FM systems that transmit two channels plus additional information on the allocated bandwidth.

Although the quality and noise immunity of the FM system is much better than the AM system, the bandwidth requirements are considerably higher. A commercial AM transmitter uses a bandwidth of slightly more than 10 kHz to transmit a maximum modulation frequency of 5 kHz. If an AM signal had a maximum baseband frequency of 15 kHz, a bandwidth of 35-40 kHz would be used for transmission. This can be compared to the 180 kHz for an FM signal calculated from Eq. (7.23) that is a very conservative estimate. It should also be noted that a SSB system requires only one-half the bandwidth of a conventional AM system; thus microwave telephone relays and some emergency services use SSB to conserve bandwidth even more.

#### Narrowband Angle Modulation

If a phase modulator has an index of modulation, $m_p$, of about 0.2 rad or less, the resulting output is referred to as a narrowband signal. This definition has some implications for an FM system based on this phase modulator. As we will see in Chapter 9, the small value of $m_p$ will lead to a correspondingly low value of frequency deviation when used in a frequency modulation system. With tone modulation, the frequency deviation of a phase-modulated signal was found to be

$$\Delta f = k_p M f_m = m_p f_m$$

whereas the index of modulation of an FM signal is

$$m_f = \frac{\Delta f}{f_m} = m_p$$

Historically, John Carson, after whom the rule is named for FM bandwidth approximation, published a paper in a leading technical journal indicating that FM was a poor method to use in radio transmission. Edwin Armstrong later demonstrated that FM was much better than AM in terms of noise immunity. Although Carson was a very able engineer with many contributions to his name, he based his conclusion on the narrowband case rather than the wideband modulation actually used in commercial FM that may have an index of modulation in the range of 3 to 10.

#### Practice Problem

**7.11** What is the required bandwidth for an FM signal with a maximum modulation frequency of 10 kHz and a maximum frequency deviation of 55 kHz? *Ans:* 130 kHz.

### 7.5.3 Phase-Shift Keying (PSK)

Although PSK is a special case of phase modulation, its importance in digital data transmission warrants a separate discussion of this modulation scheme. PSK is used in telephone modems and transmission of digital television signals. For modems, the carrier frequency is rather low, while television transmission requires a very high carrier frequency. There are several forms of PSK such as binary PSK (BPSK) and quaternary PSK (QPSK), as well as higher orders of PSK. We will confine our discussion to BPSK.

In transmitting digital data, a minimum of two levels is needed to represent binary 1 and binary 0. Given a sinusoidal reference signal, a 1 could be represented by an in-phase sinusoid while a 0 could be represented by a

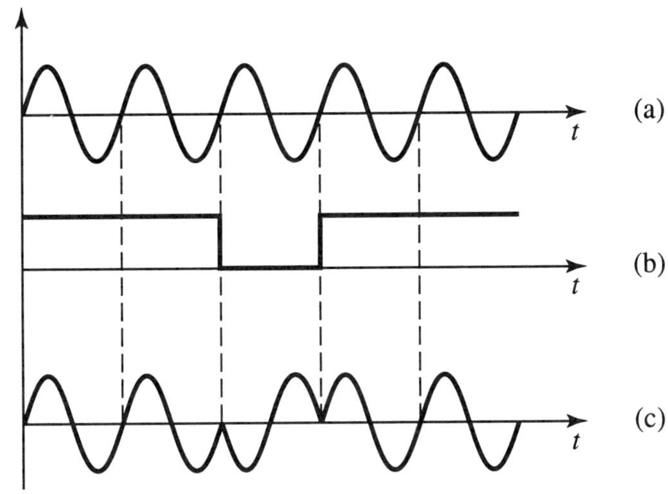


Figure 7.17: BPSK signals. (a) Reference. (b) Digital input. (c) Modulated output.

signal that is 180° out of phase with the reference. This type of representation is called BPSK and is shown in Figure 7.17.

The input data signal is typically synchronized with the reference signal so that the phase reversal occurs as a cycle of reference signal is completed.

A very simple way to create BPSK is by using a DSB-SC signal, applying the digital signal as the modulation. The digital signal uses a voltage of +1 V for binary 1 and a voltage of −1 V for binary 0. We recall that a DSBSC signal is nothing more than the product of the carrier and the modulating signals, that is,

$$v_{\text{out}} = v_m \times v_c \tag{7.24}$$

In the section on amplitude modulation, the product was formed for two sinusoidal signals. For BPSK, the carrier, $v_c$, is the sinusoidal reference signal while the digital input data is the second signal, $v_m$. The output of the multiplier is then

$$v_{\text{out}} = v_c \text{ for } v_m = +1$$

and

$$v_{\text{out}} = -v_c \text{ for } v_m = -1$$

If the digital modulation signal is synchronized with $v_c$, the signal of Figure 7.17 will result.

The BPSK system requires that the reference signal be present at the receiver for detection of the digital signal. This can be generated by various schemes from the received signal. To avoid having to generate the reference signal, differential PSK (DPSK) is sometimes used. In this form of PSK, a binary 1 may be represented by a change of phase from the previous value, and a binary 0 may be represented by no change in phase. The receiver uses the received signal as its reference and detects phase changes from this value. When the phase changes, a binary 1 is registered.

The QPSK system transmits four phases relative to a reference signal. Each phase differs by 90° from adjacent phases. This system allows four digital bits to be represented by each phase shift. Obviously, more data can be transmitted over the same bandwidth with the QPSK system. In fact, higher order systems can transmit more data, but the phase differences become smaller, resulting in a lower signal-to-noise ratio for higher order systems.

The bandwidth required to transmit BPSK is difficult to calculate, but if the data pulses are rounded appropriately, BPSK becomes quite bandwidth efficient [16].

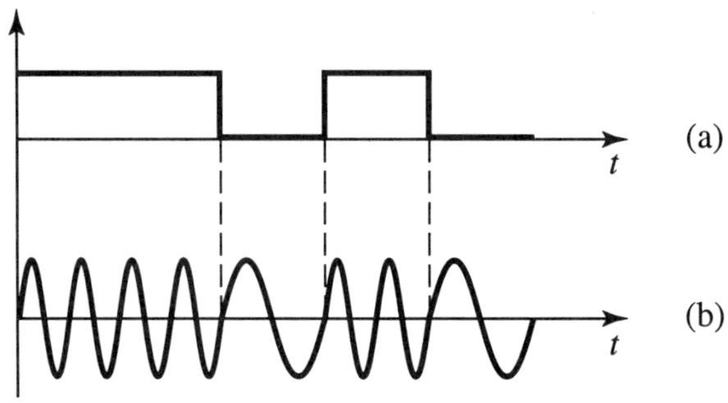


Figure 7.18: An FSK signal. (a) Digital modulating signal. (b) Output signal.

### 7.5.4 Frequency-Shift Keying (FSK)

FSK is similar to PSK except the signal changes frequency as the digital data change from 0 to 1 or 1 to 0. An FSK signal is shown in Figure 7.18.

The earliest modems were designed using FSK and allowed full-duplex operation wherein signals can be sent in both directions along the line at the same time. If a terminal was connected to a computer using a modem at both ends of the line, the terminal modem might use 1270 Hz to transmit binary 1 and 1070 Hz for binary 0. Although the computer modem recognized these frequencies in receiving data, frequencies of 2225 Hz for binary 1 and 2025 Hz for binary 0 might be used to transmit data [6]. In this way, both modems can transmit and receive simultaneously over a single pair of lines.

FSK can be generated in several ways. Two separate oscillators can be used with a switching network connecting one or the other oscillator to the output, in accordance with the modulating signal. A voltage-controlled oscillator (VCO) can directly produce the FSK signal by connecting the digital modulation to the voltage control input of the VCO.

---

#### SUMMARY

**1.** The communications field continues to expand at a rapid pace, making communication circuits an important part of the electrical engineering field.

**2.** A communication system consists of a transmitter that combines a carrier signal and a baseband signal. The transmitter sends the modulated signal over a channel to a receiver or to several receivers. The receiver amplifies the signal and recovers the baseband information.

**3.** Three important types of modulation are amplitude, phase, and frequency modulation.

**4.** In amplitude modulation, the peak value of carrier signal is varied or modulated in proportion to the value of a much lower frequency baseband signal.

**5.** Phase and frequency modulation systems generate an output signal with constant amplitude. The phase and frequency of the signal vary in accordance with the baseband signal.

---

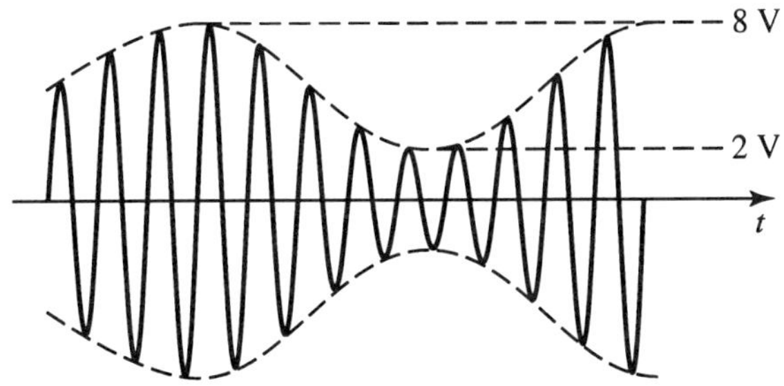


**Figure P7.6**

## PROBLEMS

### Section 7.2.2 A Communication System Example

**7.1** A transmitting antenna must have a physical dimension that is at least one-tenth the wavelength of the transmitted frequency. What would the minimum length of an antenna be to directly transmit the following frequencies?

(a) 100 Hz (b) 5 kHz
(c) 760 kHz (d) 28 MHz

**7.2** An amateur radio transmits with a carrier frequency of 27 MHz. What is an appropriate length for the transmitting antenna, assuming a $\lambda/4$ antenna is used?

**7.3** A $\lambda/4$ transmitting antenna for an AM radio station is approximately 362 feet in length. At what carrier frequency does this station operate?

**7.4*** Compare the $\lambda/4$ antenna lengths to transmit channel 2 (55.25 MHz), channel 7 (175.25 MHz), and channel 14 (471.25 MHz). Why are TV transmitting antennas placed atop very high towers, even though the actual antenna length is not large?

**7.5** Explain why TV receiving antennas have several different-length elements making up the antenna.

### Sections 7.4.1-7.4.2 The Amplitude Modulated Waveform-Index of Modulation

**7.6** Calculate the index of modulation if the applied signal is a sinusoid.

**7.7** If $m = 0.8$, calculate $V_1$ and $V_2$ assuming a sinusoidal modulation. Assume $V_c = 8$ V.

**7.8** A signal varying from 100 Hz to 5 kHz is to be transmitted by means of an AM system with a 600-kHz carrier frequency. What frequency range is required to transmit the AM signal?

**7.9*** What is the $Q_{\text{eff}}$ for a tuned circuit used in Problem 7.8 that causes a maximum of 1 dB attenuation for the transmitted frequencies?

**7.10*** Repeat Problem 7.9 for a maximum attenuation of 2 dB.

**7.11** A carrier signal with a 10-V amplitude is tone modulated to cause a maximum amplitude of 18.6 V. What is the index of modulation for this signal in percent? What is the amplitude of the sideband components?

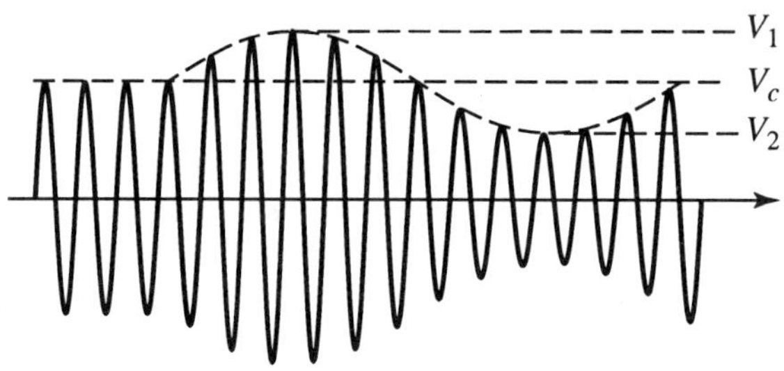


**Figure P7.7**

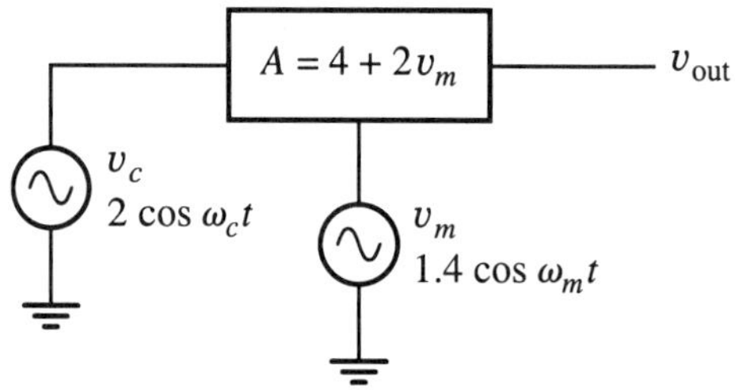


**Figure P7.16**

**7.12** A modulated signal has a maximum amplitude of 24 V. If the index of modulation is 90%, what is the amplitude of the unmodulated carrier? What is the amplitude of the sidebands?

**Section 7.4.3 Mathematical Description of the AM Waveform**

**7.13** A carrier signal of 8-V unmodulated amplitude and frequency of 1.26 MHz is tone modulated to a maximum amplitude of 14.3 V. Calculate the amplitudes and frequencies of all components of this signal.

**7.14*** A square wave, varying from -1 V to +1 V, is transmitted by an AM system. The square wave can be approximated by the first three terms of the Fourier series:

$$f(t) = \frac{4}{\pi}\left[\cos\ \omega t - \frac{1}{3}\ \cos\ 3\omega t + \frac{1}{5}\ \cos\ 5\omega t\right]$$

The period of the square wave is 0.5 ms. If the carrier frequency is 600 kHz with a 10-V amplitude, what frequency components will be present in the AM waveform? Assuming that 100% modulation is to be used, calculate the amplitudes of all components of the transmitted signal.

**7.15*** Repeat Problem 7.14 for the transmission of a sawtooth waveform that is represented by

$$f(t) = \frac{2}{\pi}\left[\sin\ \omega t - \frac{1}{2}\ \sin\ 2\omega t - \frac{1}{4}\ \sin\ 4\omega t\right]$$

The period of this sawtooth waveform is 1 ms and the waveform varies from -1 V to +1 V.

**7.16** Calculate the index of modulation for the modulator shown. Calculate the carrier output voltage and the sideband voltages.

**7.17** An AM signal is expressed as

$$v_{out} = [6 + 2\ \cos\ \omega_m t - 0.5\ \cos\ 3\omega_m t]\ \cos\ \omega_c t$$

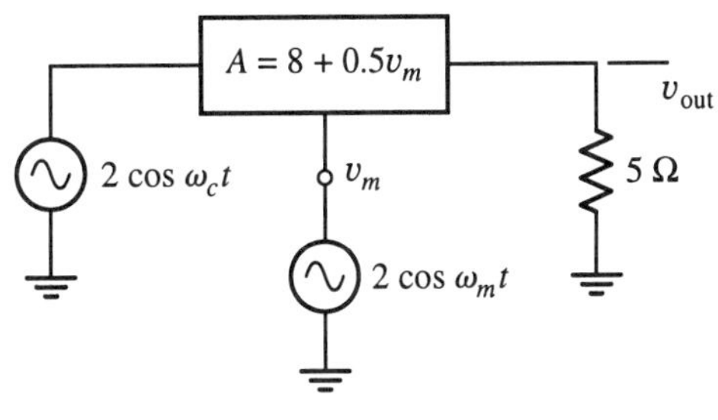


**Figure P7.20**

(a) What is the magnitude of the unmodulated carrier?
(b) What is $\Delta^+$?
(c) What is $\Delta^-$?
(d) What is $m$?

### Section 7.4.4 Power Relations in AM Waveforms

**7.18** A total of 12 W of power is transmitted in an AM waveform. If $m = 0.5$ for a sinusoidal modulating signal, calculate the carrier power and the power in each sideband.

**7.19** A total of 5 W maximum power can be transmitted using an antenna of 50-Ω impedance. Using tone modulation with $m = 1$, what is the magnitude of the largest sinusoidal signal that can be transmitted using amplitude modulation?

If SSB is used, what is the magnitude of the largest sinusoidal signal that can be transmitted?

**7.20*** For the variable-gain modulator shown,

(a) Calculate $m$, the index of modulation.
(b) Calculate the carrier power in $v_{out}$.
(c) Calculate the power in each sideband of $v_{out}$.
(d) To what peak amplitude must $v_m$ be increased to result in $m = 1$?
(e) Calculate the power in each sideband of $v_{out}$ for part (d).

### Section 7.4.8 On-Off Keying (OOK)

**7.21** An OOK system uses a carrier frequency of 1 MHz, a bit time of 0.125 ms, and a 50% duty cycle. How many frequency components must be transmitted to carry 95% of the total power of the output signal?

**7.22*** An OOK system uses a carrier frequency of 1 MHz, a bit time of 0.1 ms, and a 25% duty cycle. For a repeated one-zero (10) code, identify the smallest group of frequency components that carry at least 95% of the total power of the output signal.

### Section 7.4.10 Frequency Requirements for AM Broadcasting

**7.23** Using standard AM to transmit a 100-Hz to 5-kHz signal, what is the maximum circuit $Q$ for a narrowband circuit to attenuate the highest transmitted frequency by 0.5 dB? The carrier frequency is 1.2 MHz. What is the 3-dB bandwidth of the narrowband circuit?

**7.24** If the signal of Problem 7.19 is transmitted in a SSB system, what are the values of $Q$ and 3-dB bandwidth required?

### Section 7.5.1 Phase Modulation

**7.25** A phase modulator produces an output signal of

$$v_{\text{out}} = 10 \cos\left[2\pi \times 10^6 t + km(t)\right]$$

where $m(t)$ is the modulating signal (in volts) applied to the modulator input. The constant $k$ is 0.12 rad/V.

> **(a)** A modulation signal of $v_m = 5 \cos 2\pi \times 10^3 t$ is applied. Calculate the frequency deviation of the output signal in Hz.
>
> **(b)** Repeat part (a) if the modulating signal frequency is changed to $3 \times 10^3$ Hz.

**7.26*** A modulation voltage magnitude of 6 V is applied to a phase modulator to create a maximum phase deviation of 9°.

> **(a)** Calculate the constant, $k_p$, in rad/V from Eq. (7.16).
> **(b)** What is the maximum frequency deviation if $f_m = 100$ Hz?
> **(c)** What is the maximum frequency deviation if $f_m = 4$ kHz?
> **(d)** What is the maximum frequency deviation if $f_m = 15$ kHz?

### Section 7.5.2 Frequency Modulation

**7.27** A 10-kHz modulating signal applied to a frequency modulator causes a frequency deviation of 42 kHz. What is the required bandwidth to transmit or receive this FM signal?

**7.28*** A 1-kHz signal of amplitude 5 V applied to a frequency modulator leads to a frequency deviation of 30 kHz. The maximum magnitudes of modulating components above 1 kHz fall 6 dB/octave from the value at 1 kHz. If the maximum frequency of the modulating signal is 15 kHz, what is the bandwidth required to transmit or receive this signal?

**7.29** Repeat Problem 7.28 if the maximum magnitudes of any of the components above 1 kHz can be 5 V?

**7.30** When a 5-V modulating signal is applied to an FM modulator, a frequency deviation of 40 kHz results. The magnitudes of several components of a modulating signal applied to this modulator are tabulated below.

| $f_m$, kHz | $\lvert v_m \rvert$, V |
|---|---|
| 0.5 | 5 |
| 1 | 3 |
| 4 | 1 |
| 8 | 0.8 |
| 15 | 0.5 |

What bandwidth is required to efficiently transmit this signal?

## REFERENCES AND SUGGESTED READING

1. Abramson, A., *The History of Television, 1880 to 1941.* Jefferson, N.C.: McFarland, 1987.

2. Armstrong, E. H., "Method of reducing disturbances in radio signalling by a system of frequency modulation," *Proc. I. R. E.* Vol. 24, May 1936.

3. Braun, M. J., *AM Stereo and the FCC.* Norwood, N.J.: Ablex Publishing Company, 1994.

4. Brittain, J. E. (Ed.), *Turning Points in American Electrical History.* New York: IEEE Press, 1977.

5. Carson, J., "Notes on the theory of modulation," *Proc. I. R. E.,* Vol. 10, February 1922, pp. 57-82.

6. Couch, L. W., II, *Modern Communication Systems, Principles and Applications.* Englewood Cliffs, N.J.: Prentice-Hall, 1995.

7. Davis, H. B. O., *Electrical and Electronic Technologies.* Metuchen, N.J.: Scarecrow Press, 1983.

8. Dummer, G. W. A., *Electronic Inventions and Discoveries, 3rd edition.* Elmsford, N. Y.: Pergamon Press, 1983.

9. Farnsworth, E. G., *Distant Vision.* Salt Lake City, Utah: Pemberlykent Publishers, 1989.

10. Finn, B. S., "Controversy and the growth of the electrical art," *IEEE Spectrum,* Vol. 3, No. 1, January 1966, pp. 52-56.

11. Gregg, W. D., *Analog and Digital Communication.* New York: John Wiley and Sons, 1977.

12. Hinds, L., "It's Stratovision," *Invention and Technology.* New York: American Heritage, Inc., Spring, 1994

13. Lewis, T., *Empire of the Air* New York: HarperPerennial, 1993.

14. Shiers, G., "Early schemes for television," *IEEE Spectrum,* Vol. 7, No. 5, May 1970, pp. 24-34.

15. Taub, H. and Schilling, D., 1986, *Principles of Communication Systems, 2nd edition.* New York: McGraw-Hill Book Co., 1986.

16. Young, P. H., *Electronic Communication Techniques, 4th edition.* Englewood Cliffs, N.J.: Prentice-Hall, 1999.

# Chapter 8

# AMPLITUDE MODULATION CIRCUITS

Chapter 7 considered some of the basic ideas of AM, PM, and FM transmission. This chapter will expand the AM concepts to the point that actual circuits can then be considered to implement communication systems. Detection of modulated signals, mixing or heterodyning, and pre-emphasis of audio signals will be discussed, as well as actual circuit configurations used to accomplish these functions.

Traditionally, AM communication systems have been important in commercial radio, in the transmission of multiple telephone signals on a single pair of lines, and in the transmission of television signals. Although all three of these applications are still prominent today, the overall usage of AM methods has diminished over the past few decades. Frequency modulation is the method of choice for high-quality commercial radio. Pulse-code modulation schemes are now used much more than AM methods in the transmission of phone signals and, with the advent of high-definition television (HDTV), AM television transmission will shortly become obsolete.

A student might therefore question why an entire chapter is devoted to AM methods. Actually, many reasons justify this topic. The first is that AM is being rediscovered as the most bandwidth-efficient transmission scheme compared to FM or PM methods. The electromagnetic spectrum has very little space that is unassigned by the FCC. Schemes that use less bandwidth than others have become more important as the spectrum has become overcrowded. Recently, the FCC reallocated some FM spectrum in the range of 150 MHz to 170 MHz for AM (SSB) because of the bandwidth savings. This range of frequencies is used for emergency services, and AM allows more channels to be assigned in a given band of frequency.

Microwave links for public telephone networks have also been switching from FM to AM, again SSB, since 1980. More conversations can be relayed over a fixed total bandwidth using SSB instead of FM. The new voluntary standard for AM transmission promises to make commercial AM more competitive with commercial FM. Amateur radio operators now transmit largely SSB signals.

Because of its increasing importance in commercial applications, it is appropriate to consider AM systems. Furthermore, the creation of the DSB waveform is equivalent to the important mixing operation used in both analog and digital communication methods.

## 8.1 PRODUCTION OF AM WAVEFORMS

**Important Concepts**

**1.** For low-power systems, amplitude modulators can be based on variable gain amplifiers. The gain varies in accordance with the modulating or baseband signal. The carrier signal is amplified by this variable gain.

**2.** The modulation linearity of the transconductance multiplier is very good even for modulation indexes approaching 100%.

**3.** Class-C modulators are often applied to high-power modulation. Very high-power modulators use special purpose vacuum tubes to achieve the power levels required.

**4.** A class-D modulator can be used to produce a 50-kW amplitude modulator based on power MOSFET devices.

This section will discuss concepts and circuits designed to produce AM waveforms. The variable-gain circuit will introduce the notion of modulating the magnitude of a carrier signal although practical implementations of this simple idea are used for low-power amplitude modulators.

### 8.1.1 Variable-Gain Modulators

A conceptually simple method of achieving amplitude modulation is shown in Figure 8.1.

In this figure, the gain of the amplifier depends linearly on the modulating signal. If the modulating signal is zero, the carrier signal is amplified by a constant value. As $v_m$ becomes positive the gain increases as does the magnitude of the output signal. For negative values of $v_m$ the magnitude of the output signal decreases.

Let us assume that $v_{\text{in}}$ is a sinusoidal signal expressed by $C \cos\, \omega_c t$, while $v_m$ can be represented by $M \cos\, \omega_m t$. The output signal is given by

$$v_{\text{out}} = A_0 C \cos\, \omega_c t + A_1 M C \cos\, \omega_m t \cos\, \omega_c t \tag{8.1}$$

which can also be expressed as

$$v_{\text{out}} = A_0 C \left[1 + \frac{A_1 M}{A_0} \cos\, \omega_m t\right] \cos\, \omega_m t = V_c\, [1 + m \cos\, \omega_m t] \cos\, \omega_c t \tag{8.2}$$

where $A_0 C = V_c$ and $m = A_1 M / A_0$ for this case. The latter part of this equation represents the standard AM waveform discussed in Chapter 7.

Achieving the linear gain variation of the ideal modulator of Figure 8.1 is not a trivial matter. We will discuss two methods of implementing this variable-gain modulator. The first is the transistor modulator which now finds its major use as a *mixer*. The second is the multiplying cell that is implemented as an integrated circuit.

**Practice Problems**

**8.1** If $A_0 = 10$ V/V and $A_1 = 2$ V/V in Figure 8.1, what index of modulation will result for $v_m = 2 \cos\, \omega_m t$? *Ans:* 40%.

**8.2** If $C = 0.5$ V in Practice Problem 8.1, what are the amplitudes of the carrier output and the sidebands? *Ans:* $V_c = 5$ V, $v_{SB} = 1$ V.

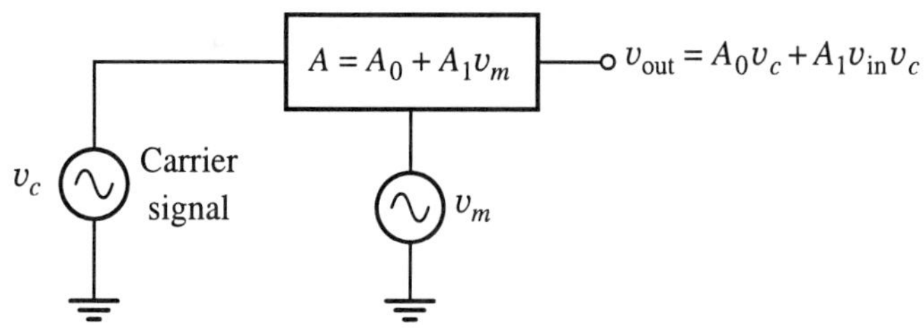


Figure 8.1: Simple amplitude modulator.

### The Transistor Modulator

The dynamic resistance of the base-emitter junction can be modulated to create a variable voltage gain in a single-stage transistor amplifier. Under proper conditions, the amplification of a carrier signal is made to vary directly with a modulating signal, producing an AM waveform at the output. This modulator corresponds closely to the basic scheme suggested in Figure 8.1.

The MOSFET can also be used as a modulator by using the modulation signal to control the transconductance, and hence the gain of the stage. We will defer the discussion of the MOSFET circuit to the section on mixers.

The BJT modulator produces rather small indexes of modulation and for that reason is rarely used as a modulator at the present time. However, it is instructional to consider the principles involved in this type of circuit. Figure 8.2 illustrates the basic circuit.

In this circuit, the capacitors $C_1$ and $C_2$ are chosen to offer low impedance to the carrier signal and high impedance to the modulating signal. This is easy to achieve if there is a large difference between carrier and modulating frequency. It will also be useful to require that $R \gg [r_x + (\beta + 1)r_e]$.

The dc source, $V_0$, and the ac modulation source, $v_m$, become a bias circuit with a slowly varying bias signal. This circuit sets and adjusts the gain of the stage for the small-input carrier signal, $v_c$.

If $v_m$ is zero, the quiescent gain of the circuit is

$$A = \frac{v_{\text{out}}}{v_c} = \frac{-\alpha R_L}{(1-\alpha)r_x + r_e} \approx -\alpha R_L r_e$$

The dc emitter current will be determined by $V_0$ and $R$ and will fix the value of $r_e$. To modulate the gain and hence the output signal, a finite value of $v_m$ is imposed to control the emitter current and $r_e$. The emitter current is now given by

$$I_E = (\beta + 1)I_B = (\beta + 1)\frac{[V + v_m]}{R} \tag{8.3}$$

where $V = V_0 - V_{BE(on)}$. Using $r_e = 26/I_E$, we find that the output signal is then

$$v_{\text{out}} = \frac{-\alpha R_L v_c}{r_e} = \frac{-\beta R_L v_c V}{0.026R}\left[1 + \frac{v_m}{V}\right] \tag{8.4}$$

Equation (8.4) includes a product term involving $v_c$ and $v_m$. As mentioned in Chapter 7, this is the term that leads to the two sidebands.

Using a carrier signal of $v_c = C \cos\, \omega_c t$ and a modulating signal of $v_m = M \cos\, \omega_m t$ gives

$$v_{\text{out}} = \frac{-\beta R_L V C}{0.026R}\left[1 + \frac{M}{V}\cos\, \omega_m t\right]\cos\, \omega_c t$$

$$= -V_c\,(1 + m \cos\, \omega_m t)\, \cos\, \omega_c t \tag{8.5}$$

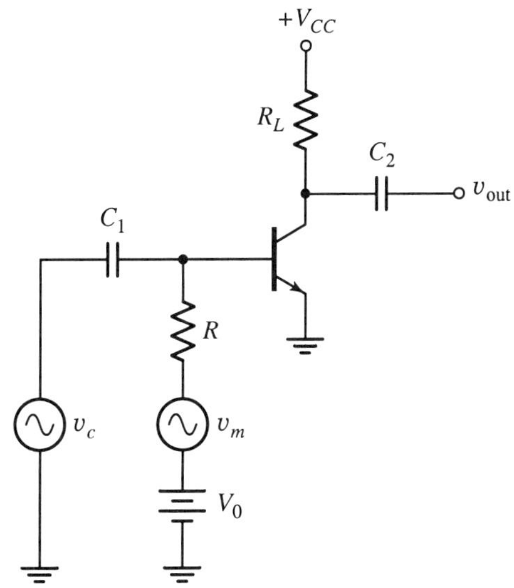


Figure 8.2: Transistor modulator using a BJT.

where

$$V_c = \frac{\beta R_L V C}{0.026R} \tag{8.6}$$

and

$$m = \frac{M}{V} \tag{8.7}$$

The transistor modulator uses the sum of the bias voltage and the modulating signal to control the large-signal value of $I_E$. This controls the amplifier gain. The carrier signal must be small enough to have a negligible effect on this large-signal value of $I_E$. The equation for dynamic diode resistance applies only when the dc bias current is much greater than the ac signal current and when the transistor remains in the active region. We can consider the bias current for the modulator to result from the voltage $V_0 + v_m$. As $v_m$ swings positive to result in a maximum bias voltage of $V_0 + M$, the transistor must not saturate. As $v_m$ swings negative to a value of $V_0 - M$, the transistor must not cut off.

**Practical Considerations**

From Eq. (8.7), 100% modulation cannot be obtained unless $M = V_0 - V_{BE(on)}$; thus the percent modulation must be limited to some value less than 100% in order to avoid cutoff. Further limiting the index of modulation is the fact that the approximation of $r_e + (1-\alpha)r_x \approx r_e$ used in developing Eq. (8.4) will not hold at high currents. The net result of these nonideal effects is that unwanted nonlinearities are introduced into the modulator. The results of these nonlinearities can be reduced by filtering in some applications, but in general they cause a limitation to be placed on the index of modulation. An index of modulation ranging from 0.1 to 0.15 is rather typical for a transistor modulator of this type. Applications for a modulator with such a low index of modulation might include a wireless intercom or some other low-quality voice circuit.

# Example 8.1

Using a transistor with $\beta = 80$, $V_{BE(\text{on})} = 0.7$ V, $r_x = 200\ \Omega$, and a 12-V power supply, design a transistor modulator with $m = 10\%$. The carrier source is $5 \cos\, \omega_c t$ mV at a frequency of 1260 kHz. The modulating signal is a sinusoidal signal at 2 kHz. The load impedance that must be driven is 10 kΩ. Specify all element values and the magnitude of the modulating signal.

**Solution** The circuit of Figure 8.2 will be used with the addition of a 10-kΩ resistance to the output, connected to $C_2$. In order to minimize loading in the output circuit, the collector resistor will be selected to be 2 kΩ.

With a 12-V power supply, $R$ can be selected to bias the stage to $V_{CEQ} = 6$ V. This requires 3 mA of collector current and a base current of

$$I_{BQ} = \frac{3}{80} = 37.5\ \mu\text{A}$$

Using $V_0 = 12$ to require a single power supply, we find the resistance $R$ to be

$$R = \frac{12 - 0.7}{0.0375} = 301\ \text{k}\Omega$$

The amplitude of the modulating signal is calculated from Eq. (8.7) as

$$M = \text{mV} = 0.1 \times (12 - 0.7) = 1.13\ \text{V}$$

The remaining elements to select are the capacitors $C_1$ and $C_2$. Both of these elements must pass the carrier frequency while attenuating the modulation frequency. The capacitor $C_1$ should have a low impedance to the carrier frequency compared to the input resistance of the transistor. The input resistance of the transistor is

$$r_{\text{in}} = r_x + (\beta + 1)r_e = 200 + 81 \times \frac{26}{3} = 902\ \Omega$$

The value of $C_1$ can be chosen to lead to an impedance of perhaps one-twentieth of this value. This gives

$$C_1 = \frac{1}{2\pi \times 1.26 \times 10^6 \times 45} = 2.8\ \text{nF}$$

The capacitor $C_2$ can be arbitrarily selected to attenuate the modulating signal by a factor of 100. This would require that the impedance of the capacitor be approximately 1.2 MΩ at a frequency of 2 kHz. A value of

$$C_2 = \frac{1}{2\pi \times 2000 \times 1.2 \times 10^6} = 66.3\ \text{pF}$$

satisfies this requirement. The impedance of this capacitor to the carrier signal is only 1.9 kΩ; thus, the carrier will not be heavily attenuated. The unmodulated carrier output will be approximately 0.96 V.

---

### Practice Problems

**8.3** Find the index of modulation for the transistor modulator if $M = \cos\,\omega_m t$ V, $V_0 = 9$ V, and $V_{BE(\text{on})} = 0.7$ V. *Ans:* 12%.

**8.4** If the unmodulated carrier output in Practice Problem 8.3 is 1 V and the transistor $\beta$ is 100, calculate the amplitude of the input carrier signal $v_c$. Assume that $R_L = 1$ kΩ and $R = 300$ kΩ. *Ans:* 9.4 mV.

### The Transconductance Multiplier

The linearity problem of the transistor modulator can be overcome by using a pair of difference amplifiers as shown in Figure 8.3.

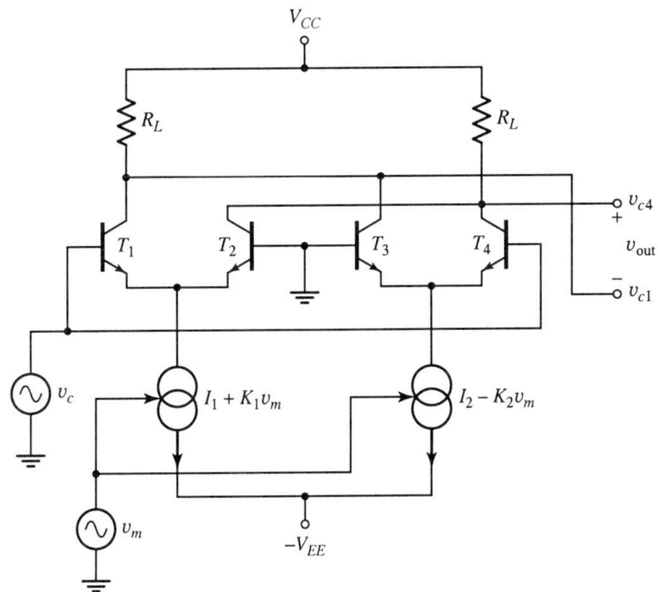


Figure 8.3: Variable transconductance modulator using the Gilbert gain cell.

This design was developed by B. Gilbert and is sometimes called the Gilbert gain cell. The two dependent current sources are controlled by the modulating signal. The gain of the carrier path from input to $v_{c4}$ can be found to be approximately

$$\frac{v_{c4}}{v_c} = -\frac{R_L}{r_{e3} + r_{e4}} + \frac{R_L}{r_{e1} + r_{e2}} = -\frac{R_L}{2r_{e4}} + \frac{R_L}{2r_{e1}}$$

Similarly, $v_{c1}/v_c$ can be written as

$$\frac{v_{c1}}{v_c} = -\frac{R_L}{r_{e1} + r_{e2}} + \frac{R_L}{r_{e3} + r_{e4}} = -\frac{R_L}{2r_{e1}} + \frac{R_L}{2r_{e4}}$$

The output voltage is the difference between $v_{c4}$ and $v_{c1}$ and is found to be

$$v_{\text{out}} = v_{c4} - v_{c1} = \left(\frac{R_L}{r_{e1}} - \frac{R_L}{r_{e4}}\right) v_c \tag{8.8}$$

The values of $r_{e1}$ and $r_{e4}$ are controlled by the current sources. We can write these resistances as

$$r_{e1} = \frac{26}{I_{E1}} = \frac{26}{\frac{1}{2}(I_1 + K_1 v_m)} = \frac{52}{I_1 + K_1 v_m}$$

and

$$r_{e4} = \frac{26}{I_{E4}} = \frac{26}{\frac{1}{2}(I_2 - K_2 v_m)} = \frac{52}{I_2 - K_2 v_m}$$

where the current source outputs are expressed in mA.

Substituting these values into Eq. (8.8) yields

$$v_{\text{out}} = \frac{R_L}{52}\left[(I_1 - I_2) + (K_1 + K_2) v_m\right] v_c \tag{8.9}$$

If sinusoids are used for $v_c$ and $v_m$ and $I_1 \neq I_2$, this equation becomes

$$v_{\text{out}} = \frac{R_L(I_1 - I_2)}{52}\left[1 + \frac{(K_1 + K_2)}{(I_1 - I_2)} M \cos \omega_m t\right] C \cos \omega_c t = V_c\,(1 + m \cos \omega_m t) \cos \omega_c t \tag{8.10}$$

In Eq. (8.10),

$$V_c = \frac{R_L C}{52}(I_1 - I_2) \tag{8.11}$$

and

$$m = \frac{M(K_1 + K_2)}{(I_1 - I_2)} \tag{8.12}$$

The unmodulated output of the transconductance multiplier is given by Eq. (8.11), while the index of modulation is calculated from Eq. (8.12).

The basic idea governing the operation of the transconductance multiplier is similar to the transistor modulator: that is, the modulating voltage controls the gain by controlling the emitter current. The significant difference, however, is that the transconductance multiplier applies a differential mode of operation. The overall gain is given by the difference of two gains, each of which is controlled by the modulating voltage. This overall gain can then go to zero when the modulating voltage has some finite value without requiring extreme values of $r_e$. In the transistor modulator, the value of $r_e$ would have to become infinite to reduce the gain to zero, which is required for 100% modulation. The resulting nonlinearities in gain with $v_m$ for such an extreme change in $r_e$ limit the index of modulation to small values for this circuit. On the other hand, the transconductance multiplier can produce 100% modulation with relatively small changes in the values of the dynamic junction resistances.

The popular MC1496/1596 modulator chip is based on the transconductance multiplier. This circuit is shown in Figure 8.4 within the dashed lines. Elements outside the dashed lines must be added externally.

An AM waveform or a suppressed carrier signal can be produced by adjusting some of the external elements. The bias networks set the bases of $T_1$, $T_2$, $T_3$, and $T_4$ at a level of approximately 6 V and the bases of $T_5$ and $T_6$ near ground. The current sources, $T_7$ and $T_8$, produce collector currents of 1 mA. The emitter currents of $T_5$ and $T_6$ depend on the current through the resistor $R_E$, as well as the collector currents of $T_7$ and $T_8$. These currents are

$$I_{E5} = I_{C7} + I_{R_E} = 1 + I_{R_E}$$

and

$$I_{E6} = I_{C8} - I_{R_E} = 1 - I_{R_E}$$

If the variable tap of the 50-kΩ potentiometer is centered so that $I_{C7} = I_{C8}$ and the current $I_{R_E}$ is zero, then $I_{E5} = I_{E6}$. In this case, the currents $I_1$ and $I_2$ are equal. From Eq. (8.11) it is seen that the unmodulated carrier output amplitude is zero. The output signal will then be a DSB-SC signal. Thus, the circuit can generate a suppressed carrier output.

In order to generate an AM waveform, the potentiometer must be unbalanced. This causes the dc bias currents, $I_1$ and $I_2$, to become unequal. The 10-kΩ resistors must also be reduced to extend the current range. A typical value is 750 Ω. Values of $I_1$ = 1.2 mA and $I_2$ = 0.8 mA result in a reasonable gain for the carrier frequency. If the carrier amplitude, $C$, is chosen as 50 mV, the unmodulated carrier output is

$$v_{\text{out}} = \frac{R_L C}{52}(I_1 - I_2) = 1.5\ V = V_C$$

The modulating voltage causes a change in $I_{R_E}$ of approximately

$$\Delta I_{R_E} = \frac{v_m}{R_E} = K_1 v_m$$

Consequently, $K_1 = K_2 = 10^{-3}$ Siemen. If 80% modulation is desired, $M$ can be found from Eq. (8.12) to be

$$M = \frac{m(I_1 - I_2)}{(K_1 + K_2)} = 0.16 \text{ V}$$

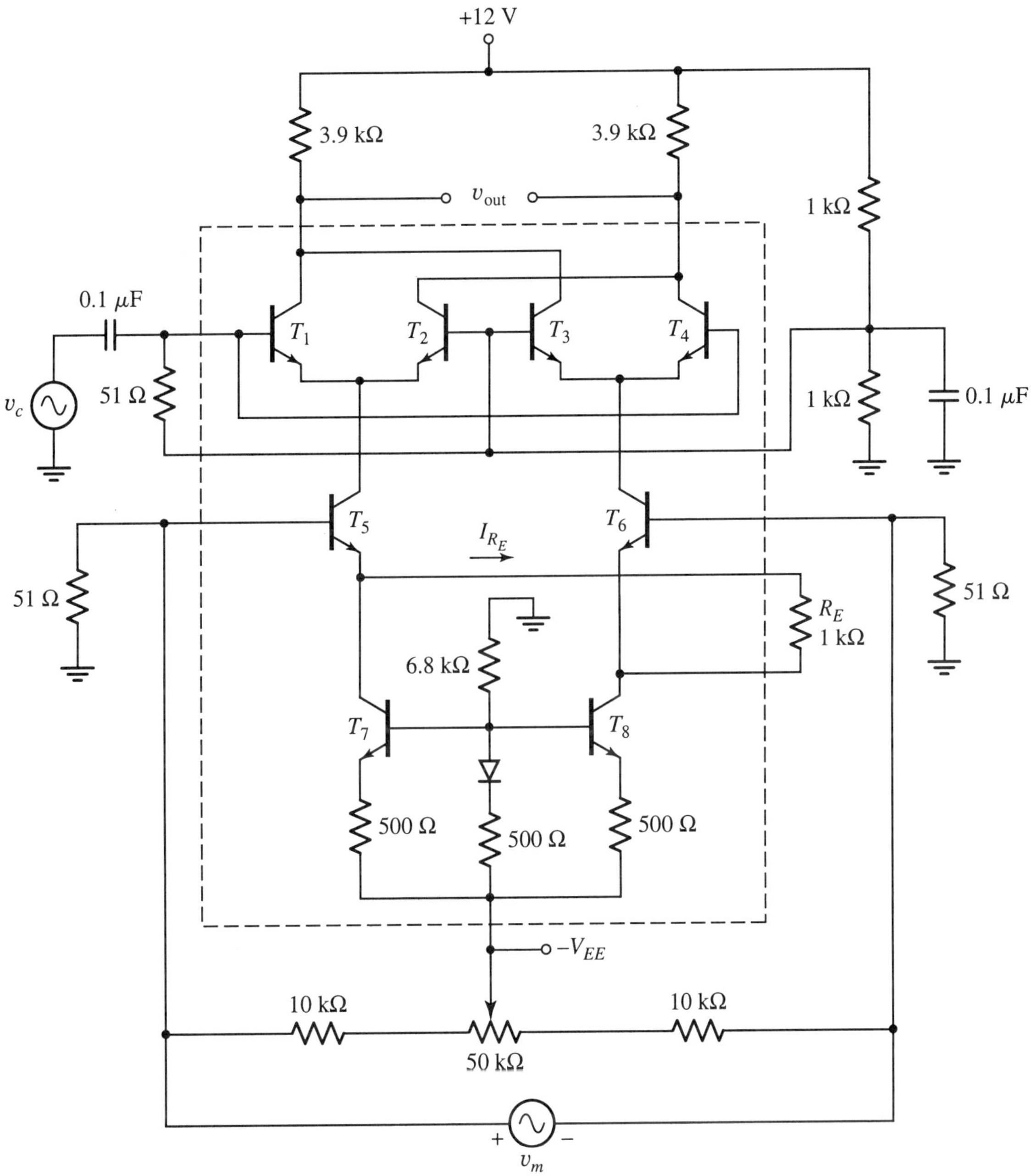


Figure 8.4: A practical transconductance multiplier circuit.

**Practical Considerations**

Although this circuit typically produces relatively small outputs, the modulated signal can be amplified to drive the output stage. However, since the amplification must be linear, it is typically done by a class-A stage and has a limited output power. The class-B stage can also be used, but will not generate the very large output powers needed in commercial transmission.

**Practice Problem**

**8.5** Select the resistance $R_E$ in Figure 8.4 to lead to an index of modulation of 50%. Assume that $I_1 = 1.2$ mA, $I_2 = 0.8$ mA, and $M = 0.2$ V. *Ans:* 2.5 kΩ.

### 8.1.2 Class-C Modulators

Although the transconductance multiplier offers good linearity with an integrated circuit, it operates in the class-A mode. For very high-power outputs with good circuit efficiency, class-C operation must be used. Class-C amplifiers do not amplify an input signal linearly and cannot be used to amplify the output of the transconductance multiplier. This stage must then simultaneously generate a high output power and produce the modulated output rather than amplifying an AM signal generated by an earlier modulator. The next few paragraphs will discuss this important mode of operation. The basic principles applying to high-power semiconductor modulators also apply to the power vacuum-tube tetrodes generally used for commercial AM transmission. Semiconductor devices are limited to maximum power outputs in the range of hundreds of watts rather than the thousands of watts transmitted by the commercial stations. In the last decade or so, 50-kW AM transmitters using transistors in the class-D configuration have become available [5].

#### Base Modulation

The basic idea of the base-modulation circuit is to control the magnitude of the collector current by the modulating voltage, which is placed in the base-emitter loop. In class-C operation, the collector current is nonsinusoidal, but a tuned collector load allows only the fundamental frequency voltage to appear at the output. Higher harmonics of the current waveform flow through a very small impedance while the tuned circuit presents a resonant impedance of $R_{sh}$ to the fundamental component.

We will first analyze an ideal base modulator and then discuss some problems that arise in a more practical situation. The circuit of Figure 8.5 represents a simple class-C base-modulator. The transistor will be assumed to have a constant value of $\beta$ when the transistor is in the active region. The value of $\beta$ will drop abruptly to zero when the voltage around the base-emitter loop is less than $V_{BE(\text{on})}$. Base and collector currents will flow only when the applied carrier voltage exceeds $V_0 + V_{BE(\text{on})}$. The collector current can be expressed for positive values as

$$i_c = \beta i_b = \frac{\beta[C \cos \omega_c t - V_0 - V_{BE(\text{on})}]}{R_B} \tag{8.13}$$

if we assume that $R_B$ is much larger than $r_x + r_\pi$. Of course, the collector current is zero for values of Eq. (8.13) that are zero or negative.

The voltage around the base-emitter loop of the circuit will be designated $v_L$. The positive peak value of this voltage is

$$v_{L\text{peak}} = C - V_0 - V_{BE(\text{on})} \tag{8.14}$$

The peak value of collector current is then

$$I_{CP} = \frac{\beta[C - V_0 - V_{BE(\text{on})}]}{R_B} \tag{8.15}$$

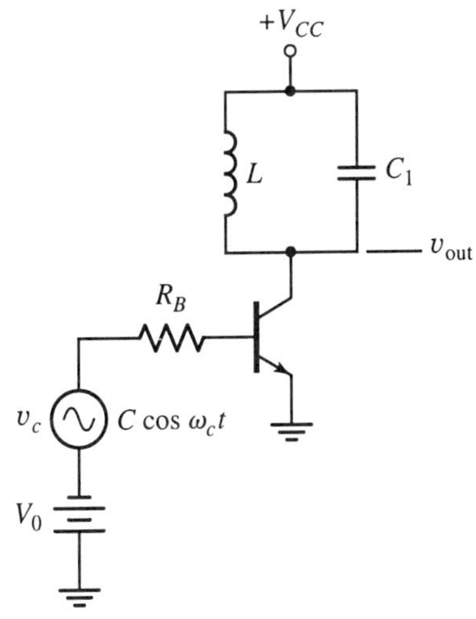


Figure 8.5: Base-modulation circuit.

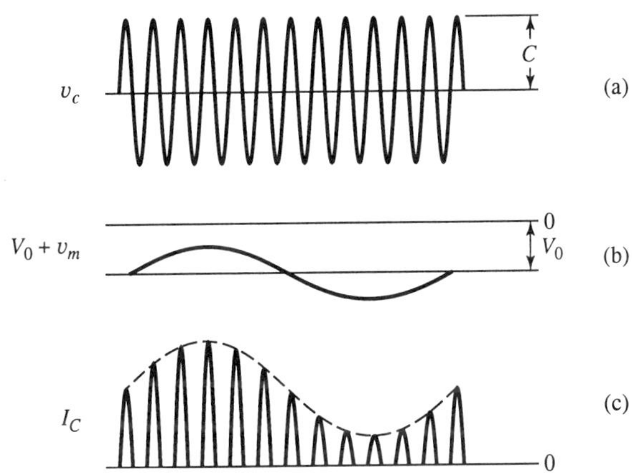


Figure 8.6: Waveforms of a class-C base modulation circuit: (b) modulating signal and dc bias; (b) collector current.

Since the collector current is repetitive, it contains a fundamental component of frequency $\omega_c$. If the tuned circuit of the collector resonates at this frequency, a sinusoidal voltage will appear at the output. This voltage will be approximately proportional to the peak value of collector current. Although higher harmonics are present in the collector-current waveform, no voltage will develop across the tuned circuit for these components. In this situation, the circuit operates as a class-C amplifier.

A modulating signal of value $V_m = M \cos \omega_m t$ can be added in series with $V_0$ to modulate the value of $I_{CP}$. As $I_{CP}$ changes with $v_m$, so also does the magnitude of the output signals. The peak collector current becomes

$$I_{CP} = \frac{\beta[C - V_0 - V_{BE(\text{on})} + M \cos \omega_m t]}{R_B} \tag{8.16}$$

The input and output waveforms are shown in Figure 8.6.

We now make an additional assumption to simplify the analysis of the modulator. We assume that the conduction angle of the collector current remains constant with $v_m$. The angle $\Theta$ that is one-half the conduction angle depends on the base-emitter loop voltage. As $v_m$ varies, so also do the values of $I_{CP}$ and $\Theta$, but we are assuming that $\Theta$ remains unaffected by $v_m$. The consequences of this assumption will be considered later.

The maximum value of $I_{CP}$ occurs at the peak of the modulating voltage and is given by

$$I_{CP\max} = \frac{\beta[C - V_0 - V_{BE(\text{on})} + M]}{R_B} \tag{8.17}$$

With the assumption that $\Theta$ remains constant with $v_m$, the fundamental frequency component of $i_c$ is proportional to the peak value of $i_c$. The unmodulated magnitude of the fundamental current is

$$I_{um} = KI_{CP} = \frac{K\beta[C - V_0 - V_{BE(\text{on})}]}{R_B}$$

where $K$ is a constant that could be evaluated using Fourier series methods. The maximum magnitude of $i_c$ is

$$I_{\max} = KI_{CP\max} = \frac{K\beta[C - V_0 - V_{BE(\text{on})} + M]}{R_B}$$

The unmodulated output voltage magnitude and the maximum output voltage magnitude can be written respectively as

$$|v_{\text{out}-um}| = \frac{K\beta R_{sh}[C - V_0 - V_{BE(\text{on})}]}{R_B} \tag{8.18}$$

and

$$|v_{\text{out}-\max}| = \frac{K\beta R_{sh}[C - V_0 - V_{BE(\text{on})} + M]}{R_B} \tag{8.19}$$

The index of modulation can be found from these two equations to be

$$m = \frac{|v_{\text{out}-\max}| - |v_{\text{out}-um}|}{|v_{\text{out}-um}|} = \frac{M}{C - V_0 - V_{BE(\text{on})}} \tag{8.20}$$

In summarizing the operation of the base modulator we can see that $I_{CP}$ varies directly with $v_m$. If we assume a constant angle $\Theta$ with $v_m$, then the fundamental component of $i_c$ varies directly with $I_{CP}$ and, hence, with $v_m$. The output voltage magnitude varies directly with the fundamental component of $i_c$; therefore, the output voltage varies directly with $v_m$.

The major departure from this ideal theory occurs as a result of the change in $\Theta$ with modulating voltage. This leads to nonlinear modulation that limits the index of modulation to about 50%. We will not consider the mathematical analysis of this case, but will proceed to a method that removes this limitation.

The output voltage can be found by evaluating $\Theta$ and then using this value to find the fundamental collector-current component. From the analysis of the class-C amplifier stage of Chapter 2, we write

$$\Theta = \cos^{-1}\left[\frac{V_0 + V_{BE(\text{on})}}{C}\right] \tag{8.21}$$

This angle is one-half the total conduction angle.

The amplitude of the fundamental component of collector current was considered in Chapter 2 as derived from Fourier series methods and is given by

$$I = \frac{\beta C}{\pi R_B}\left[\Theta(t) - \frac{\sin\, 2\Theta(t)}{2}\right] \tag{8.22}$$

If $R_{sh}$ is the resonant collector impedance, the output voltage is then

$$v_{\text{out}} = \frac{-\beta C R_{sh}}{\pi R_B}\left[\Theta(t) - \frac{\sin\, 2\Theta(t)}{2}\right] \cos\,\omega_c t \tag{8.23}$$

This represents the unmodulated output voltage of the base modulator.

The transistor must never reach saturation even on the peaks of the modulating voltage. The maximum output voltage can be found as

$$v_{\text{out-max}} = (1 + m)\, v_{\text{out}} \tag{8.24}$$

where $v_{\text{out}}$ is calculated from Eq. (8.23).

If this magnitude is smaller than the active region of the transistor collector, then saturation will not be reached. If the magnitude is greater than the size of the active region, saturation will occur, necessitating a change in the value of $R_B$ or $C$.

In many instances, the preferred configuration removes $R_B$ and inserts $R_E$ between emitter and ground. The pertinent equations can then be revised by replacing $R_B$ with $(\beta + 1)R_E$. A possible disadvantage of this arrangement is the reduced size of the active region when $R_E$ is placed in the emitter circuit.

Figure 8.7 shows two practical realizations of base-modulation circuits. In Figure 8.7(a), the capacitor $C_B$ is used to prevent carrier frequency current from passing through the modulation transformer. However, $C_B$ should present a high impedance to the modulating frequency. Figure 8.7(b) shows an alternate method of applying the modulating signal to the base of the output transistor. The emitter follower can be biased to have a zero or negative quiescent voltage at the emitter (class-B or class-C).

It is often necessary to place a diode in series with the base-emitter junction. Modern transistors produced by the diffusion process typically exhibit rather low reverse breakdown voltages. Low-power transistors often have breakdown voltages as low as two to three volts. The series diode will absorb the bulk of the reverse voltage that occurs around the base-emitter loop of a class-C stage.

# Example 8.2

The base modulator of Figure 8.8 has the following values: $L = 100\ \mu\text{H}$, $Q_{\text{coil}} = 80$, $V_{BE(\text{on})} = 0.7$ V, and $\beta = 80$. Assume that the output impedance at the collector of the transistor is infinite. The carrier frequency is 600 kHz and the modulating frequency is 2 kHz.

**(a)** Find the resonating capacitance, $C_1$.
**(b)** Find the value of $M$ to lead to $m = 0.4$.
**(c)** Calculate $R_B$ to lead to a 5-V unmodulated output amplitude.
**(d)** Select a value of emitter resistance $R_E$ to allow removal of $R_B$.

**Solution (a)** The capacitor is calculated to make the tuned circuit resonate at 600 kHz. Thus,

$$C_1 = \frac{1}{\omega_c^2 L} = \frac{1}{(2\pi \times 6 \times 10^5)^2 \times 10^{-4}} = 704 \text{ pF}$$

**(b)** The value for $M$ is found from Eq. (8.20), including the effect of $V_{BE(\text{on})}$. This gives

$$M = m[C - V_0 - V_{BE(\text{on})}] = 0.4[7 - 4.7] = 0.92 \text{ V}$$

**(c)** The resonant resistance of the tank circuit will be the parallel combination of $R_p$, the resistor representing the tank circuit losses, and the 20-kΩ load resistor. The resistor $R_p$ is

$$R_p = \omega_c L Q_{\text{coil}} = 2\pi \times 6 \times 10^5 \times 10^{-4} \times 80 = 30.2 \text{ k}\Omega$$

The resonant resistance is then

$$R_{sh} = \frac{R_p 20}{R_p + 20} = 12 \text{ k}\Omega$$

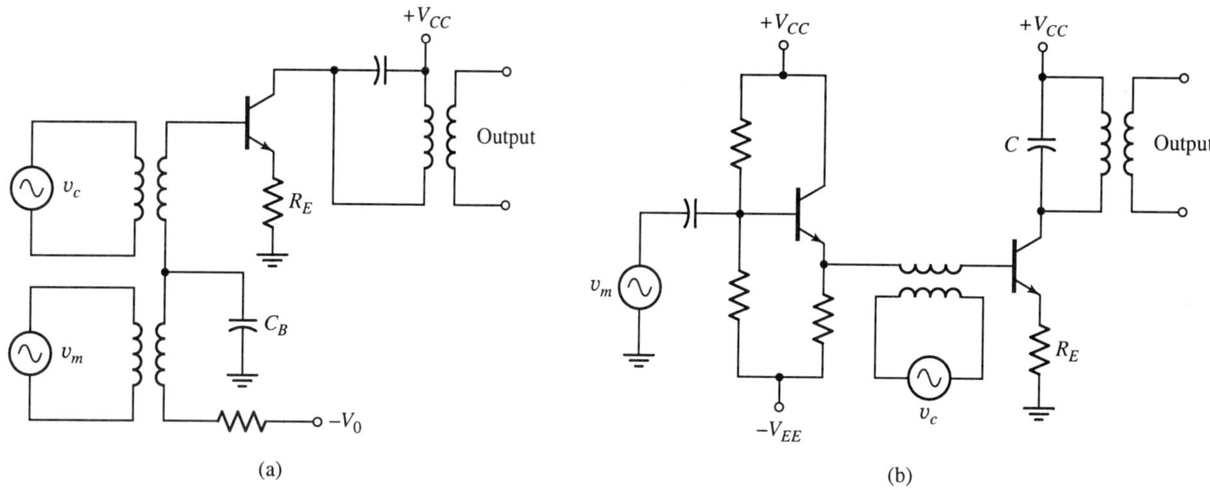


Figure 8.7: Class-C base modulation circuits. (a) Transformer coupled. (b) Emitter coupled.

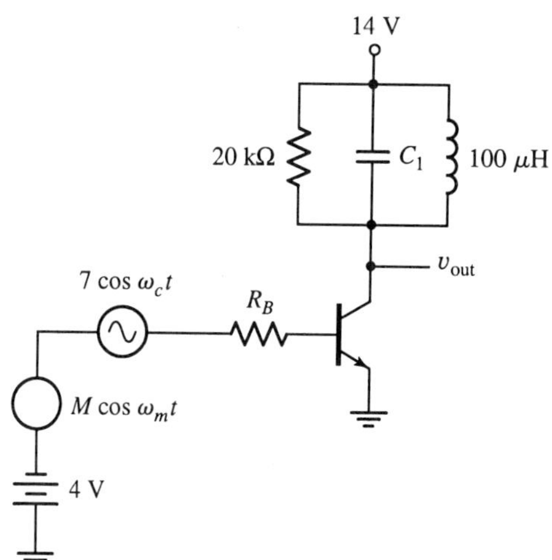


Figure 8.8: Base modulator of Example 8.2.

The value of $R_B$ to cause an output voltage of 5 V is found using Eq. (8.23). This gives

$$R_B = \frac{\beta C R_{sh}}{\pi\,|v_{\text{out}}|}\left[\Theta - \frac{\sin\,2\Theta}{2}\right]$$

The value of $\Theta$ is found to be

$$\Theta = \cos^{-1}\left[\frac{V_0 + V_{BE(\text{on})}}{C}\right] = \cos^{-1}\left[\frac{4.7}{7}\right] = 47.8^\circ = 0.835 \text{ rad}$$

Solving for $R_B$ results in a value of

$$R_B = 144 \text{ k}\Omega$$

**(d)** In order to replace $R_B$ with an emitter resistance, a value of

$$R_E = \frac{R_B}{\beta + 1} = \frac{144}{81} = 1.78 \text{ k}\Omega$$

is required. The active region size is reduced slightly when $R_E$ is placed in series with the emitter. In this case, the maximum positive applied voltage in the base-emitter loop drops across the junction and $R_E$. At the peak of the modulating voltage and the peak of the carrier input, the emitter voltage will be

$$V_{E\max} = M + C - V_0 - V_{BE(\text{on})} = 0.92 + 7 - 4 - 0.7 = 3.22 \text{ V}$$

With an output amplitude of $(1+m)\ 5\ V = 7\ V$ across the tuned circuit and 3.22 V appearing at the emitter, the collector to emitter voltage of the transistor is $14 - 7 - 3.22 = 3.78\ V$. This means that the transistor is not saturated under maximum output voltage conditions; thus, the calculated value of $R_E$ is acceptable.

---

The linearity of modulation can be improved by using a rectangular wave in place of the sinusoidal carrier frequency. This would eliminate the variation of $\Theta(t)$ with modulating voltage. Although this modification improves the linearity of the circuit, it also introduces higher amplitude harmonics in the collector current waveform. Since the tuned circuit is not ideal, these higher-amplitude components tend to add more harmonic distortion to the modulator when driven with a rectangular wave.

A rectangular wave base modulator with reasonably high efficiency might apply a carrier with a 10% duty cycle as shown in Figure 8.9.

The peak collector current is

$$I_{CP} = \frac{\beta(V_1 - V_{BE(\text{on})})}{R_B} \tag{8.25}$$

where $V_1$ is the positive peak value of the input rectangular waveform. The Fourier series for a waveform that varies between 0 and $I_{CP}$ has a period of $T$, and a duty cycle of $a/T$ is expressed as

$$i = I_{CP}\frac{a}{T}$$

$$+\frac{2I_{CP}}{\pi}\left[(\sin\ \omega_o a/2)\ \cos\ \omega_o t + \cdots + (\frac{1}{n}\ \sin\ n\omega_o a/2)\ \cos\ n\omega_o t + \cdots +\right] \tag{8.26}$$

where $n$ is the number of the harmonic. If the collector is tuned to the frequency of the carrier signal, $f = 1/T$, with a shunt impedance of $R_{sh}$, the magnitude of the output voltage is

$$|v_{\text{out}}| = \frac{\beta(V_1 - V_{BE(\text{on})})}{R_B} \times \frac{2}{\pi} \times R_{sh} \times \sin\ \omega_o a/2 \tag{8.27}$$

For a 10% duty cycle, $a = 0.1\,T$ and the magnitude of the output voltage is

$$|v_{\text{out}}| = 0.197\,\frac{\beta(V_1 - V_{BE(\text{on})})}{R_B} \times R_{sh}$$

If a modulating signal is applied in the base circuit, the peak collector current varies in proportion to $v_m$. This peak value becomes

$$I_{CP} = \frac{\beta(V_1 + v_m - V_{BE(\text{on})})}{R_B} \tag{8.28}$$

It will be left to the student to show that the index of modulation is given by an equation similar to Eq. (8.20); that is,

$$m = \frac{|v_{\text{out}-\max}| - |v_{\text{out}-um}|}{|v_{\text{out}-um}|} = \frac{M}{V_1 - V_{BE(\text{on})}} \tag{8.29}$$

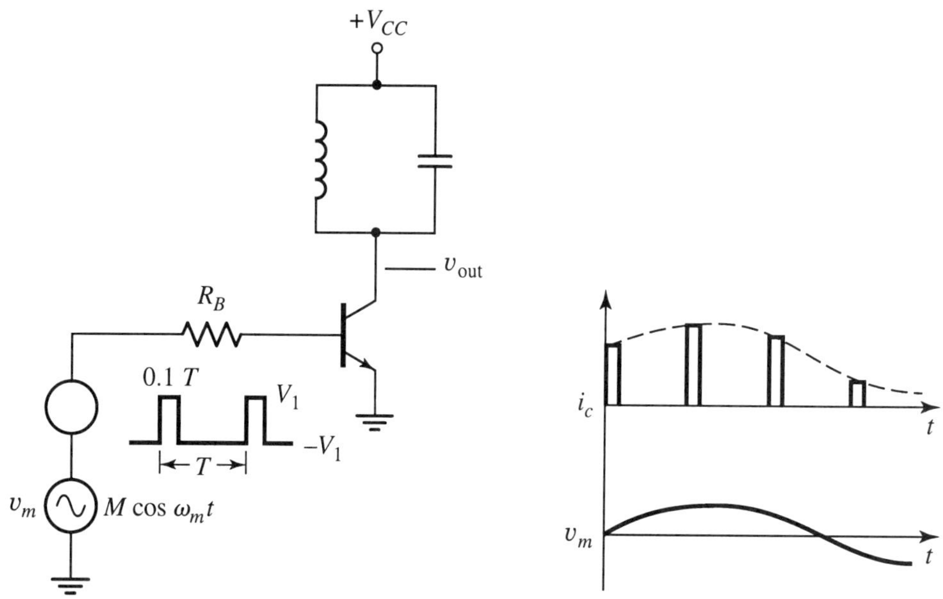


Figure 8.9: A 10% duty cycle input and resulting collector current.

Although the rectangular carrier signal leads to high circuit efficiency with a small duty cycle, the magnitudes of the third, fifth, and other odd harmonics of collector current, given by Eq. (8.26), are relatively large compared to that of the fundamental. This leads to more distortion in the output waveform. Furthermore, a much wider bandwidth is required for the transistor modulator to amplify the square wave. However, this method of base modulation is used in practical amplitude modulators [7].

### Practice Problem

**8.6** In Figure 8.9, $V_1 = 4$ V, duty cycle $= 0.2$, $V_{CC} = 12$ V, and the resonant resistance of the tuned circuit, including output impedance of the transistor, is 5 kΩ. If $V_{BE(\text{on})} = 0.8$ V and $\beta = 80$, find the value of $R_B$ to create an unmodulated output carrier amplitude of 3 V. *Ans:* 160 kΩ.

### Collector Modulators

Good linearity of modulation for large values of $m$, approaching unity, can be achieved with the collector modulator. Even though more modulation source power is required, this circuit is almost universally used in high-quality transmitters.

The collector-modulation circuit, shown in Figure 8.10, can be operated as a class-B or class-C stage. The fundamental component of collector current must develop enough voltage across the tank circuit to saturate the transistor at the negative peaks of the collector voltage. Again, the collector-current waveform is high in harmonic content, but the tank circuit develops an output voltage only for the signal of frequency $\omega_c$. The higher-order current harmonics develop negligible voltage across the tuned circuit. As discussed in Subsection 5.6.2, the amplitude of the output signal under saturation-limited conditions equals the power supply voltage. This is the key point in collector modulator operation because a change in the power supply voltage then changes the output signal proportionately.

The modulating signal is applied, often through a transformer, in series with the dc voltage supply giving a total power supply voltage of

$$v_m + V_{CC} = M \cos \omega_m t + V_{CC}$$

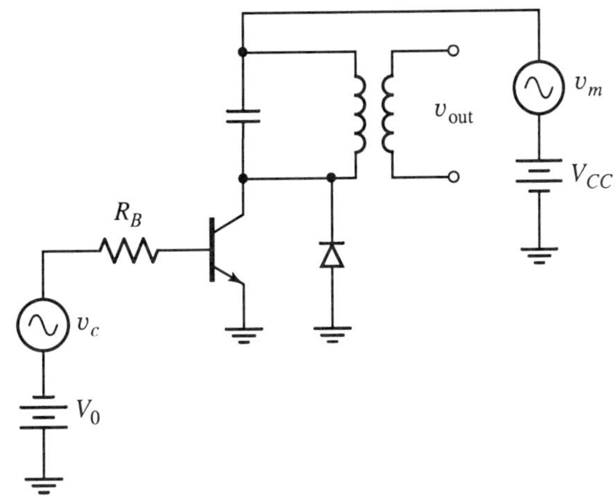


Figure 8.10: Collector modulator.

As long as the carrier component of $i_c$ is large enough to saturate the transistor on the negative peaks of the collector voltage, the collector voltage is expressed as

$$v_{c1} = V_{CC} + M \cos \omega_m t + (V_{CC} + M \cos \omega_m t) \cos \omega_c t$$

The first two terms of this equation represent the power supply voltage that appears at the collector of the transistor while the third term is the AM waveform. The tuned transformer circuit offers a resonant impedance only at $\omega_c$ and no dc voltage ($V_{CC}$) or low-frequency voltage ($M \cos \omega_m t$) appears across the tuned circuit. The output voltage at the transformer secondary, assuming a turns ratio of unity, is

$$v_{\text{out}} = (V_{CC} + M \cos \omega_m t) \cos \omega_c t$$

The unmodulated amplitude of $v_{\text{out}}$ is $V_{CC}$, and the peak change in output amplitude when modulated is equal to $M$. The index of modulation is then calculated to be

$$m = \frac{M}{V_{CC}} \tag{8.30}$$

For 100% modulation the modulating voltage amplitude $M$ must equal $V_{CC}$. The power supply voltage is then $V_{CC}(1 + \cos \omega_m t)$, and the AM waveform at the secondary is given by

$$V_{CC}(1 + \cos \omega_m t) \cos \omega_c t \tag{8.31}$$

The pertinent waveforms for the collector-modulation stage are shown in Figure 8.11 for $m = 1$.

The difference between the collector waveform in Figure 8.11(c) and the output waveform of (d) is the filtering of the dc and modulating voltages. Removing these components from the collector waveform results in the output waveform. If the transformer is a high-frequency transformer and the tuned circuit has a narrow bandwidth that does not pass the modulation frequency, proper filtering will take place. Otherwise, an appropriate filter must be designed.

An important difference between the collector-and base-modulator circuits is the constant conduction angle for collector current in the collector modulator. The angle $\Theta$ is established at a value given by

$$\Theta = \cos^{-1}\left(\frac{V_0 + V_{BE(\text{on})}}{C}\right) \tag{8.32}$$

and remains constant. In the base modulator, the modulating voltage was applied in the base-emitter loop and caused $\Theta$ to vary with $v_m$. Since $v_m$ is applied in the collector-emitter loop of the collector modulator, no change in $\Theta$ results.

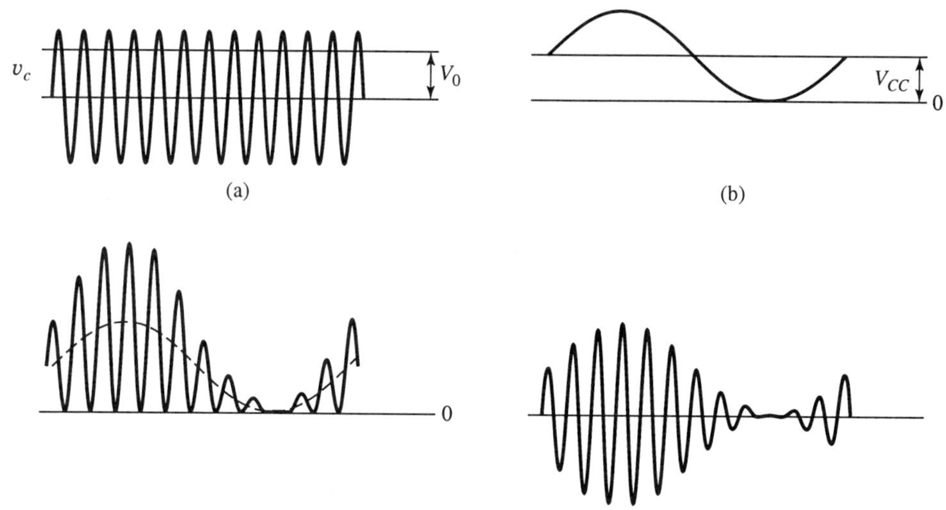


Figure 8.11: Waveforms for collector modulation. (a) Input loop voltage. (b) Collector supply voltage for 100% modulation. (c) Collector voltage. (d) Output voltage.

The fundamental collector-current component has a magnitude of

$$I = \frac{\beta C}{\pi R_B}\left[\Theta - \frac{\sin 2\Theta}{2}\right] \tag{8.33}$$

This value of current must cause saturation of the transistor for all levels of power supply voltage. The worst-case condition occurs when $v_m$ is at its positive peak. The power supply voltage can be written as

$$V_{CC} + M \cos \omega_m t = V_{CC}(1 + m \cos \omega_m t)$$

The maximum power supply voltage is $V_{CC}(1 + m)$. The maximum amplitude of the carrier must equal this maximum power supply voltage. Therefore, if the resonant collector impedance is given by $R_{sh}$, the circuit must be designed such that

$$IR_{sh} \geq V_{CC}(1 + m) \tag{8.34}$$

For 100% modulation, the output carrier amplitude is $2V_{CC}$ and thus, $IR_{sh} \geq 2V_{CC}$.

When the power supply voltage approaches its lowest value on the negative peak of $v_m$, the required saturation current is far less than the actual collector current component. Since the tank circuit does not have infinite $Q$, some distortion may show up in the output voltage as a result of the hard saturation of the transistor. Additional tuned circuits can be used to decrease the distortion due to this effect.

Equation (8.33) is valid for the nonsaturating class-C stage. This equation does not accurately apply to the continuous operation of the collector modulator as hard saturation occurs. It is of value in the limiting case, however, as the modulation voltage peaks and the transistor is only driven to the edge of saturation. For this reason, Eq. (8.33) can be used along with Eq. (8.34) to calculate the required circuit parameters.

For values of $m$ around 90%, the collector modulator is more linear than the base modulator. The modulation source must now supply collector current while it need only supply base current in the base-modulator circuit. This can be a serious problem in television transmission requiring output powers in the hundreds of kW range.

**Practical Considerations**

In order to ensure that the collector voltage limits the negative voltage swings near ground level, a diode is often added to the circuit as shown in Figure 8.10. The addition of the diode has little or no effect unless the collector voltage moves toward a negative value. When this occurs, the diode clamps this voltage to a few tenths of a volt below ground.

Another problem that arises is the lowered value of $Q$ of the tuned circuit resulting from the transistor becoming saturated. The value of $R_{sh}$ appearing across the tuned circuit determines the $Q$ and bandwidth of the modulator. Although $R_{sh}$ can be large during most of a carrier cycle, when saturation occurs, the output impedance of the transistor becomes quite small. This lowers the value of $R_{sh}$ seen by the tuned circuit. The resulting output signal has an increased bandwidth due to the lower value of $Q$ in the circuit. Although there are methods of reducing this effect on $Q$, we will not discuss them at this point.

# Example 8.3

For the circuit of Figure 8.12 $V_{BE(\text{on})} = 0.7$ V, $\beta = 60$, and $V_{CE(sat)} = 0.2$ V. If the resonant impedance of the tuned circuit at $\omega_c$ is 8 kΩ, find the value of $M$ to result in $m = 0.8$ and calculate the maximum value of $R_B$.

**Solution** The amplitude of the modulating voltage can be found by solving Eq. (8.30) for $M$. This gives

$$M = mV_{CC} = 0.8 \times 12 = 9.6 \text{ V}$$

In order that the transistor saturate for all values of $v_m$, the inequality of Eq. (8.34) must be satisfied. Accounting for a nonzero value of $V_{CE(\text{sat})}$, this requires that

$$I \geq \frac{V_{CC}(1+m) - V_{CE(\text{sat})}}{R_{sh}} = \frac{12(1+0.8) - 0.2}{8} = 2.68 \text{ mA}$$

The fundamental collector-current component must equal or exceed 2.68 mA. Equation (8.33) can be solved for $R_B$ giving

$$R_B \leq \frac{\beta C}{\pi I}\left[\Theta - \frac{\sin\, 2\Theta}{2}\right]$$

The value of $\Theta$ is first found from Eq. (8.21) to be

$$\Theta = \cos^{-1}\left(\frac{4.7}{8}\right) = 54^\circ$$

This leads to a maximum value for $R_B$ of 26.6 kΩ.

**Practice Problem**

**8.7** Rework Example 8.3 if the value of $V_{CC}$ is changed to 16 V. *Ans:* $M = 12.8$ V, $R_B = 19.9$ kΩ.

### 8.1.3 The Loop-Gain Modulator

This modulator operates on a different principle than the previous circuits discussed [1,2]. In this type of modulator, the nominal loop gain of an oscillator circuit is varied by the modulating voltage. This circuit

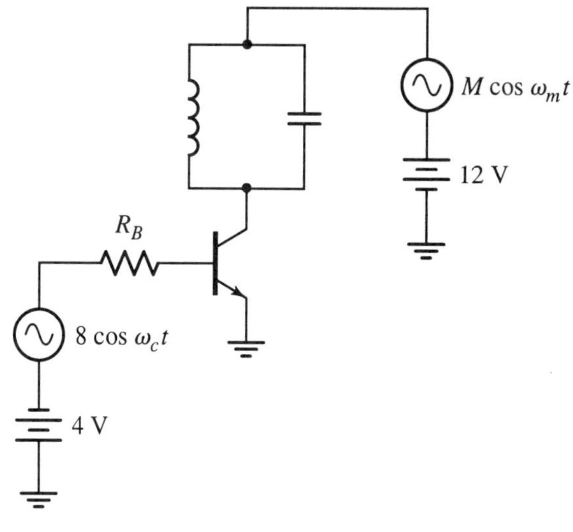


Figure 8.12: Collector modulator for Example 8.3.

exhibits very good linearity of modulation and requires little modulation source power. Both class-A and class-C modes of operation are possible.

The heart of a loop-gain modulator is the sinusoidal feedback oscillator. Figure 8.13 shows a simple model for this circuit. The forward gain of the amplifying element is represented by $A$, while $F$ represents the transfer function of the feedback network. Either $A$ or $F$ may be frequency dependent and the product $AF$ must be frequency dependent. As discussed in Chapter 6, the conditions for sinusoidal oscillation are as follow:

(1) The magnitude of the loop gain, $AF$, must be unity.

(2) The phase shift of $AF$ at the desired frequency of oscillation must be 0° or an integral multiple of 360°.

A crucial point in the operation of the loop-gain modulator is the process of output amplitude limiting that takes place in the sinusoidal oscillator. The magnitude of the loop gain is set to be slightly larger than unity at the frequency of oscillation. If no circuit parameters changed, this would result in a continually increasing output amplitude. In the actual circuit, however, as the amplitude grows, the distortion introduced by the nonlinearities of the amplifying elements increases. This distortion modifies the dc level of output voltage and current flowing through the amplifying element which in turn decreases the forward gain $A$. The output amplitude builds to the point that the magnitude of $AF$ is driven to unity. At this point, the amplitude stabilizes. In Figure 8.14, this effect is modeled by a block that generates some ill-defined dc distortion voltage, $v_d$, when driven by the sinusoidal output voltage. This dc voltage controls the gain of the amplifying element. The open-loop value of the magnitude of $AF$ is called the nominal value of loop gain. When the loop is closed, the valve of $AF$ is driven to unity as long as sinusoidal oscillations are present at the output.

The sinusoidal oscillator can be converted to a voltage-variable output if the nominal value of forward gain can be varied by an applied control voltage. Figure 8.14 shows a model for such an oscillator. The gain of the amplifying stage is now controlled by the control voltage, $v_m$, and the dc voltage produced by the output distortion. The distortion voltage is a function of the amplitude or peak value of the output voltage, $v_p$. In order to emphasize this relationship, the distortion voltage will be expressed as $v_d(v_p)$. The voltage $v_A$ is the difference between $v_m$ and $v_d(v_p)$. The amplifier gain is assumed to increase with $v_A$.

If $v_m$ is zero when power is applied to the circuit, the output amplitude, $v_p$, will be established at a value

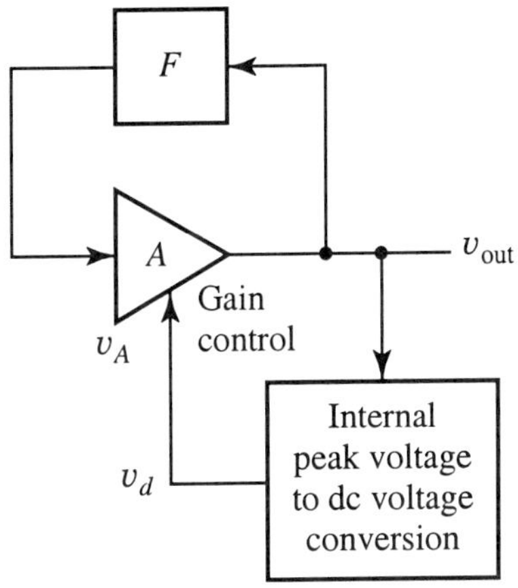


Figure 8.13: A sinusoidal oscillator.

that leads to unity loop gain. The value of $v_A$ is now

$$v_{Ao} = -v_d(v_{po})$$

where $v_{po}$ is the output voltage with zero control voltage. As $v_m$ increases from zero, the nominal value of gain increases and $v_p$ becomes larger. Distortion in the output grows and so does $v_d(v_p)$. The output stabilizes at a value that again forces the loop gain to equal unity. The control voltage must remain constant, giving

$$v_{Ao} = v_m - v_d(v_p) \tag{8.35}$$

Thus, as $v_m$ increases, $v_d(v_p)$ must increase by the same amount. If $v_d(v_p)$ is a linear function of $v_p$, this voltage also varies linearly with $v_m$. Assuming a linear variation leads to the equation

$$v_d(v_p) = k_d v_p \tag{8.36}$$

where $k_d$ is called the distortion constant. Substituting Eq. (8.36) into Eq. (8.35) gives

$$v_p = \frac{v_m}{k_d} - \frac{v_{Ao}}{k_d} \tag{8.37}$$

The second term on the right side of Eq. (8.37) is constant while the first term reflects a linear relationship between $v_m$ and $v_p$.

This variable-amplitude sinusoidal oscillator can be used as an amplitude modulator by setting the oscillation frequency to the desired carrier frequency and applying the modulation voltage as the control voltage. The linearity of modulation depends on the validity of Eq. (8.36). Whereas the distortion feedback voltage may not be perfectly linear over a large range of output voltages, it is surprisingly linear for some oscillation circuits. However, for practical circuits, the performance should not be dependent on ill-defined parameters. Consequently, the dependence of output voltage on distortion must be minimized. This can be accomplished by the addition of an amplitude-stabilizing circuit. A block diagram of the loop-gain modulator is shown in Figure 8.15.

This circuit includes a peak detector that serves an amplitude-stabilizing function. The voltage $v_A$ is now given by

$$v_A = v_m - k_p v_p - k_d v_p \tag{8.38}$$

where $k_p$ is the transfer function of the peak detecting circuit; that is,

$$k_p = \frac{v_f}{v_p} \tag{8.39}$$

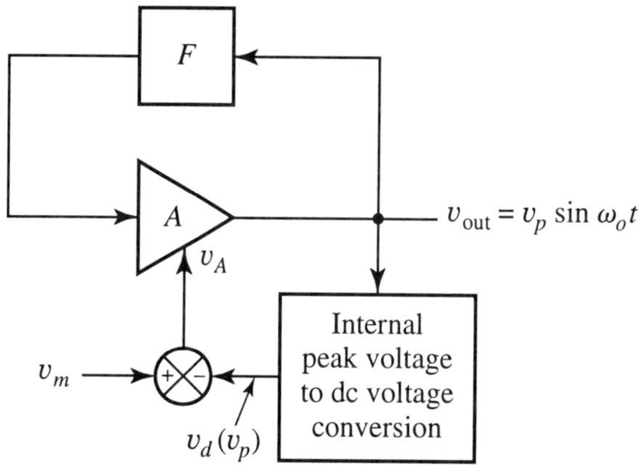


Figure 8.14: A variable-amplitude sinusoidal oscillator.

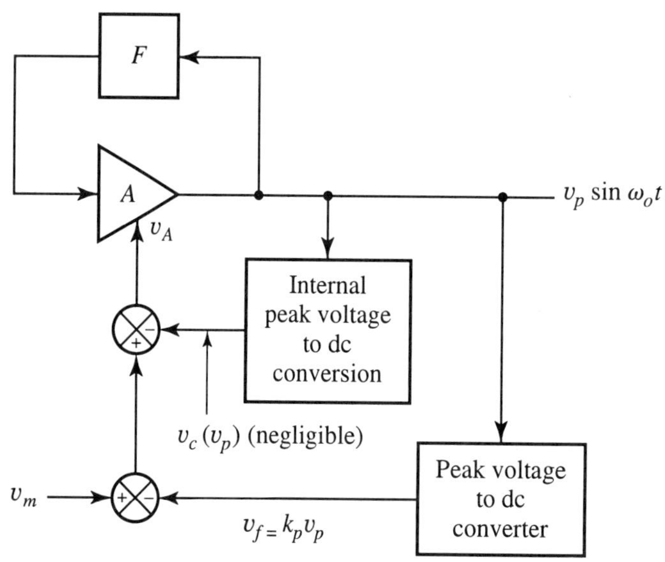


Figure 8.15: A loop-gain modulator.

It should be noted that $k_p$ relates the peak value of an ac input voltage, $v_p$, to a dc output value, $v_f$.

Although Eq. (8.38) assumes that the distortion voltage varies linearly with $v_p$, the significance of this assumption is relatively small. The constant $k_p$ is typically much greater than $k_d$; hence, the term involving $k_d$ can be neglected giving

$$v_A = v_m - k_p v_p \tag{8.40}$$

As oscillation takes place, the peak value is now expressed as

$$v_p = \frac{v_m}{k_p} - \frac{v_{Ao}}{k_p} \tag{8.41}$$

Although Eq. (8.41) is similar to Eq. (8.37), the constant $k_p$ reflects a true linear relationship rather than an approximately linear function. The voltage $v_m$ now controls the amplitude of the output in a linear fashion as long as $k_p$ is much larger than $k_d$.

Two realizations of the loop-gain modulator are shown in Figs. 8.16 and 8.17. Figure 8.16 shows a class-A modulator that requires no inductor and Figure 8.17 shows a class-C circuit with modulation voltage applied in the base circuit. Very little modulation source power is necessary for this circuit and a modulation index of 90% is obtainable with good linearity of modulation. This modulator can be applied to high-output power circuits without requiring the modulation source to supply a great amount of power, as in the collector modulator. In television transmitters with output power levels in the hundreds of kW, this circuit can significantly reduce the required power from the modulation source. The details of design for the loop-gain modulator will not be given here. The interested reader should consult [1,2].

### 8.1.4 A Class-D Amplitude Modulator

Although the transistor began replacing the vacuum tube in computers and television receivers in the 1950s, AM transmitters continued to use special-purpose tubes for many years. Power tetrodes, traveling-wave tubes, and other high-power, high-frequency tubes are still used for high-power transmission. Both FM and television stations typically transmit at considerably higher powers than a clear AM station (50,000 W). Yet it was only about 15 years ago that 50,000 W semiconductor AM transmitters were developed [5]. This section will give an overview of a class-D or pulse-width modulated AM transmitter. Figure 8.18 shows a block diagram for this system.

The modulating signal is applied to a pulse-width modulation (PWM) circuit that varies the on duty cycle of a fixed period signal. The frequency of the PWM is 75 kHz. The PWM drives a switching circuit that applies the power supply to an amplifier when the PWM is in the on state and removes the power supply voltage when the PWM is in the off state. The carrier signal is applied to the amplifier input, but will be amplified only when the PWM is in the on state. The output of the amplifier for various modulation voltage levels is shown in Figure 8.19.

When power is applied to the amplifier, the output consists of the amplified carrier signal with a constant amplitude. As the power switches to the off state, no output signal is present. The number of cycles of the carrier output that occurs during the on state is proportional to the duration of the on state. Therefore, the number of cycles of the carrier is also proportional to the amplitude of the modulating signal.

A Fourier analysis of a waveform consisting of $n$ cycles of a carrier sinusoid that then goes to zero for the remainder of the period of repetition shows that the amplitude of the carrier frequency is proportional to the number of cycles of the carrier signal. In other words, a fixed period that begins with several cycles of a carrier signal and then goes to zero for the remainder of the period, will contain a frequency component at the carrier frequency, with an amplitude proportional to the number of cycles of the applied carrier signal. Since the number of cycles of the carrier signal is proportional to the amplitude of the modulating signal, the output of the amplifier contains a carrier component with an amplitude proportional to the modulating signal. A tuned circuit filter follows the amplifier to select this single component at the carrier frequency.

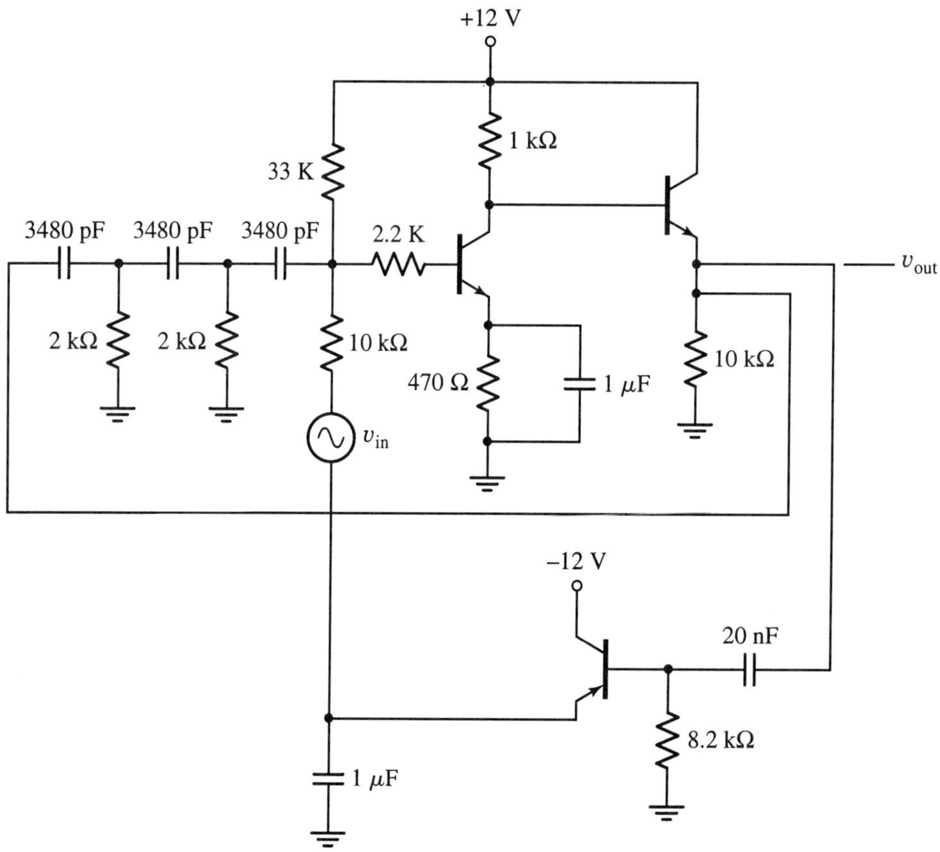


Figure 8.16: Class-A modulator based on an RC oscillator.

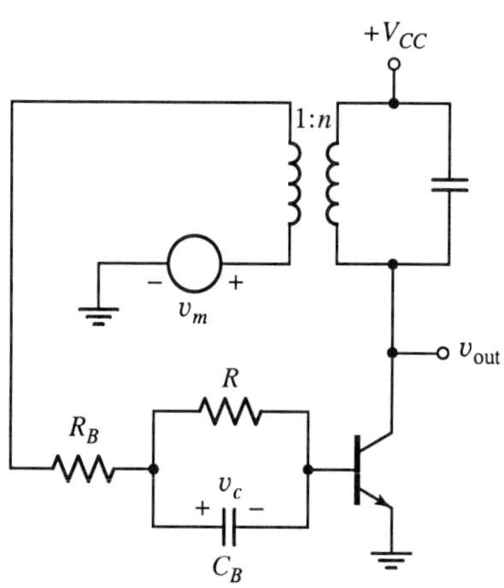


Figure 8.17: A class-C loop-gain modulator.

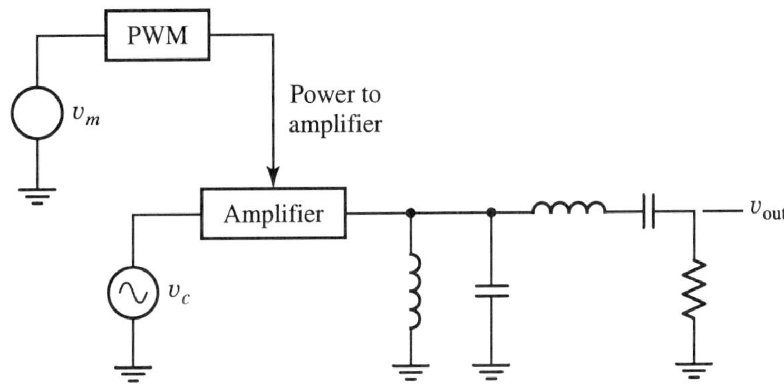


Figure 8.18: A class-D amplitude modulator.

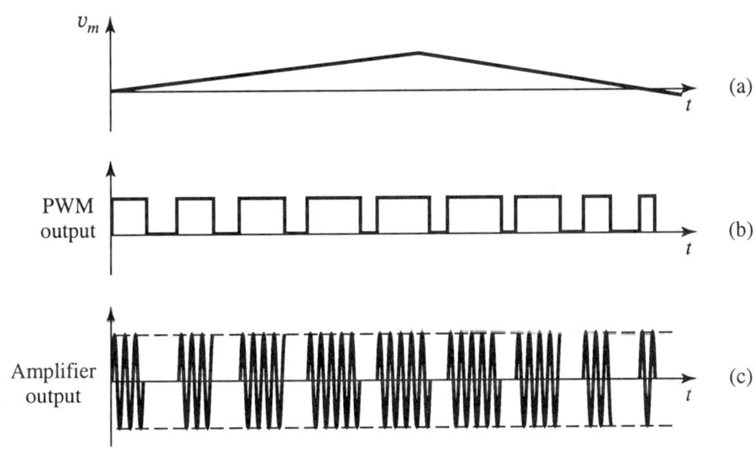


Figure 8.19: Transmitter signals. (a) Modulating levels. (b) Power to amplifier. (c) Amplifier output.

A 50-kW transmitter using power MOSFET devices was reported in 1989 [5] and several clear stations now use this type of transmitter. Modulating frequencies up to 10 kHz allow this circuit to be used by stations observing the voluntary AM standard.

## 8.2 PRODUCTION OF SUPPRESSED-CARRIER WAVEFORMS

**Important Concepts**

**1.** A suppressed carrier signal can be produced with a multiplying circuit. The MC1496 chip uses this approach.

**2.** Suppressed carrier signals can also be produced with amplitude modulators and difference circuits. These types of systems are used for higher power modulation systems.

**3.** The suppressed carrier circuit is often used as a building block for a single-sideband system.

A DSB-SC modulator is variously named a *balanced modulator* or a *mixer*. We will discuss the origins of these names as we consider two different circuits. Carrier suppression is conveniently accomplished by a balanced modulator such as that shown in Figure 8.20. The individual amplitude modulators are assumed to be identical, hence the term balanced modulator for the overall system. Assuming identical modulators with equal gains, the same carrier signal is applied to the carrier input of each modulator. The baseband signal is applied to the modulation input of one modulator, whereas the inverted baseband signal is applied to the modulation input of the other modulator. If the modulators are considered to be of the variable gain variety, the output of the upper modulator is

$$v_1 = Av_c = (A_0 + A_1 v_m)v_c = A_0 v_c + A_1 v_m v_c$$

The lower modulator is similar, but the modulation signal is inverted; thus,

$$v_2 = (A_0 - A_1 v_m)v_c = A_0 v_c - A_1 v_m v_c$$

Each of these signals is a normal AM signal with a carrier along with two sidebands resulting from the product term $A_1 v_m v_c$. A difference circuit subtracts the output of the lower modulator from the output of the upper modulator to produce the term $2A_1 v_m v_c$. The carrier signal $A_0 v_c$ appearing in both modulator outputs is canceled perfectly if the two modulators are perfectly balanced.

The output signal of the difference circuit is a product term involving the multiplication of the modulating signal and the carrier signal. If each of these signals is sinusoidal, that is, $v_m = M \cos \omega_m t$ and $v_c = C \cos \omega_c t$, the output signal is

$$v_{\text{out}} = 2MC \cos \omega_m t \cos \omega_c t = A_1 MC[\cos (\omega_c - \omega_m)t + \cos (\omega_c + \omega_m)t] \tag{8.42}$$

This signal contains only the two sidebands with no carrier signal. In practice, the two modulators may not be perfectly balanced, and the difference circuit may not be perfect. This results in a small carrier signal, but in actual DSB-SC systems, this signal may be 40 to 60 dB below the sideband signals and may become negligible.

We emphasize that the output of the balanced modulator is the product of the carrier signal and the modulating signal. This product can also be formed by the transconductance multiplier discussed in the last section. The output voltage of this circuit from Eq. (8.9) is

$$v_{\text{out}} = \frac{R_L}{52}\left[(I_1 - I_2) + (K_1 + K_2)v_m\right] v_c$$

This signal consists of a carrier given by

$$V_c = \frac{R_L}{52}(I_1 - I_2)v_c$$

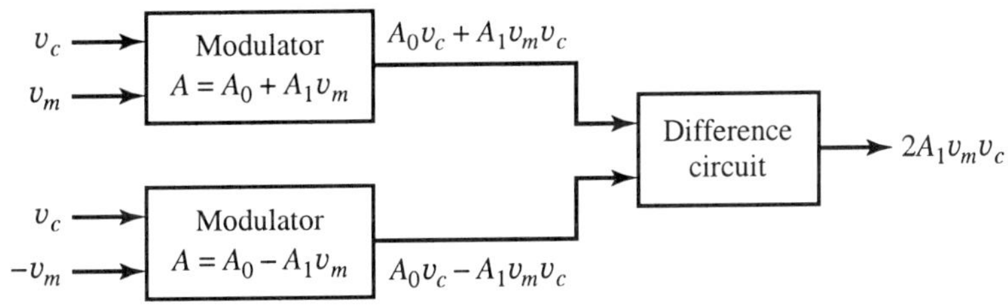


Figure 8.20: Block diagram of a balanced modulator.

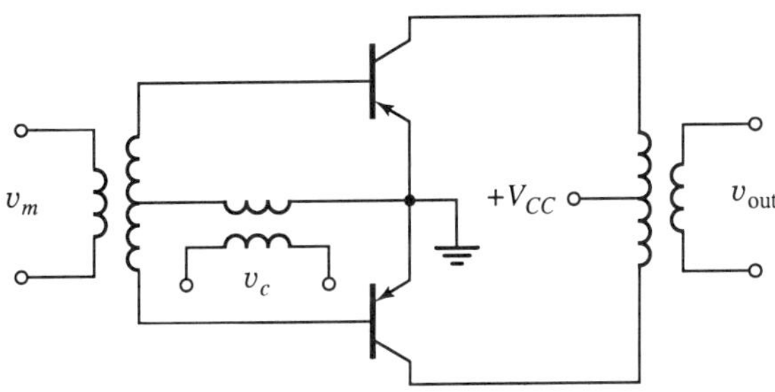


Figure 8.21: Balanced modulator circuit.

and a product term

$$\frac{R_L}{52}(K_1 + K_2)v_m v_c$$

If the two currents $I_1$ and $I_2$ are made to be equal, the carrier signal disappears and only the product signal remains in the output. We have discussed this MC1496 chip that can generate a DSB-SC signal earlier and will not go into more detail at this point. This chip specifies a carrier suppression of over 50 dB, but produces a low-power output signal. Although this signal can be amplified, the next circuit to be discussed can generate a higher output power for the DSB-SC signal.

Figure 8.21 shows a possible high-power circuit for the balanced modulator. The carrier signal is through a transformer between the center tap and ground of the secondary of a second transformer. This applies the carrier signal in the same phase to both transistor bases simultaneously. Applying the modulating signal to the primary of the second transformer splits the modulating signal into two equal-amplitude signals applied to the two bases of the transistors. The phase of the signal applied to one base has a 180° phase difference from the signal applied to the other base. The nonlinearity of the base-emitter junctions modulates the carrier signal. The resulting collector currents flow through the output transformer in such a way as to generate a difference voltage across the secondary winding. This circuit implements the block diagram of Figure 8.20.

The modulator of Figure 8.21 can produce high levels of output power, but suffers from several shortcomings that make the circuit complex and expensive. The nonlinearity of the base-emitter junctions introduces several unwanted components into the output signal. In order to minimize these components, additional filtering is required and the index of modulation is limited. Special bias networks must be used to remove crossover distortion and achieve perfect balancing of both transistors. This circuit is now rarely used because of the good performance of the transconductance multiplier chip. For low-power DSB-SC signal production, the transconductance multiplier of Figure 8.4 is often the circuit of choice.

## 8.3 PRODUCTION OF SINGLE-SIDEBAND WAVEFORMS

**Important Concepts**

**1.** Filtering may be used to eliminate one sideband and the carrier from an AM waveform in order to obtain a single-sideband filter. The required sharpness of the filter in many applications makes this method difficult in practice.

**2.** Crystal filters are used in some applications, whereas a two-stage modulation method can be used in other instances.

**3.** Phase-shift methods can be applied to achieve single-sideband signals without severe filtering requirements. The Weaver method is often used for this purpose.

### 8.3.1 Filtering to Obtain SSB

The simplest method of single-sideband generation uses filtering to eliminate one sideband from a suppressed carrier signal. A telephone company may wish to transmit the upper sideband of a signal resulting from modulating a 60-kHz carrier by a signal varying from 300 to 3400 Hz. Using a balanced modulator results in a lower sideband extending from 56.6 to 59.7 kHz. A bandpass filter with a passband extending from 60.3 to 63.4 kHz (the upper sideband range) follows the modulator. The filter is designed to fall sharply as frequency decreases from 60.3 to 59.7 kHz. Typically a 40- to 60-dB difference in response at these frequencies must result at the filter output. A rather sharp filter is required as the frequency change is only 1% in this case, but such filters are available.

With higher carrier frequencies, it becomes very difficult to use the simple method just described due to even more severe filter requirements. For example, if the same modulating signal is to be transmitted with a 10-MHz carrier and 40-dB attenuation between sidebands, a very complex filter will be required. This filter would have to provide 40-dB attenuation within 600 Hz at a frequency of 10 MHz, a 0.006% frequency change. This would require an excessively sharp filter such as a crystal filter.

As shown in Figure 8.22, two stages of modulation may be used to ease the filtering problem [6]. The second filter must now provide 40-dB attenuation between the frequencies 9.8997 and 10.1003 MHz, a 200.6-kHz or two-percent difference. Use of this two-step method requires more bandwidth per information channel, but some applications are concerned more with obtaining the single-sideband signal than with the channel spacing. Crystal filters are generally used to achieve the required sharpness.

### 8.3.2 The Phase-Shift Method

Two other methods for generating SSB waveforms without sharp filters are sometimes used. The phase-shift method [6] is shown in Figure 8.23.

This method requires a 90° phase shift of the modulating signal. The modulating signal with no phase shift is applied to the upper balanced modulator while the modulating signal with a 90° phase shift is applied to the lower balanced modulator. The carrier signal is also applied to the upper modulator with no phase shift and to the lower modulator with a 90° phase shift. The output of the upper modulator is

$$v_1 = \frac{mV_c}{2}\left[\cos\,(\omega_c - \omega_m)t + \cos\,(\omega_c + \omega_m)t\right]$$

The output of the lower modulator is

$$v_2 = \frac{mV_c}{2}\left[\cos\,(\omega_c - \omega_m)t - \cos\,(\omega_c + \omega_m)t\right]$$

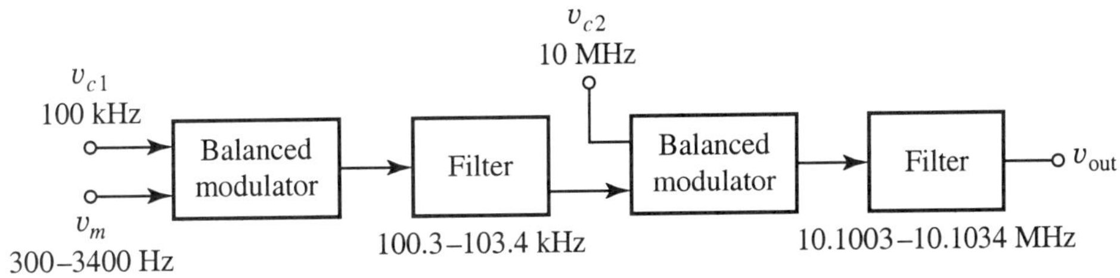


Figure 8.22: Single-sideband generation using two modulators.

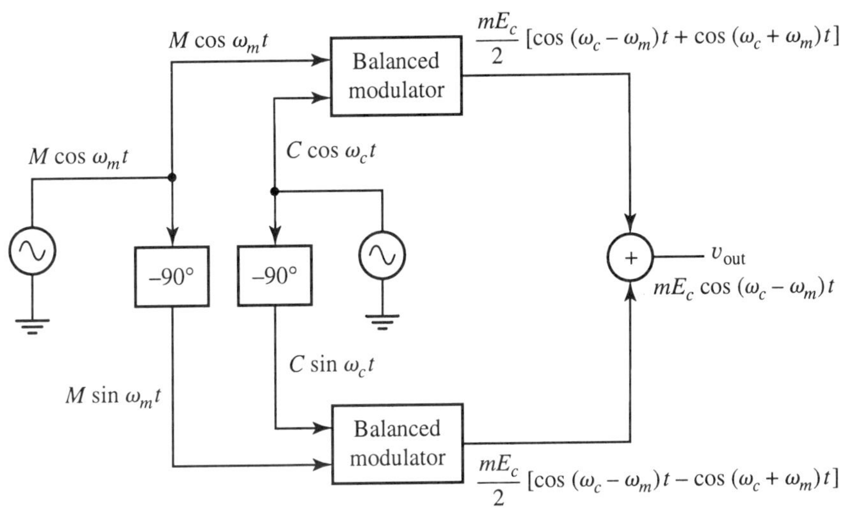


Figure 8.23: Phase-shift method of SSB generation.

If a summing circuit is used to add these signals, the result is

$$v_{out} = mV_c \cos (\omega_c - \omega_m)t \tag{8.43}$$

which is the desired SSB signal.

One problem with this method is the need for a 90° phase shift for all components of the modulating voltage. It is rather easy to shift the phase of a constant-frequency signal such as the carrier by 90°. The modulating signal may contain components that vary in frequency by a factor of ten or more. A single phase-shift circuit cannot shift the phase of all components by exactly 90°. In the practical case, two allpass networks with a phase shift that varies with frequency implements the wideband phase-shift network. The difference of the phase shifts of the two all-pass networks approximates the required 90°. However, the purity of the SSB signal deteriorates rapidly if the phase shift of either the carrier or the modulating signal departs from 90°. The ratio of amplitude of the unwanted sideband to that of the desired sideband for a given single modulating frequency is equal to $(\tan \frac{\Delta}{2})$ where $\Delta$ is the departure from 90°. A difference of only 2° results in an unwanted sideband that is 35 dB below the desired one.

### 8.3.3 The Weaver Method

A method that is used more often than the phase-shift approach is referred to as the Weaver method. The block diagram that implements this approach is shown in Figure 8.24. This approach requires a 90° phase shift of the carrier in two sections of the system, but these phase shifts are easily and accurately obtainable for the

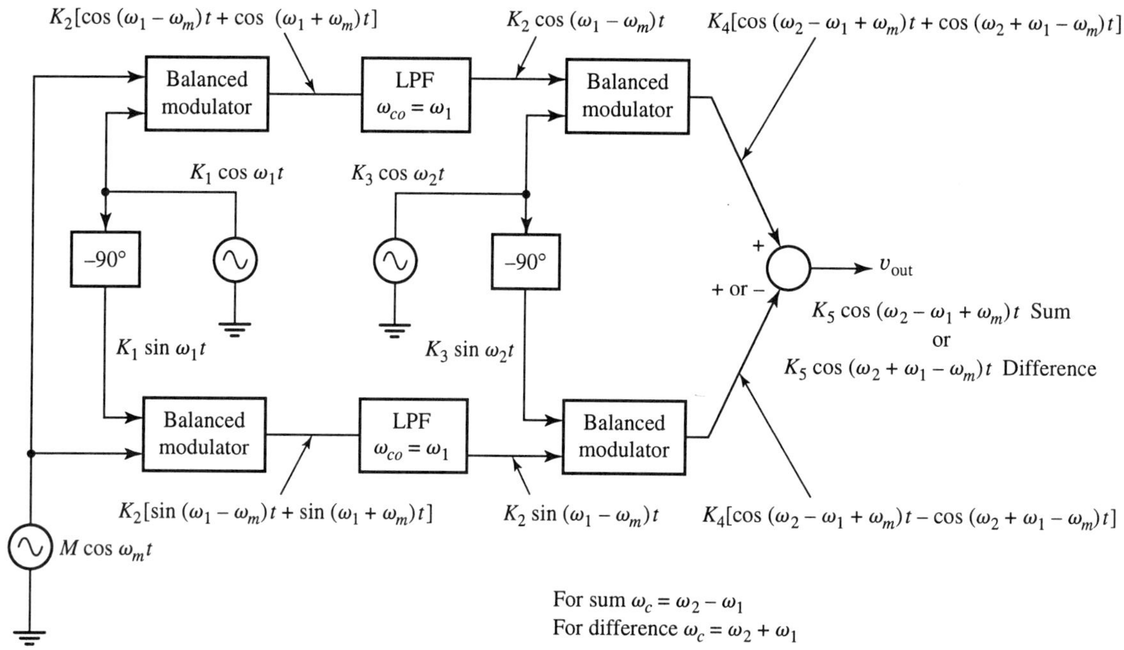


Figure 8.24: Weaver method of SSB generation.

fixed carrier frequency. We note that the lowpass filters must separate an intermediate lower sideband from an upper sideband frequency. This can often be done without excessively sharp filters. It is convenient to require that the frequency $\omega_1$ and $\omega_2$ be synchronized. This eliminates any phase difference between the two signals and simplifies the mathematics. The simplest method of accomplishing this is to multiply $\omega_1$ by a frequency multiplier to obtain $\omega_2$. The output carrier frequency is equal to the sum or difference of $\omega_1$ and $\omega_2$ depending on whether the difference or sum of the two signals is formed at the output. In either of the two methods available for creating SSB signals using phase shifts, the gains of the balanced modulators must be closely matched to accurately cancel the unwanted component.

## 8.4 MIXERS

**Important Concepts**

**1.** Mixers are used to translate a carrier frequency with its sidebands to a new intermediate frequency with sidebands.

**2.** A circuit that multiplies two input signals can function as a mixer.

**3.** A device that obeys the square law forms the product of two input signals, but also contains other unwanted components in the output signal.

**4.** Double conversion is used to remove image frequencies from the mixed signal.

### 8.4.1 Simple Mixers

A mixer is used in all radio and television receivers. This important circuit is very similar to a modulator, but is typically designed to be driven by two input signals with a frequency ratio of two or three to one. Most modulators are driven by a carrier frequency that is at least 100 times the maximum modulating frequency. A mixer in an AM radio receiver might have two applied signals of frequency 980 kHz and 525 kHz. With two such input frequencies, the resulting output frequencies are widely separated. It is then easy to filter out all but the desired component because of the wide frequency separation. The mixing operation is also called *heterodyning.*

Although this process is basically a modulation process, it is usually performed on an AM waveform without destroying the information. In effect, mixing translates the frequency of the AM waveform in such a way that the sidebands are centered about a new carrier frequency. One very useful application of mixing occurs in all standard commercial radio receivers. Information is transmitted with a carrier frequency falling in the 540- to 1700-kHz range. The desired signal is translated from the incoming carrier frequency to a carrier of 455 kHz carrying the same modulation information. A fixed-frequency tuned amplifier is then used to amplify this signal. Other signals can be amplified with this same amplifier simply by translating each AM waveform to this 455-kHz carrier frequency.

This idea of a fixed intermediate-frequency amplifier is useful in a receiver that is to receive any one of several different signals within a specified frequency range. The tuned amplifiers of the tuned stages can be carefully tuned or aligned to the intermediate frequency and never require tuning to another frequency. A variable frequency oscillator, called the *local oscillator*, is applied to the mixer along with the incoming AM signals of several stations. The local oscillator frequency is adjusted to have a difference from the desired carrier frequency of 455 kHz. This new AM waveform with a 455 kHz carrier frequency contains the same information as that contained by the original AM waveform. After passing through the IF strip, any extraneous frequencies will be removed, while the translated carrier and sidebands of the desired frequency will be amplified.

Mixing is also useful in single-sideband detectors either to reinsert a carrier frequency or to recover the original modulating or baseband signal. This application will be discussed in more detail in the next section. Mixing can be performed by means of a balanced modulator, although a simple nonlinear element such as a bipolar-junction transistor or a field-effect transistor is also used in some less stringent applications. Let us assume that we apply a constant-amplitude component $v_1 = D\cos\,\omega_1 t$ to the modulating input of a balanced modulator or a multiplier. To the other input is applied an AM waveform

$$v = V_c\,\cos\,\omega_c t + \frac{mV_c}{2}\,\cos\,(\omega_c - \omega_m)t + \frac{mV_c}{2}\,\cos\,(\omega_c + \omega_m)t$$

where $m$ is the index of modulation of the AM waveform. The resulting signal is

$$v_h = \frac{kDV_c}{2}\,\left[\cos\,(\omega_1 - \omega_c)t + \frac{m}{2}\,\cos\,(\omega_1 - \omega_c - \omega_m)t + \frac{m}{2}\,\cos\,(\omega_1 - \omega_c + \omega_m)t\right]$$

$$+\frac{kDV_c}{2}\,\left[\cos\,(\omega_1 + \omega_c)t + \frac{m}{2}\,\cos\,(\omega_1 + \omega_c - \omega_m)t + \frac{m}{2}\,\cos\,(\omega_1 + \omega_c + \omega_m)t\right] \tag{8.44}$$

where $k$ is a constant of the balanced modulator. The voltage of Eq. (8.44) represents two AM waveforms, one centered about a carrier of frequency $(\omega_1 - \omega_c)$ and the other centered about $(\omega_1 + \omega_c)$. Under usual conditions both $\omega_1$ and $\omega_c$ are much larger than the maximum modulation frequency; thus, the sidebands associated with this waveform do not overlap. A tuned amplifier with a bandwidth slightly larger than $2\omega_m$ can be used to separate either of the frequency ranges from the other. In commercial applications the signal with the frequency $(\omega_1 - \omega_c)$ is generally amplified as the intermediate frequency. Figure 8.25 shows the spectra for both input and output waveforms, assuming tone modulation.

We note that if $\omega_1 = \omega_c$, the lower frequency AM signal contains a dc term and a term equal to

$$\frac{kDmV_c}{2}\,\cos\,\omega_m t$$

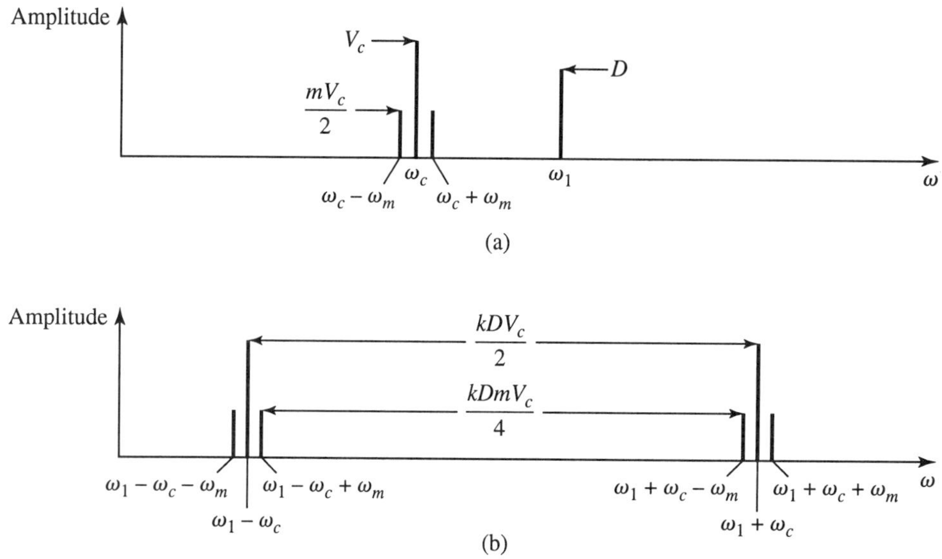


Figure 8.25: Amplitude spectra for mixer. (a) Input spectrum. (b) Output spectrum.

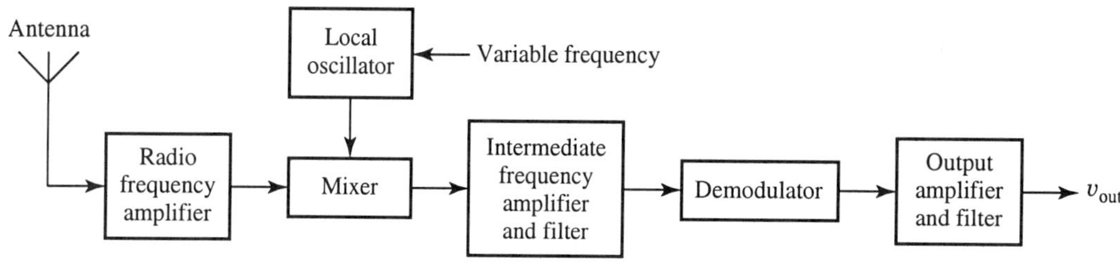


Figure 8.26: Superheterodyne receiver.

This signal is proportional to the original modulating signal, $M \cos \omega_c t$. Although it may be difficult to synchronize the frequencies $\omega_1$ and $\omega_c$, some special-purpose receivers use this basic idea to recover the transmitted information.

The mixer in most commercial receivers uses a simple transistor mixer (BJT or FET) rather than a perfect multiplier, but additional output frequencies are far out of the intermediate-frequency band and present no problems to the heterodyne process. The block diagram of a typical superheterodyne receiver is shown in Figure 8.26.

The AM waveform is received by the antenna and amplified by the RF amplifier. Some filtering can be done in this amplifier, but the selectivity or ability to discriminate against other off-carrier frequencies is determined primarily by the IF section. The local oscillator is tuned so that the oscillation frequency minus the carrier frequency of the desired signal equals the intermediate frequency. In a superheterodyne system, the oscillator frequency is always greater than the RF carrier frequency. The IF section passes the signal centered at the intermediate frequency with a small bandwidth and discriminates heavily against signals of other frequencies. The signal is then demodulated, amplified, and filtered, if necessary, to produce the output signal. This signal is then used to drive a speaker or other output device.

### 8.4.2 A Perfect Square-Law Multiplier

Some electronic devices approximate square-law operation. We will first look at a perfect square-law multiplier before proceeding to multipliers realized with real, nonideal elements. A perfect square-law device does not result in an ideal multiplier. The perfect square-law device accepts the sum of two signals at its single input

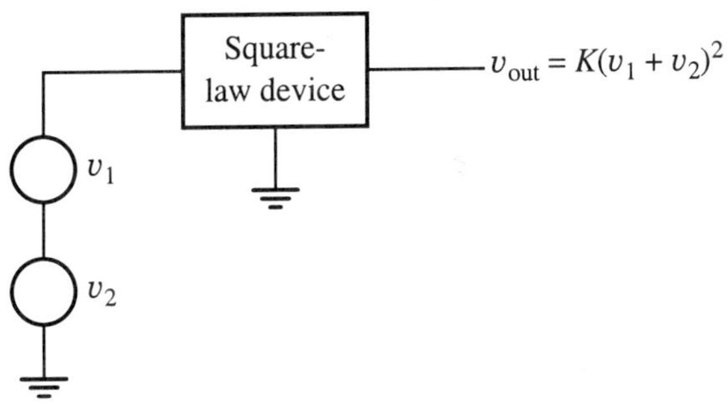


Figure 8.27: A square-law device.

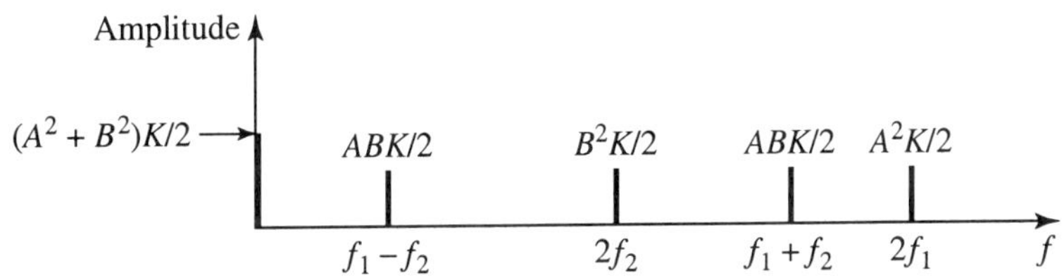


Figure 8.28: Spectral components of a square-law circuit output.

and produces the squared value of this sum at the output. The squared value of the sum of the two signals contains several terms including one that is the product of the two input signals. The block diagram of Figure 8.27 represents this situation.

With an input given by

$$v_{in} = v_1 + v_2 = A\cos\,\omega_1 t + B\cos\,\omega_2 t$$

The output becomes

$$v_{out} = K(A\cos\,\omega_1 t + B\cos\,\omega_2 t)^2 = K(A^2\cos^2\,\omega_1 t + 2AB\cos\,\omega_1 t\,\cos\,\omega_2 t + B^2\cos\,\omega_2 t)$$

Again using trigonometric identities, this output signal can be expressed in terms of pure sinusoidal components as

$$v_{out} = \frac{K}{2}\left(A^2 + B^2 + B^2\cos\,2\omega_2 t + A^2\cos\,2\omega_1 t\right) + \frac{K}{2}\left(AB\cos\,(\omega_1 - \omega_2)t + AB\cos\,(\omega_1 + \omega_2)t\right)$$

The two components with frequencies that are the sum and difference of the two input values represent the multiplied signal. All other components are undesirable and must be filtered from the system to achieve the heterodyned or mixed signal. Figure 8.28 shows the spectral components of this output signal.

Depending on the relative values of $\omega_1$ and $\omega_2$, it might be difficult to filter all unwanted signals from the desired components. For example, if $f_1 = 1000$ kHz and $f_2 = 333.33$ kHz, the desired component $f_1 - f_2$ and the unwanted component $2f_2$ both equal 666.66 kHz. It is impossible to filter one of these signals without eliminating the other.

On the other hand, if $f_1 = 1000$ kHz while $f_2 = 200$ kHz, the output signal will contain desired components at 800 kHz and 1200 kHz and undesired components at zero frequency (dc), 400 kHz, and 2000 kHz. These undesired components can be easily filtered with tuned circuits.

We may therefore conclude that an ideal multiplier circuit can be implemented by a square-law circuit followed by effective filters, assuming that the input frequencies are restricted to proper values or ranges. With inappropriate frequency inputs, the square-law circuit cannot implement the multiplier or DSB circuit.

### 8.4.3 Practical Multipliers

Two practical multiplier circuits will be considered here. The first is often called a transconductance multiplier. One version of this chip is the Motorola MC1496/1596 discussed earlier in the text. This chip produces an output given by Eq. (8.45)

$$v_{\text{out}} = \frac{R_L}{50}\left[(I_1 - I_2) + (K_1 + K_2)v_m\right]v_c \tag{8.45}$$

This AM waveform becomes a DSB-SC waveform if $I_1 = I_2$. The chip provides for an external resistive tuning circuit, allowing this condition to be satisfied. Of course, the cancellation of two currents can never be perfect over a range of temperatures or with limited resolution of the tuning network, so the manufacturer guarantees that the suppressed carrier magnitude will be much smaller than the magnitude of the input carrier signal. A typical value might be carrier suppression of −65 dB at 0.5 MHz or −50 dB at 10 MHz for the MC1496 chip. This chip, while a good approximation to a multiplier circuit, is limited in frequency performance.

When input frequency ranges are such that a square-law circuit and filter can be used, the MOSFET is the device of choice. This component can have very good frequency response and approximates a square-law variation surprisingly well. To a first approximation, the MOSFET with channel lengths of 1 $\mu$ or more obeys a drain current relationship in the active region given by

$$I_D = \frac{\mu C_{\text{ox}} W}{2L}(V_{GS} - V_T)^2 \tag{8.46}$$

where $\mu$ is the carrier mobility, $C_{\text{ox}}$ is the channel capacitance per unit area, $W$ is the channel width, $L$ is the channel length, and $V_T$ is the threshold voltage.

When this current flows through a real impedance, an output voltage is developed that is proportional to the square of the gate-to-source voltage. Second-order effects such as channel-length modulation tend to detract from the ideal behavior of the device. A simple MOSFET mixer is shown in Figure 8.29. For this circuit, the input voltage is equal to the gate-to-source voltage or

$$V_{\text{in}} = V_{GS} = v_c + v_d + V_1$$

where $V_1$ is the bias voltage needed to ensure operation in the active region. The signals $v_c$ and $v_d$ can be written as

$$v_c = C\cos\omega_c t \qquad v_d = D\cos\omega_d t$$

The drain current is then

$$I_D = K[v_c + v_d + V_{\text{eff}}]^2 \tag{8.47}$$

where $K = \mu C_{\text{ox}} W/2L$ and $V_{\text{eff}} = V_1 - V_T$.

Using sinusoidal quantities for $v_c$ and $v_d$ and a few trigonometric identities results in

$$\begin{aligned} I_D = K&\left[V_{\text{eff}}^2 + \frac{C^2}{2} + \frac{D^2}{2}\right] \\ &+ \left[2V_{\text{eff}}C\cos\omega_c t + 2V_{\text{eff}}D\cos\omega_d t + \frac{C^2}{2}\cos 2\omega_c t + \frac{D^2}{2}\cos 2\omega_d t\right] \\ &+ \left[CD\cos(\omega_c - \omega_d)t + CD\cos(\omega_c + \omega_d)t\right] \end{aligned} \tag{8.48}$$

The first few terms of this equation represent the unwanted dc component. The next terms represent unwanted components at frequencies of $\omega_c$, $\omega_d$, $2\omega_c$, and $2\omega_d$. The last two terms are the desired sideband terms. Note that the coefficients of the unwanted terms can be comparable to the coefficients of the desired terms.

In order to filter the unwanted terms out of the output signal, a tank circuit can form the drain load as shown in Figure 8.29. Of course, the frequencies $f_1$ and $f_2$ must have values that allow filtering without eliminating the desired components.

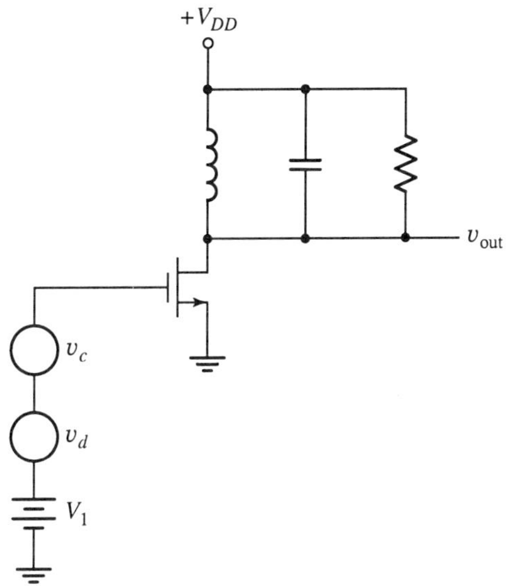


Figure 8.29: A MOSFET mixer.

### 8.4.4 Double Conversion

Earlier in this section we discussed the use of the multiplier circuit in heterodyning to achieve an intermediate frequency for further amplification and filtering. Although a simple mixer can be used in AM applications, most receivers now use two mixing operations in a process known as *double conversion.*

The double converter is used to minimize a problem due to the *image* frequency. To understand the origin of an image frequency, let us assume we are trying to tune an AM receiver to 580 kHz. The local oscillator frequency used to convert this incoming frequency to the intermediate frequency, $f_{\text{if}} = 455$ kHz, could be 1035 kHz so that

$$f_{\text{osc}} - f_c = 1035 - 580 = 455 \text{ kHz}$$

The image frequency arises from another carrier frequency of 1490 kHz since the difference between the local oscillator frequency and this frequency is also 455 kHz. Hence, two different carrier frequencies are heterodyned to the intermediate frequency. The equation for the image frequency is given by

$$f_i = f_c + 2f_{\text{if}}$$

One method used on all receivers to minimize the effect of the image frequency is to use a tuned circuit on the first RF stage that precedes the mixer stage. This circuit would be tuned to the desired carrier of 580 kHz and should discriminate against the 1490-kHz signal. If both signal strengths were of equal size, the 1490-kHz signal would be essentially eliminated. However, if the station generating the 580-kHz signal were located some distance away from the receiver while the station generating the 1490-kHz signal was nearby, the signal strength might be radically different. In addition, the station generating the 1490-kHz signal may have a "clear" rating while the other does not; thus, the clear station may generate much more power than the other station. The tuned circuit on the RF stage may not be able to discriminate against the unwanted strong signal in such cases. This leads to interference and noise in the detected signal.

A second method popular in AM receivers applies a *double conversion* approach. The first mixer stage uses a relatively high oscillator frequency. For example, $f_{\text{osc1}}$ might tune over the range of 10,540 kHz to 11,700 kHz to convert the desired incoming carrier frequency to 10,000 kHz. If $f_{\text{osc1}} = 10,580$ kHz, the 580 kHz signal is converted to $10,000$ kHz, but the 1490 kHz signal is converted to 9090 kHz. A tuned filter follows this first conversion, but it need not be a high-$Q$ filter. This signal is then mixed with the output of a second local oscillator with a fixed frequency of 10,455 kHz. The resulting intermediate frequency is 455 kHz, which can be amplified and filtered by a 455-kHz IF strip.

## 8.5 DETECTORS

**Important Concepts**

**1.** Detectors are used to recover the baseband information transmitted over a modulated signal.

**2.** The peak detector is a nonsynchronous detector that is simple and inexpensive. It has been used in commercial AM receivers for many decades.

**3.** The synchronous detector is often used to demodulate a single-sideband signal.

After a radio signal has been received, converted to an IF signal, separated from all extraneous signals, and amplified, the modulation signal must be recovered. This process is called *demodulation* or *detection.* Two types of detectors are used in AM communication, depending on the type of transmission used. These detectors are classified as *nonsynchronous* or *synchronous.* The synchronous detector is often used for a SSB system and uses a mixer to translate the incoming sideband to the correct modulating frequency. The nonsynchronous detector is simpler and will only function when driven by an AM signal. It cannot detect DSB-SC or SSB signals without additional circuitry.

### 8.5.1 The Nonsynchronous Detector

The information transmitted by a simple AM wave is carried by the peak amplitude of the waveform. Thus a peak-detecting circuit should recover this information. The simple circuit of Figure 8.30(a) will detect peaks as long as the peak amplitude does not decrease. This detector will not follow the peaks of decreasing amplitude unless a resistor is used to slightly discharge the capacitor between cycles, as shown in Figure 8.30(b).

The output voltage of the detector contains three components: a high-frequency ripple voltage; a much lower frequency voltage that is proportional to the amplitude of the peaks of the AM waveform; and a dc voltage equal to the amplitude of the unmodulated carrier signal. The signal that contains the transmitted information is the one that is proportional to the peaks of the waveform. The dc signal and the high-frequency signal are undesirable and must be removed. The detector voltage might ultimately be used to activate a loudspeaker; hence, a coupling capacitor should be used to remove the dc component. The high-frequency component can be filtered or even ignored in some applications, since it is far above the audio range. The signal shown in Figure 8.30 does not demonstrate the true frequency difference between the IF signal and the modulating signal. We must bear in mind that the intermediate frequency is several hundred times greater than the modulating frequency.

The time constant of the RC-combination is chosen to be much larger than the period of the carrier and much smaller than the period of the modulating signal. A reasonable value for this simple circuit is

$$RC_1 = \sqrt{1/\omega_m \omega_{\text{if}}} \tag{8.49}$$

Equation (8.49) gives a time constant which is the geometric mean of the two periods divided by $2\pi$. The discharge between cycles of the carrier signal should be negligible, but the discharge rate must be high enough to follow accurately the decrease in amplitude of the peaks.

At least two practical problems are associated with the simple detector of Figure 8.30. One shortcoming relates to the voltage necessary to cause current conduction in the diode. A silicon diode might drop 0.6 or 0.7 V before allowing significant current to flow. This drop can be minimized by using a germanium diode, but the drop may still be too large. The second problem is that very small input signals may not be reproduced accurately. The discharge rate of a capacitor depends on the time constant, the voltage across the capacitor, and the target voltage to which the capacitor is discharging. The initial discharge rate is given by

$$\frac{dv_{\text{cap}}}{dt} = \frac{v_i - v_t}{\tau}$$

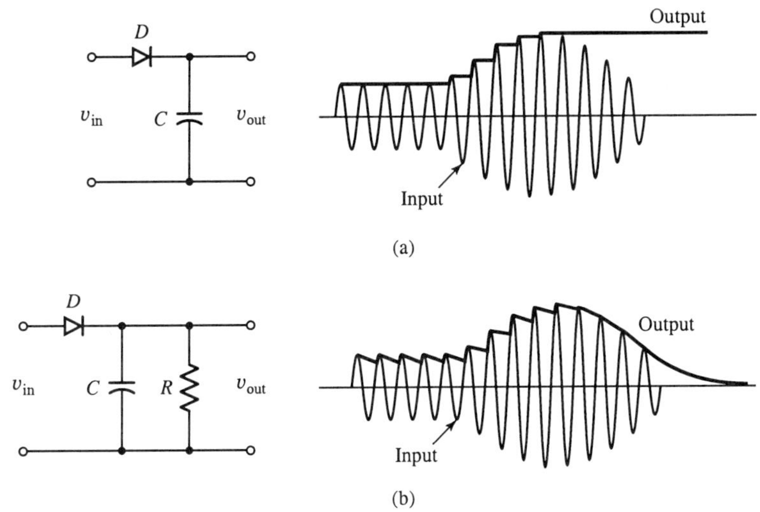


Figure 8.30: Simple diode detector. (a) Peak detector. (b) Discharge resistor added.

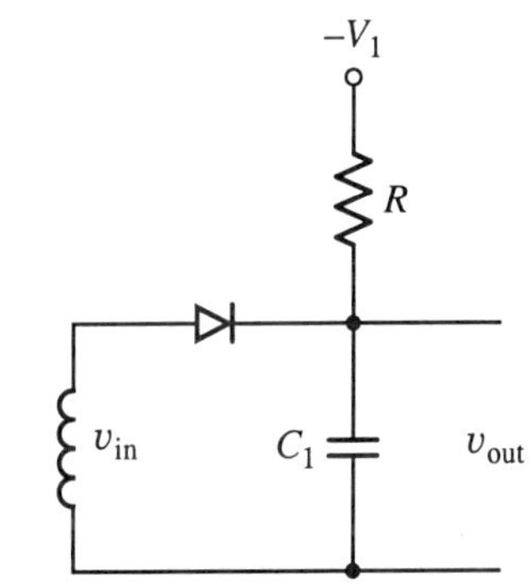


Figure 8.31: An improved detector.

where $v_{cap}$ is the capacitor voltage, $v_i$ is the initial capacitor voltage, $v_t$ is the target voltage, and $\tau$ is the time constant. For small values of input signal, the capacitor voltage is small and the target voltage is zero. The output signal may not respond rapidly enough to follow the changes in the modulating signal.

The detector of Figure 8.31 solves both of these problems. The negative voltage provides a forward bias for the diode at all times. In addition, a target voltage of $-V_1$ exists in this circuit rather than a voltage of zero. The capacitor discharge rate is nearly constant for all input signal levels, providing that $V_1$ is considerably larger than the maximum input level. The diode can be replaced by the base-emitter junction of an emitter-follower stage to achieve current gain. Figure 8.32 shows a detector followed by a highpass filter.

The coupling capacitor, $C_c$, removes the dc component of the demodulated signal from the output voltage while passing the modulation frequency. The discharge rate of the capacitor, $C_1$, can be selected to follow the largest expected change of modulating signal. For a simple tone-modulated signal, assuming 100% modulation, the greatest slope will be $D\omega_m$, where $D$ is the magnitude of the unmodulated IF signal.

In Figure 8.32, $R_i$ might represent the input resistance to an amplifying stage. This resistance is typically much larger than $R$ to minimize loading. If an unmodulated carrier is applied to the input of the receiver, the IF signal will also be unmodulated. the voltage $v_d$ will be a dc signal equal to the amplitude of the IF signal (neglecting the diode drop). The voltage $v_{out}$ will equal zero. The average current drawn from the source is then

$$I_{in} = \frac{D + V_1}{R} \tag{8.50}$$

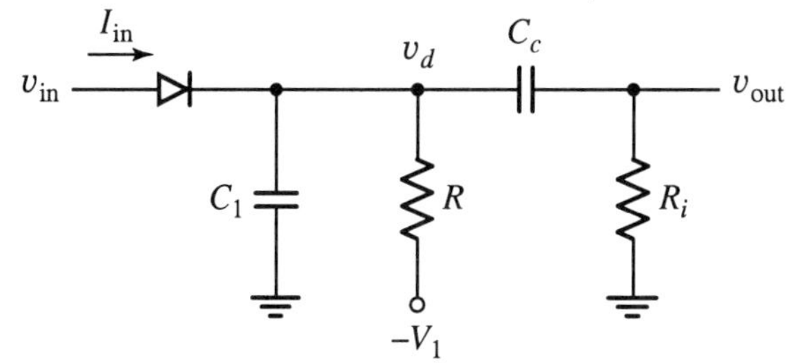


Figure 8.32: A practical detector.

The capacitor discharge rate between cycles of the carrier signal is given by

$$\frac{\Delta v_d}{\Delta t} = \frac{I_{in}}{C_1} = \frac{D + V_1}{RC_1}$$

which should equal or exceed the greatest slope of the modulating signal. This leads to the requirement

$$\frac{D + V_1}{RC_1} \geq mD\omega_m$$

or

$$RC_1 \leq \frac{1}{m\omega_m}\left(1 + \frac{V_1}{D}\right)$$

The peak detector is used in all commercial AM receivers because it is relatively inexpensive.

### 8.5.2 Synchronous Detection

An incoming signal may be AM, DSB, or SSB. Commercial radio stations in the 540- to 1700-kHz frequency range are AM waveforms due partly to tradition and to the low cost of peak detection circuitry.

Amateur radio operators can generally transmit AM or SSB. Older telephone transmission schemes, many of which are still in service, use SSB with a pilot frequency to aid in the demodulation. Microwave relay stations have been returning to SSB to preserve bandwidth. Mobile services use SSB in some frequency ranges, again owing to the efficient bandwidth usage of SSB.

The popular peak detector used for AM signals will not work directly for SSB or DSB signals. It is possible to reinsert a carrier to convert the DSB signal to an AM signal and then use a peak detector for demodulation.

Synchronous or coherent detection can be used to demodulate either of the three types of amplitude-modulated signals. In this process, the incoming signal is mixed with a local signal that has phase and frequency coherence with the original carrier signal. The coherence can be achieved in various ways.

One method of achieving coherence is through constantly tuning the local oscillator as the strength of the detected signal is optimized. This is the method most often used by amateur radio operators when receiving an SSB signal.

Long-distance phone lines using SSB also send a pilot frequency when transmitting multiple conversations over a single pair of lines. This pilot frequency is coherent with each carrier frequency used to create the individual signals at the transmitter. When recovered and frequency multiplied by appropriate values, these signals can be used for synchronous detection of the baseband signals.

The synchronous detector is also called a *coherent* detector and always contains a mixer. Two inputs are applied to this circuit: (1) the signal to be detected, often SSB, and (2) the signal having a mixing frequency designed to translate the SSB signal to a baseband signal. The mixing frequency must equal the carrier frequency of the signal to be detected. Because the mixing frequency is the same as the carrier frequency, the signals are said to be synchronized or coherent.

For an AM waveform, the nonsynchronous envelope detector is the least expensive demodulator that can be used. This detector is always used with AM waveforms such as those transmitted over the commercial AM band. The major use of the synchronous detector is in the recovery of the baseband signal from an SSB or VSB waveform.

If the mixing signal is synchronized in both frequency and phase, the two mixer inputs are

$$v_{usb} = S \cos(\omega_c + \omega_m)t$$

and

$$v_1 = D \cos \omega_c t$$

The mixer output is then

$$\begin{aligned} v_{\text{mix}} &= kSD \cos \omega_c t \cos(\omega_c + \omega_m)t \\ &= \frac{kSD}{2}[\cos \omega_m t + \cos(2\omega_c + \omega_m)t] \end{aligned}$$

The higher frequency component is easily removed by filtering, leaving

$$v_{\text{out}} = \frac{kDS}{2} \cos \omega_m t \tag{8.51}$$

This signal is proportional, through the value of $S$, to the original baseband signal and has the same frequency.

It would appear that the synchronous detector can be implemented by a balanced modulator that forms the product of the two input signals and by a lowpass filter. The above results were derived based on the assumptions that the two input signals were synchronized in both phase and frequency.

In most practical systems, the mixing signal will have either a phase or frequency difference from the SSB signal, or both. If the transmitter sends only the SSB signal, the receiver must generate the mixing signal. It is difficult, if not impossible, to produce a mixing signal that is perfectly synchronized with the carrier of the SSB signal. A system that produces its own mixing signal is called a *self-synchronous* system. The Class-D citizens band (CB) radio transceivers transmit standard AM signals. However, the more versatile models also allow the transmission and reception of SSB signals. By concentrating the maximum allowable 5 W of average power into a single sideband, the range of communication is greatly extended. A peak envelope power of 12 W is allowed for SSB transmission. These systems use a beat frequency oscillator (BFO) to produce an appropriate mixing signal. The resulting distortion due to imperfect synchronization allows audio signals to be recovered with acceptable quality only because the human ear is very tolerant of this type of distortion.

Another method used to produce a mixing signal is the transmission of a pilot tone along with the SSB or VSB signal. The pilot signal can then be utilized to generate a synchronized mixing signal. This method is implemented in the FDM system employed in long-distance telephone transmission. A pilot tone, synchronized with the carriers used to generate the various SSB signals, is transmitted along with the SSB signals. This pilot can be used to produce the necessary mixing signals to demodulate the SSB signals. The transmission of TV signals is done with VSB and includes the transmission of an attenuated carrier or pilot signal to be used in the demodulation process.

In order to understand the effects of a nonsynchronous mixing signal, we will assume that an upper sideband signal of frequency $(\omega_c + \omega_m)$ is applied to a mixer, along with a mixing signal having a frequency of $\omega_1$ and a phase of $\phi$. After lowpass filtering the output of the mixer, the resulting signal is

$$v_{\text{out}} = \frac{kDS}{2} \cos[(\omega_c - \omega_1 + \omega_m)t - \phi] \tag{8.52}$$

If the mixing frequency is equal to $\omega_c$, but $\phi$ is nonzero, $v_{\text{out}}$ becomes

$$v_{\text{out}} = \frac{kDS}{2} \cos[\omega_m t - \phi]$$

Using a trigonometric identity, we can write this as

$$v_{\text{out}} = \frac{kDS}{2}[\cos \phi \cos \omega_m t + \sin \phi \sin \omega_m t]$$

The magnitude of the detected signal is

$$|v_{\text{out}}| = \frac{kDS}{2}\sqrt{\cos^2\,\phi + \sin^2\,\phi} = \frac{kDS}{2}$$

This magnitude is completely independent of the phase difference. The phase of the signal equals $\phi$, which in the practical case will vary with time.

If the mixing signal has a different frequency, but no phase difference at some beginning time, the output of the detector is

$$v_{\text{out}} = \frac{kSD}{2}\cos\,(\omega_m - \Delta\omega)t$$

where $\Delta\omega = \omega_1 - \omega_c$. This signal can again be shown to have a magnitude that is independent of $\Delta\omega$, but a phase difference of $\Delta\omega t$. This signal can also be considered to have a frequency of $\omega_m - \Delta\omega$. Thus the detected signal will have a different frequency than the original baseband signal.

Because the human ear is basically insensitive to phase shift of the various components of speech or music, a BFO may be used to generate the mixing signal in a self-synchronous detector. The mixing frequency must be relatively close to the carrier or harmonic distortion takes place. For example, if a sound is transmitted that has frequency components at 200 Hz, 400 Hz, and 600 Hz, a frequency difference of $\Delta f = 50$ Hz will result in detected components of 150 Hz, 350 Hz, and 550 Hz. These components are no longer harmonically related since 350 Hz is not the second harmonic of 150 Hz and 550 Hz is not the third harmonic of 150 Hz. It has been found for voice transmission using SSB that a frequency difference of 30 to 40 Hz is acceptable for reasonable intelligibility. The CB units that receive SSB generally allow for fine tuning to maximize the intelligibility.

## 8.6 AUTOMATIC GAIN CONTROL

**Important Concepts**

**1.** The signal strength of a radio signal can vary drastically with time, especially in mobile receivers.

**2.** The automatic gain control circuit is used to minimize the effect of signal strength variation at the receiver.

The signal strength received by a radio receiver can vary dramatically and rapidly. If the receiver is an automobile radio, the signal strength varies as the transmission path from transmitter to automobile changes. When mountains, buildings, and other obstacles are near the vehicle, attenuation and reflection of the signal can easily take place. In order to minimize the effect of varying signal strength on output volume, most receivers have *automatic gain control* (AGC). AGC receivers increase the gain of the IF amplifiers when the signal strength drops and decreases the gain when the signal strength increases.

The signal strength is measured at the demodulator output and integrated by a simple RC lowpass circuit. The time constant of the RC circuit is long compared to the minimum modulation frequency. The average value of the AM waveform is then passed to the capacitor. This is proportional to the amplitude of the unmodulated carrier signal. An increase in this value represents a stronger received signal, and the higher capacitor voltage is used to decrease the gain of the IF stages. Generally, this is done by controlling the dynamic base-emitter junction resistance, $r_e$, for BJT stages or by controlling the value of $g_m$ for MOSFET stages. The circuit of Figure 8.33 shows a simplified version of an AGC, with gain control applied to only one transistor.

The value of $V_{CC}$ for this circuit is negative. The detector diode is forward biased through resistors $R_1$ and $R_2$. When a signal is received and detected by the demodulator, the average value of the signal is fed through $R_2$ to $C_2$. The modulation signal recovered by the detector will not be passed to $C_2$ because the cutoff frequency of the RC network is too low. If the average value at the detector increases, the capacitor charges to a more positive value. This increase in base voltage decreases the forward bias of the base-emitter junction. The resulting decrease in emitter current leads to a larger value of $r_e$, and this lowers the gain of the tuned stage. The time constant of the AGC circuit is generally in the 30- to 50-ms range.

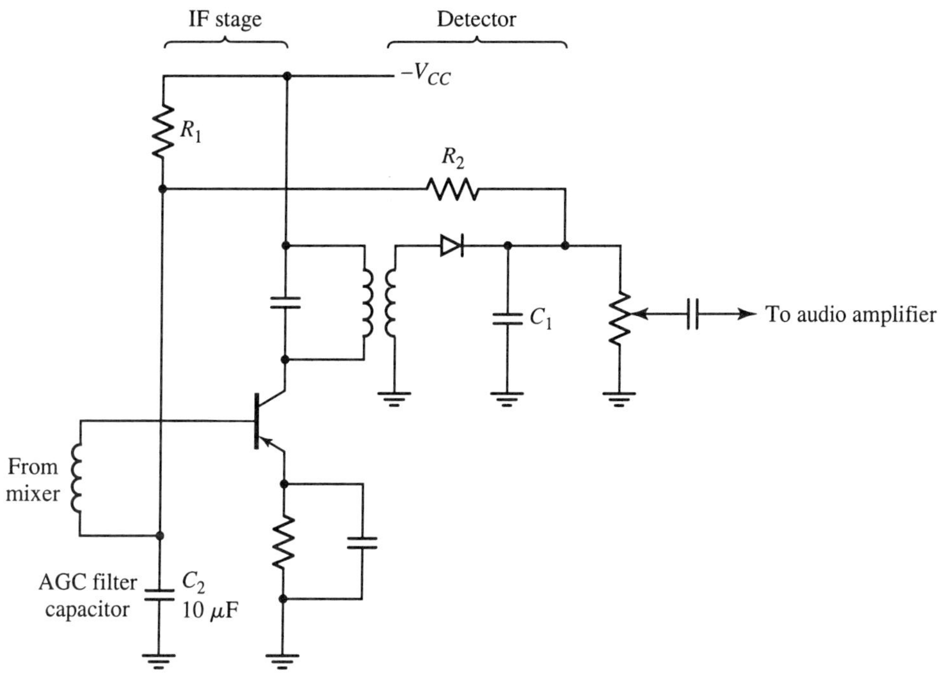


Figure 8.33: A simplified AGC circuit.

The potentiometer following the detector is the volume control for the audio signal. The AGC attempts to drive this potentiometer with a constant input voltage, but does not accomplish this objective. However, applying the AGC signal to all IF stages minimizes the variation in output voltage as signal strength varies.

## SUMMARY

**1.** Although AM is one of the oldest forms of transmission of information, it is still used today. This is due partly to the efficient use of bandwidth in this modulation method.

**2.** A variable-gain amplifier can be used to create the AM signal. This method is used in low-power applications.

**3.** Class-C modulators are important in higher power applications of AM transmission. Traditionally, special-purpose vacuum tubes are utilized in very high-power transmitters.

**4.** A class-D modulator allows semiconductor circuits to deliver 50,000 W of power to the transmitting antenna.

**5.** A DSB-SC signal can be produced by a balanced modulator that cancels the carrier signal by subtraction.

**6.** An SSB signal can be generated by filtering a sideband from the DSB-SC signal or by using phase-shift methods to cancel a sideband by subtraction.

**7.** Mixers are used to translate an incoming carrier frequency to an intermediate frequency. The process of mixing is accomplished by multiplication of the two incoming signals to form a product.

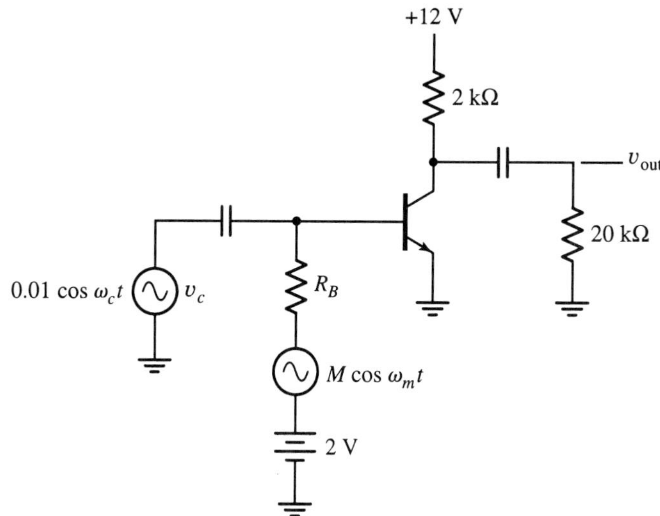


**Figure P8.3**

**8.** Detectors are used to recover the modulation or baseband information from the transmitted signal. Detectors can be nonsynchronous or synchronous.

---

## PROBLEMS

### Section 8.1.1 Variable-Gain Modulators

**D8.1** Design a transistor modulator such as that shown in Figure 8.2 to give 10% modulation. The unmodulated carrier output should have a 0.5-V amplitude. Assume that $\beta = 100$, $V_{CC} = 18$ V, $f_c = 600$ kHz, and $f_m = 5$ kHz.

**D8.2** In the circuit of Figure 8.2, $R_L = 1$ kΩ, $V_0 = 10$ V, $v_c = 0.01 \cos \omega t$, $\beta = 100$, and $V_{BE(\text{on})} = 0.7$ V.

(**a**) Choose $R$ to set the unmodulated carrier amplitude to 0.6 V.
(**b**) If a modulation voltage of $v_m = 1.0 \cos \omega t$ is applied, what is the resulting index of modulation?
(**c**) If a 4-Ω resistance is inserted between emitter and ground, calculate the new value of unmodulated output amplitude if $v_c$ remains constant.
(**d**) Calculate the index of modulation after the emitter resistor is added if both $v_c$ and $v_m$ remain constant.

**8.3** For the transistor shown in Figure P8.3, $\beta = 150$ and $V_{BE(\text{on})} = 0.7$ V. If $v_{\text{out}} = 1.0 \cos \omega_c t$ V (unmodulated) and $m = 0.1$, calculate

(**a**) The amplitude of the modulating voltage, $M$
(**b**) The resistance $R_B$

Assume that the capacitors are open circuits to the modulating signal and short circuits to the carrier signal.

**D8.4** The transistor in the circuit of Figure P8.4 has a $\beta = 120$ and $V_{BE(\text{on})} = 0.6$ V.

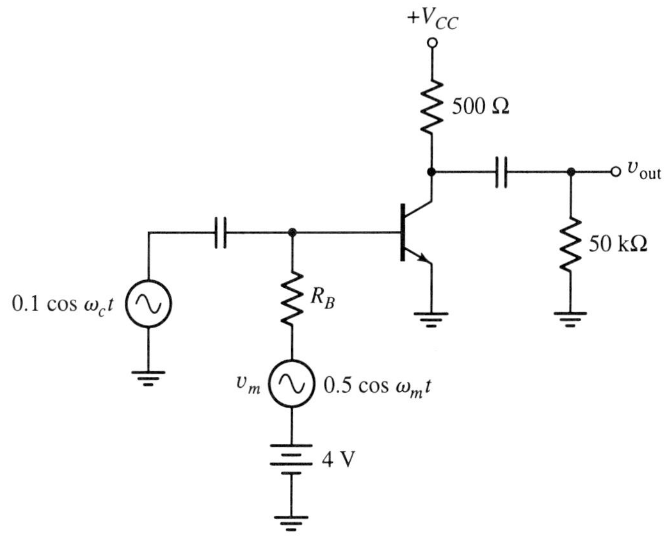


**Figure P8.4**

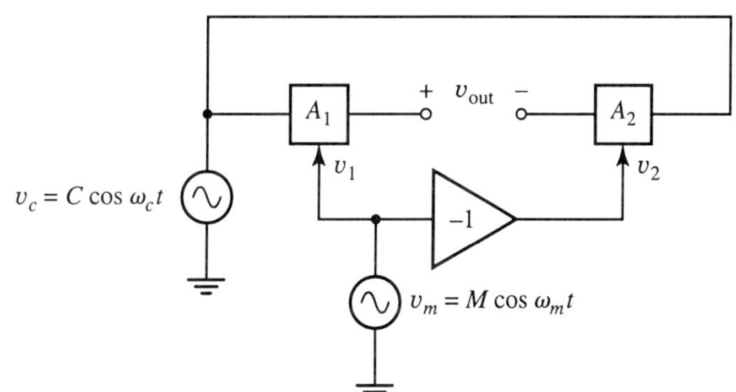


**Figure P8.5**

**(a)** Calculate the value of $R_B$ to set the unmodulated carrier output magnitude to 1.6 V.
**(b)** Calculate the maximum value of $v_{\text{out}}$ when modulation is applied.

Assume that the capacitors are open circuits to the modulating signal and short circuits to the carrier signal.

**8.5*** In Figure P8.5, the gain of stage 1 is $A_1 = 10 + 0.2v_1$ and the gain of stage 2 is $A_2 = 6 + 0.2v_2$.

**(a)** Calculate the value of $M$ required to result in an index of modulation of $m = 0.8$.
**(b)** Calculate $C$ to result in an unmodulated output voltage with an 8-V magnitude.
**(c)** If $v_{\text{out}}$ is applied to a 50-Ω resistance, calculate the power carried by the carrier and each sideband for 80% modulation of the 8-V carrier.

**8.6** Assume that $\beta = 80$ and $V_{BE(\text{on})} = 0.7$ V for both transistors in Figure P8.6. The input carrier signal is $v_c = 0.05 \cos \omega_c t$,

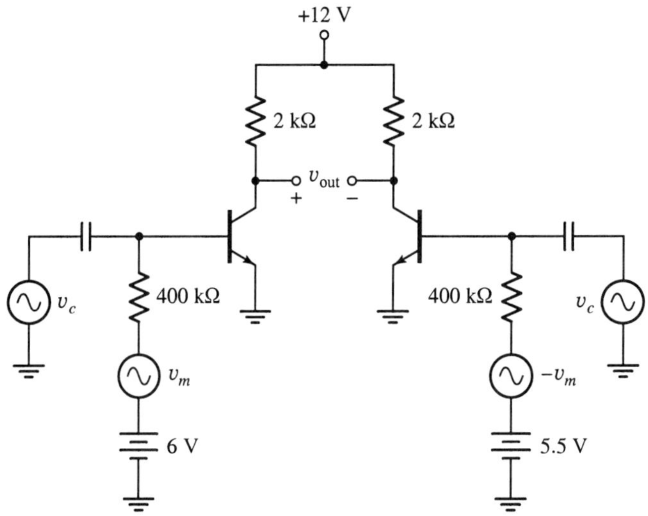


**Figure P8.6**

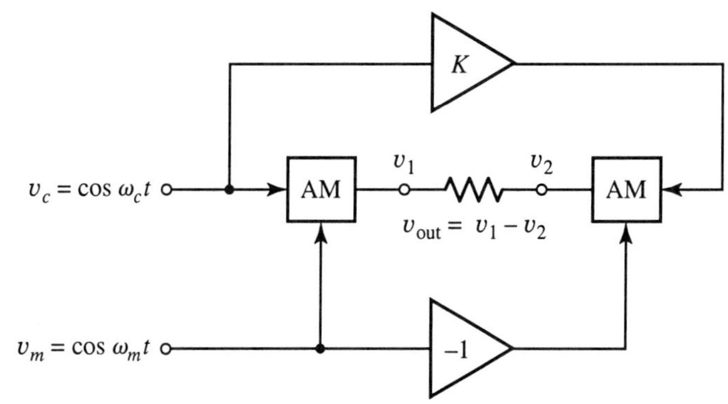


**Figure P8.7**

**(a)** Calculate the magnitude of the unmodulated carrier output.
**(b)** What is the index of modulation for $v_m = 0.2 \cos \omega_m t$?
**(c)** What peak value of $v_m$ causes 100% modulation?

**8.7*** The two modulator circuits of Figure P8.7 are identical and produce output signals of $v_{am} = 6v_c + 1.2v_m v_c$.

**(a)** What is the index of modulation, $m$, for the single-ended output signal $v_1$?
**(b)** Select the gain $K$ to result in a double-ended output $v_{out}$ that has an unmodulated amplitude of $V_c = 3$ V.
**(c)** Calculate $m$ for the output signal of part (b).
**(d)** Choose a new value of $K$ to lead to $m = 1$.
**(e)** If $K = 1$, what is the output carrier amplitude?
**(f)** If $K = 1$, what is the amplitude of the upper sideband?

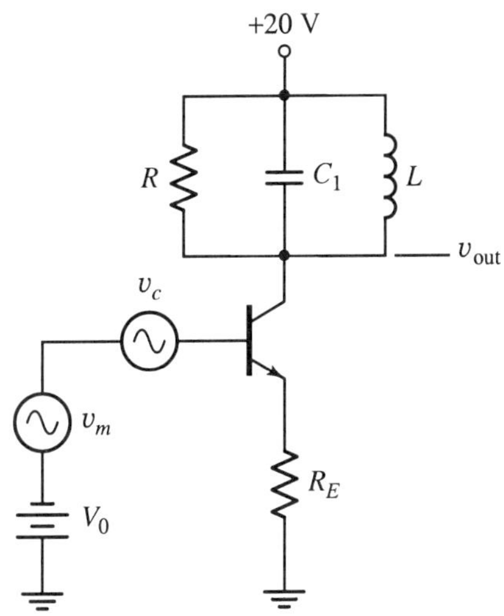


**Figure P8.10**

### Sec. 8.1.2 Class-C Modulators

**D8.8*** A class-C base modulation circuit is to be used to produce an AM waveform. The carrier frequency is 1 MHz and the highest modulation frequency is 5 kHz. A 10-$\mu$H coil with $Q_o = 120$ is used in the resonant circuit. Calculate the required value of $C_1$ and the shunt resistance that must be added to the tank circuit. *Hint:* The highest modulation frequency should be transmitted with no more than a 1-dB attenuation compared to that of the carrier frequency.

**8.9** If the carrier amplitude applied to the circuit of Figure 8.5 is 10 V and $V_0 = 2.5$ V what is the magnitude of modulating signal required for 30% modulation? Assume that $V_{BE(\text{on})} = 0.7$ V.

**8.10** A carrier signal of 5-V amplitude and 0.71-MHz frequency is applied to the circuit shown in Figure P8.10. The modulating frequency is 5 kHz. Assuming that the output resistance of the transistor is 100 k$\Omega$ and that $L = 40\ \mu$H with $Q_o = 120$, calculate all other element values required. Specify the amplitude of $v_m$ to give 40% modulation if $V_0 = 2$ V and $\beta_o = 80$.

**8.11** Assume that $V_{BE(\text{on})} = 0.7$ V, $V_{CE(sat)} = 0$ *V*, and $\beta = 60$ in the circuit of Figure P8.11. The resonant impedance at $f_c$ presented from collector to signal ground is 8 k$\Omega$.

(a) Calculate $M$ to result in $m = 40\%$.
(b) Calculate the minimum allowable value of $R_S$.

**D8.12** Assume that $\beta = 100$, $V_{BE(\text{on})} = 0.7$ V, and the resonant resistance of the tuned circuit is 6 k$\Omega$ in the circuit of Figure P8.12. The modulation voltage consists of two harmonically related sinusoids and is expressed as

$$v_m = 2 \cos \omega_m t + 0.5 \cos 3\omega_m t \text{ V}$$

(a) Select the input carrier amplitude $C$ to lead to an index of modulation of $m = 40\%$.
(b) If $C$ is chosen to be 9 V, calculate $R_B$ to lead to an unmodulated output carrier amplitude of 6 V.

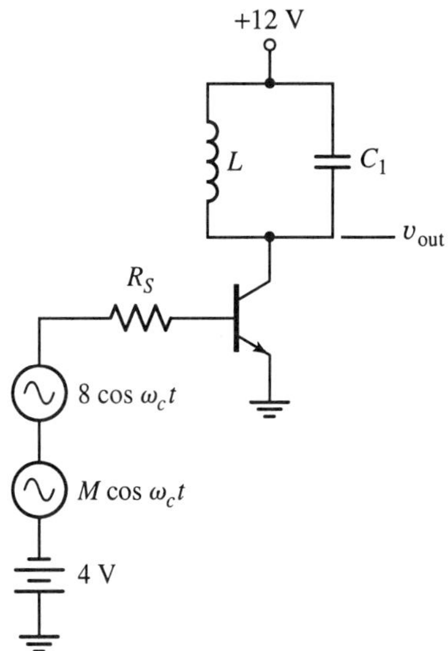


**Figure P8.11**

(c) For $C = 9$ V, what is the largest amplitude of $v_{\text{out}}$ when the given $v_m$ is applied?

**D8.13** Assume that $\beta = 100$, $V_{BE(\text{on})} = 0.7$ V, and the resonant resistance of the tuned circuit at $\omega_1$ is 5 k$\Omega$ in the circuit of Figure P8.13. A square wave of current with peak value $I_P$ can be represented by the Fourier series

$$i = \frac{I_P}{2} + \frac{2I_P}{\pi}\left[\cos \omega_1 t - \frac{1}{3}\cos 3\omega_1 t + \frac{1}{5}\cos 5\omega_1 t - \cdots\right]$$

(a) Calculate $M$ to give an index of modulation of 0.8.
(b) Calculate the maximum value of $R_B$ for proper operation.

**D8.14** Design a collector-modulation stage using the same coil and transistor, as well as carrier and modulating frequencies as in Problem 8.10. The modulating voltage now appears in series with the power supply, and 80% modulation is desired. The carrier amplitude is now 12 V while $V_0 = 2$ V.

**D8.15** If the modulating voltage in Problem 8.11 is moved to the collector circuit to create a collector modulator, calculate the maximum possible value of $R_S$ and the value of $M$ for 80% modulation.

**8.16** Using the same transistor and resonant circuit found in Problem 8.11 in the circuit of Figure P8.16, calculate the minimum value of carrier amplitude, $C$, for proper operation. What is the index of modulation for the circuit?

**8.17** Evaluate the carrier power, the upper sideband power, and the lower sideband power in Problems 8.10 and 8.16.

**8.18** Evaluate the power supplied by the modulator in Problems 8.10 and 8.14.

### Section 8.3 Production of Single-Sideband Waveforms

**8.19*** A 30-MHz carrier is applied to a balanced modulator. If the lowest baseband frequency to be passed is 1 kHz, how many dB/octave must a filter attenuate between the maximum lower sideband frequency and the

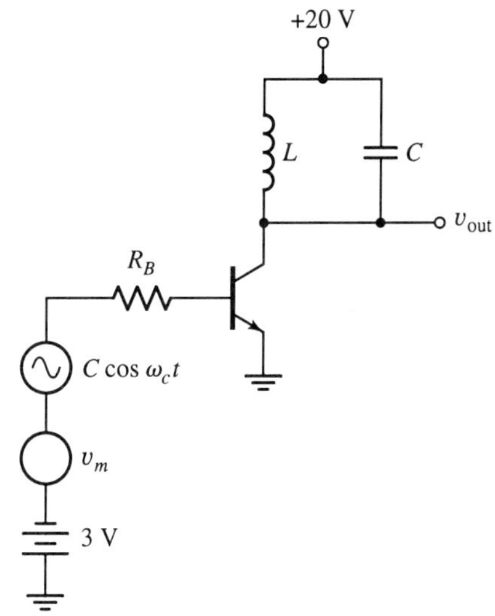


**Figure P8.12**

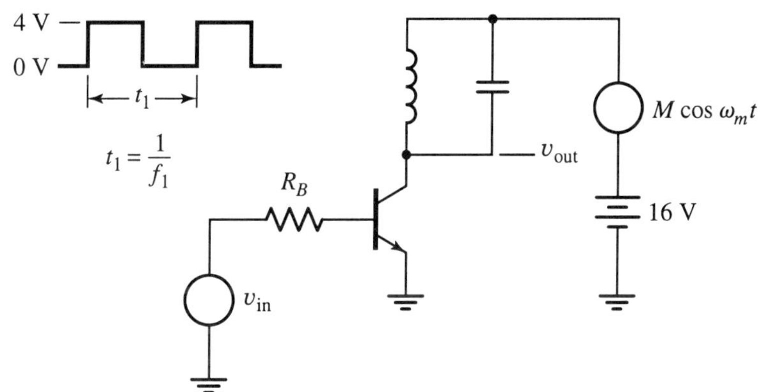


**Figure P8.13**

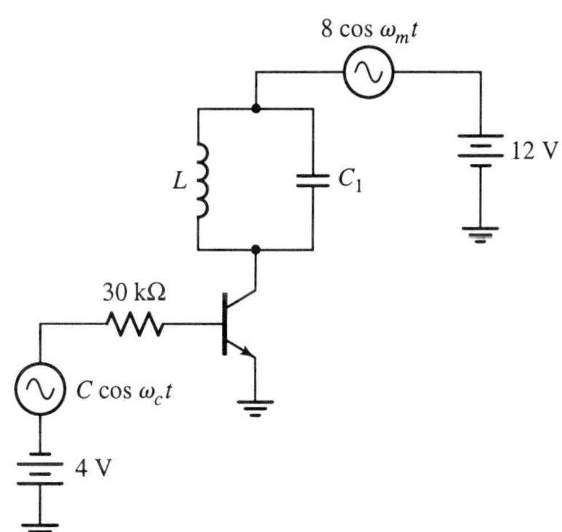


**Figure P8.16**

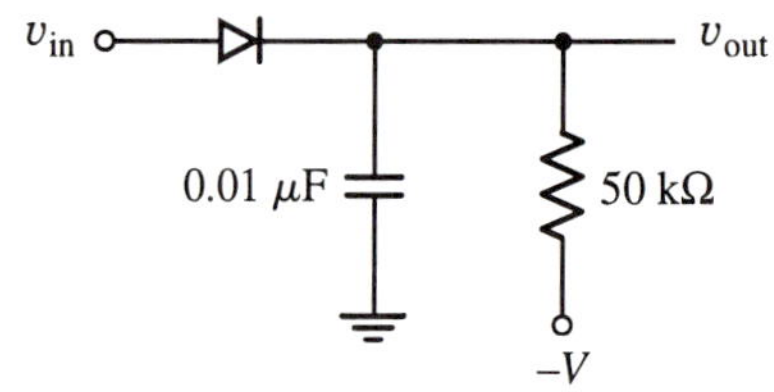


**Figure P8.24**

minimum upper sideband frequency to be down by 40 dB?

**8.20** A phase-shift method is used to create a SSB waveform. If the phase-shift circuit for the modulating signal varies from 84° to 94° over the baseband bandwidth, how many dB is the undesired sideband down compared to the desired sideband at the extreme baseband frequencies?

**Section 8.4 Mixers**

**8.21*** An AM signal is transmitted on an 800-kHz carrier. The baseband frequency range extends from 100 Hz to 5 kHz. This signal is to be detected by a superheterodyne receiver with a 455-kHz intermediate frequency.

(a) What is the frequency of the local oscillator?
(b) What must be the $Q$ of the IF stages if the 1-dB points are to correspond to the maximum and minimum frequencies of the IF signal?

**8.22** A perfect square-law device is used to translate a carrier frequency of 800 kHz to an IF frequency of 455 kHz. The two signals applied in series to the input of the mixer are

$$v_c = 2\cos\omega_c t \;\; and \;\; v_1 = 4\cos\omega_1 t$$

The difference of $f_c$ and $f_1$ is to be 455 kHz.

Select an appropriate circuit $Q$ to cause 40 dB of attenuation of the components nearest the IF signal.

**Section 8.5 Detectors**

**8.23** Select values for the diode detector of Figure 8.32 if $R_i = 10$ kΩ. The modulation frequency range extends from 100 Hz to 5 kHz and the carrier is 800 kHz. The index of modulation will never exceed 0.8, and the unmodulated carrier amplitude is 1.8 V.

**8.24** If the forward drop across the diode in Figure P8.24 is 0.7 V and the input signal is

$$v_{in} = [4 + 3\cos(2\pi \times 2000\, t)]\cos(2\pi \times 455{,}000\, t)$$

(a) What is the largest change in peak value of $v_{in}$ between successive peaks?
(b) Select $V$ to lead to a maximum error of 2 mV between $v_{out}$ and the positive peak value of $v_{in}$.

## REFERENCES AND SUGGESTED READING

1. Comer, D. J., Christiansen, M. L., and Fox, D. K., "The Loop-Gain Modulator: A New Class of Linear Amplitude Modulators," *IEEE Transactions on Circuits and Systems.* Vol. CAS-34, No. 10, October 1987, pp.

1161-1164.

2. Comer, D. J. and Herron, R., "A Class-C Loop-Gain Amplitude Modulator," *IEEE Transactions on Circuits and Systems.* Vol. CAS-36, No. 1, January 1989, pp. 159-161.

3. Couch, L. W. II, *Digital and Analog Communication Systems, 5th edition,* New York: MacMillan Publishing Co., 1997

4. Couch, L. W. II, *Modern Communication Systems, Principles and Applications.* Englewood Cliffs, N.J.: Prentice-Hall, 1995.

5. Fink, D. G. and Christiansen, D. (Eds.) *Electronics Engineers' Handbook, 3rd Ed.* New York: McGraw Hill Book Co., 1989.

6. Taub, H. and Schilling, D., 1986, *Principles of Communication Systems, 2nd edition.* New York: McGraw-Hill Book Co., 1986.

7. P. H. Young, P. H., *Electronic Communication Techniques, 4th Ed.* Upper Saddle River, N.J.: Prentice-Hall, 1999.

# Chapter 9

# ANGLE MODULATION CIRCUITS

Angle modulation refers to a modification of the argument of a carrier signal in accordance with the baseband or modulating signal. Phase modulation and frequency modulation are examples of angle modulation. Prior to the 1980s, a major purpose of a phase modulator was to generate frequency modulation for commercial FM transmission. In the last 25 years, phase modulation has become extremely important in the transmission of digital data. Frequency modulation has also expanded to digital transmission systems, and these two forms of modulation are very significant in the twenty-first century.

Digital phase and frequency modulation systems are often based on discrete systems that do not require continuous-phase modulators such as those used in analog systems. This chapter will discuss continuous-angle modulators, leaving the digital versions to the digital communications classes.

## 9.1 PHASE MODULATORS

**Important Concepts**

**1.** A phase modulator has a carrier input and a modulation input. The output has a constant-amplitude signal with a phase angle that varies proportionally to the modulating signal.

**2.** A phase modulator can be based on an amplitude modulator. A balanced modulator or DSB-SC system extends the linearity of the phase modulation.

As noted in Chapter 7,the phase modulator produces a constant amplitude signal with a changing phase. This signal can be expressed as

$$v_{\text{out}} = C \cos(\omega_c t + \phi_0 + k_p M \sin \omega_m t)$$

The phase varies linearly with the applied modulating signal. The maximum phase shift is given by the index of phase modulation, which is

$$m_p = k_p M_{\max}$$

This chapter considers methods of generating this ideal phase-modulated signal.

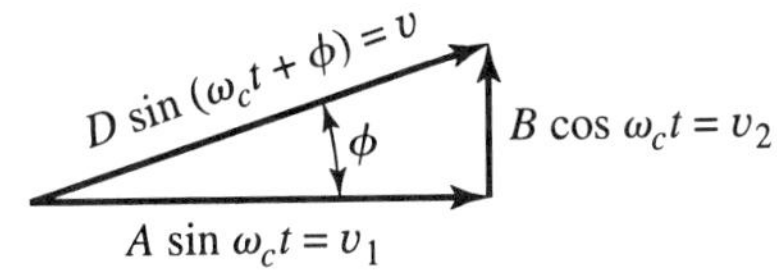


Figure 9.1: Quadrature signals of equal frequency.

### 9.1.1 Phase Modulators Based on Amplitude Modulators

A popular method of phase modulation applies the principle that two quadrature sinusoids of the same frequency create a resultant sinusoid with a phase shift. This is demonstrated in Figure 9.1.

The diagram is a phasor representation of the signal $A \sin \omega_c t$ and $B \cos \omega_c t$, the latter signal leading the former by 90°. The resulting signal is designated $v$ and can be expressed as

$$v = D \sin(\omega_c t + \phi) \tag{9.1}$$

where

$$D = \sqrt{A^2 + B^2} \tag{9.2}$$

and

$$\phi = \tan^{-1} \frac{B}{A} \tag{9.3}$$

This signal leads the signal $A \sin \omega_c t$ by the angle $\phi$.

In order to use this principle to construct a phase modulator, the angle $\phi$ is modulated by modulating the amplitude of the signal $B \cos \omega_c t$. If this signal is replaced by an amplitude modulated signal given by

$$v_2 = [V_c (1 + m \cos \omega_m t)] \cos \omega_c t$$

or

$$v_2 = [V_c (1 + kM \cos \omega_m t)] \cos \omega_c t$$

the phase angle will be modulated in accordance with $M \cos \omega_m t$. The constant $k$ is a function of the amplitude modulator and $kM$ equals $m$, the index of modulation.

The system of Figure 9.2(a) shows an implementation of this scheme, and Figure 9.2(b) indicates the phasor representation of the quadrature signals.

The phase of the carrier is shifted by −90° and added to a much smaller AM waveform having the same carrier frequency. The phase shift can easily be done using an allpass network. The resulting waveform at the adder output is the vector sum of the two components, as shown in Figure 9.2(b). Since the AM waveform varies in amplitude, the adder output voltage, $v$, varies in phase and slightly in amplitude. If the phase deviation is limited to a small value, a reasonably linear variation of phase with modulating voltage results. The limiter is used to keep the amplitude variations of the output waveform at a minimum level. More linearity is achieved if a balanced modulator is used in place of the amplitude modulator. This leads to an unmodulated phase of 0°, and a sinusoidal modulating voltage leads to equal positive and negative phase deviations. The phase index is generally limited to about 0.25 rad (14.3°) to minimize nonlinearities.

The amplitude of $v_2$, the quadrature signal, corresponds to the amplitude $B$ shown in Figure 9.1 and can be expressed by

$$|v_2| = B = V_c (1 + kM \cos \omega_m t)$$

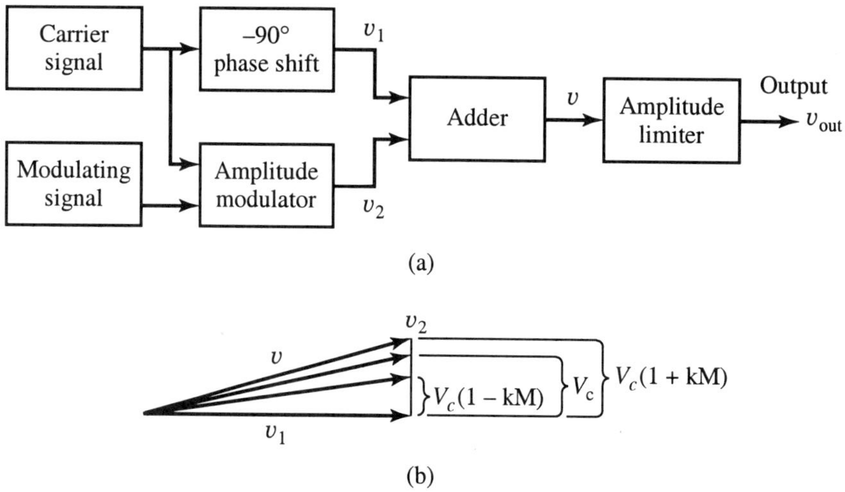


Figure 9.2: (a) A phase modulator. (b) Phasor representation of signals.

If this amplitude is substituted into Eq. (9.3) for $\phi$, the result can be written as

$$\phi = \tan^{-1} \frac{V_c\ (1 + kM\ \cos\ \omega_m t)}{A} \tag{9.4}$$

A phase modulator must exhibit a linear relationship between the phase angle and the modulating signal. The previous equation will represent a linear relationship between $M\ \cos\ \omega_m t$ and $\phi$ only for small angles such that

$$\tan^{-1}\ \phi \approx \phi$$

When this approximation is valid, the phase angle is expressed as

$$\phi = \frac{V_c\ (1 + kM\ \cos\ \omega_m t)}{A} \tag{9.5}$$

The resulting expression for the sum of $v_1$ and $v_2$ is now

$$v = D\ \sin\ (\omega_c t + \phi_0 + m_p\ \cos\ \omega_m t) \tag{9.6}$$

where

$$\phi_0 = V_c/A \tag{9.7}$$

and

$$m_p = kMV_c/A \tag{9.8}$$

This equation reflects a phase angle that varies linearly with the modulating voltage.

In order to keep the phase angle small enough to validate the earlier approximation, the amplitude of $v_2$ must be small compared to $A$. When this is true, the value of $D$ varies little as $v_2$ changes in amplitude. The voltage $v_2$ in Figure 9.2(b) varies linearly with the modulating signal. This causes the tangent of the angle $\phi$ to vary approximately linearly, but only over a small range of values, as can be seen in Table 9.1.

**Table 9.1 Nonlinearity of Phase Modulation.**

| $B/A$ | tan $\phi$ | $\phi$ Degrees | $\phi$ Radians | Error Percent |
|---|---|---|---|---|
| 0.0400 | 0.0400 | 2.29 | 0.0400 | 0.00 |
| 0.0800 | 0.0800 | 4.57 | 0.0798 | 0.21 |
| 0.1200 | 0.1200 | 6.84 | 0.1194 | 0.48 |
| 0.1600 | 0.1600 | 9.09 | 0.1586 | 0.84 |
| 0.2000 | 0.2000 | 11.31 | 0.1974 | 1.30 |
| 0.2400 | 0.2400 | 13.50 | 0.2355 | 1.86 |
| 0.2800 | 0.2800 | 15.64 | 0.2730 | 2.50 |
| 0.3200 | 0.3200 | 17.74 | 0.3097 | 3.22 |

A linearity error of 2% or less is generally acceptable in communication applications. This error figure results for an angle of 14°. If a conventional amplitude modulator is used, the unmodulated phase shift is set to be 7°. An index of modulation for the AM signal of 100% then results in a phase deviation of ±7° to a maximum of 14° and a minimum of 0°. If the modulator is replaced by a balanced modulator, phase shifts of ±14° can result.

**Practice Problems**

**9.1** If $v_1$ in Figure 9.2 is 16 sin $\omega_c t$ and $v_2$ is an AM signal with an unmodulated amplitude of 1.4 V and an index of modulation of 0.8, calculate the unmodulated phase shift, $\phi_0$, and the maximum phase deviation of the adder output signal, $m_p$. *Ans:* $\phi_0 = 5°$, $m_p = 4°$.

**9.2** If $v_1$ in Figure 9.2 is 16 sin $\omega_c t$ and a balanced modulator that produces an output of $v_2 = 1.5 \cos \omega_m t \cos \omega_c t$, calculate the maximum phase deviation of the adder output signal. *Ans:* $m_p = 5.4°$.

### 9.1.2 A Large-Deviation Phase Modulator

The circuit of Figure 9.3 shows a large-deviation phase modulator [3] using a varistor (silicon-carbide voltage-variable resistor). The modulator allows linear phase shifts of 50° or greater and does not require an amplitude limiter. Unfortunately, the varistor resistance has a fairly large temperature coefficient.

When the modulating voltage is zero, the circuit reduces to a phase-shift circuit similar to that shown in Subsection 6.2.2. If the amplitude of the carrier is fairly small, it will pass through the phase-shift circuit with constant amplitude and a phase shift of

$$\phi = 2 \tan^{-1} \frac{1}{\omega_c C_1 r_v} \tag{9.9}$$

For phase shifts up to 100°, Eq. (9.9) can be approximated by

$$\phi = \frac{2}{\omega_c C_1 r_v} \tag{9.10}$$

The phase shift can be varied by changing $r_v$, but the variation is an inverse one. The voltage variable resistor (VVR) is a device that has a V-I relationship expressed by

$$I = kV^2 \tag{9.11}$$

The exponent can be made to lie anywhere from 2 to 7, depending on the construction process, but for our purposes a value of 2 is required. The incremental or small signal resistance of the VVR is

$$r_v = \frac{dV}{dI} = \frac{K}{V} \tag{9.12}$$

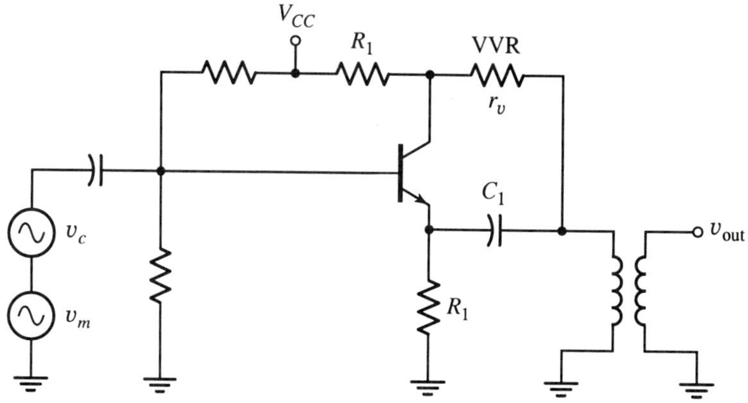


Figure 9.3: Large-deviation phase modulator.

where $K = 1/2k$. The dynamic resistance of the VVR depends inversely on the voltage applied to the device. The phase shift of the carrier can then be related to the quiescent collector voltage as

$$\phi_{CQ} = \frac{2}{\omega_c C_1 r_v} = \frac{2}{\omega_c C_1 K} V_{CQ}$$

If the modulating voltage has a much larger amplitude than the carrier signal, the resistance $r_v$ will be

$$\frac{K}{V_{CQ} + M \sin \omega_m t}$$

and the phase shift of the carrier is

$$\phi = \frac{2}{\omega_c C_1 K} [V_{CQ} + M \sin \omega_m t] \tag{9.13}$$

The phase deviates linearly with the amplitude of $v_m$. The transformer chosen must offer a low impedance to the modulating signal and a high impedance to the carrier. The transformer can be replaced with an inductor.

The output signal is then given by

$$v_{\text{out}} = D \cos \left( \omega_c t + \frac{2V_{CQ}}{\omega_c C_1 K} + \frac{2M}{\omega_c C_1 K} \sin \omega_m t \right) \tag{9.14}$$

Comparing Eq. (9.14) to the phase-modulated waveform of Eq. (9.6) results in the identifications

$$\phi_0 = \frac{2V_{CQ}}{\omega_c C_1 K}$$

and

$$m_p = \frac{2M}{\omega_c C_1 K}$$

This circuit can be used in automatic phase correction systems, but it lacks the inherent precision needed in most communication applications.

# 9.2 FREQUENCY MODULATORS

**Important Concepts**

**1.** A simple frequency modulator should produce a frequency deviation that is a function of the amplitude of the modulating signal, but is independent of modulation frequency.

**2.** The phase modulator produces an output with a frequency that is dependent on both modulation signal amplitude and frequency.

**3.** If a modulating signal is integrated and then applied to a phase modulator, a frequency-modulated signal results, with a frequency deviation that is not a function of modulation frequency.

**4.** Commercial FM transmitters generate a frequency deviation that increases at higher frequencies to emphasize the small amplitudes of these components in voice and music.

## 9.2.1 The Indirect Method of Frequency Modulation

As mentioned in Chapter 7, it should be possible to base a frequency modulator on a phase modulator. To review the motivation behind this method, we recall that the frequency deviation of a phase-modulated waveform is

$$\Delta\omega = k_p M \omega_m = m_p \omega_m$$

This signal has a frequency deviation that depends on the magnitude of the modulating signal as well as the frequency of this signal. The frequency of this signal is given by

$$f = f_c + k_p M f_m \cos \omega_m t$$

A frequency-modulated waveform should have a constant frequency deviation with $\omega_m$ or $f_m$. The expression for frequency of an FM signal is given by Eq. (7.20) as

$$f = f_c + k_f M \cos \omega_m t$$

This frequency is a function of the magnitude of the modulating signal, but not a function of the modulation frequency.

The dependence of frequency deviation on $\omega_m$ in a phase modulator can be removed by applying a modulating signal of

$$\frac{M_1}{\omega_m} \sin \omega_m t$$

to the phase modulator. The frequency deviation can be found from Eq. (7.19) by replacing $M$ by $M_1/\omega_m$. This gives a frequency deviation of

$$\Delta\omega - k_p \frac{M_1}{\omega_m} \omega_m = k_p M_1$$

The result is a frequency deviation that is a function of $M_1$, but does not depend on $\omega_m$.

Before this method is validated, we must consider two important issues. One relates to the method of producing a modulating signal with an amplitude that varies inversely with $\omega_m$, that is, we must make $M = M_1/\omega_m$. The other issue relates to whether the phase modulator with its limited index of modulation can produce a sufficient frequency deviation to be useful in a practical system.

The first issue is easily resolved by applying a constant-amplitude modulating signal to an integrator. If $M \cos \omega_m t$ is the input to an integrator, the output is

$$e_I = k_I \int M \cos \omega_m t dt = \frac{k_I M}{\omega_m} \sin \omega_m t \tag{9.15}$$

where $k_I$ is a constant determined by the integrating circuit. The constant $M_1$ is then given by

$$M_1 = k_I M \tag{9.16}$$

# Example 9.1

A phase modulator with a carrier frequency of 100 kHz is used to create a frequency modulator. When a modulating signal of 5-V amplitude is applied to the phase modulator, the index of phase modulation reaches its maximum value of 0.2 rad. Beyond this value of applied voltage, the modulator becomes excessively nonlinear. Determine the largest value of $k_I$ that can be used for the integrating circuit. Assume that the frequency range of the modulating signal is 100 Hz to 8 kHz and the modulating voltage amplitude is 5 V. Calculate the frequency deviation for the resulting circuit. Note that this index of phase modulation is on the border between narrowband and wideband angle modulation.

**Solution** The magnitude of the output voltage of the integrator is

$$|e_I| = \frac{k_I M}{\omega_m}$$

Since the output of the integrator will be applied to the modulator, this value must not exceed 5 V. The maximum magnitude occurs at the lowest modulation frequency, so we can solve for $k_I$ from the equation

$$\frac{k_I M}{2\pi \times 100} = 5$$

This gives $k_I = 628$.

The resulting frequency deviation can be found from Eq. (7.22) after evaluating $k_p$. From Eq. (7.17), $k_p$ can be found to be

$$k_p = \frac{m_p}{M_{\text{max}}} = \frac{0.2}{5} = 0.04 \frac{\text{rad}}{\text{volt}}$$

The frequency deviation is now calculated as

$$\Delta f = \frac{k_I k_p M}{2\pi} = \frac{628 \times 0.04 \times 5}{6.28} = 20 \text{ Hz}$$

---

The very small frequency deviation calculated in Example 9.1 is typical of an FM signal generated by a phase modulator. Note also that the small value of $m_p$ in this modulator leads to a small value of $\Delta f$: hence, the term *narrowband angle modulation* is appropriate. Although $\Delta f$ may appear to be too small to be practical, such is not the case. Using mixers and frequency multipliers, this signal can be converted to a practical FM waveform.

Before we discuss methods of increasing the frequency deviation of the narrowband circuit, let us consider a practical integrator circuit. One of the better integrator circuits for lower frequencies is the op amp circuit of Figure 9.4. The transfer function of this circuit can be written as

$$-\frac{1}{sRC} = -\frac{1}{j\omega RC} \tag{9.17}$$

In the Laplace domain, integration is indicated by the function $1/s$. Thus, the function represented by Eq. (9.17) is an inverting integrator with an integration constant of

$$k_i = -\frac{1}{RC} \tag{9.18}$$

In Example 9.1, the integration constant was required to be $k_i = 628$. The sign of this constant is unimportant in determining the frequency deviation of the circuit. The op amp integrator could be used to produce this constant by the proper selection of the $RC$ time constant in Eq. (9.18).

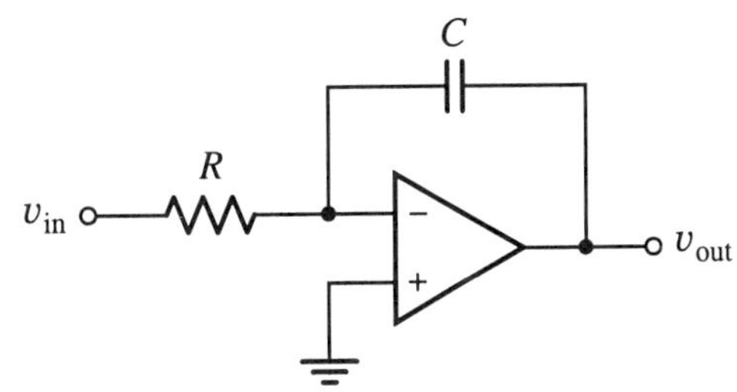


Figure 9.4: An integrator circuit.

Of course, op amps have some frequency limitations, but for commercial FM or television audio, the modulation frequency is limited to relatively low frequencies. Commercial FM, for example, limits the modulating frequency to 15 kHz. Consequently, the op amp integrator can be very useful in such applications.

The frequency deviation and the carrier frequency of a signal can both be increased by frequency multiplication. If the signal has a frequency given by

$$f_c + \Delta \cos \omega_m t$$

a frequency multiplication of 20 will result in a frequency of

$$f_{\text{new}} = 20[f_c + \Delta \cos \omega_m t] = 20f_c + 20\Delta \cos \omega_c t$$

The new carrier frequency is $20f_c$ and the new frequency deviation is $20\Delta$.

It is also possible to change the carrier frequency while leaving the frequency deviation unchanged by using a mixing circuit. If a local oscillator frequency of $f_{\text{osc}}$ is applied to one input of a mixer and a signal of frequency $f_c + \Delta \cos \omega_m t$ is applied to the other input, the output will have a frequency of

$$f_c - f_{\text{osc}} + \Delta \cos \omega_m t$$

The new carrier frequency is now $f_c - f_{\text{osc}}$ and the frequency deviation remains at $\Delta$.

### Frequency Multipliers

It is possible to apply a sinusoid of a given frequency to a class-C stage and generate an output that has a frequency two or three times the input value. The output current of the stage is nonsinusoidal, but periodic, and contains the fundamental component and higher order harmonics. The amplitude of these harmonics drops off at higher frequencies. The resonant circuit is tuned to the desired harmonic. Often, the stage is driven hard enough to saturate and limit the amplitude of the output voltage to a peak value equal to the power supply voltage, even if the input voltage varies slightly in magnitude. This same principle is applied to the limiter employed in the phase modulator of Figure 9.2, except that the tuned circuit is resonant at the fundamental frequency instead of a harmonic.

Since the amplitudes of the higher order harmonics in the collector-current waveform decrease with frequency, it is difficult in practice to select frequencies above the third harmonic. Consequently, each class-C stage of a multiplier generates a factor of either two or three. This limits the overall multiplication to a number that contains these factors.

The frequency multiplier should not be confused with the mixer, which is a multiplying or product circuit. The mixer multiplies two input signals to form the product. The output product contains a frequency that is the difference of the two input frequencies. This circuit translates the frequency of one input signal by an amount equal to the other input frequency. The output amplitude is a function of the two input amplitudes. The frequency multiplier, on the other hand, has only one input signal and may produce an output amplitude that is not a function of the input amplitude due to the limiting of the last stage of the circuit. However, the frequency of the output signal is some multiple of the input frequency.

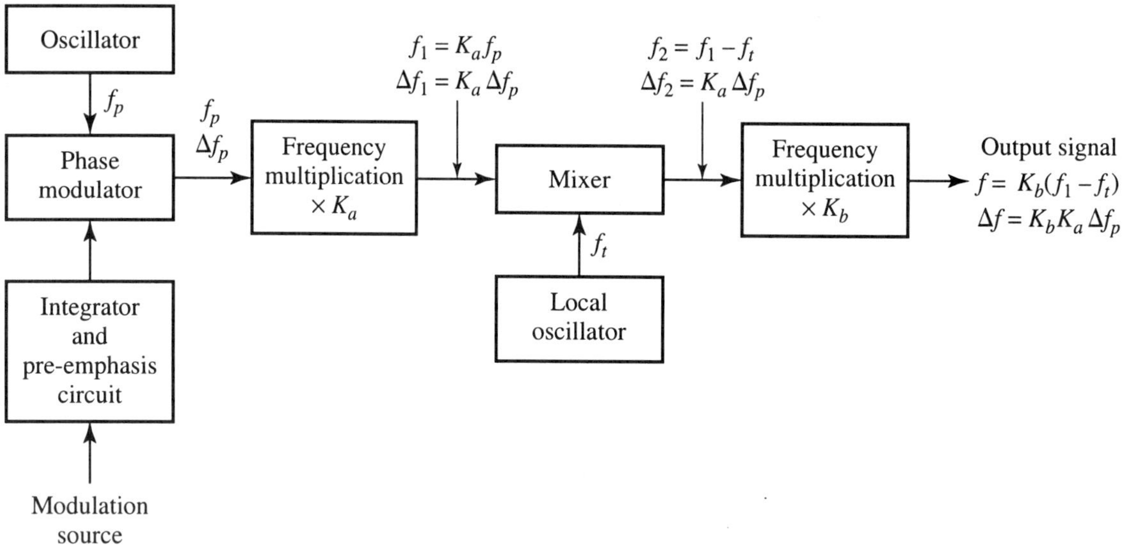


Figure 9.5: Indirect method of FM waveform generation.

### The Armstrong or Indirect Method of FM Waveform Production

This method, reported by Armstrong in 1936 [1], has been used with slight variations for over 60 years. The basic scheme is depicted in Figure 9.5. The frequency multiplication could be done in one multiplier, but using two frequency multipliers allows the mixer to operate in the MHz range, which is a reasonably practical frequency range. The frequency multipliers multiply both the carrier frequency and the frequency deviation by the same factor. The mixer translates the carrier frequency to a new value, but the frequency deviation is unaffected.

The FCC specifies the allowed values of the output frequency, $f$, and deviation, $\Delta f$. The design of the phase modulator and integrator determines $f_p$ and $\Delta f_p$. The frequency multiplying factors, $K_a$ and $K_b$, along with the local oscillator frequency, $f_t$ must then be selected properly to match the known values. An example will demonstrate this procedure.

## Example 9.2

Using the integrator and phase modulator of Example 9.1, select $K_a$, $K_b$, and $f_t$ in Figure 9.5 to create an output signal with $f = 100$ MHz and $\Delta f = 75$ kHz.

**Solution:** The total frequency multiplication required can be calculated from

$$K_{mul} = \frac{\Delta f}{\Delta f_p} = \frac{75,000}{20} = 3750$$

To simplify the selection of $K_a$ and $K_b$, we will assume that these factors must be powers of two only. The nearest power of two that leads to a multiplication exceeding 3750 is 12; that is,

$$2^{12} = 4096 > 3750$$

If this multiplication factor is used, the output deviation will exceed the specified value of 75 kHz. However, if the 5-V amplitude of the applied modulating voltage is reduced, the value of $\Delta f_p$ can be reduced to the required value. If $\Delta f_p$ is reduced from 20 to

$$20 \times \frac{3750}{4096} = 18.31 \text{ Hz}$$

a frequency multiplication of 4096 will lead to an output deviation of exactly 75 kHz. This value of $\Delta f_p$ can be achieved by dropping the modulation voltage amplitude to

$$5 \times \frac{3750}{4096} = 4.578\ V$$

The selection of $K_a$ and $K_b$ is somewhat arbitrary as long as each is a power of two and the product equals 4096. We will select values of $K_a = K_b = 64$. This will produce a value of

$$f_1 = K_a f_p = 64 \times 100 \text{ kHz} = 6.4 \text{ MHz}$$

In order for the second frequency multiplier to result in an output carrier frequency of 100 MHz, the value of $f_2$ must be

$$f_2 = \frac{100 \text{ MHz}}{64} = 1.5625 \text{ MHz}$$

The mixer must then translate the 6.4-MHz signal to a 1.5625-MHz signal which requires that

$$f_t = 6.4 - 1.5625 = 4.8375 \text{ MHz}$$

The resulting FM output signal now satisfies the original specifications.

---

**Practice Problems**

**9.3** A frequency is to be multiplied by a factor of 76. If multiplier circuits with frequency multiplication factors of two or three are available, what is the minimum number of stages needed to equal or exceed this requirement? *Ans:* Four stages, each with a factor of three.

**9.4** A frequency deviation of 24 Hz must be converted to a deviation of 50 kHz by frequency multiplication. What is the minimum number of stages required to equal or exceed this requirement? Assume that only stages with a multiplication factor of two are available. *Ans:* 12 stages.

### 9.2.2 The Direct Method of FM Generation

While the Armstrong circuit is still used in some commercial FM transmitters, many newer transmitters now use the *automatic frequency control* (AFC) method to create the FM waveform. This approach is also referred to as the direct method of frequency modulation. The FM signal is generated at a low-power level in this method and is then amplified by class-C stages before being applied to the transmitting antenna. (The student should understand how it is possible to do class-C amplification of an FM signal while AM signals cannot be amplified by class-C stages.) The block diagram of Figure 9.6 shows the general configuration for the direct method.

The frequency reference signal is often produced by a temperature-controlled crystal oscillator to achieve the required stability for the carrier signal to be transmitted. Incidentally, the FCC specification on stability specifies that the carrier frequency always be within 2 kHz of the specified carrier frequency. For a 100.1-MHz carrier, this amounts to a stability of 0.002%.

The actual FM oscillator frequency is continually compared to the reference frequency and a dc error signal, which is a function of the frequency difference, is generated. If the two frequencies are equal, the error signal is zero. If the frequencies are unequal, the error signal drives the FM oscillator toward the reference frequency. The FM oscillator is a *voltage-controlled oscillator* (VCO). The VCO and the frequency comparator circuit can be combined in a phase-locked loop, to be discussed in a later section.

An astute student might question the validity of the frequency comparison since the output of the FM oscillator varies in both directions about its carrier frequency by up to 75 kHz. Yet this signal is compared to a

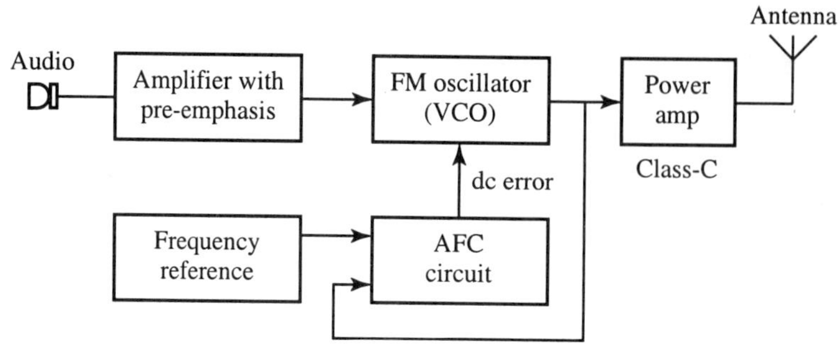


Figure 9.6: The direct method of FM signal production.

reference signal, and the carrier of the FM oscillator must fall within 2 kHz of the reference frequency. This is accomplished by making the frequency comparator respond very slowly to the frequency difference. In fact, this difference is averaged over a long time period so that the instantaneous frequency deviation becomes unimportant. The average value of the modulating signal is clearly zero; consequently, the average value of the frequency of the FM oscillator must equal its carrier frequency.

### The Tuned-Collector Modulator

One method of producing a VCO for a narrow range of frequencies is to utilize the dependence of the depletion capacitance of a *pn* junction on voltage applied to the junction. Figure 9.7 shows an FM modulator based on the tuned-collector oscillator that uses this concept.

If the collector depletion-region capacitance is neglected, the reverse-biased diode determines the tank circuit capacitance. Assuming an abrupt-junction diode, we find that this capacitance varies with the reverse-bias voltage across the junction, as shown by

$$C_{dl} = \frac{k}{\sqrt{\psi + V}} \approx \frac{k}{\sqrt{V}}$$

where $\psi$ is the barrier voltage of the junction. When $V$ is much larger than the barrier voltage, the above approximation is valid. The resonant frequency of the tuned circuit, neglecting stray capacitance is

$$\omega_0 = \frac{1}{\sqrt{LC_{dl}}} = \frac{V^{1/4}}{\sqrt{Lk}} \tag{9.19}$$

The voltage $V$ is the total reverse-biased voltage across the diode and is equal to $V_1 + M \cos \omega_m t$ where $V_1$ is the dc bias voltage. Equation (9.19) can then be expressed as

$$\omega_0 = \frac{V_1^{1/4}[1 + (M/V_1) \cos \omega_m t]^{1/4}}{\sqrt{Lk}}$$

Using the binomial expansion for the numerator term, we obtain a frequency of

$$\omega_0 = \frac{V_1^{1/4}}{\sqrt{Lk}} \left(1 + \frac{1}{4}\frac{M}{V_1} \cos \omega_m t - \frac{3}{32}\frac{M^2}{V_{1^2}} \cos^2 \omega_m t + \cdots \right)$$

Applying the identity $\cos^2 \omega_m t = \frac{1}{2} + \frac{1}{2} \cos 2\,\omega_m t$ allows this equation to be written as

$$\omega_0 = \frac{V_1^{1/4}}{\sqrt{Lk}} \left(1 - \frac{3}{64}\frac{M^2}{V_1^2} + \frac{1}{4}\frac{M}{V_1} \cos \omega_m t - \frac{3}{64}\frac{M^2}{V_1^2} \cos 2\omega_m t\right) \tag{9.20}$$

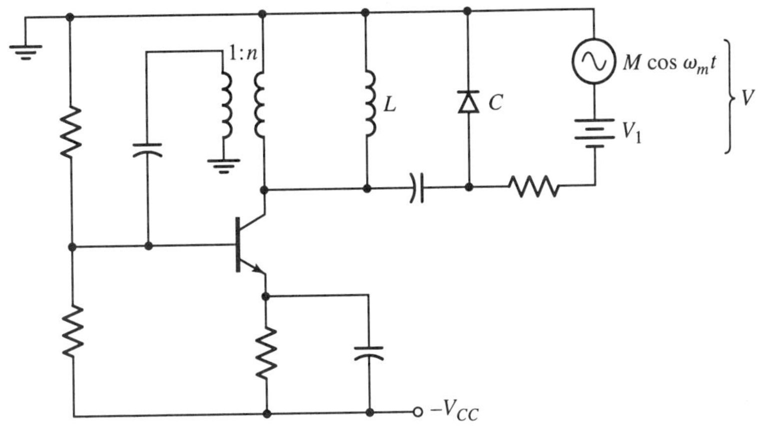


Figure 9.7: A tuned-collector frequency modulator.

If

$$\frac{V_1^{1/4}}{\sqrt{Lk}}\left[1-\frac{3}{64}\frac{M^2}{V_1^2}\right]\approx\frac{V_1^{1/4}}{\sqrt{Lk}}$$

is equated to the carrier frequency and

$$\frac{V_1^{1/4}}{\sqrt{Lk}}\times\frac{1}{4V_1}=2\pi k_f$$

then the first two terms in Eq. (9.20) describe the frequency of an FM waveform as given by Eq. (7.20). Unfortunately, there is a third term that represents a frequency error proportional to the second harmonic of the modulation frequency. Comparing the amplitude of the second harmonic term to that of the fundamental shows that the percent distortion is

$$\frac{3M}{16V_1}\times 100 \tag{9.21}$$

In order to keep this distortion at a negligible level, $M$ must be somewhat smaller than $V_1$. This requirement imposed on $M$ limits the linear frequency deviation that can be obtained with this circuit. This type of circuit can allow frequency deviations of 2 to 5% of the carrier frequency without serious distortion. For commercial FM systems, this deviation is more than enough to meet FCC specifications.

In practice, the depletion layer capacitance of the collector-base junction is used to resonate the tank circuit. The modulating voltage is then applied to the transistor base and the diode is eliminated from the tank circuit. More flexibility results, however, when the diode is replaced by a varactor diode, which offers a wide range of capacitance values. This circuit can serve as the VCO in the direct method of FM generation. The voltage $V_1$ is adjusted in accordance with the error signal to force the VCO frequency to equal the reference frequency.

### 9.2.3 Pre-Emphasis in FM Transmitters

Every electrical system is subject to noise contamination; especially communication systems that use the Earth's atmosphere as the channel. Without going into detail relative to the effects of noise, we will note that noise causes spurious amplitude and phase deviations in the transmitted signal. White noise, which is often the major type of noise occurring in electrical systems, has a uniform distribution of energy or amplitude at all frequencies.

The AM waveform is affected more by noise-induced amplitude variations than by phase variations since AM demodulators are peak amplitude detectors. The bandwidth of the detected standard AM signal from a commercial station is only about 5 kHz; thus, noise above this frequency has little effect on the signal. The noise

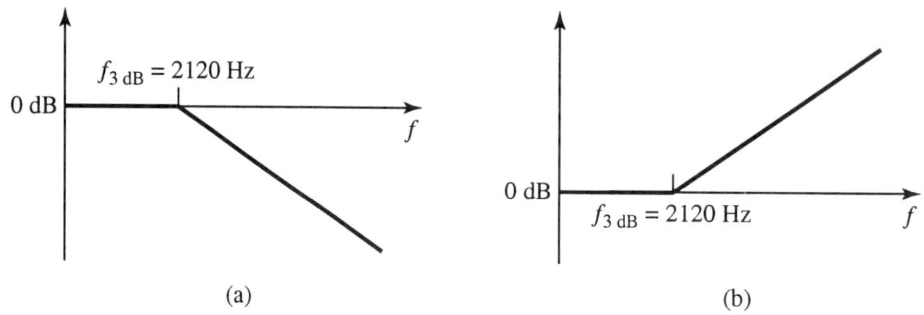


Figure 9.8: (a) Response of de-emphasis filter. (b) Response of pre-emphasis filter.

within this 5-kHz bandwidth does degrade the quality of the detected signal, especially if the receiver is several miles from the transmitter.

An FM receiver is made to be insensitive to amplitude variations of the incoming signal. This is done by placing an amplitude limiter before the detector or by employing a detector that ignores amplitude variations such as a phase-locked loop. Unfortunately, the spurious phase variations lead to spurious frequency deviations to which the detector responds. The frequency deviations due to noise increase with higher frequencies even though the phase noise is uniformly distributed over frequency. This increase occurs because more rapid phase deviations lead to larger frequency deviations since frequency deviation is a function of the derivative of the instantaneous phase. Furthermore, the bandwidth of the audio modulating signal in commercial FM is 15 kHz rather than the 5-kHz bandwidth of commercial AM. Because noise degrades the quality of the FM signal through phase variations rather than amplitude variations, the overall noise degradation of the FM signal is less than that of the AM waveform. On the other hand, FM transmission is intended to have much better fidelity than AM transmission, so noise problems should be minimized.

Audio signals have much higher energies at frequencies below 1000 Hz than at higher frequencies. Any noise at low frequencies will be much smaller than the signal components and will generally have a negligible effect on these components. At higher frequencies, the audio signal energy continues to decrease. The noise due to frequency deviations increases with frequency and at higher frequencies can approach the level of the signal components. In order to attenuate these noise effects, the output of the FM detector is passed through a lowpass filter that has a corner frequency of 2120 Hz as shown in Figure 9.8(a). This filter has an asymptotic falloff of 6 dB/octave above the corner frequency. Whereas the output of the detector due to noise increases with frequency, the noise output of the filter is constant with frequency. This filter is called a de-emphasis circuit.

As a result of using this de-emphasis filter in the receiver the actual audio signal components above 2120 Hz are also attenuated. This problem is solved by using a highpass filter with the response shown in Figure 9.8(b) in the transmitter. This filter can be inserted between the modulation source and the integrator to emphasize the higher frequencies prior to modulation. Under these conditions, a 10-kHz component will cause a frequency deviation that is twice that caused by an equal amplitude 5-kHz component. This circuit is called a pre-emphasis filter and offsets the high-frequency attenuation of the de-emphasis filter in the receiver. The FCC requires commercial FM transmitters to generate this pre-emphasized signal.

The pre-emphasis function can be combined with the integrator circuit of the Armstrong transmitter rather than using a separate filter. It will be recalled that integration of a modulation signal results in a falloff of -6 dB/octave. This signal that falls with frequency is applied to the phase modulator and the resulting output frequency deviation is flat with frequency. We conclude that the frequency deviation varies with frequency with a slope that is 6 dB/octave above the slope of the input signal amplitude with frequency. This arises from the fact that the frequency deviation of a phase-modulated signal is proportional to the derivative of the phase change, which, in turn, is proportional to the amplitude of the applied signal.

If a frequency deviation is to be generated that varies with frequency in the manner indicated by Figure 9.8(b), the signal applied to the phase modulator must have a slope with frequency that is 6 dB/oct less than that of the figure. The frequency response shown in Figure 9.9 produces the modulating signal needed for the phase modulator that leads to the pre-emphasized output frequency deviation desired. This response can be

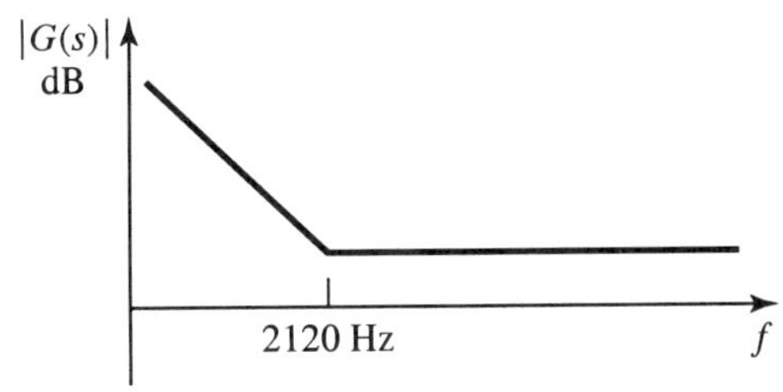


Figure 9.9: Desired response of pre-emphasis circuit.

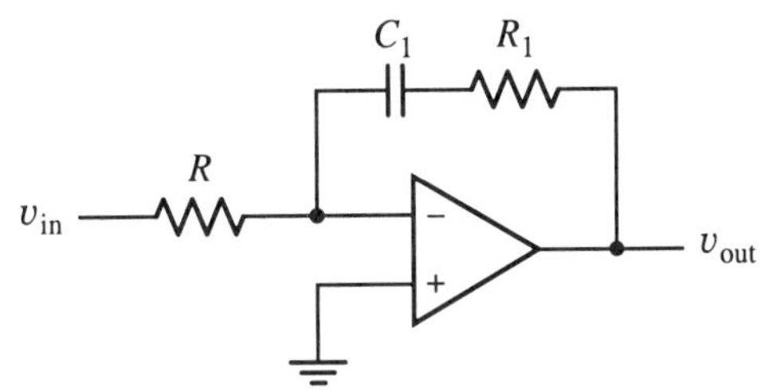


Figure 9.10: An op amp pre-emphasis circuit.

achieved by modifying the integration circuit used in the Armstrong modulator.

The response of Figure 9.9 can be expressed as

$$G(s) = \frac{(s\tau_1 + 1)}{s\tau_2} \tag{9.22}$$

where $\tau_1 = 1/2\pi f_1 = 1/2\pi \times 2120$. The constant $\tau_2$ is a gain constant that shifts the entire curve in the vertical direction.

The op amp circuit of Figure 9.10 has a transfer function of

$$G_1(s) = -\frac{1}{sC_1R}(sC_1R_1 + 1) \tag{9.23}$$

Ignoring the phase inversion, the magnitudes of $G_1(s)$ and $G(s)$ can be made equal if

$$k_I = \frac{1}{\tau_2} = \frac{1}{C_1R} \tag{9.24}$$

and

$$f_1 = \frac{1}{2\pi\tau_1} - \frac{1}{2\pi C_1 R_1} \tag{9.25}$$

Inserting the circuit of Figure 9.10 between the modulating source and the phase modulator accomplishes the pre-emphasis function required by commercial FM transmitters. This circuit behaves as an integrator at frequencies well below 2120 Hz and as a fixed-gain amplifier at frequencies well above 2120 Hz.

Although the frequency deviation at higher frequencies is increased by the pre-emphasis scheme, it will not exceed the maximum allowed deviation of 75 kHz. This follows from the fact that the higher frequency components of an audio signal are quite small. Amplifying these components by a small factor results in only a slight difference in overall frequency deviation.

In the direct method of FM signal generation, the pre-emphasis filter is generally added to the modulation amplifier. The transfer function in this case has the response of that shown in Figure 9.8(b).

**Practice Problems**

**9.5** In an FM transmitter with pre-emphasis, the frequency deviation for a modulation frequency of 3 kHz is 800 Hz. What is the frequency deviation for a 7.9 kHz signal of the same amplitude. *Ans:* 2107 Hz.

**9.6** If $C_1 = 0.025\ \mu$F in Figure 9.10, select $R$ and $R_1$ to result in a standard pre-emphasis circuit with a value of $k_I = 1000$. *Ans:* $R = 40$ k$\Omega$, $R_1 = 3$ k$\Omega$.

# 9.3 FM DETECTORS

**Important Concepts**

1. One method of FM signal detection converts the frequency variation of the signal into an amplitude variation and then applies an AM detector.

2. Configurations have been developed to make the conversion from frequency to amplitude be as linear as possible.

3. The phase-locked loop is an important element in electronic circuits and can be used for FM detection if an accurate frequency reference is available.

Several different types of FM detectors are available. The traditional detectors that have been used for over 50 years convert the frequency deviation of the waveform to a corresponding amplitude change. The ratio detector, based on the Foster-Seeley discriminator, uses this approach. Even today, with several attractive alternative circuits, the ratio detector dominates the FM receiver market.

A second type of detector converts the frequency deviation into a phase change and then converts this phase change back into the modulating voltage waveform. This type of detector has been used for years in TV receivers to recover the audio information. In this so-called quadrature detector, the quality of the demodulation is not as high as that of the ratio detector. The simplicity and low cost of the quadrature detector has appealed to TV manufacturers since the days of the vacuum tube TV receiver.

A third type of detector measures zero-axis crossings of the FM waveform and reconstructs the modulating voltage based on the time deviation from the axis crossings of the carrier.

The phase-locked loop represents a fourth type of FM detector that has gained popularity in recent years. This circuit is also useful in several other communication applications and will be discussed in detail later in this section.

Although the last two types of detector are less sensitive to amplitude variations than the first two, an FM receiver generally contains an amplitude-limiting circuit that precedes the detector. The first type of detector implements two distinct sections. The initial section converts the constant-amplitude FM signal into a signal having an amplitude that varies with frequency. This section essentially converts the FM signal into an AM signal. The second section is an AM detector that demodulates the newly created AM waveform. While this is the traditional approach to FM detectors, newer circuits operate on different principles. These will be considered later in this section.

## 9.3.1 The Foster-Seeley Discriminator

A resonant circuit tuned to a slightly different frequency than the carrier of the FM signal could be used for the first section of the traditional detector. If a current proportional to the FM signal enters the tank circuit, the output-voltage amplitude can be found by considering the impedance curve of the tank circuit shown in Figure 9.11. As the frequency varies about the carrier frequency, the output voltage varies from the unmodulated level. One problem with this circuit is that the output is linear with frequency deviation only for small frequency

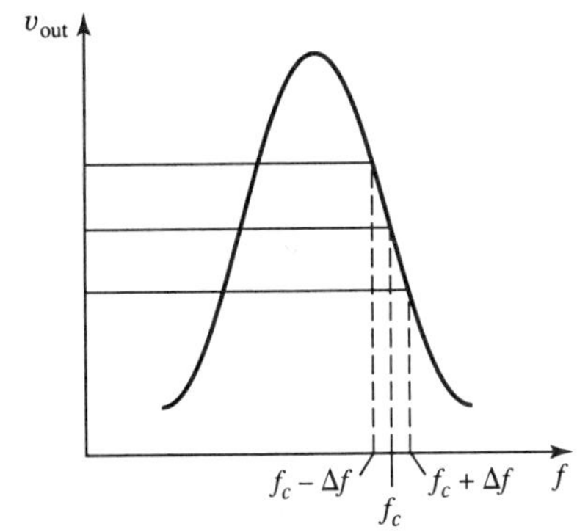


Figure 9.11: Response of a tuned circuit to an FM waveform.

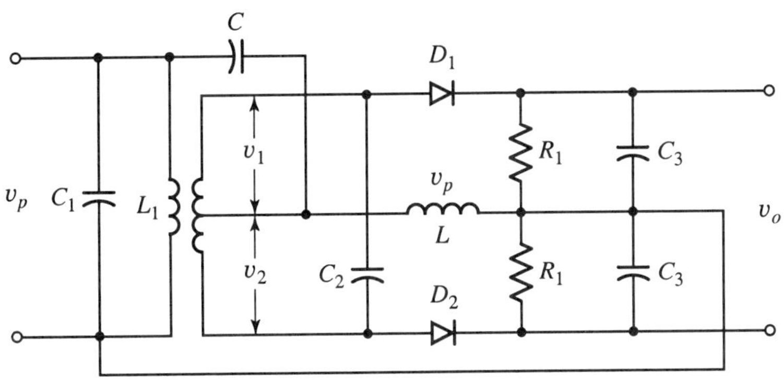


Figure 9.12: Foster-Seeley discriminator.

deviations.

Other types of detectors can be used to improve the linearity. The Foster-Seeley discriminator and the ratio detector are quite linear FM detectors. The Foster-Seeley discriminator is shown in Figure 9.12. Both primary and secondary are tuned to resonance at the detector input frequency. In fact, the input frequency is the intermediate frequency of the FM system. For commercial FM, this intermediate frequency is 10.7 MHz.

The current induced in the secondary is in phase with the input primary voltage. This current flows through the capacitor, $C_2$, developing a voltage that leads the input voltage by 90°. This total voltage equals $v_1 + v_2$, where $v_1 = v_2$ because of the center-tapped secondary.

The primary voltage is also applied at the center tap of the secondary through the coupling capacitor $C$ and the RF choke, $L$. The coupling capacitor is selected to offer a small impedance and the RF choke is selected to offer a large impedance to the resonant frequency. The voltage applied to the diode circuit that includes $D_1$ is $v_{1a} = v_p + v_1$ while the voltage applied to the other diode circuit is $v_{2a} = v_p - v_2$. A phasor representation of these voltages is shown in Figure 9.13.

When the frequency varies from resonance, the phases of $v_1$ and $v_2$ change. Voltage $v_{1a}$ increases and voltage $v_{2a}$ decreases as the frequency varies in one direction from resonance. As the input frequency varies in the opposite direction from resonance, $v_{1a}$ decreases and $v_{2a}$ increases. The diode circuits are peak detectors; thus, the voltage across the output capacitors represents the magnitude of the voltages $v_{1a}$ and $v_{2a}$. The output voltage then equals the difference between $|v_{1a}|$ and $|v_{2a}|$, and this voltage varies linearly with frequency deviation over a sizable range of frequency.

The ratio detector extends the Foster-Seeley method to make the detector less sensitive to amplitude variations of the applied FM waveform. Details of operation can be found in [6].

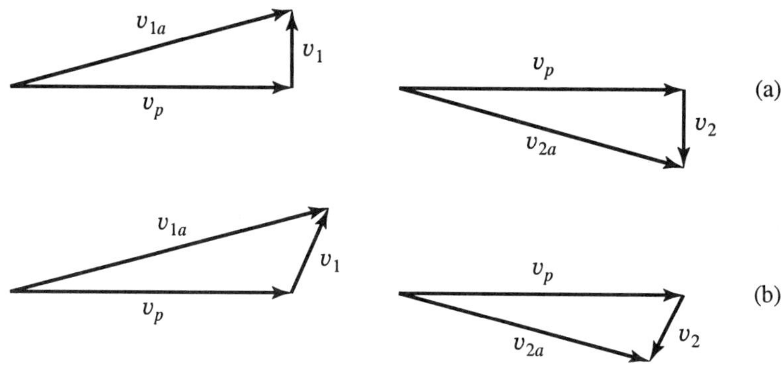


Figure 9.13: Voltage relations in the FM discriminator: (a) at resonance (b) off resonance.

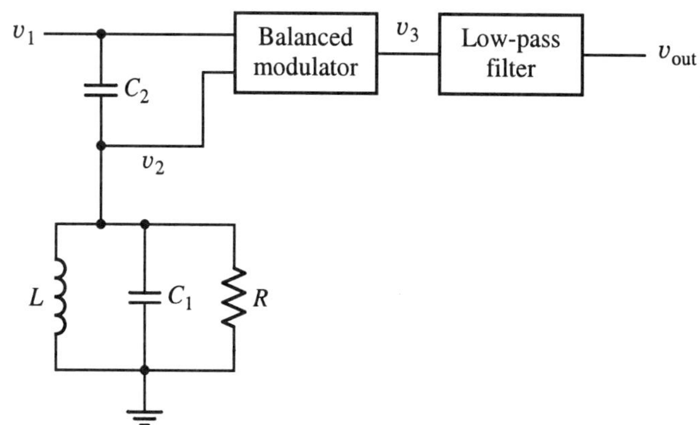


Figure 9.14: Quadrature FM detector.

### 9.3.2 The Quadrature Detector

The circuit of Figure 9.14 demonstrates the basic principle of the quadrature detector. The tuned circuit and capacitor, $C_2$, cause a phase shift between $v_1$ and $v_2$. If $G(\omega)$ is the transfer function of the network, then $v_2$ can be written as

$$v_2 = v_1 G(\omega) = v_1 G_1 \text{ at an angle } \Theta$$

where $G_1$ is the magnitude of $G(\omega)$ and $\Theta$ is the phase of $G(\omega)$. If a sinusoid is applied to the circuit, the inputs to the balanced modulator are

$$v_1 = V_1 \cos \omega t$$

and

$$v_2 = V_1 G_1 \cos (\omega t + \Theta)$$

The balanced modulator forms the product of $v_1$ and $v_2$ giving an output of

$$v_1 = V_1^2 G_1 \cos (\omega t + \Theta) \cos \omega t$$

A trigonometric identity gives

$$v_1 = \frac{V_1^2 G_1}{2} [\cos (2\omega t + \Theta) + \cos \Theta] \tag{9.26}$$

Assuming no attenuation of the low-frequency component of the modulator output by the lowpass filter, the output voltage of the detector is

$$v_{\text{out}} = \frac{V_1^2 G_1}{2} \cos \Theta \tag{9.27}$$

For good demodulation, the output voltage should vary linearly with frequency. Both $\Theta$ and $G_1$ depend on $\omega$, but it can be shown that $v_{\text{out}}$ approximates a linear dependence on $\omega$. At one time it was conventional to select circuit parameters such that $\Theta$ was near 90° at the input carrier frequency—hence the name quadrature detector. Other values of $\Theta$ are often used, however, with reasonable linearity.

The relationship between $v_{\text{out}}$ and $\omega$ can be found by writing $G(\omega)$ and then evaluating $\Theta$ and $G_1$ for use in Eq. (9.27). The transfer function is

$$G(\omega) = \frac{\omega^2 C_2 LR}{\omega^2 LR[C_1 + C_2] - R - j\omega L} \tag{9.28}$$

The magnitude of this function is

$$\frac{\omega^2 C_2 LR}{\left[\omega^4 (LR[C_1 + C_2])^2 + \omega^2 (L^2 - 2R^2 L[C_1 + C_2]) + R^2\right]^{1/2}} \tag{9.29}$$

and the phase is

$$\Theta = \tan^{-1}\left(\frac{\omega L}{\omega^2 LR[C_1 + C_2] - R}\right) \tag{9.30}$$

It is difficult to calculate the linearity of $v_{\text{out}}$ as a function of $\omega$, but typical circuit values can be selected and the values of $v_{\text{out}}$ versus $\omega$ can be tabulated. This is done in Table 9.2 for a nominal phase shift of 45° and an input magnitude of 6 V. A typical TV audio IF of 4.5 MHz and a maximum frequency deviation of 25 kHz are assumed. The necessary circuit element values are $R = 884\ \Omega$, $C_1 = C_2 = 40$ pF, and $L = 31.27\ \mu$H. The output voltage in Table 9.2 has a linearity error ranging from -3% to +3% over the extreme values of frequency deviation.

**Table 9.2 Output of Quadrature Detector**

| **Input Frequency** kHz | $G_1$ | $\cos\Theta$ | $v_{\text{out}}$ V |
|---|---|---|---|
| 4475 | 0.70902 | 0.70103 | 8.9467 |
| 4480 | 0.70862 | 0.70224 | 8.9572 |
| 4490 | 0.70784 | 0.70464 | 8.9778 |
| 4500 | 0.70705 | 0.70701 | 8.9980 |
| 4510 | 0.70627 | 0.70935 | 9.0178 |
| 4520 | 0.70548 | 0.71167 | 9.0373 |
| 4525 | 0.70509 | 0.71282 | 9.0468 |

The sensitivity of this detector to frequency change is rather small, about 2 mV per kHz. A significant amount of amplification is required before a speaker can be driven. Furthermore, the detector output is dependent on the square of the input amplitude, requiring a good limiter circuit to be placed before the detector. For TV-quality sound, the quadrature detector is sufficient. For FM radio, this detector is generally not used unless it is important to minimize receiver size.

The quadrature detector has been implemented using a combination of analog and digital circuitry, making the output less dependent on input amplitude. This detector is available on an IC chip (Motorola MC1357) that requires only the tuned circuit and $C_2$ as external elements. This type of IC quadrature detector is used in some FM receivers.

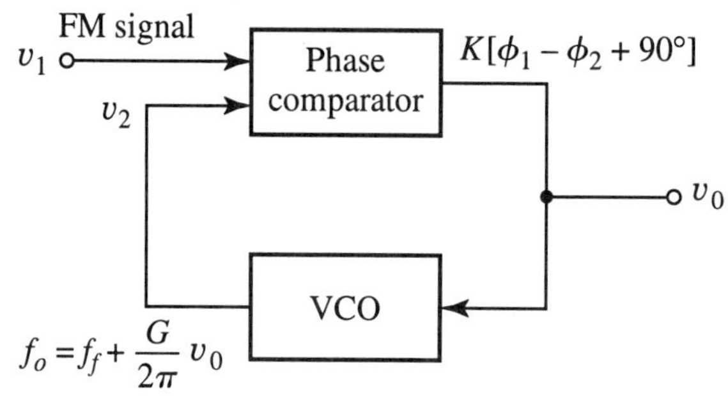


Figure 9.15: A phase-locked loop used as an FM detector.

### 9.3.3 The Phase-Locked Loop

The phase-locked loop (PLL) can be used as an FM demodulator. In this application, a precision voltage-controlled oscillator (VCO) with a stable equilibrium frequency is required. The PLL also finds application as a frequency multiplier or divider, as a local oscillator, and as an FM stereo decoder. This circuit consists of a phase comparator and a voltage-controlled oscillator (VCO), as shown in Figure 9.15.

Two types of phase comparator can be used in the PLL. One is a multiplier circuit or balanced modulator that forms the product of the two input signals. If the inputs are given by $v_1 = C \cos(\omega_c t + \phi_1)$ and $v_2 = D \cos(\omega_c t + \phi_2)$, the product is

$$CD \cos(\omega_c t + \phi_1) \cos(\omega_c t + \phi_2) = \frac{CD}{2}[\cos(2\omega_c t + \phi_1 + \phi_2) + \cos(\phi_1 - \phi_2)]$$

Passing the output of the multiplier through a lowpass filter results in an output signal of

$$v_0 = \frac{CD}{2}[\cos(\phi_1 - \phi_2)] \tag{9.31}$$

When the phase difference is $\pm 90°$, the output of the comparator is zero. As this difference becomes greater than $90°$, the output becomes negative while a phase difference less than $90°$ leads to a positive output.

Another type of comparator converts the input signals to square-wave, logic-level signals by amplification and limiting. The phases of these two logic signals are then compared. An exclusive-OR or EXOR gate can be used for this comparison. When both inputs are at logic 0 or logic 1, the EXOR generates a 0 output. When only one signal is 0 and the other is at 1, the EXOR produces a logic 1 output. If the EXOR generates an output voltage of $+V$ for logic 1 and $-V$ for logic 0, the waveforms will appear as in Figure 9.16.

The output of the EXOR gate is either integrated or lowpass filtered to generate the final error signal. The area under the EXOR output waveform will be zero when the two input signals are $90°$ out of phase as shown in Figure 9.16(a). As the relative phase shift changes from this $90°$ value, the area varies linearly with phase change. The error signal is then linearly proportional to the phase change.

In the development that follows, we will assume a linear variation of comparator output with phase shift. The assumed comparator output is then

$$v_0 = K[\phi_1 - \phi_2 + 90°] \text{ for } -180° \leq (\phi_1 - \phi_2) \leq 0° \tag{9.32}$$

and

$$v_0 = -K[\phi_1 - \phi_2 - 90°] \text{ for } 0° \leq (\phi_1 - \phi_2) \leq 180° \tag{9.33}$$

Figure 9.17 shows the comparator output as a function of phase difference.

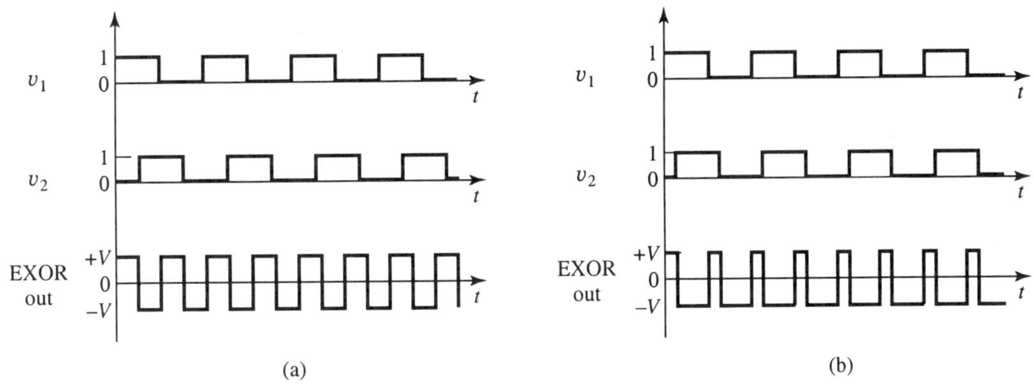


Figure 9.16: Outputs of the EXOR phase comparator. (a) 90° phase difference. (b) 60° phase difference.

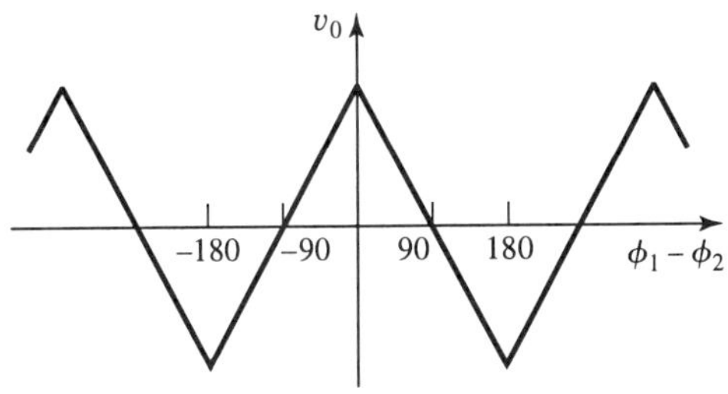


Figure 9.17: Comparator output as a function of phase difference.

The VCO is an oscillator with an output frequency linearly controlled by an applied voltage. This circuit is basically an FM modulator with a frequency deviation at least as large as that of the input signal. The VCO is driven by the phase comparator output voltage, $v_0$. The frequency of the VCO when $v_0$ is zero is called the free-running frequency and is designated $f_f$.

For use as an FM demodulator the PLL is so designed that the VCO oscillates at the frequency of the incoming or IF signal when no control voltage is applied; that is, $f_f = f_{\text{if}}$. For nonzero values of $v_0$, the VCO frequency is expressed as $f_o = f_f + \frac{G}{2\pi}v_0$. This relation is shown in Figure 9.18.

An equilibrium is reached when the phase difference between the FM input and the VCO signal is $-90°$. A constant phase difference implies that the input frequency and the VCO frequency are equal; otherwise, a continual change in phase difference would result.

Let us assume that the system initially exhibits a phase difference of only $-60°$; that is, the VCO output signal leads the FM carrier by 60° or $\phi_1 - \phi_2 = -60°$. A positive voltage will be generated by the comparator according to Eq. (9.32). This voltage will increase the output frequency of the VCO or the frequency of the comparator

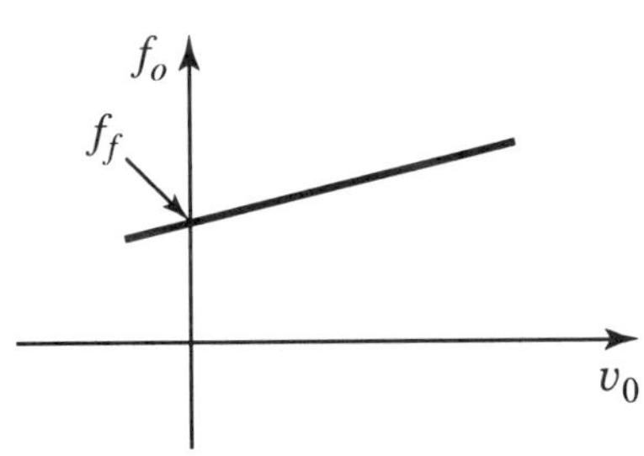


Figure 9.18: VCO frequency vs. control voltage.

input signal $v_2$. With a slightly higher VCO frequency, the angle $\phi_2$ will increase and the absolute phase difference will increase over each cycle. As this difference approaches $-90^\circ$, the comparator output approaches zero, and the VCO frequency approaches the carrier frequency. Other values of initial negative phase difference also lead to the same equilibrium point. Positive values of phase difference between $0^\circ$ and $180^\circ$ lead to an equilibrium point of $+90^\circ$.

If the input frequency and the VCO frequency when $v_0 = 0$ V are equal, a phase difference other than $\pm 90^\circ$ leads to temporary nonzero values of $v_0$ as the phase is modified to reach this point. When this occurs, the PLL quickly comes into lock. A change in frequency of the input leads to the generation of a nonzero value of $v_0$ as the VCO frequency attempts to reach a constant phase difference.

Because

$$d\phi_1/dt = \Delta\omega_1$$

an abrupt variation of input frequency from $f_{\text{if}}$ to $f_{\text{if}} + \Delta f_1$ leads to a phase change in $\phi_1$, referenced to the old value of $\phi_1$, of

$$\Delta\phi_1 = \Delta\omega_1 t$$

This phase change generates a positive voltage at the comparator output that increases the frequency of the VCO until a new equilibrium point has been established. The new value of $\phi_1 - \phi_2$ for equilibrium is no longer $\pm 90^\circ$. Obviously, this new VCO frequency must equal the FM intermediate frequency for equilibrium, and $v_o$ is proportional to the difference between the incoming frequency and $f_f$ or $f_{if}$.

We will now derive the response of the PLL to an FM input signal. In doing so, we will assume the use of a PLL operating in the mode that obeys Eq. (9.32); that is,

$$v_0 = K[\phi_1 - \phi_2 + 90^\circ]$$

We will also again assume that the FM signal to be demodulated is applied to input number 1 of the phase comparator. The output of the VCO is applied to the second input. We will further assume that the PLL is operating in an equilibrium situation with the free-running frequency of the VCO equal to the incoming IF frequency. This results in an output voltage from the phase comparator of 0 V when $\phi_1 - \phi_2 = -90^\circ$. A time reference will be selected such that $\phi_1 = 0^\circ$ and $\phi_2 = 90^\circ$.

The frequency of the FM signal may be represented by

$$\omega = \omega_{\text{if}} + m(t)$$

where $m(t)$ is the frequency deviation due to the modulation signal at time $t$. With no modulating signal the value of $\phi_1 = 0^\circ$. When $m(t)$ is finite, causing a deviation in frequency from the carrier frequency, $\phi_1$ will change.

Since the instantaneous phase of an FM signal, expressed by $v_1 = C \cos \Theta(t)$, is related to the frequency of the signal by

$$\omega = \frac{d\Theta}{dt}$$

we can integrate this expression to find $\Theta(t)$. This leads to

$$\Theta(t) = \int \omega dt = \int [\omega_{\text{if}} + m(t)]dt = \omega_{\text{if}} t + \int m(t)dt$$

If $\phi_1$ is $0^\circ$ at $t = 0$ or when the modulation signal is applied, $\phi_1$ will change from $0^\circ$ to

$$\phi_1 = \int m(t)dt$$

As this angle changes, the phase comparator generates an output voltage that modifies the VCO frequency and thereby modifies the phase angle $\phi_2$. This angle was initially $90^\circ$ at $t = 0$ and now becomes

$$\phi_2 = 90^\circ + \int G v_0 dt$$

From Eq. (9.32) we can now write that

$$v_0 = K[\phi_1 - \phi_2 + 90°] = K[\int m(t)dt - \int Gv_0 dt]$$

We note that the unknown value of output voltage, for which we need to solve, appears on both sides of this integral equation. We therefore differentiate both sides of the equation to get the first-order differential equation

$$\frac{dv_0}{dt} + KGv_0 = Km(t) \tag{9.34}$$

This equation applies to the output voltage of the PLL for an FM input waveform that has a radian frequency change given by $m(t)$.

### Step Function Change in Frequency

Let us first consider the case of an abrupt change in modulating voltage that causes a change in frequency from zero to $m_1$ or $m(t) = m_1 u(t)$ where $u(t)$ is the unit step function. Solving the differential equation given by Eq. (9.34) results in

$$v_0 = \frac{m_1}{G}[1 - e^{-t/\tau}] \tag{9.35}$$

where

$$\tau = 1/GK \tag{9.36}$$

The output voltage reaches the desired voltage exponentially. If the time constant $\tau$ is small, the output voltage rapidly reaches a value that is proportional to the modulating signal. The constants $G$ and $K$ can easily be chosen by selecting an appropriate PLL to result in a suitable time constant for FM signals with a baseband limitation of 15 kHz.

### A Tone-Modulated FM Signal

Now let us assume that the input frequency is given by

$$f = f_{\text{if}} + \frac{m_1}{2\pi} \sin \omega_m t$$

which is the frequency of a tone-modulated FM signal. The differential equation for $v_0$ becomes

$$\frac{dv_0}{dt} + KGv_0 = Km_1 \sin \omega_m t$$

Solution of this differential equation gives

$$v_0 = \frac{K\tau m_1}{1 + \omega_m^2 \tau^2} \sin \omega_m t + \frac{K\tau^2 \omega_m m_1}{1 + \omega_m^2 \tau^2}(e^{-t/\tau} - \cos \omega_m t)$$

If we require the time constant $\tau$ to be much smaller than the period of the modulating signal, so that $\omega_m^2 \tau^2 \ll 1$, this equation reduces to

$$v_0 = K\tau m_1 \sin \omega_m t = \frac{m_1 \sin \omega_m t}{G} \tag{9.37}$$

The output voltage of the PLL is proportional to the sinusoidal modulating signal. Obviously, the PLL can then act as a demodulator for an FM tone-modulated signal. Again, the PLL can be chosen with appropriate values of $G$ and $K$ to demodulate any tone up to the 15 kHz limit of the baseband signal. It should be obvious that the PLL can be used to demodulate commercial FM.

One matter that is not so obvious is the effect of drift of the free-running frequency on the demodulation of the FM signal. Typically, the VCO of the PLL is not a precision oscillator. The FCC requires that the carrier frequency of commercial FM be stable to within ±2 kHz of the carrier frequency. In order that the VCO also stay within this tolerance, a more accurate reference oscillator must be used that allows comparison of the VCO frequency to this precision oscillator frequency. Any differences apply an additional control voltage to the VCO to maintain a precise average frequency value. This automatic frequency control (AFC) circuit adds cost to the detector circuitry and indicates why other detectors are often used in commercial FM receivers.

The input amplitude does not appear in any of the equations. Depending on the type of phase comparator used, the value of $K$ may depend on the input amplitude. For those comparators that convert the input signal to a square wave, the amplitude limiting is a "built-in" feature of the PLL. For the product-type phase comparator, amplitude limiting is necessary to create a fairly constant value of $K$.

The PLL is available as an IC chip from several manufacturers. Some of the parameters of the PLL that specify performance are as follow:

*Free-running frequency.* This is the frequency of the VCO when the phase comparator output voltage is zero. This parameter can be determined by external circuitry within some predefined limits for some PLLs. This parameter was designated $f_f$ in the previous discussion.

*Lock range.* Once in lock, this parameter specifies the range of frequency over which the PLL will remain in lock. Often, the free-running frequency is near the midpoint of the lock range.

*Capture range.* This is the frequency range over which the PLL can acquire lock if not initially locked. This parameter can never be larger than the lock range and is often smaller.

*Phase comparator conversion gain.* This constant relates the output voltage of the phase comparator to the phase error of the PLL when the loop is locked. This constant was designated $K$ in the previous discussion and is usually specified in units of volts per radian.

*Conversion gain.* This constant relates the VCO frequency change to the VCO control voltage. The units are radians per second per volt, and G is the symbol used for this constant.

**Practice Problem**

**9.7** The free-running frequency of the VCO in a PLL is 10.0 MHz. A PLL output signal of 1.8-V magnitude shifts this frequency to 10.14 MHz. Calculate the value of $G$ for this VCO. *Ans:* 0.489 MHz/V.

## 9.4 FM RECEIVERS

**Important Concepts**

**1.** The monaural FM receiver is similar to the AM receiver in that a mixer to an intermediate frequency is used, followed by an IF strip consisting of several tuned stages. The detector must demodulate frequency changes rather than peak value changes as in AM.

**2.** The FM stereo system transmits one composite signal consisting of the left channel signal plus the right channel signal. A second composite signal consists of the left channel signal minus the right channel signal. A third signal is the pilot signal indicating stereo is being transmitted.

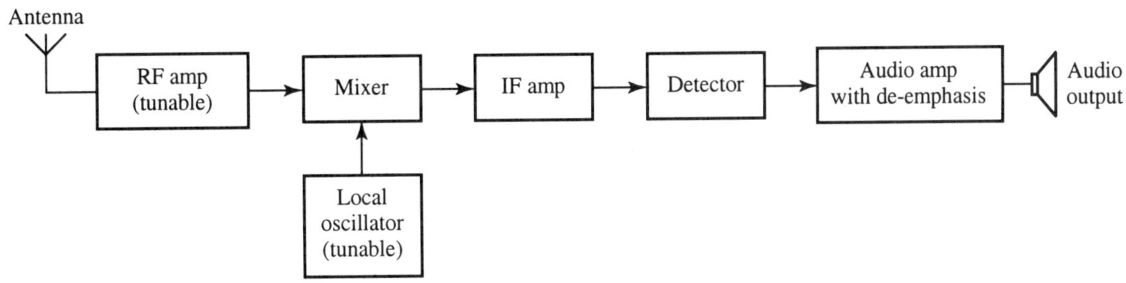


Figure 9.19: An FM receiver.

### 9.4.1 A Monaural FM Receiver

The block diagram of a commercial monaural FM receiver is shown in Figure 9.19.

The first section of the receiver is an RF stage that is tunable over the frequency range of 88 MHz to 108 MHz. The mixer then creates an intermediate frequency of 10.7 MHz. The local oscillator employs automatic frequency tuning to maintain the correct frequency in the presence of parameter drift due to temperature changes. The local oscillator allows voltage fine tuning and uses the dc output level of the detector for a drive signal. The dc output of the detector when the local oscillator is tuned properly should be zero. This value will change as the local oscillator frequency drifts.

The IF amplifier may consist of three or four tuned stages that result in a 200-kHz bandwidth. Stagger-tuning of the tuned stages is used to achieve a sharp falloff for frequencies outside the desired bandwidth, along with a relatively flat response inside the passband. In many respects, the FM receiver is similar to the AM receiver discussed in Chapter 8.

### 9.4.2 FM Stereo Transmission

The FCC issued standards for stereo transmission in 1961 [2]. This standard allows two channels of information to be transmitted simultaneously over the same channel. For creation of a "wide-stage" effect, a left microphone and a right microphone, along with the associated amplifiers are used at the sound source. These microphones are placed to pick up the same sounds that would be heard by each of the human ears, if the human were present at the source. A concert can be recorded by a left microphone that is directed toward the sound arising from the left portion of the stage and a right microphone that is directed toward the sound arising from the right portion of the stage. If these two signals or channels are directed separately to properly placed speakers, the reproduced sound will be very similar to the actual sound in the concert hall. One of the two speakers is placed to the left of the listener, while the other is placed to the right.

For FM transmission, it would be a simple matter to transmit the sound signal produced by the two microphones separately. However, the FCC has approved stereo transmission on the basis that a monaural receiver must be able to receive the stereo transmission. Of course, the monaural receiver cannot produce a stereo output, but it must be able to respond to both left and right channel information. In fact, the monaural system must respond to a signal that is the sum of the left and right information channels.

Because of this required compatibility with a monaural receiver, the system of Figure 9.20 is used to generate the FM stereo system.

The audio information from the left and right channels is added together to form the L + R signal. It is also subtracted to form the L - R signal. The audio bandwidth of these signals is 15 kHz. The L - R signal is then applied to one input of a balanced modulator that has a 38-kHz carrier applied to the other input. The output of the modulator, which is a DSB-SC signal, is applied to one input of a summing network. The L + R signal is directly applied to the second input of the summer. A 19-kHz signal, synchronized with the 38 kHz carrier, is applied to the third input of the summer. The output of the summer is the baseband signal applied to an FM

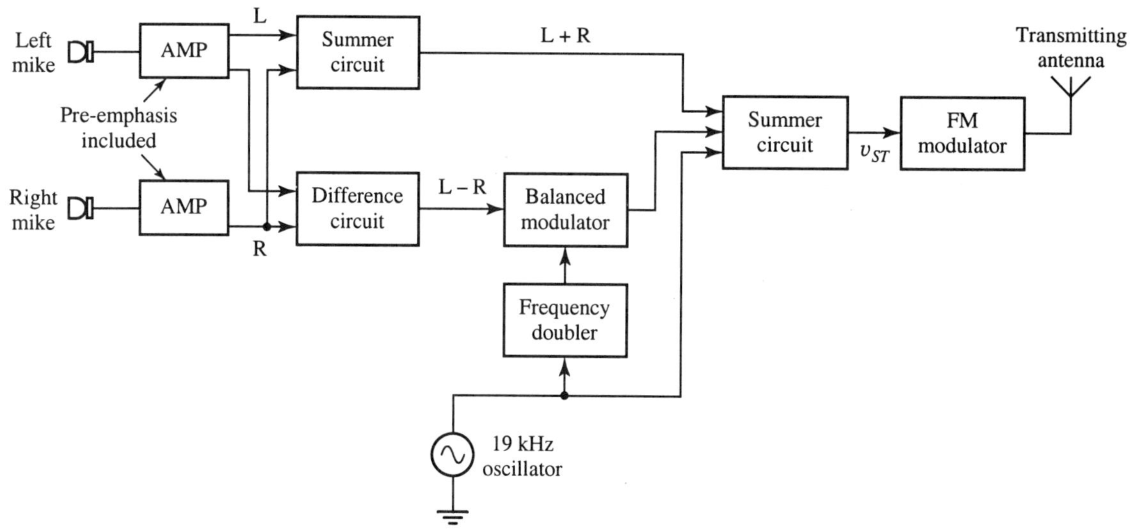


Figure 9.20: An FM stereo transmitter.

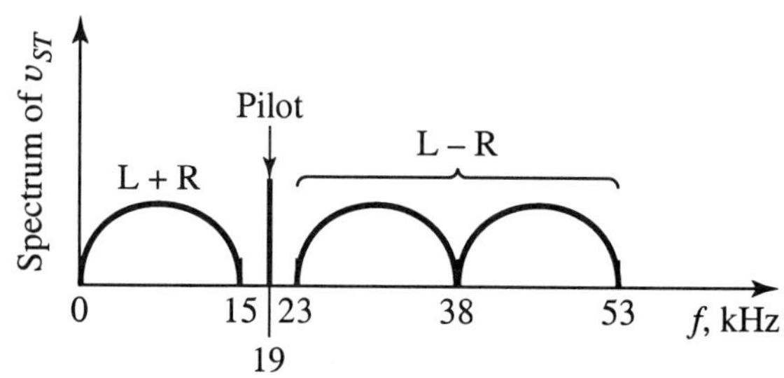


Figure 9.21: Spectrum of baseband signal applied to FM modulator.

modulator. The frequency spectrum of this signal is indicated in Figure 9.21.

Some FM stations may be authorized to broadcast a second program that is added to the baseband signal. This is called a Subsidiary Communications Authorization (SCA) signal. This signal is centered about a frequency that is higher than the 53-kHz maximum sideband frequency of the L - R channel. The SCA signal requires a special decoder to receive and background music for businesses such as banks and medical/dental offices is transmitted. We will ignore this service in our discussion of stereo.

The bandwidth of a stereo receiver must be somewhat higher than that of a monaural receiver, as indicated by Eq. (7.23). The bandwidth should be at least

$$BW = 2[\Delta f + f_m(\text{max})] = 2[75 + 53] \text{ kHz} = 256 \text{ kHz}$$

This result is far too conservative and predicts a larger required bandwidth than is actually necessary. The SCA channel is a rather small signal that leads to a frequency deviation that may be only 10% of the total value of allowed frequency deviation. The resulting bandwidth required is only slightly larger than that required by a monaural system.

When a monaural receiver sends the FM stereo signal through its IF stages, information pertaining to the L - R channel is lost. Any remaining information from this channel is eliminated by the 15-kHz bandwidth of the detector and audio circuitry. Only the L + R channel is demodulated and used to activate the speaker.

The block diagram of Figure 9.22 is that of a stereo receiver. The RF and IF sections are similar to the monaural receiver, but have a wider bandwidth. The L + R signal appears as a component of the FM detector output, while the L - R signal must be demodulated by an AM demodulator. The 19 kHz pilot signal is separated

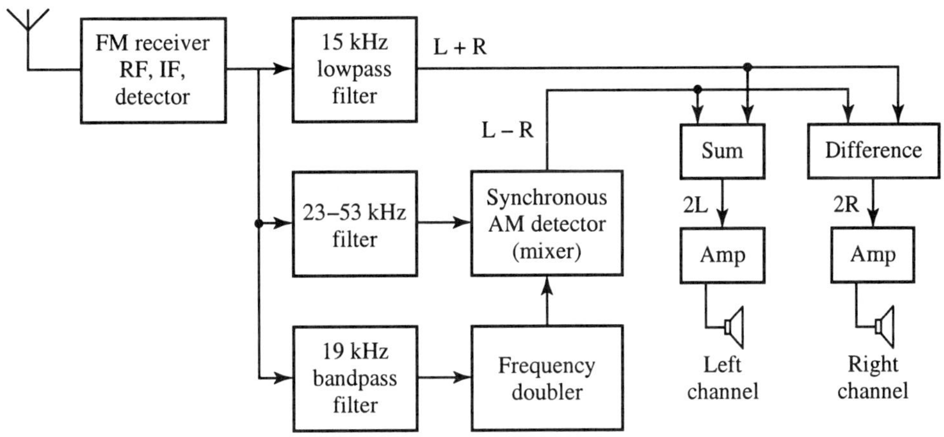


Figure 9.22: An FM stereo receiver.

and applied to a frequency doubler. This signal is synchronous with the original carrier signal and can be used by the synchronous detector or mixer to demodulate the L - R channel.

The L + R information is then added to the L - R information to produce a signal proportional to the original left channel. The two signals are subtracted to produce a signal proportional to the right channel.

## SUMMARY

**1.** A phase modulator changes the phase of the output signal in direct proportion to the instantaneous magnitude of the modulating signal.

**2.** A phase modulator can be based on an amplitude modulator or a balanced modulator.

**3.** The frequency modulator produces a frequency deviation in the output signal in direct proportion to the instantaneous magnitude of the modulating signal.

**4.** The phase modulator can be used to create a frequency modulator by integrating the baseband signal prior to applying it to the modulation input of the phase modulator.

**5.** A direct method of generating an FM signal is based on an automatic frequency control circuit.

**6.** FM detectors must produce an output voltage that varies in proportion to the frequency changes of the incoming signal.

## PROBLEMS

### Section 9.1.1 Phase Modulators Based on Amplitude Modulators

**9.1** The output of the amplitude modulator in Figure P9.1 is given by

$$V_c(1 + 0.2M \cos \omega_m t) \cos \omega_c t$$

Choose the values of $V_c$ and $M$ that will lead to a phase-modulated output signal of

$$v_{\text{out}} = 20 \cos (\omega_c t + \phi_0 + k_p M \cos \omega_m t)$$

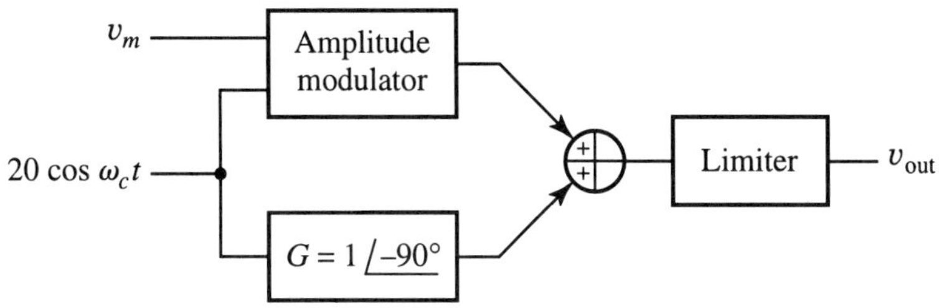


**Figure P9.1**

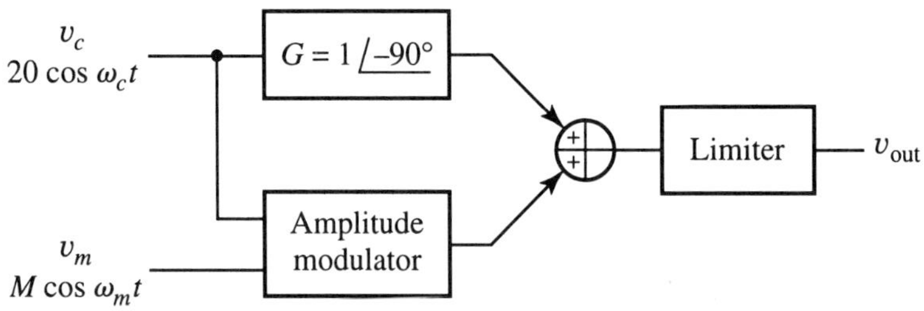


**Figure P9.2**

where $\phi_0 = 8°$ and the phase deviation is 4°.

**9.2** If the output of the amplitude modulator in Figure P9.2 is

$$2.5 \cos \omega_c t + 0.1M \cos (\omega_c - \omega_m)t + 0.1M \cos (\omega_c + \omega_m)t$$

select the magnitude of the input modulating signal, $M$, to create a phase deviation of 5°.

**9.3** A phase-modulated signal is expressed as $v_{\text{out}} = 6 \cos (\omega_c t + K \sin \omega_m t)$. The frequency deviation of this signal is 120 Hz when $f_m = 4$ kHz.

**(a)** Find the value of $K$.
**(b)** What is $\Delta f$ at $f_m = 2.5$ kHz?

**9.4*** The modulating signal shown is applied to a phase modulator with $k_p = 0.01$ rad/V to produce an output signal of $v_{\text{out}} = A \cos (\omega_c t + k_p v_m)$.

**(a)** What is the maximum phase deviation in degrees for the output signal?
**(b)** What is the maximum frequency deviation in Hz for the output signal?
**(c)** If the half-period, $T/2$, is changed to 75 $\mu$s and the amplitude of the modulating signal is changed to 4 V, what is the new maximum frequency deviation?

**9.5** The output of a phase modulator is given by

$$v_{\text{out}} = 10 \cos (\omega_c t + 0.02M \sin \omega_m t)$$

**(a)** Calculate the value of $M$ required to cause a frequency deviation of 10 Hz when $f_m = 100$.
**(b)** Calculate $M$ to cause a frequency deviation of 100 Hz when $f_m = 5$ kHz.

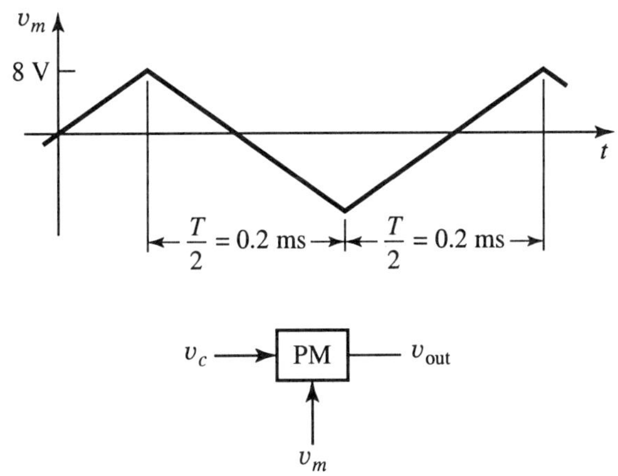


**Figure P9.4**

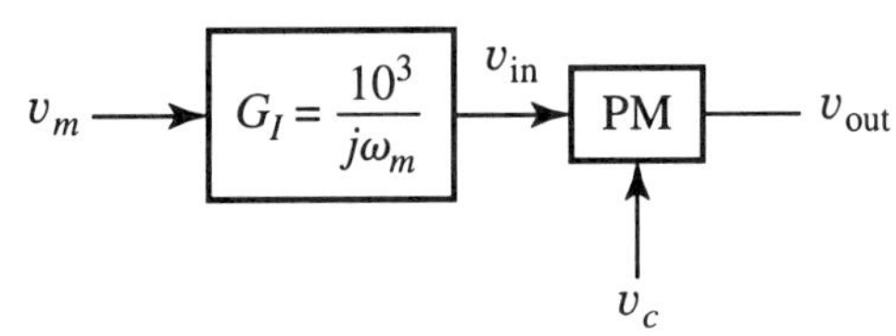


**Figure P9.8**

### Section 9.1.2 A Large-Deviation Phase Modulator

**D9.6*** Design a phase modulator using the circuit of Figure 9.3. The carrier signal is 0.1 cos $\omega_c t$,where $f_c = 100$ kHz, and the modulating signal is $v_m = 3$ sin $\omega_m t$ where $f_m = 200$ Hz. Assume that $r_v = 90$ k$\Omega$ with 1 V appearing across the VVR. The quiescent phase shift should be 70°. Calculate $m_p$ for your circuit.

### Section 9.2.1 The Indirect Method of Frequency Modulation

**9.7** Assume that a frequency modulator is used with a 1-MHz carrier and a 0.2% frequency deviation. If a 96-MHz carrier with a 75-kHz deviation is required for transmission, show how to mix and multiply this waveform to obtain the desired signal.

**9.8** If $v_m = 10$ cos $\omega_m t$, select a value of $k_p$ for the phase modulator shown to make $\Delta f = 40$ Hz for all values of $f_m$ varying between 100 Hz and 15 kHz.

**9.9*** A phase modulator produces an output signal of 10 cos $[2\pi \times 10^6 t + km(t)]$, where $m(t)$ is the modulating signal in volts applied to the phase modulator input and $k = 0.12$ rad/V.

**(a)** A modulating signal of 5 cos $2\pi \times 10^3 t$ V is applied to the phase modulator input. Calculate the frequency deviation of the output signal in Hz.
**(b)** Repeat part (a) if the modulating signal frequency is changed to $3 \times 10^3$ Hz.
**(c)** The signal of part (a) is now applied to an integrator with $k_I = 5000$. The output of the integrator is connected to the phase modulator input. Calculate the frequency deviation of the output signal.
**(d)** Repeat part (c) if the modulating signal frequency is changed to $3 \times 10^3$ Hz.

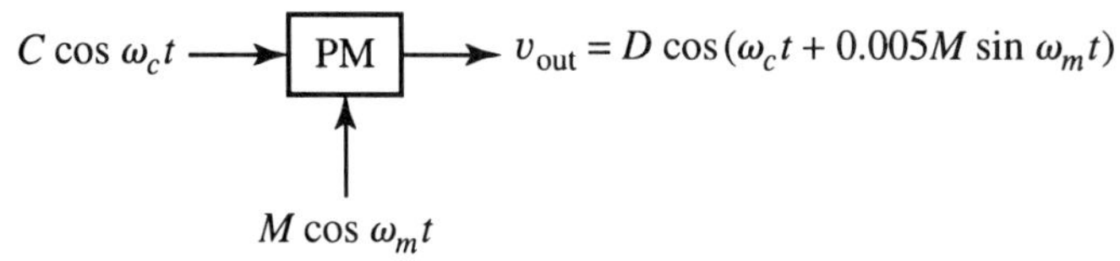


**Figure P9.10**

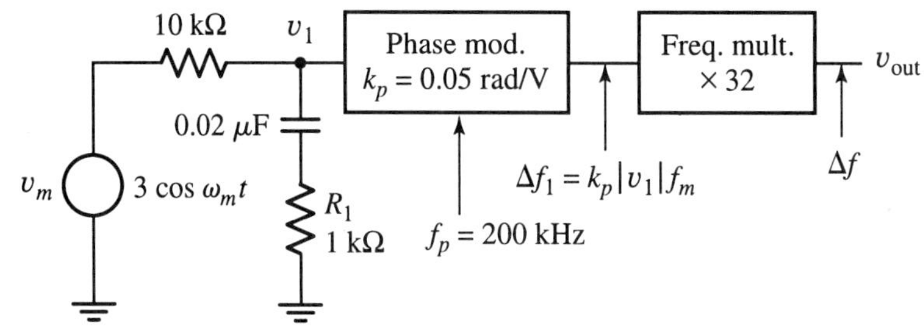


**Figure P9.11**

**9.10** Calculate $M$ in the circuit shown to lead to a maximum frequency deviation of $\Delta f = 20$ Hz for $f_m = 2$ kHz and $\Delta f = 1.5$ kHz for $f_m = 150$ kHz.

**9.11*** Plot the frequency deviation, $\Delta f$, at the output versus the input modulating frequency as $f_m$ varies from 100 Hz to 15 kHz for the modulator of Figure P9.11. Do an asymptotic plot similar to a typical frequency response. Show the values of any corner frequencies and key values of $\Delta f$, including values at $f_m = 100$ Hz and $f_m = 15$ kHz.

### Section 9.2.2 The Direct Method of FM Generation

**9.12** Design a frequency modulator using the tuned-collector modulator of Figure 9.7. The carrier frequency is to be 1 MHz, and a 5% second-harmonic distortion is allowed. What is the maximum frequency deviation that can be obtained with your circuit? State any assumptions made.

### Section 9.2.3 Pre-Emphasis in FM Transmitters

**9.13** Design an integrator/pre-emphasis circuit for the frequency modulator of Example 9.2.

**D9.14*** When a 4-V peak modulating signal with $f_m = 1$ kHz is applied to a phase modulator, the frequency deviation is $\Delta f = 4$ kHz. Show the asymptotic frequency response of a circuit that can be inserted between the modulating signal and the phase modulator to create the asymptotic output frequency deviation shown. Label the magnitude of the transfer function, $|G|$, at the critical corner frequencies.

### Section 9.3.3 The Phase-Locked Loop

**9.15** The maximum frequency of a modulating signal is 10 kHz. Select an appropriate product for GK of the phase-locked loop used to detect this signal.

**9.16** Assume that the changes in frequency of the incoming signal of Figure P9.11 are perfectly abrupt; that is, the two changes are step function changes. Plot the output of the PLL if $G = 10^4$ rad/V-s and $K = 10$ V/rad.

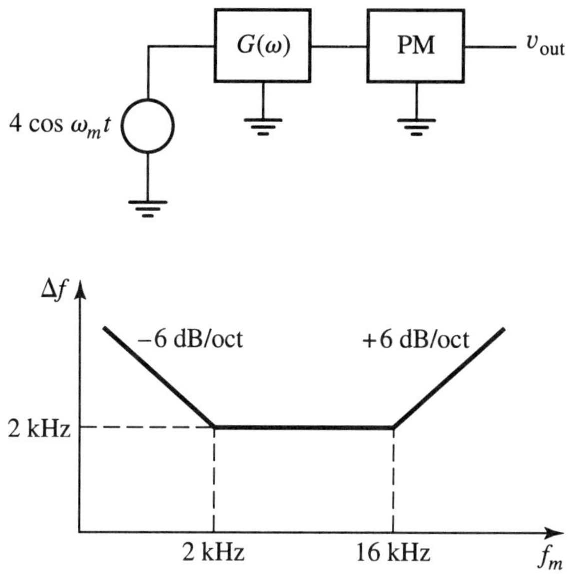


**Figure P9.14**

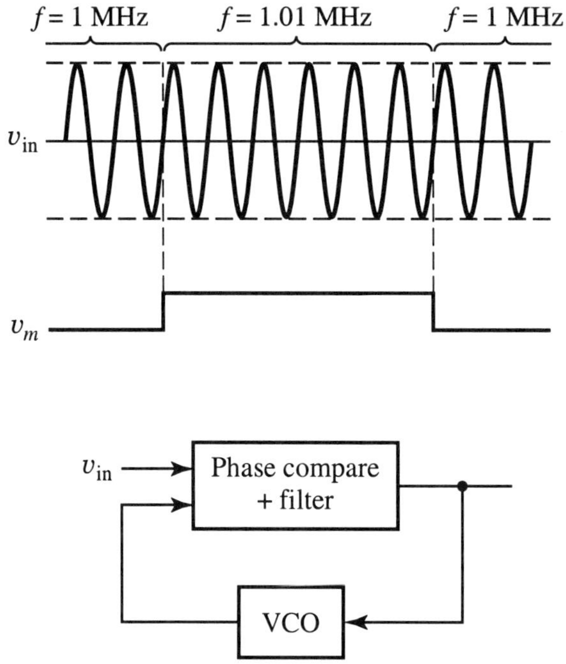


Free running frequency = 1 MHz

**Figure P9.16**

## REFERENCES AND SUGGESTED READING

1. Armstrong, E. H., "Method of reducing disturbances in radio signalling by a system of frequency modulation," *Proc. I. R. E.* Vol. 24, May 1936.

2. Braun, M. J., *AM Stereo and the FCC.* Norwood, N. J.: Ablex Publishing Company, 1994.

3. Comer, D. J., "Large-deviation phase and frequency modulators," *Electronic Engineering.* Vol. 39, No. 474, August 1967, pp. 495-497.

4. Couch, L. W. II, *Digital and Analog Communication Systems, 5th edition,* New York: MacMillan Publishing Co., 1997.

5. Couch, L. W. II, *Modern Communication Systems, Principles and Applications.* Englewood Cliffs, N.J.: Prentice Hall, 1995.

6. Young, P. H., *Electronic Communication Techniques, 4th edition* Upper Saddle River, N.J.: Prentice Hall, 1999.

# Chapter 10

# MIXED-SIGNAL INTERFACING CIRCUITS

A mixed-signal system involves both analog and digital signals within the system. The input signal might be a small analog signal that is amplified and converted to a digital code or word at specified time intervals. This stream of digital words may then be processed using some digital signal processing algorithm, with the results sent to a storage register at prescribed intervals. The output of the register might be converted back to a single analog voltage and used to activate some mechanism.

A well-known, but highly distributed mixed-signal system is that of Direct TV©. This system converts an input signal consisting of the composite analog TV signal to digital data and transmits this information to a satellite using phase-shift keying. After the signal is transmitted back to the various receiving antennas on Earth, it is amplified and converted back to an analog signal to drive the analog TV. This system that occupies a huge geographic area can be contrasted to a mixed-signal system on an IC chip. A microcontroller chip may consist of a circuit that converts an analog input signal to a digital signal that is processed and converted back into some analog control signal to drive some mechanism. In this latter case, the entire system is implemented on a single chip.

## 10.1 INTRODUCTION TO MIXED-SIGNAL SYSTEMS

**Important Concepts**

**1.** Analog circuits and digital circuits are now placed on the same IC chip.

**2.** Converter circuits are used on the chip to allow communication between analog and digital sections.

**3.** Circuits that deal with both analog and digital signals are called mixed-signal circuits.

In early electronic systems, digital circuits and analog circuits were generally separate. The digital computer accepted punched tape or punched card input and produced numerical output. This input or output is represented by discrete data as opposed to continuous-time analog information. Communication systems such as AM radio, FM radio, and conventional TV are analog systems. The signal processing required for these signals is accomplished in a continuous manner.

In the 1960s, digital data acquisition systems gained in popularity. The naturally-occurring analog signals from various transducers were converted to digital words by analog-to-digital converters (ADCs) for direct input to the digital computer. If conversion back to analog was required for the processed signals, digital-to-analog converters (DACs) were used. These data acquisition systems were quite expensive, but were often supplied as integral parts of the digital computing systems. In the late 1970s and early 1980s, VLSI circuits advanced to the point that data acquisition systems were implemented as microcontrollers on a single IC chip. The early chips were rather rudimentary, but continued to improve over the next few decades. At present, the capabilities of mixed-signal circuits on a chip are quite advanced. These chips may include several channels of analog-to-digital conversion, a central processing unit (CPU), digital signal processing circuits, and some digital-to-analog conversion capability. Both analog and digital circuitry is implemented on this chip.

The architecture of these and other mixed-signal chips must be considered in terms of the appropriate partitioning between analog and digital functions. Furthermore, the division between analog and digital sections changes as more capability in either area increases. The huge number of transistors now available on a chip influences both analog and digital circuitry, and a designer should be knowledgeable in both areas.

As a system designer attempts to design a mixed-signal system, interfacing elements are required to communicate between the digital and analog sections. Many manufacturers have standard cells designed to be used for interfacing purposes on the chip. Table 10.1 lists several such functioning cells.

**Table 10.1 Typical Cells Used in Mixed-Signal Chips**

| | |
|---|---|
| 1. D/A converter (DAC) | 7. Op amp |
| 2. A/D converter (ADC) | 8. Precision voltage source |
| 3. Comparator | 9. Logic gates |
| 4. Sample/hold circuit | 10. Flip-flops and latches |
| 5. Window comparator | 11. Tri-state drivers |
| 6. Analog switch | 12. Register and counter cells |

The first six circuits in the table are mixed-signal circuits that deal with both analog and digital signals. The op amp and precision voltage source are analog devices whereas the gates, flip-flops, tri-state drivers, and registers/counters are digital devices. The following sections will consider the basic system specifications and operation of the mixed-signal circuits of Table 10.1. Some of these circuits are used in conjunction with the conversion process and will be discussed prior to a consideration of converter circuits.

# 10.2 CONVERTER INTERFACE CIRCUITS

**Important Concepts**

**1.** The sample-and-hold circuit samples the value of the analog input when instructed by the digital control signal. Whatever analog value was present at the time of the sample is held constant and becomes the output signal.

**2.** The comparator circuit outputs a high-level or a low-level signal depending on which of two inputs is the most positive.

**3.** The analog switch presents the input signal to the output terminal of the switch in one mode and presents an open circuit to the output terminal in the other mode. A digital control signal determines the mode of operation.

## 10.2.1 Sample-and-Hold Circuit

The sample-and-hold (S/H) circuit is a mixed-signal component that has two modes of operation. In the *sample* or *track* mode, the output signal follows the analog input precisely. In the *hold* mode, the output voltage is held

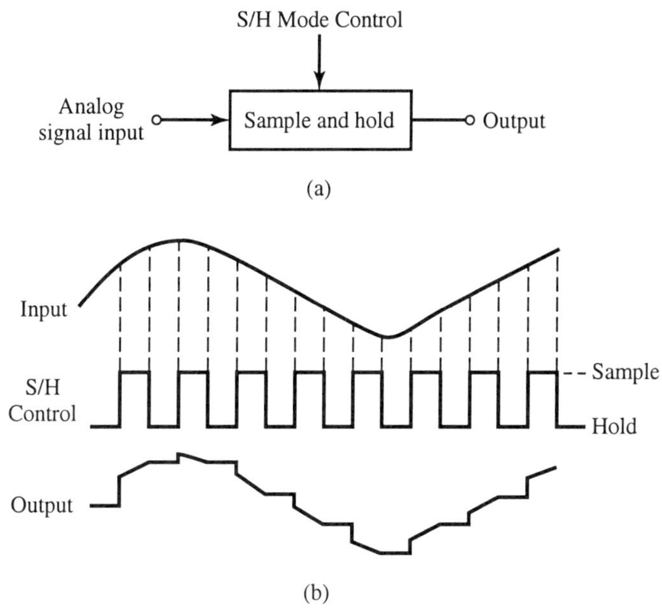


Figure 10.1: The S/H circuit. (a) Block diagram. (b) Typical waveforms.

at the value that existed at the instant this mode was initiated. This constant signal is maintained as long as the hold mode is in effect. A digital control signal determines whether the S/H circuit is in the sample or the hold mode. Figure 10.1 shows a block diagram representation of an S/H circuit along with typical input, output, and control waveforms. For this circuit, a high control signal dictates the sample function and a low control signal puts the circuit into the hold mode.

The S/H circuit is often used to hold an analog value constant while an ADC converts this value to a digital code. If a binary ADC converts an input signal to an $N$-bit code, the largest output code of the ADC, 11 1111 1111 ($1023_{10}$), should occur for the largest input signal. If the largest input signal is 6.00 V, for example, the code for $1023_{10}$ will appear at the output. The smallest change in input signal that leads to a change of code to the adjacent value or the input signal leading to an output of 00 0000 0001 is called a quantum interval, $q$, or a least-significant bit designated $LSB$. This interval is given by

$$LSB = \frac{v_{\text{in}}(\max)}{2^N - 1} \tag{10.1}$$

This equation is often approximated for large $N$ as

$$LSB \approx \frac{v_{\text{in}}(\max)}{2^N} \tag{10.2}$$

For a 10-bit ADC that converts a maximum signal of 6 V, the quantum interval is

$$LSB = \frac{6}{1023} = 5.865 \text{ mV}$$

This parameter will be discussed in more detail in connection with the ADC circuit at the end of this chapter. At the present time, it is introduced to define the maximum error that can be allowed by the S/H circuit. Typically, this circuit should never introduce more error than one-half of a quantum interval or 1/2 $LSB$.

### When to Use the S/H Circuit

When used in applications that require ADC conversions at fixed, periodic intervals, the S/H module precedes the ADC if the converter speed is too slow. If the converter can generate output codes more quickly than the input can change by an *LSB*, the S/H circuit is not required. The equation that calculates the ADC conversion time necessary to avoid using the S/H circuit can be expressed in terms of the maximum input signal slope. If the input must change less than 1 *LSB* between conversions, we can write

$$t_c \times \frac{dv_{\text{in}}}{dt} \leq LSB \approx \frac{v_{\text{in}}(\text{max})}{2^N} \tag{10.3}$$

## Example 10.1

A 10-bit binary ADC has a conversion time of 1 $\mu$s. The maximum slope of the analog signal presented to the ADC is 0.04 V/s and the maximum input is 10 V. Determine whether an S/H circuit is required.

**Solution** The left side of Eq. (10.3) is found to be

$$t_c \times \frac{dv_{\text{in}}}{dt} = 10^{-6} \times 4 \times 10^4 = 4 \times 10^{-2}\ \text{V}$$

The right side of the equation is

$$\frac{v_{\text{in}}(\text{max})}{2^N} = \frac{10}{2^{10}} = 9.77 \times 10^{-3}\text{V}$$

The inequality of Eq. (10.3) is not satisfied; therefore, an S/H module should precede the ADC. If the ADC had a faster conversion time, the S/H module could be eliminated. This could occur if the conversion time satisfied

$$t_c \leq \frac{LSB}{\frac{dv_{\text{in}}}{dt}} = \frac{9.77 \times 10^{-3}}{4 \times 10^4} = 0.244\ \mu\text{s}$$

---

There are several practical implementations of the S/H circuit. One method for lower frequency applications employs two op amps connected as unity-gain buffers, as depicted in Fig. 10.2. The first buffer, $A_1$, feeds the *hold* capacitor, $C_H$, through an analog switch. The voltage across the hold capacitor is transmitted to the output through the second buffer, $A_2$. The analog switch status is determined by the S/H control signal. This switch may be implemented by an MOS switch that is discussed in a later section. The switch is in the closed state in the sample mode and is in the open state during the hold mode. When the switch opens, the holding capacitor is decoupled from $A_1$.

If the switch has an infinite open impedance and the input impedance of $A_2$ is also infinite, the capacitor holds the value reached when the switch opened. When the switch closes, a zero output impedance of $A_1$ will change the capacitor voltage instantaneously to the value dictated by the input voltage. In practice, these ideal impedances and finite switching times will cause the performance of the S/H circuit to be degraded. The performance of the circuit in Fig. 10.2 will deviate from the ideal due to the impedance and switching effects of both the op amps and the analog switch. These effects will be considered later in terms of the S/H parameters defined in the following subsection.

### Sample-and-Hold Parameters

**Aperture Time:** The aperture time is the time that elapses between the application of the hold command and the actual opening of the analog switch. This time is generally in the nanosecond range for most IC S/H circuits.

**Acquisition Time:** This parameter is the minimum time between a sample command and the time after a hold command that the output voltage is acquired to within a prescribed accuracy. Acquisition time depends

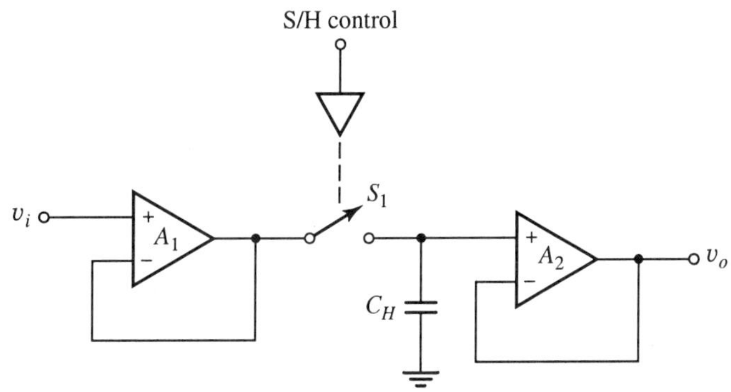


Figure 10.2: S/H circuit using op amps.

on several factors. This time is often dominated by the time to *slew* the hold capacitor to the desired level and the required time to settle to the desired accuracy. Acquisition time is typically specified as the time needed to acquire a *full-scale* voltage step and settle to within a prescribed percentage of the final value.

**Droop Rate:** This parameter specifies the rate of drift of the output during the hold mode. The droop rate is expressed in units of volts per second and is determined by the ratio $I/C_H$ where $I$ is the discharge current during the hold mode. This current is made up of

1. Leakage across the S/H mode control switch
2. Input bias current of the second buffer amplifier
3. Parasitic leakage to the bias supplies through active circuit elements or isolation diodes

**Gain Error:** This parameter is a measure of the difference between input and output signals when in the sample mode. Ideally, the gain would be unity, but the buffer stages generally differ very slightly from this value. The gain for the first stage can be expressed as

$$G_1 = \frac{A}{1+AF} = \frac{A_1}{1+A_1}$$

since the feedback factor is unity. The difference between a gain of unity and the actual gain is the error and is found to be

$$er_1 = 1 - \frac{A_1}{1+A_1} = \frac{1}{1+A_1} \tag{10.4}$$

For the high-gain op amp buffers shown in Fig. 10.2, the gain errors of the stages are approximated as

$$er_1 = \frac{1}{A_1} \tag{10.5}$$

and

$$er_2 = \frac{1}{A_2} \tag{10.6}$$

The overall gain error is

$$er_o = \frac{A_1 + A_2}{A_1 A_2} \tag{10.7}$$

**Sample-to-Hold Offset Error:** This error is sometimes referred to as the *pedestal error* or *charge-injection error*. It is the difference in output levels instantaneously before the mode control switches from sample to hold and instantaneously after switching. This error is considered to be a step function error caused by the injection of charge into the hold capacitor resulting from the input amplifier output circuit and the switch circuit during the sample-to-hold transition time.

**DC Offset Voltage Error:** This error is defined as the output voltage measured with an input voltage of zero when in the sample mode. The offset voltages of the two amplifiers and the offset introduced by the sample switch contribute to the DC offset voltage error.

**Zero-Scale Error:** This error is defined as the difference between output and input voltages instantaneously after the mode control has switched from sample to hold. It is a combination of dc offset and sample-to-hold offset errors.

Table 10.2 lists parameters for a commercial application-specific integrated circuit or ASIC S/H module.

**Table 10.2 Typical ASIC S/H Parameters**

| Parameter | Value |
|---|---|
| Aperture time | 50 ns |
| Acquisition time (10 V step) | 5 $\mu$s (0.01% of final value) |
| Droop rate | 50 nV/ms ($C_H = 0.005\ \mu$F) |
| Voltage gain | 0.99950 |
| Charge transfer | 5 pC |
| S/H offset | 1 mV ($C_H = 0.005\ \mu$F) |
| Zero-scale error | 2 mV |

**Practice Problems**

**10.1** The leakage current in the hold mode from a 0.01 $\mu$F hold capacitor is 2 nA. What is the droop rate of the output voltage? *Ans:* 0.2 V/s.

**10.2** For the S/H circuit of Fig. 10.2, the amplifiers have gains of $A_1 = 1200$ V/V and $A_2 = 2000$ V/V. What is the overall gain error of the system? *Ans:* 0.00133 V/V.

## S/H Circuit Applications

Sample-and-hold circuits are widely used in data acquisition systems to hold the value of an analog signal while a conversion to digital code is accomplished using an ADC. Another common application for S/H circuits is in data recovery or signal reconstruction filters when it is required to create a continuous waveform from a train of finite samples processed from the original signal. The S/H circuit fills in the waveform between samples with constant values. Figure 10.3 shows the reconstruction of a portion of a sine wave with a simple S/H circuit driven by an ADC that accepts digital input values.

It is possible to rearrange the elements of an S/H circuit to create a *peak detector* circuit. The function of the positive peak detector is to capture and hold the largest instantaneous value of an analog waveform applied during some interval of interest. Figure 10.4 shows a block diagram of a positive peak detector along with input and output waveforms.

The peak detector uses the same elements as the S/H circuit with the addition of a diode. The switch is arranged to short the hold capacitor to ground when the reset command is asserted. The output signal from $A_2$ is fed to the inverting input of $A_1$, creating negative feedback around the entire circuit. This forces the output signal to equal the input within the offset and gain errors of the amplifier $A_1$.

With the arrangement of Fig. 10.4, the output of the circuit will change only when the capacitor is charged through the diode. This takes place when the input signal exceeds the value that was stored on $C_H$ from a

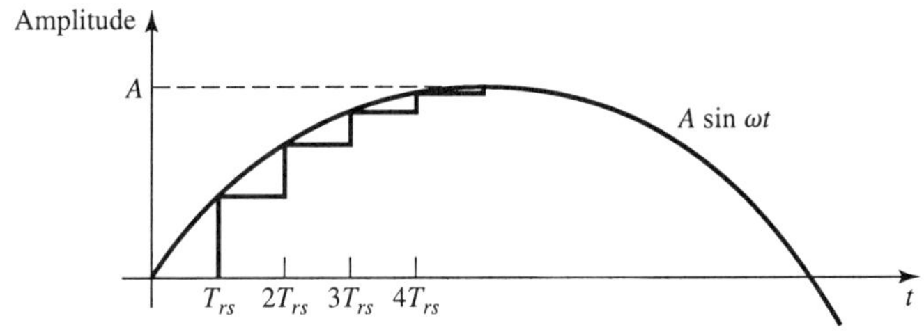


Figure 10.3: Reconstruction of a sine wave using the S/H circuit.

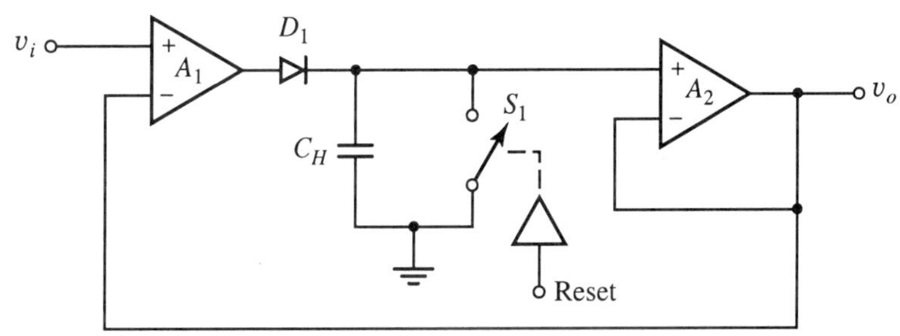


(a) Block diagram

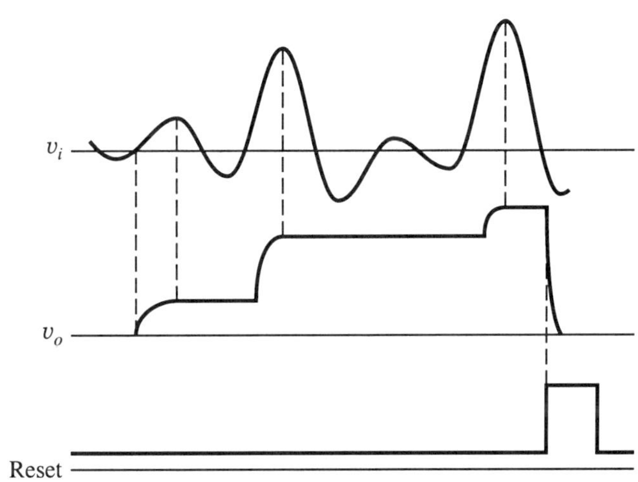


(b) Typical waveforms

Figure 10.4: (a) Positive peak detector. (b) Waveforms.

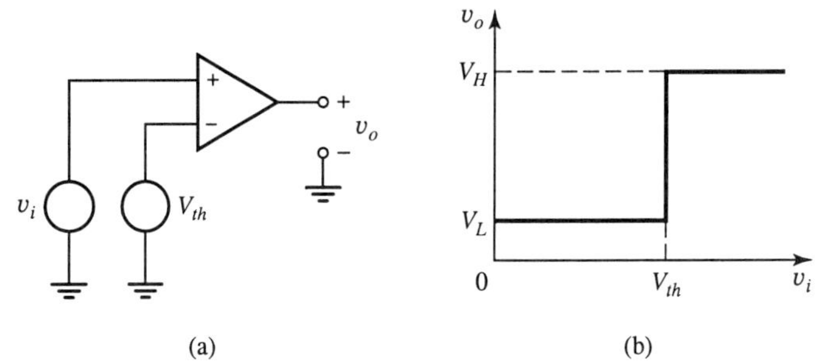


Figure 10.5: (a) Comparator circuit. (b) Ideal transfer function.

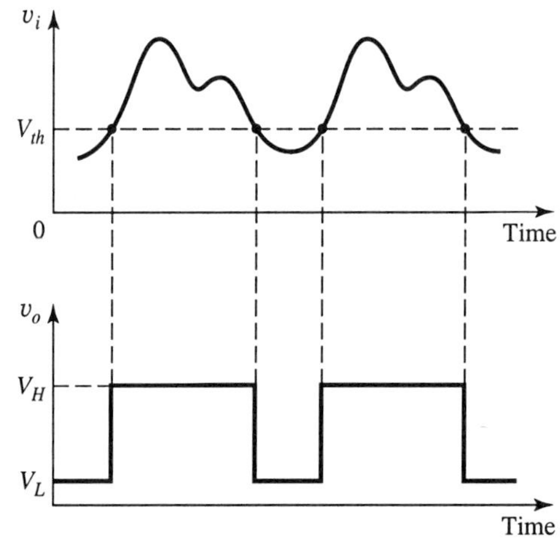


Figure 10.6: Comparator response.

previous peak. The output level of the circuit will then be the largest peak value of the input that has occurred from the last reset to the present time. By asserting the reset control, the capacitor can be discharged to zero and the detection process repeated.

### 10.2.2 The Comparator Circuit

The function of a comparator circuit is to compare the magnitudes of two analog input signals and generate a digital output signal that indicates which of the two inputs is the largest. When one of the inputs is at a constant voltage level, the circuit becomes a *threshold detector* that indicates when the input signal exceeds this constant threshold level. Figure 10.5 shows a block diagram of the comparator and the ideal transfer function for the device.

For the circuit shown, the threshold level, $V_{th}$, is applied to the inverting input and an input signal, $v_i$, is applied to the noninverting terminal. The output level will either be a high level, $V_H$, if $v_i$ exceeds $V_{th}$ or a low level, $V_L$, if $v_i$ is less than $V_{th}$. The low- and high-output levels can be defined as logic 0 and logic 1. The comparator can then compare the instantaneous value of $v_i$ to the threshold level and produce a logic 1 when $v_i$ is more positive than the threshold level. Figure 10.6 illustrates the output waveform resulting from an arbitrary input waveform applied to the input.

The comparator can be used as part of an ADC as will be discussed later. It can also be used as a pulse height discriminator, zero-crossing detector, or as a digital receiver with adjustable noise immunity. One useful application of the comparator is that of turning a nonrectangular waveform into a rectangular waveform. If the

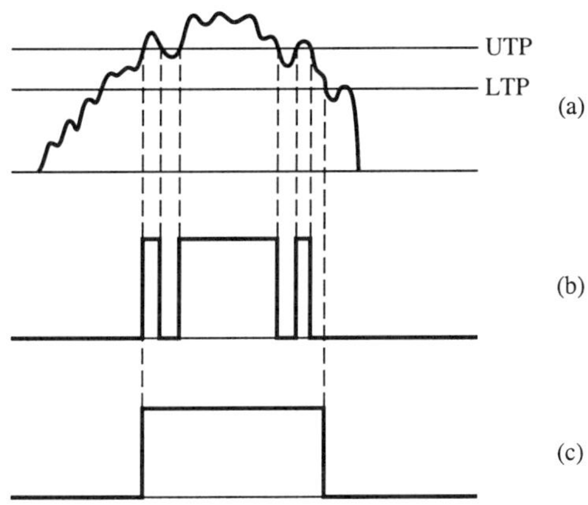


Figure 10.7: Effect of hysteresis: (a) input signal with noise; (b) output of comparator with no hysteresis; (c) output of comparator with hysteresis.

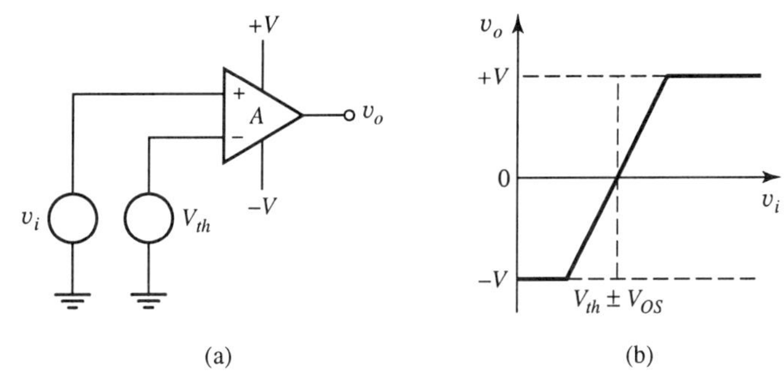


Figure 10.8: (a) Op amp comparator. (b) Transfer characteristics.

threshold is set to zero volts, the comparator makes a transition from negative to positive each time an axis crossing with a particular slope occurs. If an axis crossing with the opposite slope occurs, the comparator output switches from positive to negative.

Some situations require a squaring function or conversion to a rectangular waveform even in the presence of noise. This conversion can be accomplished by introducing *hysteresis* into the comparing operation. It follows that there must be two trip points of the comparator-an *upper trip point* (UTP) and a *lower trip point* (LTP). The output switches to the positive state as the input signal exceeds the UTP. After switching to this state, the trip point changes to the value given by the LTP. When the input drops below this new trip point, the comparator output changes to its original state and the trip point returns to the value given by the UTP. This feature discriminates against noise, as demonstrated in Fig. 10.7.

With no hysteresis, a comparator can produce *chatter* at the output as the single threshold level is crossed. The output rapidly switches between alternate states as shown in part (b) of the figure. With hysteresis, the chatter is eliminated as indicated in part (c). In order to eliminate chatter, the hysteresis or difference between the UTP and LTP should exceed the peak amplitude of the noise signal.

### Comparator Design

It is possible to construct a comparator using a general-purpose op amp as connected in Fig. 10.8. If equal positive and negative power supplies are used for the op amp, the output will swing between $+V_1$ and $-V_1$ where $V_1$ is about a volt less than the power supply voltage. With no feedback around the op amp, the voltage gain is very high. When the input voltage is just below the threshold voltage, the output sets at $-V_1$. As the input

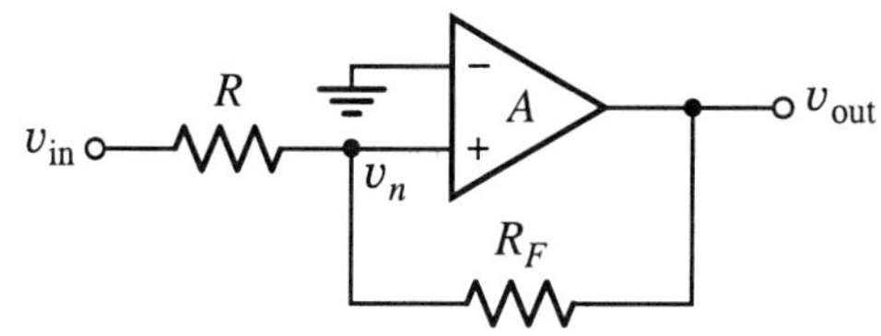


Figure 10.9: A comparator with hysteresis.

voltage increases slightly above the threshold voltage, the output is driven positive to $+V_1$. If the difference in voltage between the input and the threshold voltage is $\Delta V$, the output is driven toward $A\Delta V$ where $A$ is the op amp gain. For gains of 200,000 or $300,000$ V/V, the output would exceed $+V_1$ if not limited by the edge of the active region of the op amp. Thus, the transfer characteristics of Fig. 10.8(b) result. If $V_1 = 14$ V, the input voltage will need to exceed the threshold voltage by only

$$\Delta v_i = \frac{14}{200,000} = 70\ \mu\text{V}$$

to lead to an output of 14 V. A comparator circuit with hysteresis is shown in Fig. 10.9.

When the input voltage of the comparator in Fig. 10.9 is less than the LTP, the output will be at the low level of $-V_1$. The output will swing to $+V_1$ when the input exceeds the UTP. The two trip points can be calculated by finding the input voltages required to cause the noninverting node voltage, $v_n$, to equal zero volts. This voltage can be expressed as

$$v_n = v_{\text{in}} \frac{R_F}{R + R_F} + v_{\text{out}} \frac{R}{R + R_F} \tag{10.8}$$

The UTP is found by equating $v_{\text{out}}$ to $-V_1$, setting $v_n = 0$, and solving for $v_{\text{in}}$. This yields

$$UTP = \frac{R}{R_F} V_1 \tag{10.9}$$

When the UTP is exceeded, the output switches to $+V_1$ and remains there until $v_{\text{in}}$ drops below the LTP. This value is found from Eq. (10.8) by equating $v_{\text{out}}$ to $+V_1$, setting $v_n = 0$, and again solving for $v_{\text{in}}$. The LTP is

$$LTP = -\frac{R}{R_F} V_1 \tag{10.10}$$

If the inverting terminal is connected to a positive voltage, $V_r$, as shown in Fig. 10.10, both trip points can be made to be positive. The UTP is

$$UTP = \frac{R}{R_F} V_1 + \left(1 + \frac{R}{R_F}\right) V_r \tag{10.11}$$

The LTP is

$$LTP = -\frac{R}{R_F} V_1 + \left(1 + \frac{R}{R_F}\right) V_r \tag{10.12}$$

The voltage $V_r$ can be selected to set the LTP, whereas the ratio $R/R_F$ can be selected to set the difference between the UTP and the LTP.

# Example 10.2

Design a comparator with a UTP of 1.4 V and an LTP of 0.8 V. The op amp used has an output that swings

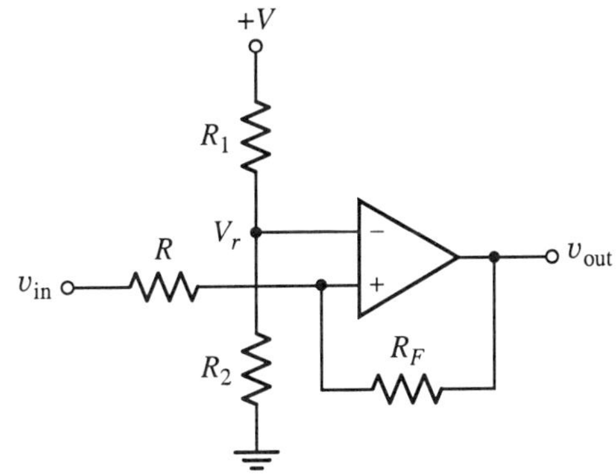


Figure 10.10: A comparator with positive trip points.

between +10-V and -10-V limits and $V = 12$ V.

**Solution** The difference between the desired trip points is $1.4 - 0.8 = 0.6$ V. From Eqs. (10.11) and (10.12) this difference is expressed as

$$UTP - LTP = 2V_1 \frac{R}{R_F}$$

For this circuit, this equation becomes

$$0.6 = 20 \frac{R}{R_F}$$

giving $R/R_F = 3 \times 10^{-2}$. If $R$ is selected to be 3 kΩ and $R_F$ to be 100 kΩ, a 0.6-V difference in trip points results. With these values, Eq. (10.12) can be used to find that

$$LTP = 0.8 = -\,0.3 + 1.03\ V_r$$

Solving for $V_r$ leads to a value of $V_r = 1.07$ V. Values of $R_1 = 20.4$ kΩ and $R_2 = 2$ kΩ set $V_r$ at this value.

---

**Practice Problem**

**10.3** If $R_2 = 2$ kΩ and $R_F = 100$ kΩ in Example 10.2, select values of $R$ and $R_1$ to lead to a UTP of 1.2 V and an LTP of 0.9 V. *Ans:* $R = 1.5$ kΩ, $R_1 = 17.4$ kΩ.

Whereas the op amp offers an accurate comparison of an input signal to a threshold voltage, it has at least two limitations that prevent its universal application. The op amp has a relatively slow large-signal response and is therefore limited to low- or medium-frequency uses. Generally, the output levels are not directly compatible with logic levels, but can be interfaced to convert to logic levels.

In IC designs, comparators are based on lower gain amplifiers that can operate at much higher speeds than comparators based on op amps. Two typical IC comparators are shown in Fig. 10.11.

The input stage of Fig. 10.11(a) is a buffered differential pair with a current mirror load. Devices $Q_7$ and $Q_8$ are common-emitter gain stages with an open-collector output. The MOS comparator is similar with a differential input stage and current mirror load followed by a gain stage ($M_5$) and an inverting amplifier ($M_6$ and $M_7$) as the output stage.

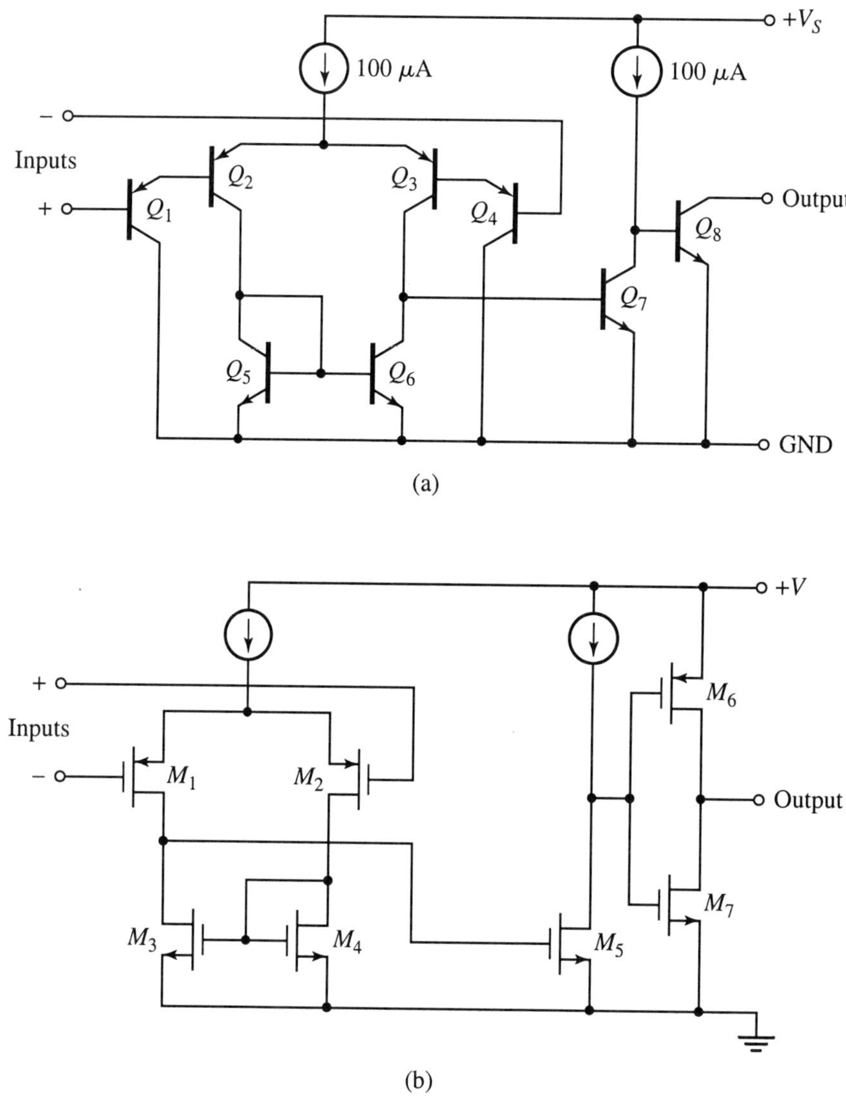


Figure 10.11: (a) BJT comparator with open-collector output. (b) MOS comparator.

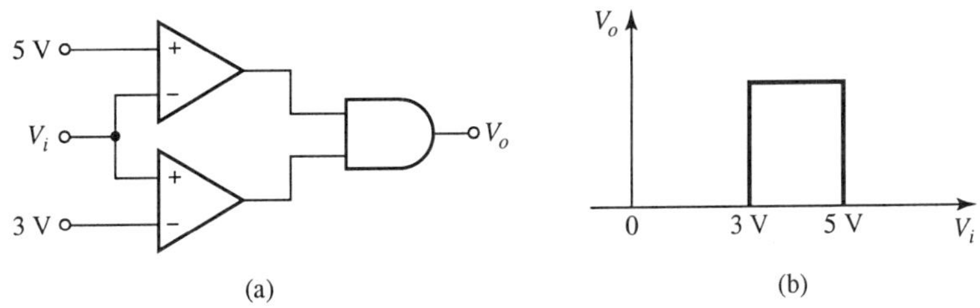


Figure 10.12: A window comparator. (a) Circuit. (b) Transfer characteristics.

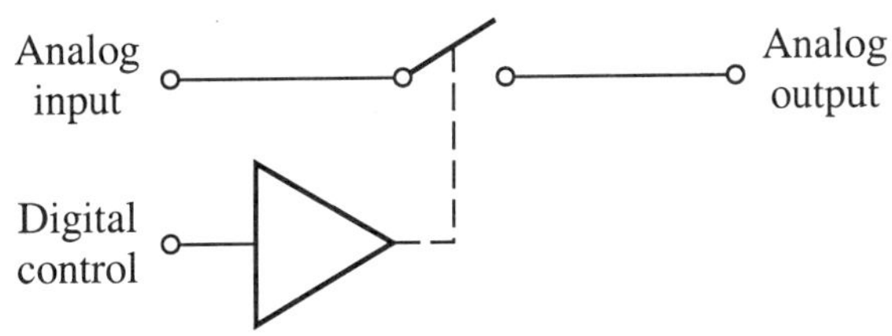


Figure 10.13: An ideal analog switch.

**The Window Comparator**

Some applications require a logic level that signals when an analog input voltage or current is within a certain range or *window*. For example, an automatic test system may be designed to accept all devices that exhibit an output between 3 and 5 V for specified drive conditions. Output levels lower than 3 or higher than 5 V are considered out of specification. Such a measurement can be done by a window comparator that is constructed from two standard comparators as illustrated in Fig. 10.12.

The lower comparator generates an output when the input exceeds 3 V and the upper comparator generates an output when the input is less than 5 V. Both comparator outputs must be high to assert the AND gate. The transfer characteristics show an output only when the input falls between 3 and 5 V.

## 10.2.3 The Analog Switch

A block diagram of an analog switch is shown in Fig. 10.13. This switch is controlled by a digital input signal that determines whether the switch is open or closed. If open, the input analog signal does not influence the output. When the switch is closed, the analog input signal becomes the output signal. One application of the analog switch is in the S/H circuit. Other applications include multiplexing, modulation, and switched-capacitor/resistor emulation. In many analog applications, nonideal elements of the analog switch, inherent to its implementation, become significant. This section will consider this nonideal behavior. A general model including nonideal parasitic elements is indicated in Fig. 10.14.

The switch can be represented by three nodes plus a ground node. Nodes $a$ and $b$ are the input and output nodes, and node $c$ is the control node. Parasitic capacitances between each pair of nodes are designated $C_{ab}$, $C_{ac}$, and $C_{bc}$. In addition, both input and output nodes have parasitic capacitance between these terminals and ground. These elements are designated $C_a$ and $C_b$. These capacitances not only affect the speed of operation of the switch, but indicate that *feedthrough* of the digital signal to the analog input or output nodes also takes place.

The resistive elements, $R_{\text{on}}$ and $R_{\text{off}}$, are included to model the resistance when the switch is closed and when it is open. Instead of an ideal short circuit, some small resistance, $R_{\text{on}}$, is present when the switch is on. The resistance $R_{\text{off}}$ is a large resistance that models the open resistance of the switch. Even though a well-designed switch may have a value of $R_{\text{on}}$ in the range of ohms and $R_{\text{off}}$ may be several megohms, they may have some

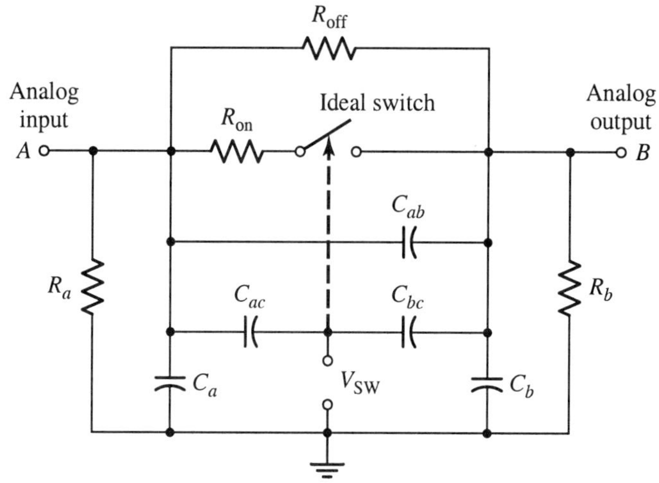


Figure 10.14: General analog switch model.

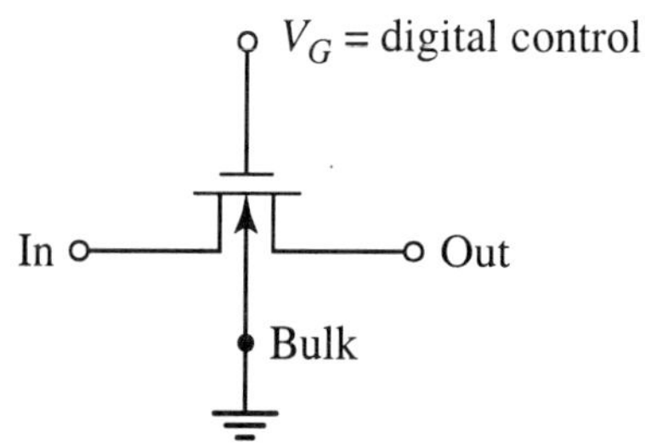


Figure 10.15: nMOS analog switch.

effect on performance. Resistors $R_a$ and $R_b$ represent conductive paths to ground or to the substrate of an IC.

MOSFET devices possess characteristics that make them more suitable than BJTs for the implementation of analog switches. MOSFET devices are inherently symmetrical between source and drain terminals and are capable of transmitting analog signals of positive and negative polarity. Bipolar devices suffer from junction offsets and are only capable of conducting current in a single direction. The MOSFET can be used in a straightforward manner to implement an analog switch, using the source and drain for the input and output nodes and the gate as the control node.

### MOSFET Analog Switches

For many applications, the enhancement-mode device, available in standard CMOS processes, is used for the analog switch. Either pMOS, nMOS, or the complementary switch may be fabricated, depending on the requirements of the switch. Figure 10.15 shows an nMOS device arranged as an analog switch.

When the switch is in the *on* state, the resistance between source and drain should be small, thus, the device must operate in the triode region. If drain current is small, the drain current is expressed as

$$I_D = \frac{\mu C_{\text{ox}} W}{L}(V_{GS} - V_T)V_{DS} \tag{10.13}$$

The resistance can be found from this equation from

$$R_{\text{on}} = \frac{1}{\frac{\partial I_D}{\partial V_{DS}}} = \frac{L}{\mu C_{\text{ox}} W (V_{GS} - V_T)} \tag{10.14}$$

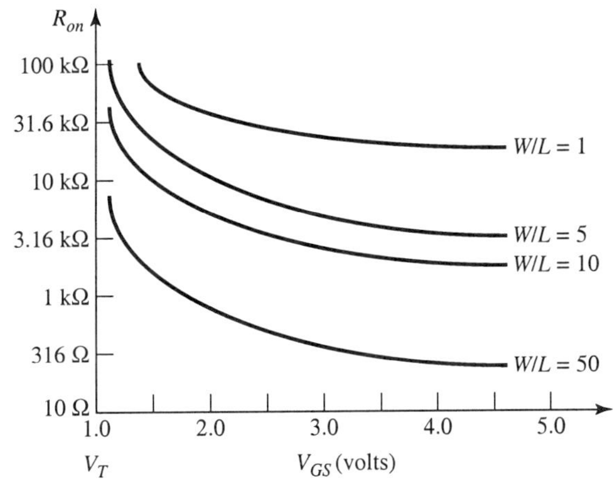


Figure 10.16: $R_{\text{on}}$ as a function of $V_{GS}$ for various scaling ratios.

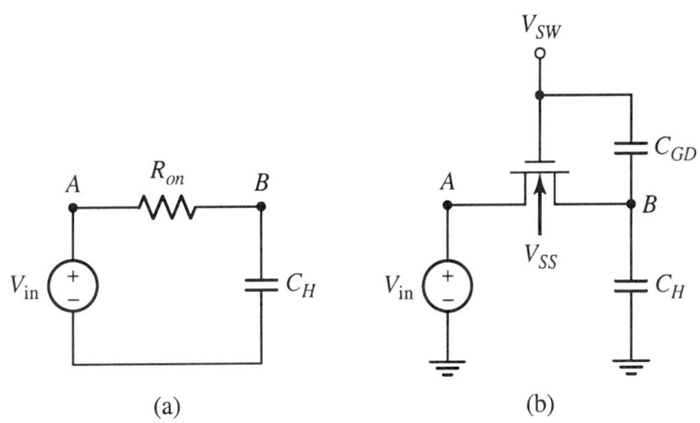


Figure 10.17: nMOS switch in the S/H circuit. (a) Sample mode model. (b) Dynamic switch model.

From Eq. (10.14) we see that the value of $R_{on}$ becomes smaller if the ratio $L/W$ decreases and if $V_{GS}$ is large compared to the threshold voltage. This information is plotted in Fig. 10.16.

The minimum length, $L$, will generally be fixed by the process, but $W$ can be varied to lower the value of $R_{\text{on}}$. Larger values of $W$, however, lead to more parasitic capacitance and can degrade the dynamic performance of the switch. Often a compromise must be made between good dynamic performance and good static characteristics. Of course, if $V_{GS}$ can be driven to a high *on* voltage, the value of $R_{\text{on}}$ will be decreased, but this is limited by available power supply voltages.

The parasitic capacitances of the switch in Fig. 10.15 can be easily identified with those of the general switch model of Fig. 10.14. The source-to-bulk and drain-to-bulk capacitances equate to $C_a$ and $C_b$. The gate-to-source and gate-to-drain capacitances equate to $C_{ac}$ and $C_{bc}$ and source-to-drain capacitance equates to $C_{ab}$.

The importance of low values for both $R_{\text{on}}$ and parasitic capacitance can be illustrated by considering the S/H circuit that uses the analog switch. A high value of $R_{\text{on}}$ slows the charging of the hold capacitor during acquisition of the input signal in the sample mode and limits the acquisition speed. A large value of parasitic capacitance between gate and source will create a large charge injection when the control terminal switches to the hold mode. These two effects can be visualized by means of the models in Fig. 10.17.

Obviously, the value of $R_{\text{on}}$ directly affects the time constant of acquisition as seen in part (a) of the figure. Charge injection errors can be calculated from the model of Fig. 10.17(b). If the feedthrough capacitance consists

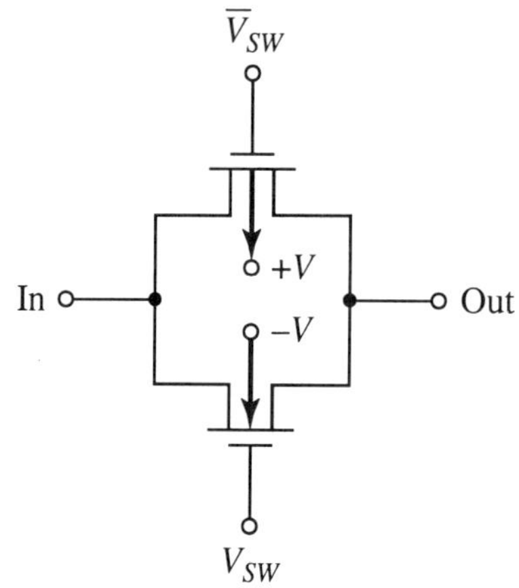


Figure 10.18: A CMOS switch.

primarily of gate-to-drain capacitance, the size of the injected error, $\Delta V$, can be found as

$$\Delta V = \frac{C_{gd}}{(C_H + C_{gd})} \Delta V_{SW} \tag{10.15}$$

where $\Delta V_{SW}$ is the voltage swing of the control voltage at the gate of the switch.

It is possible to improve the acquisition time and decrease the charge injection error by using a CMOS switch, as shown in Fig. 10.18.

The gates of the two devices are driven in opposite directions to control the state of the switch. The CMOS switch has a greater range of analog input signals than the nMOS switch over which $R_{\text{on}}$ remains low. As the analog input signal swings in either direction from ground level, the gate-to-source voltage of one of the devices increases. From Eq. (10.14), the value of $R_{\text{on}}$ is lowered as $V_{GS}$ increases. In addition to this improvement in $R_{\text{on}}$, the charge injection problem is also minimized by the CMOS configuration. Since the nMOS and pMOS devices require opposite polarity control signals applied to the gates, the charge injection feedthrough is diminished by cancellation. As one device removes charge from the holding capacitor, the other device replaces charge. Although the CMOS switch needs additional control circuitry to generate both polarities of control voltage, this is the preferred configuration in most applications.

### BJT Analog Switches

A BJT circuit used for several years as an analog switch is illustrated in Fig. 10.19. This circuit is referred to as a diode bridge switch.

The switch is *on* when the control signals $V_{SW1}$ and $V_{SW2}$ are high. Devices $Q_2$ and $Q_4$ are then conducting and the diodes are forward biased as indicated in Fig. 10.20(a).

Each diode has a forward current of $I/2$ through it and a forward voltage drop near 0.7 V. The output voltage can be written as

$$v_o = v_i + V_{d1} - V_{d2} \tag{10.16}$$

Since equal currents flow through each diode, the voltages $V_{d1}$ and $V_{d2}$ can be matched very precisely so that the output voltage equals the input voltage.

When the control voltages switch to their low levels, devices $Q_2$ and $Q_4$ go to the nonconducting state and the switch is *off*. The diode circuit appears as in Fig. 10.20(b). With no current to drive the diodes, they are in the nonconducting state and the output terminal is isolated from the input voltage. Although both control

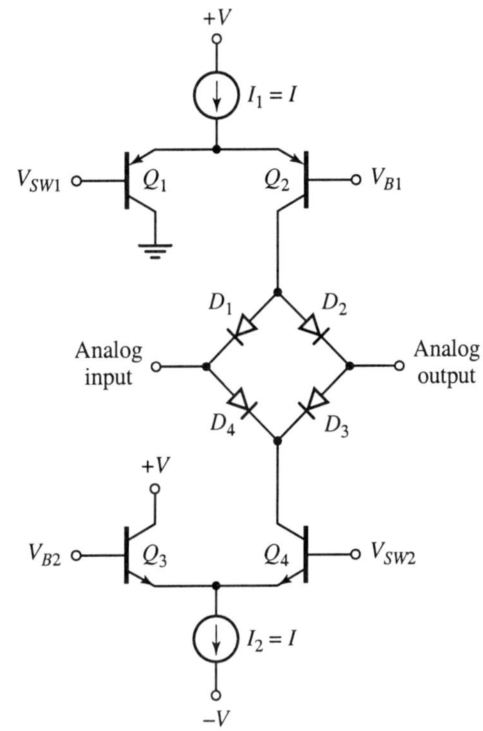


Figure 10.19: Diode bridge switch.

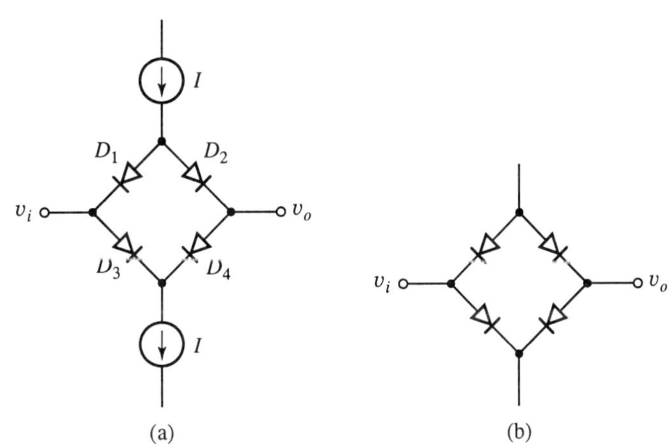


Figure 10.20: (a) Diode bridge in *on* state. (b) Diode bridge in *off* state.

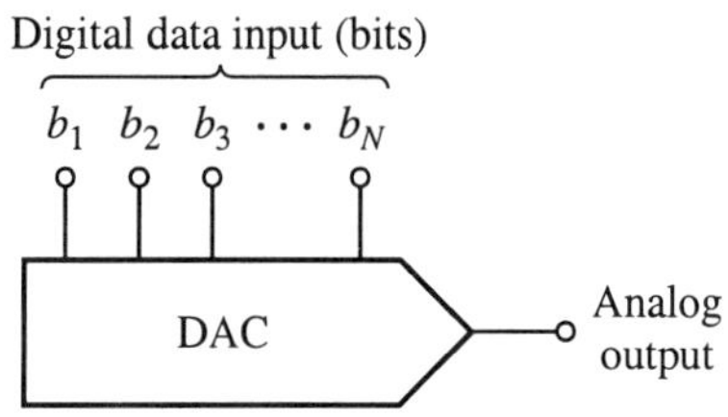


Figure 10.21: The DAC.

voltages, $V_{SW1}$ and $V_{SW2}$, switch in the same directions to control the switch, they are not equal in value. Since $Q_2$ must remain in its active region as $v_i$ moves to its highest level, the voltage $V_{B1}$ is set as close to the power supply voltage $+V$ as possible. For the same reason, $V_{B2}$ is chosen to be near the negative supply voltage, $-V$. The two control voltages move in the same direction to control the switch, but there must be a dc level shift between the two signals. For carefully designed IC switches, the offsets between input and output voltages are approximately 1 mV and the charge injection error may be around 5 pC.

## Example 10.3

The diode bridge of Fig. 10.19 is driven by current sources that supply 5 mA of current when the switch is on. The input can vary from -5 V to +5 V. What capacitance is needed for this switch to have an acquisition time of 50 ns?

**Solution** The longest charge time of the capacitor will occur when the input changes from -5 V to +5 V between samples. This gives a voltage change on the capacitor of $\Delta v_c = 10$ V. The charge on the capacitor will be supplied by a constant current of 5 mA. Using the equation for voltage change of a capacitor with constant current charge gives

$$\Delta v_c = \frac{1}{C} \times I\Delta t$$

We solve for $C$ to find that

$$C \leq \frac{I\Delta t}{\Delta v_c} = \frac{5 \times 10^{-3} \times 50 \times 10^{-9}}{10} = 25 \text{ pF}$$

Although this value is small, it is appropriate for integrated circuit construction.

## 10.3 THE DIGITAL-TO-ANALOG CONVERTER

**Important Concepts**

1. The DAC converts a digital input code or number to a corresponding analog output value.
2. Quantization errors are inherent in the operation of a DAC. Other errors result from nonlinearities and offsets.
3. Three types of DAC are classified as current-scaling, voltage-scaling, or charge-scaling circuits.

Figure 10.21 shows a block diagram of a digital-to-analog converter, a D/A converter, or a DAC.

The input to the DAC is a digital word of a prescribed length or number of bits, $N$, and the output is an analog signal related uniquely to the input word. In a typical situation, a series of $N$-bit digital words is

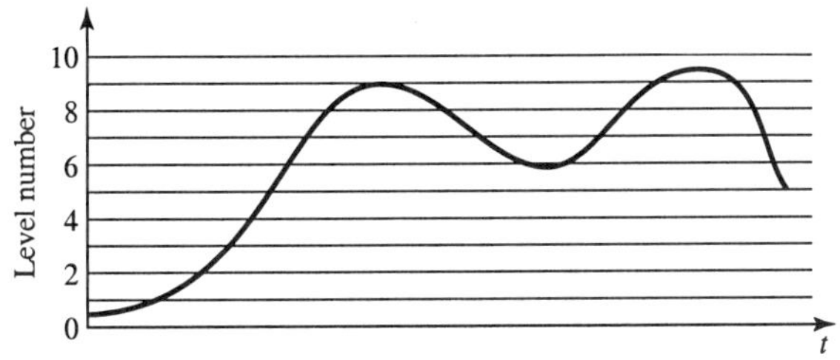


Figure 10.22: An analog signal along with quantization levels.

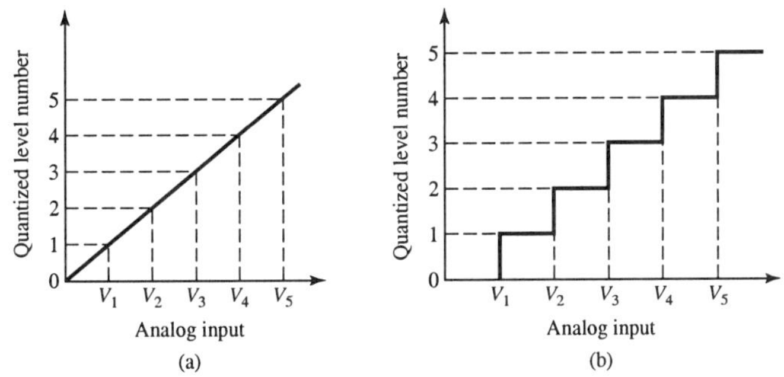


Figure 10.23: (a) Ideal analog output. (b) Actual output of DAC.

presented to the digital inputs at intervals of time and the output generates voltage levels corresponding to the input codes at each time interval.

Several sources of error in a DAC limit its overall accuracy of conversion. The major errors are *quantization errors*, converter nonlinearities, errors due to finite switching times, and offset errors.

### Quantization Errors

A finite-length digital code can represent only a finite number of levels. On the other hand, an analog signal has an infinite number of values. When an analog signal is to be generated by a digital code, the number of discrete levels produced is limited. The division of an analog range into a number of discrete levels is called *quantization*. Figure 10.22 shows an analog voltage quantized into equal-width intervals. The DAC can only generate output signal levels corresponding to the horizontal lines. Figure 10.23 helps to define this problem for a DAC that generates an output voltage.

Part (a) of the figure shows the ideal output of the converter and part (b) shows the actual output of a converter. As the ideal signal varies from 0 to $V_1$, the DAC generates a signal of zero volts. For an ideal output between $V_1$ and $V_2$, the signal moves to level 1 to generate an output voltage of $V_1$. The output of the DAC moves to level 2 with a voltage of $V_2$ as the ideal voltage reaches and exceeds $V_2$. The width of these levels is usually chosen to be equal and is designated as the quantum interval, $q$. This width is also called the least-significant bit or $LSB$.

The quantization error is calculated as the difference between the ideal analog signal and the actual converter output. This error is shown in Fig. 10.24 and varies with the value of ideal signal. The greatest possible quantization error is 1 $LSB$.

Additional errors are introduced by the conversion circuit. Slightly different values of quantum interval at different values of output signal lead to nonlinear operation. Finite switching times may introduce errors as high conversion speeds are approached. Parameters that characterize these effects are considered in the following subsection.

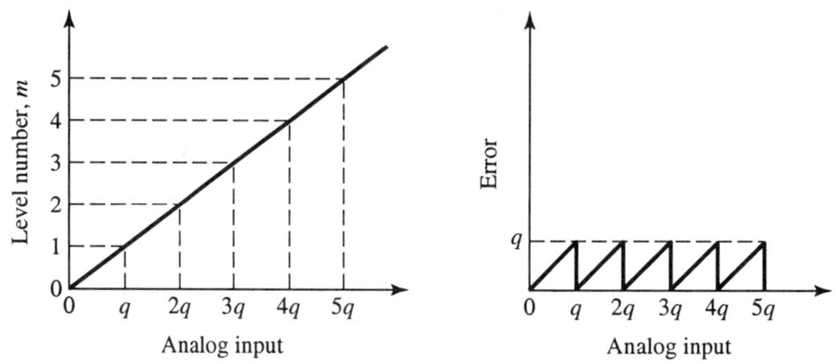


Figure 10.24: Quantization error.

### 10.3.1 Performance Parameters of the DAC

Several parameters influence the selection of a DAC in a particular application.

**Resolution:** The number of bits the DAC can convert is an important consideration. The *resolution* of a DAC relates to the number of bits. The resolution is the smallest increment of output signal that the DAC can produce. An 8-bit DAC is said to have a resolution of 8 bits or 1 part in $2^8$. In percent, this is $1/256 \times 100 = 0.391\%$. Although the resolution specification generally corresponds to the total number of bits, this may not be true for larger DACs. A 14-bit DAC may only have 12-bit resolution. This implies that a change in the thirteenth or fourteenth bit of the input code may result in an output change that is less than the change due to nonlinearity or temperature drift.

**Linearity or Accuracy:** This parameter is sometimes referred to as nonlinearity or integral nonlinearity (INL). It is the maximum deviation of the DAC output from the ideal straight line drawn between the zero-scale and full-scale outputs. This parameter may be given in percent or in terms of the fraction of an $LSB$. An 8-bit DAC with $\pm 1/2\ LSB$ linearity corresponds to $\pm 0.195\%$ linearity.

**Monotonicity:** A DAC is said to be *monotonic* if each digital code increment produces an output that is equal to or larger than the preceding code output.

**Settling Time:** The *settling time* is measured from the instant of a digital input code change to the time that the analog output reaches its corresponding new value to within a specified error. Often the error is $\pm 1/2\ LSB$. Worst-case settling time is measured between zero-scale and full-scale codes.

**Differential Nonlinearity:** The maximum deviation from the ideal difference between any two adjacent outputs is the *differential nonlinearity* or DNL of the DAC. It is expressed as a percent of full scale or in terms of an LSB value. Table 10.3 shows a set of specifications for an IC DAC.

**Table 10.3 Eight-Bit DAC Specifications**

| Parameter | Specification |
|---|---|
| Resolution | 8 bits |
| Differential nonlinearity | $\pm 1/2\ LSB$ |
| Linearity (% of full scale) | $\pm 0.1\%$ (max.) |
| Settling time to $\pm 1/2\ LSB$ | 135 ns (max.) |
| Full-scale current (nominal) | 2 mA |
| Temp. coef. of $I_{FS}$ with ext. ref. | 10 ppm/$^\circ C$ |
| Output impedance | $\geq 20$ MΩ |
| Power dissipation | 174 mW (max.) |
| Technology used | Bipolar with diffused resistors |

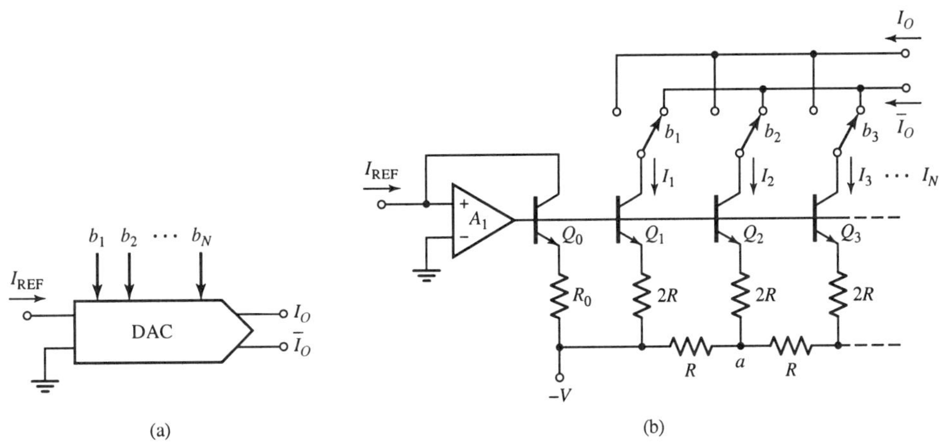


Figure 10.25: Current-scaling DAC. (a) Block diagram. (b) Schematic.

There are three popular types of converters used in IC design:

1. Current-scaling circuits
2. Voltage-scaling circuits
3. Charge-scaling circuits

In current-scaling circuits, a reference voltage is converted into binary weighted currents. The input code determines the appropriate weighted currents to sum together to either form an output current or convert this current to an output voltage. The BJT DAC often uses current scaling.

The voltage-scaling circuit divides the reference voltage into $2^N$ voltages and, depending on the input code, connects the proper voltage to the output. The MOSFET DAC often uses voltage scaling.

Charge-scaling circuits use capacitors to divide the charge appropriately. Low-power MOSFET DACs often apply charge scaling.

### 10.3.2 The Current-Scaling DAC

Figure 10.25 shows the heart of a current-scaling, $N$-bit DAC. The reference current may be created by a reference voltage and the output current may be converted to a voltage output.

The *reference amplifier* consists of an op amp and transistor $Q_0$. The inversion of the signal from base to collector on transistor $Q_0$ creates a high-gain, inverting amplifier resulting in a virtual ground at the noninverting input of the op amp. A reference voltage in series with a resistance connected to the noninverting terminal of the op amp can create $I_{\text{REF}}$. The collector current of $Q_0$ equals $I_{\text{REF}}$ and the emitter current is $I_{\text{REF}}/\alpha$. The voltage at the base of $Q_0$ equals the base voltage of all other transistors and the emitter voltages of all transistors are approximately equal as well at $V_E = V_B - 0.7$ V.

The emitter current of $Q_1$ is given by

$$I_{E1} = \frac{V_E + V}{2R} \tag{10.17}$$

Each of the emitter currents of $Q_2$, $Q_3$, $\cdots$ $Q_N$ are made to be one-half that of the adjacent, lower-numbered device. The sum of the currents $I_{E2} + I_{E3} + \cdots + I_{EN}$ must equal the current $I_{E1}$ for large values of $N$. This follows from the fact that the series

$$I_{E1}\left[\frac{1}{2} + \frac{1}{4} + \frac{1}{8} + \cdots + \frac{1}{2^{N-1}}\right] \rightarrow I_{E1}$$

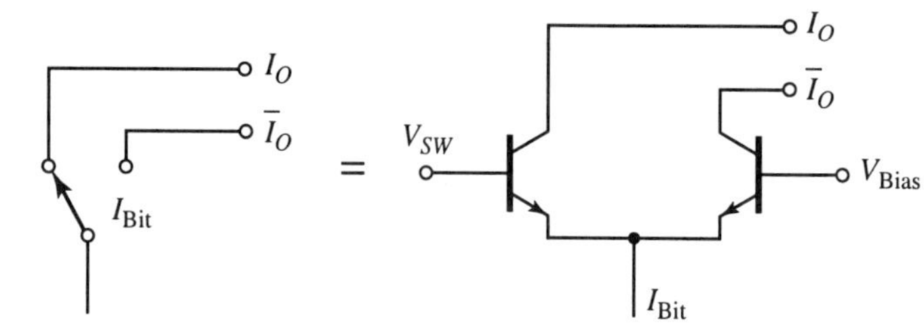


Figure 10.26: A differential switch.

as

$$N \rightarrow \infty$$

The sum of all emitter currents to the right of $Q_1$ flows through the resistance $R$. The voltage at point $a$ is found using Ohm's law. The current $I_{E2}$ can be found as

$$I_{E2} = \frac{V_E + V - RI_{E1}}{2R} = I_{E1} - \frac{I_{E1}}{2} = \frac{I_{E1}}{2} \tag{10.18}$$

In a similar manner, it can be shown that each successive transistor has an emitter current that is one-half that of the adjacent, lower-numbered transistor as a result of the $R - 2R$ ladder. The collector current of $Q_1$ can be made equal to or precisely related to, $I_{\text{REF}}$ by proper choice of $R$ and the emitter area scaling of the device.

The circuit of Fig. 10.25(b) provides an output current, $I_O$, and its complement, $\overline{I}_O$, where

$$I_O + \overline{I}_O = (2^N - 1)I_{LSB} \approx 2^N I_{LSB} \tag{10.19}$$

A differential switch such as that illustrated in Fig. 10.26 can be used to supply the true and complementary value of current.

Although several types of coding are used in converters, the most common is binary coding, as assumed in the circuit of Fig. 10.25. For this converter, a nominal full-scale current, $I_{FS}$, can be defined as

$$I_{FS} = KI_{\text{REF}} \tag{10.20}$$

where $K$ is a scaling factor determined by $R$ and the emitter area scaling. The current $I_1$, designated the most-significant bit (MSB) current, is chosen to be one-half the nominal full-scale current. Thus $I_1$ can be written as

$$I_1 = \frac{KI_{\text{REF}}}{2} = \frac{I_{FS}}{2} \tag{10.21}$$

The output current can be expressed as

$$I_O = I_{FS}\left[\frac{b_1}{2^1} + \frac{b_2}{2^2} + \cdots + \frac{b_N}{2^N}\right] \tag{10.22}$$

or

$$I_O = I_{FS}\sum_{i=1}^{N}\frac{B_i}{2^i} \tag{10.23}$$

An alternative expression for Eq. (10.22) using the decimal equivalent of the input binary code, $n_{10}$, is

$$I_O = I_{FS}\frac{n_{10}}{2^N} \tag{10.24}$$

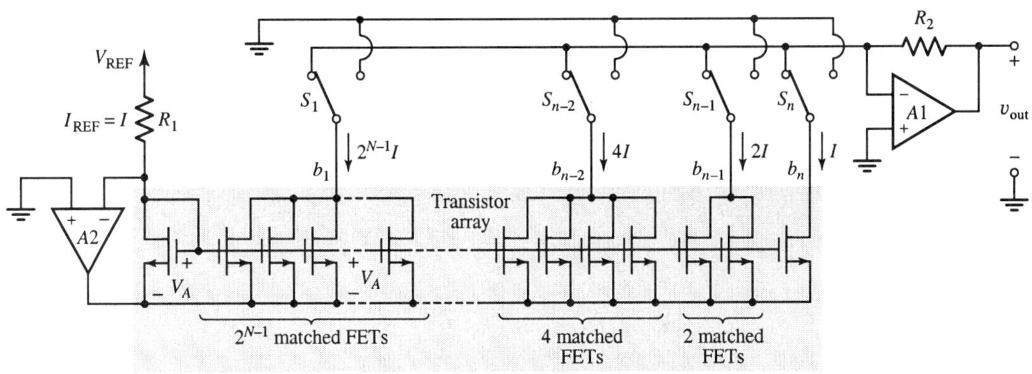


Figure 10.27: A current-scaling MOSFET DAC.

A full-scale input code produces an output current that is 1 $LSB$ less than the nominal full-scale current, $I_{FS}$. If $N$ is large enough to result in

$$2^N - 1 \approx 2^N$$

the actual output current for a full-scale input code will be very close to the nominal value.

# Example 10.4

An 8-bit binary-coded DAC with a nominal full-scale current of 2 mA has an input code of [1011 0000]. Calculate the output current, $I_O$.

**Solution** From Eq. (10.22), we write

$$I_O = 2\left[\frac{1}{2} + 0 + \frac{1}{8} + \frac{1}{16} + 0 + 0 + 0 + 0\right] = 1.375 \text{ mA}$$

If the input code is converted to a decimal number, this calculation can be made using Eq. (10.24). The input code has a decimal equivalent of 176 giving an output current of

$$I_O = \frac{2 \times 176}{256} = 1.375\, mA$$

A MOSFET op amp is shown in Fig. 10.27.

Each group of MOSFET devices could be replaced by properly scaled single devices on the chip. The reference current is established by the reference voltage and the resistance connected to the virtual ground of the input amplifier. Starting from the right side of the current-weighting network, the first group puts out a current equal to the LSB value. Moving to the left, each group puts out twice the current of its adjacent group to the right. The current is then summed and converted to a voltage by the output amplifier. If the two resistors shown in the circuit are properly chosen, the output voltage can be expressed as

$$V_O = V_{\text{REF}}\left[\frac{b_1}{2} + \frac{b_2}{4} + \cdots + \frac{b_n}{2^N}\right] \tag{10.25}$$

### 10.3.3 The Voltage-Output DAC

Figure 10.25 shows a current-output DAC and Fig. 10.27 illustrates a voltage-output DAC that converts the output current to voltage. A current-scaling DAC can convert the output current to voltage by the method

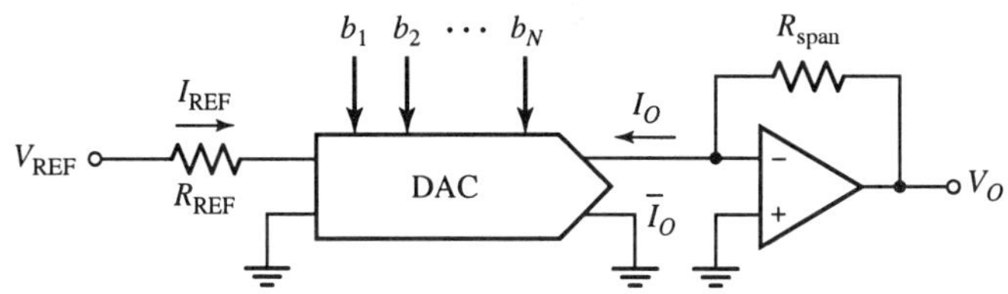


Figure 10.28: A voltage-output DAC.

depicted in Fig. 10.28.

The output current of the DAC is channeled through a *span resistor*, designated $R_{span}$, by the op amp. This develops an output voltage of

$$V_O = I_O R_{span} \tag{10.26}$$

The output current for the circuit of Fig. 10.25 can be expressed as

$$I_O = \frac{n_{10} K V_{REF}}{R_{REF} 2^N} \tag{10.27}$$

where $K$ is determined by the reference amplifier and emitter area scaling. Often $K$ is chosen to be

$$K = \frac{2R_0}{R} \tag{10.28}$$

leading to an output voltage for the circuit of Fig. 10.28 of

$$V_O = 2V_{REF}\left(\frac{R_0}{R}\right)\left(\frac{R_{span}}{R_{REF}}\right)\left(\frac{n_{10}}{2^N}\right) \tag{10.29}$$

**Practice Problems**

**10.4** A nominal full-scale voltage of 10 V should be produced in the circuit of Fig. 10.28. If the LSB current for the 8-bit DAC is 4 $\mu$A, choose the value of $R_{span}$. *Ans:* 9.766 kΩ.

**10.5** For the DAC of Fig. 10.28, $R_0 = R$ and $R_{span} = R_{REF}$. If $V_{REF} = 5$ V, calculate the *LSB* output voltage. *Ans:* 39.1 mV.

### 10.3.4 Ratiometric Design

Equation (10.29) demonstrates a relationship that depends on the ratios of two pairs of resistors. One ratio is $R_0/R$ and the other is $R_{span}/R_{REF}$. When an important output parameter depends on ratios of resistors rather than absolute values, it is referred to as a *ratiometric design.* In IC design, it is difficult to control the absolute values of resistors, but the ratio of resistors can be controlled relatively accurately. If $R_{span}$ and $R_{REF}$ need to have nominal values of 5 kΩ, they can be placed side by side on the IC layout and can be expected to vary identically with process and photolithographic variations. If both resistors are 10% high, the ratio of the two will still be unity.

Resistor ratios with values other than unity can be constructed from identically sized resistor *links.* If nominal values are $R_{span} = 5$ kΩ and $R_{REF} = 10$ kΩ, then a 5–kΩ link construction using two links would be appropriate, as shown in Fig. 10.29.

If resistors are intended to match with a specified ratio, it is generally expected that this matching will be maintained over the operating temperature range. It follows that the resistors should have a layout that

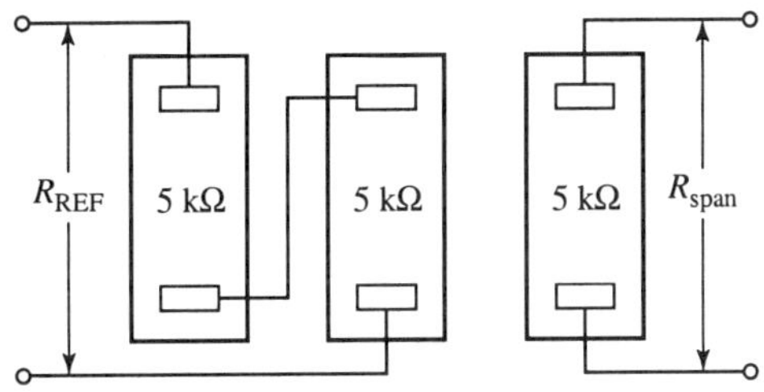


Figure 10.29: Identical link construction of ratio-matched resistors.

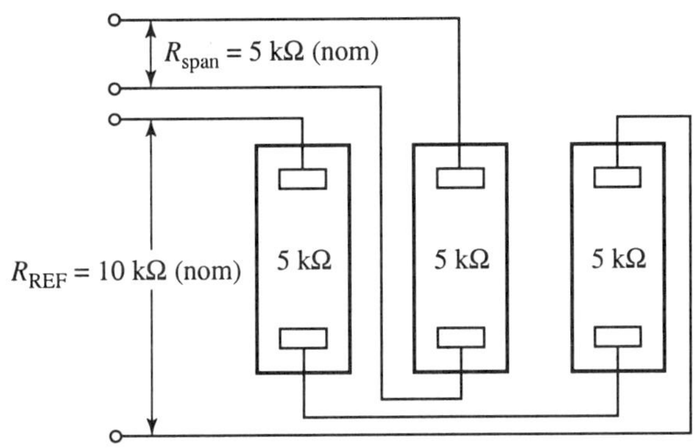


Figure 10.30: Placement of resistor links to optimize matching and tracking ratios.

experiences similar temperature gradients. In IC layouts, the resistors are intended not only to match, but to track over temperature. One way to ensure this is to place them in close proximity on the chip and surround them with similar environments. This sometimes leads to the requirement of *dummy* resistors with no electrical connection to surround the actual resistors with identical environments. Placement is important in the DAC of Fig. 10.28 because the self-heating of $R_{span}$ varies as the output changes with varying input codes, but $R_{REF}$ experiences a constant current flow. A layout that would optimize temperature tracking is illustrated in Fig. 10.30.

Placing the links as indicated in Fig. 10.30 causes both resistors to experience similar temperature gradients that may arise from self-heating of $R_{span}$.

Although the matching and tracking requirements are discussed in connection with a DAC, the example can be generalized to include any system application that demands precision. Furthermore, a system designer should avoid the requirement of matching between external components and those on an IC chip.

## 10.4 THE ANALOG-TO-DIGITAL CONVERTER

**Important Concepts**

**1.** The analog-to-digital converter (ADC) accepts an analog input signal and produces a digital output code that corresponds to some characteristic of the input signal. Often this characteristic is the magnitude of the input signal at the time of the conversion.

**2.** The range or span of the input signal is typically divided into some number of equal quantum intervals. These levels are numbered consecutively. The output of the ADC is a coded equivalent of this level number.

**3.** Among the several types of ADC are the parallel or flash converter, the successive approximation converter, and the ramp converter.

The function of an ADC is to measure an analog input signal and provide a quantized representation of the signal in the form of a digital output code. The DAC is considered a decoding device whereas the ADC can be considered an encoding device that translates between the analog signal and the digital word. Figure 10.31 shows a block diagram that represents an ADC.

In addition to the analog input signal to be converted, $V_i$, a precision dc voltage reference, $V_{\mathrm{REF}}$ is applied to the ADC. The output is an $N$-bit digital word for each conversion. This output may be generated as a serial bit stream or as a parallel word as shown in Fig. 10.31. The bit $b_1$ represents the $MSB$ and bit $b_N$ represents the $LSB$ of the output word. An input control signal is often provided to initiate the conversion and an output signal indicates when the conversion is completed. These optional signals are shown as dashed lines in Fig. 10.31.

If the largest input voltage to be converted is $V_{FS}$, the input can be quantized into intervals of width

$$q = \frac{V_{FS}}{2^N - 1} \approx \frac{V_{FS}}{2^N} \tag{10.30}$$

for large $N$. The quantum levels can be numbered from 0 to $n_{10}(\max)$ where

$$n_{10}(\max) = 2^N - 1$$

If $n_{10}$ represents the number of the highest quantum level that does not exceed the input when a conversion is made, the output of the ADC should represent this number, that is,

$$n_{10} = \frac{V_i}{q} = \frac{V_i}{V_{FS}/2^N} = \frac{V_i 2^N}{V_{FS}} \tag{10.31}$$

A binary-coded ADC will generate an output binary code that corresponds to the number $n$. The full-scale voltage of the ADC is related to the reference voltage by

$$V_{FS} = kV_{\mathrm{REF}} \tag{10.32}$$

where $k$ is a constant scale factor. In terms of the bits of the binary output, the ADC must satisfy

$$\left[\frac{b_1}{2^1} + \frac{b_2}{2^2} + \cdots + \frac{b_N}{2^N}\right] = \frac{V_i}{kV_{\mathrm{REF}}} \tag{10.33}$$

where the bits will have values of either 0 or 1.

### 10.4.1 Types of ADC

Various methods of converting an analog signal to a digital code can be applied. The following list indicates several possible types of ADC.

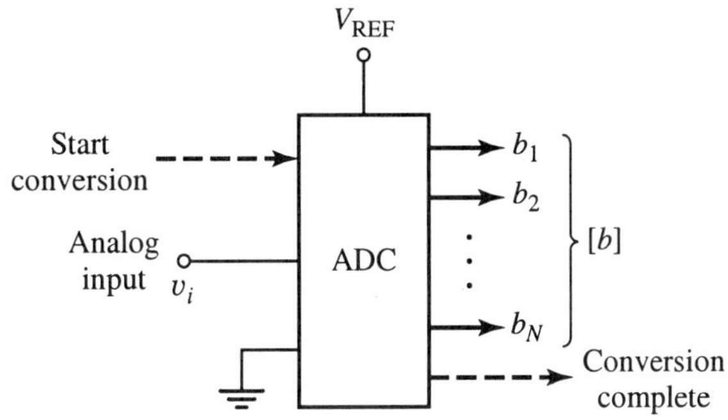


Figure 10.31: Block diagram of an ADC.

1. Parallel or flash converter
2. Successive approximation converter
3. Iterative converter
4. Staircase or counting converter
5. Tracking converter
6. Integrating or dual slope converter

The following material will consider the staircase converter, the successive approximation converter, and the parallel converter. The parallel converter is the fastest ADC, but often requires the most circuitry and the most precise components. The successive approximation ADC is simpler and slower than the parallel ADC. The staircase converter is very slow, but requires simple circuits for implementation.

### The Staircase Converter

A simplified block diagram for the staircase or counting ADC is shown in Fig. 10.32.

In addition to the dc reference voltage, $V_{\mathrm{REF}}$, a square-wave clock signal of proper frequency must also be provided. The Start Conversion control signal resets the counter and sets the $D$ latch. This latch drives the S/H circuit to the hold mode to capture the value of input signal existing at this time. This value is designated $V_A$. Setting the latch also opens the clock gate to drive the input of the binary counter. This counter, previously reset to a count of zero, now begins counting in a binary sequence. The binary output of the counter drives a DAC, generating a staircase increase of output current $I_O$. The voltage $V_x$ will initially be positive when the conversion of a positive input voltage begins. As the counter output increases, $I_O$ increases and $V_x$ can be expressed as

$$V_x = V_A - RI_O \tag{10.34}$$

When the value of $RI_O$ slightly exceeds $V_A$, the comparator input voltage becomes negative and the comparator output moves to the low state. This will reset the latch and cause the conversion complete line to go high and the clock gate to close. The binary number stored in the counter at this time corresponds to the conversion of the input voltage at the beginning of the conversion cycle.

At the end of conversion, the counter holds the binary code that makes

$$RI_O \cong V_A \tag{10.35}$$

We found earlier, in connection with the DAC, that the output current can be expressed in terms of Eq. (10.27) as

$$I_O = \frac{n_{10} K V_{\mathrm{REF}}}{R_{\mathrm{REF}} 2^N}$$

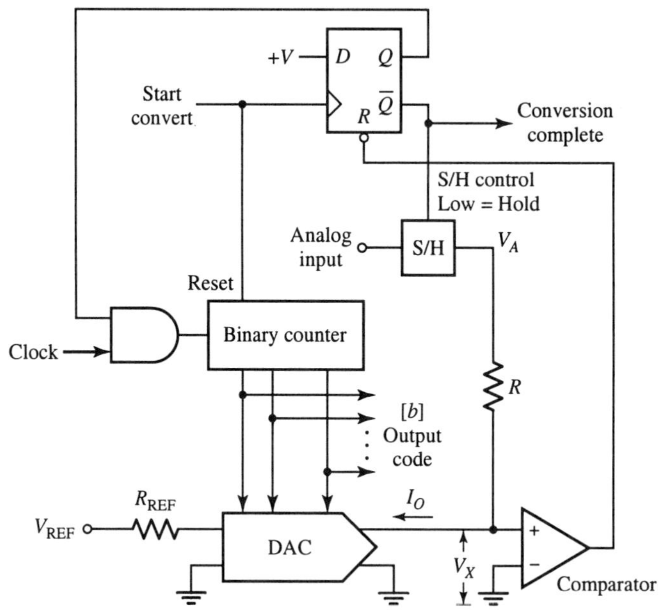


Figure 10.32: The staircase ADC.

where $n_{10}$ is the decimal equivalent of the input code to the DAC. Substituting this expression into Eq. (10.35) leads to

$$R\frac{n_{10}KV_{\text{REF}}}{R_{\text{REF}}2^N} \cong V_A \tag{10.36}$$

If $K$ is assumed to be unity and $R$ is chosen to equal $R_{\text{REF}}$, this equation tells us that

$$n_{10} \cong \frac{V_A 2^N}{V_{\text{REF}}} \tag{10.37}$$

From Eq. (10.31), this result is seen to be the correct output for the ADC.

The counter of the staircase converter is always reset to zero as the conversion begins. For large input voltages, this counter must count consecutively to some high value to complete the conversion. The conversion time is variable, depending on the value of the input signal, and can be quite long for higher input voltages.

The tracking converter represents a variation of the staircase converter that has shorter conversion times. A block diagram of the tracking converter is shown in Fig. 10.33.

The simple binary counter of the staircase converter is replaced by an *up/down counter* in the tracking ADC. This counter is directed to count up when the DAC output is less than the input analog signal. The counter counts down when the DAC output is greater than the input signal. Additional control circuitry determines when to start the conversion and the conversion is complete when the comparator output changes state. In the tracking converter, the counter is not reset between conversions. For large input signals with small changes between conversions, the counter requires little time to reach the converted value.

### The Successive Approximation ADC

The successive approximation ADC is relatively fast and has a fixed conversion time. It is based on an intelligent trial-and-error method. If a person must guess some number that falls within a given range and has feedback to know if each guess is high or low, each guess should split the known range. Given a range of 0 to 100, the first guess is 50. If this guess is higher than the number, the new range is known to be 0 to 50. The second guess should be 25. If this guess is low, the new range is known to be 25 to 50. The next guess should

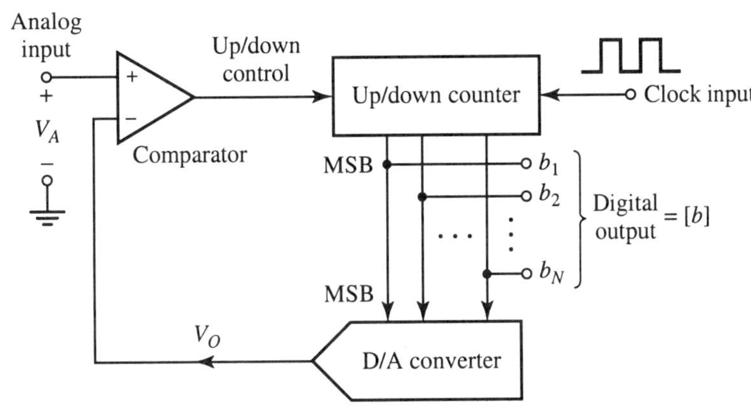


Figure 10.33: Tracking ADC.

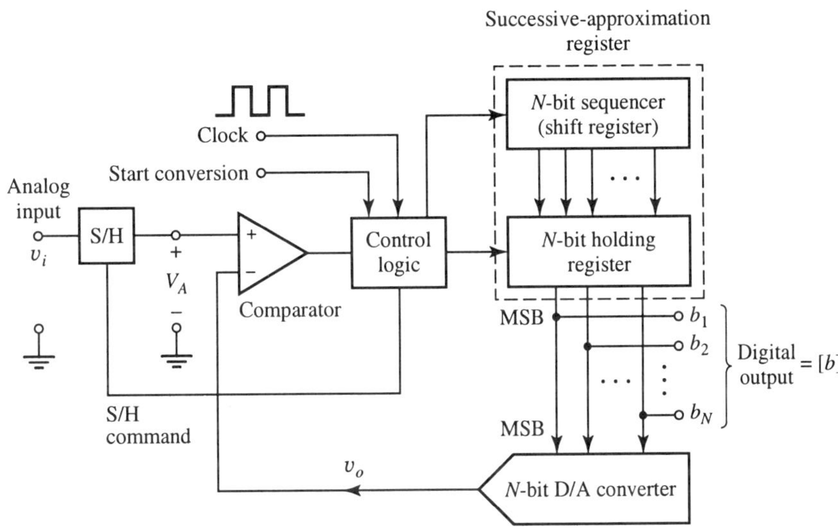


Figure 10.34: A successive approximation ADC.

be 37 or 38. By successively limiting the range by a factor of approximately one-half, the final answer is reached in a reasonably small number of guesses. This scheme is applied in the successive approximation ADC shown in Fig. 10.34.

For an $N$-bit word, the successive approximation ADC completes a conversion in $N$ clock cycles, whereas the staircase converter may require as many as 256 cycles. When the Start Conversion input is asserted, the control logic instructs the sequencer to place a 1 in the MSB position of the holding register. The DAC converts the digital guess of the holding register to an analog value and compares it to the sampled value of the input, $V_A$. The quantum interval of the DAC must equal that of the ADC for proper operation. Since the largest digital code that the holding register can contain is $2^N - 1 \approx 2^N$, the first guess should be one-half this value. A 1 bit in the MSB position with all other positions being 0 leads to a value of

$$2^{N-1} = \frac{1}{2} \times 2^N$$

Thus, this is a proper first guess.

The comparator now informs the control logic whether this guess is larger or smaller than $V_A$. If it is smaller than $V_A$, the first trial bit remains and the sequencer is instructed to insert a 1 in the next-most-significant bit of the holding register. If the output of the DAC after the first guess is larger than $V_A$, this trial bit is reset back to 0 before setting a 1 in the adjacent position of the holding register. The result of the second guess is compared to $V_A$ to determine the value of the third guess. This procedure is repeated until $N$ approximations have been made. Each comparison can be made in one clock period. At the end of $N$ clock cycles, the conversion is complete. The algorithm used in the successive approximation ADC can be demonstrated by the following example.

# Example 10.5

Assume that an 8-bit ADC is driven by an input signal of 6.825 V. The nominal full-scale input voltage is 10 V. Determine the sequence of trials that will be done to reach the final result. Express this final converted value in terms of the output binary code.

**Solution** We can write the value of binary code that should equal the input voltage from Eq. (10.33) as

$$\left[\frac{b_1}{2^1} + \frac{b_2}{2^2} + \cdots \frac{b_N}{2^N}\right] = \frac{V_i}{kV_{\text{REF}}} = \frac{V_i}{V_{FS}} = 0.6825 \text{ V}$$

The sequencer first places a 1 in the MSB of the holding register. This code of 1000 0000 generates a DAC output voltage that is one-half the full-scale voltage or 5 V. Since this value is less than the value of input voltage, the MSB bit is left at a value of 1 and another 1 bit is inserted into the next position of the holding register. This leads to a value of 7.5 V at the DAC output. This value is larger than the input signal, so this second bit is reset to 0 and the third position of the holding register is set to 1. This code of 1010 0000 leads to a DAC output voltage of 6.25 V. This value is less than the input value, so this bit remains at 1 and a 1 bit is set into the fourth position. The holding register now contains 1011 0000. This value gives a DAC output of 6.875 V. This value exceeds the input voltage, so the fourth position is reset to 0 and the fifth position is set to 1. Continuing in this manner results in a final code of 1010 1110. The DAC output is 6.797 V, giving an error of 0.028 V. The quantum interval or $LSB$ for this converter is

$$LSB = q \approx \frac{10}{2^8} = 0.0391 \text{ V}$$

The resulting error is within an $LSB$.

### The Parallel Converter

The block diagram of a simple parallel converter is shown in Fig. 10.35. This device is also called a flash converter indicating a very rapid conversion for this ADC. No other method used for conventional ADCs results in conversion speeds approaching those of the parallel method.

The parallel converter establishes $2^N - 1$ voltage levels, each differing from the next by the quantum interval $q$. The lowest level is established at $q/2$, with succeeding levels of $3q/2$, $5q/2$, $7q/2$, $\cdots$, $(2^{N+1} - 3)q/2$. These levels are generally established by a string of series resistors connected to a reference voltage.

One comparator is required for each quantum level, resulting in $2^N - 1$ comparators for an $N$-bit ADC. Circuit complexity increases rapidly with $N$, but 8-bit ADCs have been available in VLSI chips for several years. Certain modifications to this basic concept lead to simpler configurations for the parallel ADC.

The first or lowest comparator of the figure switches to an output of $+V_H$, the high logic level, when the analog input exceeds a voltage of $q/2$. Each succeeding comparator switches from a low to a high level as the analog input increases by $q$ over the switching point of the previous comparator. If $V_{FS}$ is the largest input voltage to be converted, the quantum interval is calculated from

$$q = \frac{V_{FS}}{2^N - 3/2} \tag{10.38}$$

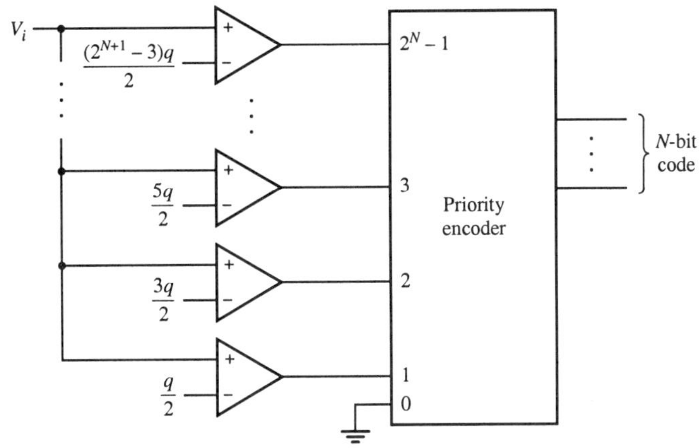


Figure 10.35: A parallel converter.

The comparator outputs drive the input of a priority encoder logic circuit. This device has $2^N$ input lines, numbered successively from 0 to $2^N - 1$. If a single input line is asserted, the encoder generates a binary output code corresponding to the number of the asserted line. If several inputs are asserted simultaneously, the priority encoder generates an output code corresponding to the highest-numbered asserted input line.

For each value of input voltage, several adjacent comparators will be asserting input lines to the encoder. The reference voltages driving each of these comparators will be less than $V_i$. All other comparators with reference voltages exceeding $V_i$ will not assert the encoder inputs. The priority encoder will then generate a binary code corresponding to the highest-numbered comparator that asserts an input line. This code is proportional to the value of $V_i$.

As the input voltage changes to a different level, the output code responds after a short period of time. This response time is determined by the propagation delay time of the comparators plus the propagation delay time of the priority encoder. Since the precise time that the input signal crosses a quantum level is undetermined, a Start Conversion signal is normally included to reference the time of conversion. This still allows the output code to be changing at the time some following device is reading the code. A transient, incorrect code could be registered under these circumstances.

In order to eliminate problems of transient output codes, a S/H circuit is inserted between the analog input signal and the input to the ADC. A Start Conversion pulse puts the S/H circuit into the hold mode to keep this signal constant until an end of conversion signal is asserted. Some chips have added latching comparators, but this method becomes quite complex as $N$ increases.

---

## SUMMARY

**1.** Analog circuits and digital circuits are now placed on the same IC chip. The circuits that interface between these circuits are called mixed-signal circuits.

**2.** Examples of mixed-signal circuits are the sample-and-hold circuit, the comparator, the analog switch, the digital-to-analog converter and the analog-to-digital converter.

**3.** The DAC converts a digital input code or number to a corresponding analog output value. Three important types of DAC are the current-scaling, voltage-scaling, and charge-scaling circuits.

**4.** The analog-to-digital converter or ADC accepts an analog input signal and produces a digital output code that corresponds to some characteristic of the input signal.

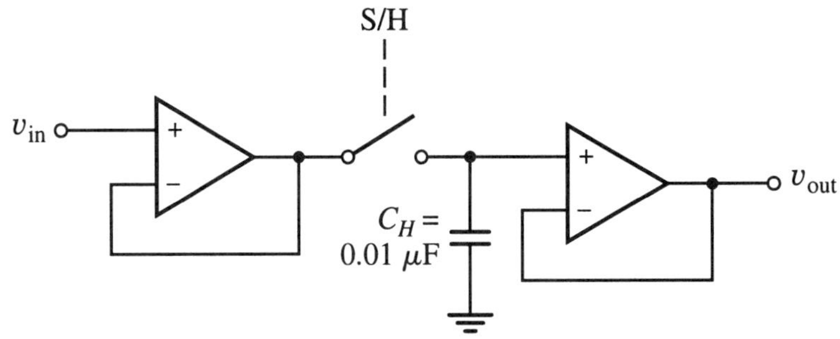


**Figure P10.3**

**5.** Three types of ADC are the staircase converter for low-frequency applications, the successive approximation converter for medium speed applications, and the parallel or flash converter for high-speed applications.

---

## Problems

### Section 10.2.1 Sample-and-Hold Circuit

**10.1** An S/H circuit can have a droop of 4.88 mV before it leads to an error. If the droop rate is 0.05 V/s, what is the maximum hold time before an error occurs?

**10.2** Derive Eq. (10.7) for overall gain error in a two-stage S/H circuit.

**10.3** The op amps in the S/H circuit shown have the following parameters:

$V_{OS} = \pm 0.5$ mV
Open-loop gain $= 20{,}000$ V/V
Common-mode gain $= 0.0005$ V/V
Input bias current $= 50$ nA (max)

The switch has a charge injection of $\pm 10$ pC. The S/H circuit should accommodate a full-scale signal of 10 V.

**(a)** What is the droop rate?
**(b)** What is the sample-to-hold offset?
**(c)** What is the worst-case zero-scale error?
**(d)** What is the worst-case full-scale error?

**10.4** An S/H module having a droop rate of $\pm 10\ \mu V/\mu s$ is to be used to hold signals ranging from 0 to 5 V. The acquisition time including the 0.005% settling time is 75 $\mu s$. Recommend and justify an appropriate number of bits to use for the digital data words. What is the maximum sample frequency that might be used?

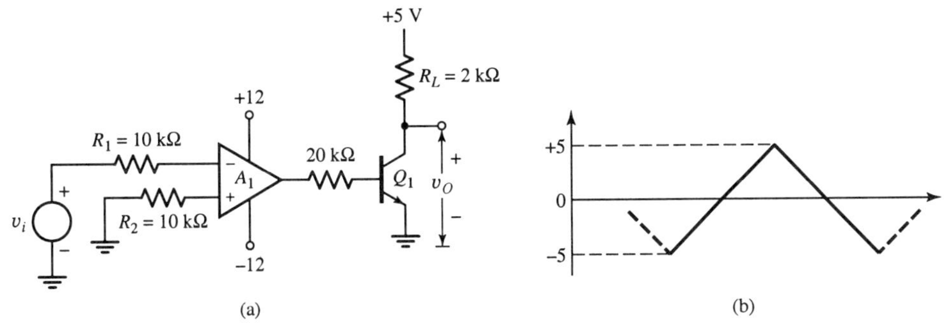


**Figure P10.6**

### Section 10.2.2 The Comparator

**10.5** Derive the equations for UTP and LTP for the circuit of Fig. 10.9 if the output swings from 0 V to $V_1$ V.

**10.6** For the comparator circuit shown,

**(a)** plot the output voltage, $V_O$ vs. the input voltage, $V_i$.
**(b)** For the triangular $V_i$ shown, sketch the output waveform.

**10.7** Assume that the op amp in Problem 10.7 is a 741 op amp and the transistor is a 2N2222 device. Simulate the circuit with a square pulse input to determine its response to leading- and trailing-edge inputs.

**10.8** Add hysteresis to the circuit of Problem 10.6 by placing a resistor between the output and the inverting input of the op amp. Calculate the value of resistance to produce a hysteresis of 0.5 V.

**10.9** A diode bridge switch such as that of Fig. 10.19 has currents of $I_1 = I_2 = 5$ mA. In switching from the *on* to the *off* condition, the current $I_1$ is switched off in 90 ns and the current $I_2$ switches off in 70 ns. Calculate the magnitude and polarity of the charge injection error.

### Section 10.3 The Digital-to-Analog Converter

**10.10** Calculate the maximum output of a 12-bit DAC with a quantum interval of 0.0040 V. Calculate the output for a code of 1000 0000 0000.

**10.11** The quantum interval in an 8-bit DAC is 0.0040 V when $V_{\text{REF}} = 12$ V. Calculate the change in $V_{\text{REF}}$ to cause a $\pm 1/2$ LSB change in output voltage.

**10.12** A 6-bit, binary-coded, current-scaling DAC has a nominal full-scale output current of 1 mA and an input code of 101 101.

**(a)** What is the decimal equivalent, $n_{10}$, of the input code?
**(b)** Compute the output current for this code.
**(c)** Compute the output current for a code corresponding to $n_{10} = 63$.
**(d)** What is the binary code for $n_{10} = 63$?

**10.13** How do the resolution and accuracy of a DAC differ? Can you think of any applications that would require a 4-bit resolution with an 8-bit accuracy?

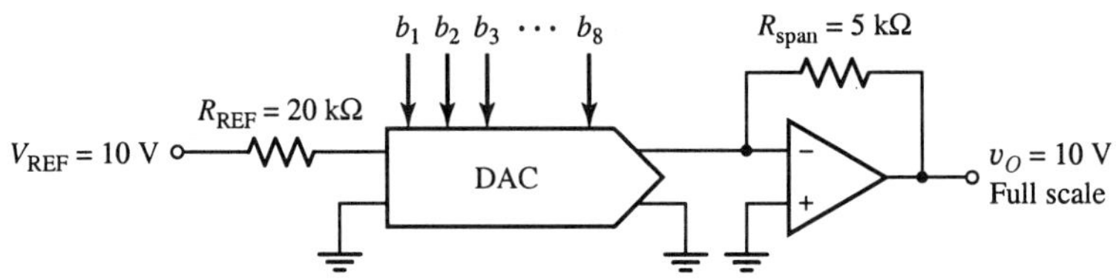


**Figure P10.16**

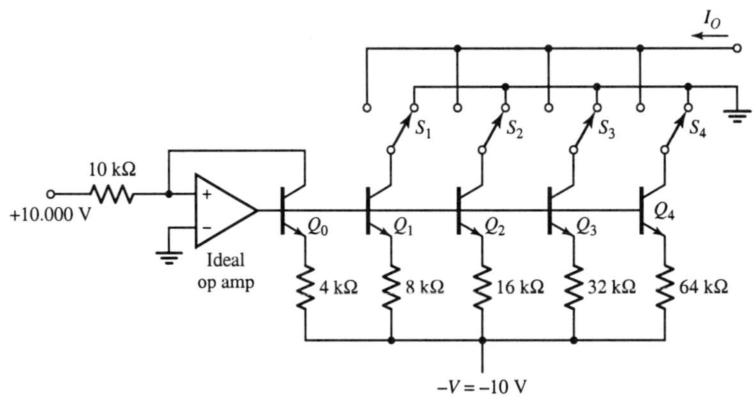


**Figure P10.18**

**10.14** A 4-bit DAC is driven by a binary counter that is driven by a clock signal with period $T$. The nominal full-scale output voltage is 8 V. Show a plot of DAC output vs. time.

**10.15** An 8-bit DAC drives a linear positioning system. The DAC linearity is $\pm 1/2\ LSB$. To what accuracy in percent can the linear positioning be controlled?

**10.16** The resistors $R_{\text{REF}}$ and $R_{\text{span}}$ have a temperature coefficient of $\pm 200$ ppm/°C and a tracking coefficient of $\pm 4$ ppm/°C. What drift in full-scale output voltage will occur as temperature changes from 0°C to 70°C, assuming all drift is due to $R_{\text{REF}}$ and $R_{\text{span}}$.

**10.17** For the DAC of Problem 10.16, what is the nominal or room temperature value of $V_O$ for the input code 1001 0111?

**10.18** The emitters of the DAC shown have been scaled so that all $V_{BE}$ drops are identical. Calculate

(a) $I_{MSB}$
(b) $I_{LSB}$
(c) The actual full-scale output current

## Section 10.4 The Analog-to-Digital Converter

**10.19** An 8-bit ADC with a 10.000-V reference has a scale factor of $k = 1$. For an analog input of 7.685 V, calculate the resulting digital code and its decimal equivalent, $n_{10}$.

**10.20** Explain why it is generally necessary to include an S/H circuit with a staircase or successive approximation ADC.

**10.21** In the staircase ADC of Fig. 10.32, $V_{\rm REF} = 10$ V, $R_{\rm REF} = 10$ k$\Omega$, and the DAC scale factor, $K = 2$. Calculate the value of $R$ that will produce a full-scale output voltage of 5 V for the ADC. To what value of $k$ does this correspond in Eq. (10.33)?

**10.22** An 8-bit successive approximation ADC has a quantum interval of 0.020 V. If an input voltage of 3.184 V is converted, list all the codes through which the holding register will sequence as the conversion is made.

## REFERENCES

1. Allen, P. E. and Holberg, D. R., *CMOS Analog Circuit Design, 2nd Edition.* New York: Oxford University Press, 2002, Chapter 10.

2. Comer, D. T., *Introduction to Mixed Signal VLSI.* Provo, Utah: Array Publishing Company, 1994, Chapter 11.

3. Miner, G. F. and Comer, D. J., *Physical Data Acquisition for Signal Processing.* Englewood Cliffs, NJ: Prentice Hall, 1992, Chapters 6-8.

# Chapter 11

# BASIC CONCEPTS IN FILTER DESIGN

In modern society, the field of communications influences our lives in rather significant ways. Television, radio, telephone, and computer communications are examples of important applications that touch our lives on a daily basis. It would be difficult to imagine going through a single day without having access to these communication devices.

Although the communications field encompasses several subfields of electrical engineering, the topic of filtering is indispensable to the operation of many types of communication hardware. A basic requirement in the communications field is a circuit that can pass a signal falling within a small range of frequencies to the output while blocking all signals with frequencies outside this range. Such a circuit is an example of an electronic filter.

In the introductory section, we will define the term *electronic filter* in terms of its frequency response. One classification of electronic filters is based on the frequency response of the filter. Several useful frequency responses, required in modern electronic systems, will be considered. The applications of filters will be discussed to emphasize the importance of these circuits in popular electronic systems.

The final portion of the section will discuss another classification scheme for filters, based on methods of implementation or on methods of fabrication. Filters may be implemented as analog circuits or digital circuits. Furthermore, they may be made from discrete circuits or integrated circuits or combinations of the two technologies.

# 11.1 ELECTRONIC FILTERS

**Important Concepts**

**1.** Most electronic filters are designed to pass the desired signals with little attenuation while attenuating the unwanted signals (noise) severely.

**2.** A popular method of classifying filters is by their frequency responses. Lowpass, highpass, and bandpass filters broadly describe the frequency range over which little attenuation occurs.

**3.** Filters are used to improve the quality of audio signals, to separate one signal (channel) from many communication signals, and to transmit and separate several signals over a single transmission medium.

**4.** Filters can be made with passive elements (passive filters), with passive elements plus amplifiers (active filters), and on integrated circuit chips (active IC filters).

## 11.1.1 Definition of an Electronic Filter

One definition states that a filter

*is a device that removes impurities or unwanted particles from a liquid or gas passed through it.*

Basically, a filter receives a mixture of substances and passes some purer substance or material while removing impurities or undesired material.

An electronic filter is very similar in that it is designed to receive an impure electrical signal and pass a desired electrical signal while removing undesired components, often referred to as noise signals.

Noise comes from many sources. Large current changes in a conductor can generate electromagnetic interference (EMI) by propagating electromagnetic energy into the atmosphere. The amount of energy depends on the magnitude of the current change, the speed of the change, the dimensions of the current conductor (antenna effect), and other factors such as shielding.

Unintentionally generated noise comes from electric arc welders, spark plugs of internal combustion engines, high-voltage ac transmission lines, large electric motors, especially with commutators, and relay or solenoid closures/openings. Line noise that may come from ac to dc conversion in power supplies also is a troublesome source in electronic circuits. The first-mentioned sources are often random in nature while power supply hum is 60 Hz or some multiple of 60 Hz.

Intentionally generated noise comes from radio waves, including TV, AM, FM, CB, and emergency services, satellite transmission, cell phones and base stations, and microwave telephone links.

Both unintentionally and intentionally generated noise can be troublesome. A research group at IBM was working on a 2000-word vocabulary voice-controlled typewriter. The desired signals were the input words with a frequency range of about 10 Hz to 12 kHz. The group found some very sharp noise spikes of perhaps 100 mV occurring on the dc power lines. They were very random, but large enough to be troublesome. After several days, they discovered that the source of the noise was the solenoids of card punch machines in an adjacent data-entry room.

On another occasion, a company that designed high-frequency instrumentation tape recorders, operating up to 100 kHz, encountered a very strange high-frequency signal on chassis ground and at other points in the circuit. The frequency was around 100 MHz and it was nearly sinusoidal. In this case, the noise source was an FM station about one-half block away that was transmitting several kilowatts of power.

One important source of noise is a switching circuit on the same chip substrate or the same printed circuit board as an analog circuit. As current switches, this signal can be coupled to an analog circuit through stray capacitance of leads (crosstalk), through the power supply, or other means. Designers of mixed-signal circuits must be extremely careful in circuit layout to minimize such effects.

In the electronic circuit itself, noise is generated by resistors, transistors, and other sources. Consequently, even if external noise does not contaminate a signal, internal noise may still appear at the output, leading to an impure signal.

Noise is generally the most troublesome problem in electronic circuit and system design. Since filters can be designed to eliminate or minimize noise problems, this area is rather important. The topic of filter design is the subject of many textbooks. This chapter will cover some basic ideas in active filter synthesis.

One of the more popular methods of specifying the behavior of a filter involves writing the transfer function for the filter or specifying its magnitude and/or phase response as a function of frequency. The transfer function responses of Fig. 11.1 indicate five of the more popular filter responses used at present. The lowpass filter passes frequencies below some cutoff frequency and attenuates frequencies far above this cutoff frequency as shown in Fig. 11.1(a). The bandpass filter of Fig. 11.1(c) is indispensable to the communications field. The highpass response of part (b) of Fig. 11.1, the band-reject response of part (d), and the allpass response of part (e) are not used as often as the lowpass and bandpass responses.

### 11.1.2 Applications of Filters

Some applications require precise, carefully designed filters; others use filters that are somewhat approximate; still others use simple circuits that may not be considered as filters, but actually perform a filtering function.

#### Lowpass Filter

The lowpass filter is often used in audio applications. A prime example of this filter is in telephone circuits. Voice signals carry much of their energy in the range of a few hundred Hz to 1 or 2 kHz. Smaller frequency bands of energy exist in the 2 to 3 kHz range, while an even smaller band occurs from about 4 to 6 kHz. In order to transmit an intelligible (understandable) conversation over phone lines, the upper energy band is eliminated. A lowpass filter that passes 0 Hz to about 3.3 kHz is used to pass only the lower energy bands. Actually, while voice sounds do not contain frequencies below the fundamental frequency, which varies from around 200 Hz to 400 Hz for different people, all frequencies below 300 Hz may also be removed from the voice signal. This reduces any unwanted noise that may occur in this lower range of signals that could result in a distorted signal.

The resulting frequency response of the telephone filters may be lowpass, from 0 Hz to 3.3 kHz, or bandpass, from 300 Hz to 3.3 kHz. The telephone company translates the voice signal output from the filter into either pulse-code modulation or single-sideband modulation. The older scheme uses single-sideband methods and many outlying areas have still not been converted to the newer pulse-code technique. These methods allow several phone conversations to be transmitted over a single transmission line simultaneously. If not for filtering, far fewer conversations could be transmitted over the single line.

As a matter of interest, it is very difficult to distinguish between the phonemes "s" and "f" over a phone line, because both of these unvoiced sounds generate only small amounts of energy in the 4 kHz to 6 kHz range of frequency. These sounds are greatly attenuated by the filter, and only the context of a word or sentence allows the listener's brain to automatically restore these sounds.

Another lowpass circuit is produced when a "decoupling" capacitor is added to the power supply pins of an op amp chip or a logic circuit. In practical circuits, these capacitors are added to every power supply pin of every chip on a printed circuit board. These capacitors combine with the output resistance of the power supply plus the wiring resistance to create a lowpass filter. In this case, only the dc or zero-frequency signal is to be passed while higher frequency components are to be blocked. This is a very simple filter that requires little accuracy or precision.

All dc power supplies require lowpass filters to eliminate frequency components related to the ac input signal. Older supplies used RC or L-section filters, while newer switched power supplies translate a voltage to a higher frequency and thereby make the lowpass filtering problem easier to solve.

#### Highpass Filter

Highpass filters are not used to the extent of lowpass filters. One application is in the driving circuitry for the high-frequency speakers of a high-fidelity music system. At least one speaker is designed to reproduce lower

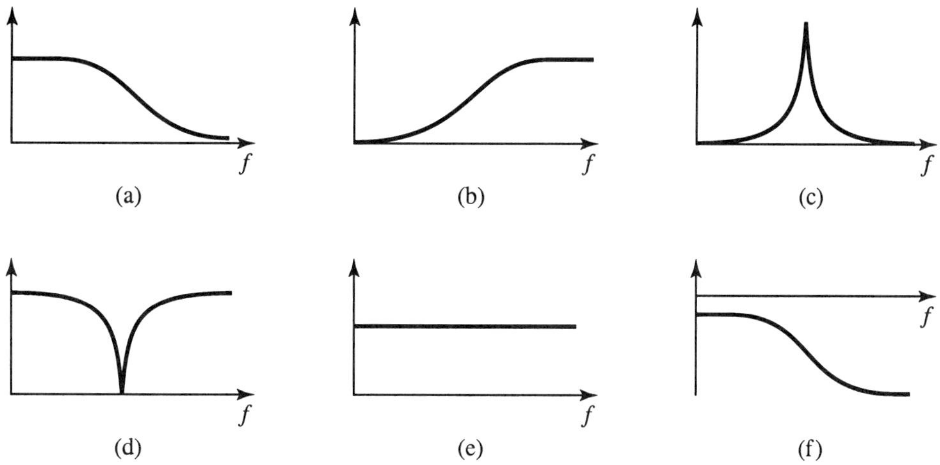


Figure 11.1: Filter transfer function responses. (a) Magnitude response of a lowpass filter. (b) Magnitude response of a highpass filter. (c) Magnitude response of a bandpass filter. (d) Magnitude response of a band-reject filter (e) Magnitude response of an allpass filter. (f) Phase response of an allpass filter.

frequency sounds. This speaker, called a "wolfer," provides excellent bass sounds. A lowpass circuit is used to eliminate higher frequencies from the entire audio signal before being applied to the speaker. The high-frequency speaker, called a "tweeter," amplifies the upper frequencies. A highpass circuit is used to remove low frequencies from the audio signal before applying the remaining signal to this speaker.

### Bandpass Filter

This type of filter is necessary to enable the communications industry as explained in earlier chapters. Frequency-division multiplexing systems that transmit multiple information channels over the same medium require bandpass filters to separate these channels in the receiver. Prime examples of these systems are AM and FM radio that transmit multiple signals through the atmosphere. Television also transmits several channels in a given geographic area through the atmosphere. The telephone companies transmit multiple voice channels over a single microwave link or over a single coaxial cable. At the receiving end of the telephone link, the various channels must be separated before recovery of the original voice signals. Bandpass filters are used in all these systems to isolate individual channels for subsequent demodulation.

Cellular phone systems and satellite communication systems also require bandpass filters. Cable television, digital television, and high-definition television systems also depend on narrowband bandpass filters for their operation. The bandpass filter is one of the most well-used and important filters in the electronics industry.

### Band-Reject Filter

This filter is not often used. It is typically employed to eliminate one troublesome frequency component or frequency range from a wideband signal. The 60-Hz ac power source often injects an unwanted 60-Hz signal, often called "hum," into an audio or other desired signal. A 60-Hz band-reject filter can be used to greatly attenuate the hum without adversely affecting any components of the audio signal.

### Allpass Filter

This filter is used in phase equalization circuits. A large percentage of filter design is based on the realization of the magnitude response of the desired function. The resulting phase response may be very undesirable. An allpass filter is sometimes cascaded with the lowpass filter to obtain an improved phase response without altering the lowpass magnitude response.

Another use of the allpass filter is in the production of other types of filters using active circuits. Specifically, the allpass circuit can be used to implement bandpass, lowpass, highpass, or band-reject filters on integrated circuit chips.

The classification of the preceding filters is based on frequency or phase response. Although frequency is not the only variable used to specify filters, it is an important one. However, there are situations in which the transient response of a filter is very important. Thus, we will expand the definition to include this case also.

> *A filter is a circuit or system that can modify the transient response, the frequency response, or the phase response of an electronic signal in a predetermined way.*

### 11.1.3 Other Classifications of Filters

Filters can be classified in many ways. Classification by response is obviously an important consideration. Another classification depends on the types of circuit components used to construct the filter, while another depends on the type of system that implements the filter.

We will first discuss the type of system that implements the filter. This leads to classification of filters into three types: the digital filter, the switched-circuit filter, and the continuous-time filter.

**1.** Digital filters. These filters require some type of computing power to implement. Although the precision and variety of responses are excellent, they cannot filter in real time at high frequencies. In addition, digital filters typically require a sample and hold circuit, an analog-to-digital converter, a pre-aliasing filter, a digital-to-analog converter, and a smoothing filter as well as a computer to do the processing.

**2.** Switched-capacitor filters. The switched-capacitor filter uses electronic switching to connect capacitors to different nodes periodically. This method makes time constants and critical frequencies become a function of switching frequency. An external source can control the switching frequency and can adjust the critical performance parameters of the filter response. If element values used to construct the filter are not precise, the external frequency can be tuned to achieve a very accurate filter response.

**3.** Continuous-time filters. This filter is also referred to as an analog filter. The analog input signal is a continuous signal in time that is not quantized or digitized. The elements of the filter are linear time-invariant components that shape the signal properly to produce an analog output signal.

Continuous-time filters are further classified according to the components used to implement the circuit. A passive filter uses only passive components such as resistors, capacitors, inductors, and transformers. An active filter uses amplifier circuits as well as passive elements to implement the desired response. These filters may use op amps and discrete passive elements or they may be implemented on an integrated circuit chip, resulting in an IC filter.

#### Passive Filters

The synthesis of passive filters is based on well-known theory, much of which was developed from the 1920s to the 1960s. The spectrum of specifications that can be implemented with passive filters is very broad. As Van Valkenburg points out [7], the large ranges of practical element values available to use in passive filters result in a broad range of specifications that can be achieved. Approximate ranges of values for resistors, capacitors, and inductors are shown in Table 11.1.

**Table 11.1 Practical Discrete Passive Component Values**

| Element | Minimum | Maximum |
|---|---|---|
| Resistor | 0.1 Ω | 500 kΩ |
| Inductor | 1 nH | 10 mH |
| Capacitor | 0.2 pF | 10 μF |

A disadvantage of passive circuits in some applications is the size or volume of the finished filter. For low-frequency filters, the inductors or capacitors may be quite large in value and in physical size. Passive filters continue to be important in applications that do not require volume minimization.

### Active Filters

Developments in active filter synthesis began in the 1940s, initially to minimize the size of passive components needed. Through active filter methods, inductors can be eliminated and capacitor values reduced in size. The vacuum tube had to be used as the active element, thus size reduction was somewhat limited because of the size of the vacuum tube. A classic article by Sallen and Key in 1955 still applied vacuum tubes rather than transistors as the amplifying element [4]. Further improvement was made when the bipolar transistor replaced the vacuum tube in the late 1950s.

The introduction of integrated circuits, especially the IC op amp around 1965, led to further reduction in size, but also limited the maximum frequency at which the filter would operate. The initial op amps had rather poor frequency response and led to a limitation on the frequency range of filter performance. While op amps have improved over the years, they still tend to limit frequency range in active filters, especially when higher gains are required.

### IC Filters

Building the entire filter on a chip presents several challenging problems. First of all, the range of possible element values is limited. Although each fabrication process may differ, a typical process may result in component ranges, as shown in Table 11.2.

**Table 11.2 Component Ranges for an IC Process**

| Element | Minimum | Maximum |
|---|---|---|
| Resistor | 0.1 Ω | 50 kΩ |
| Inductor | 0.1 nH | 10 nH |
| Capacitor | 0.1 pF | 100 pF |

A 100-pF capacitance occupies a great deal of space and may allow no more area on the chip for other capacitors. Thus, this value of capacitance may not be practical for circuits involving several capacitors. Spiral inductors now use copper wiring to achieve reasonable quality factors for inductors and cannot be used in a standard IC process. In addition to the limited range of values possible, it is expensive to include an inductor on a chip. In future years, this component will become more common, but early in the new millennium, it is not expected to have significant general usage. Most IC filters are limited to resistors and capacitors for passive elements.

Another problem with IC elements is the accuracy of component value fabrication. Resistors and capacitors can be fabricated with absolute accuracies of somewhere between 10% and 20%. While this is almost as accurate as some discrete components, a bad value cannot be trimmed by replacing or padding the inaccurate component. In a discrete circuit, a trimming capability can be included at an increased cost. It is possible to use voltage-controlled elements and tune IC components using a voltage. This method is used in automatic tuning circuits for IC filters and will be discussed later in the book.

The ratio of two resistors or two capacitors can be made accurate to near 0.1%; thus, accuracy can be achieved if performance depends on these ratios rather than on absolute values.

Integrated circuit filters can be based on op amps and passive RC circuits, just as active filters using discrete components. However, there is a paradox when combining the op amp and RC networks on a chip. Since capacitor and resistor values are limited, conventional techniques can only be used for higher frequency filters. Although high-frequency active circuits could be constructed on a chip with practical values of RC products, the op amp will not perform at these high frequencies. For the range of frequencies in which the op amp will perform efficiently, it is difficult to create the large RC products required. Conventional active circuit synthesis methods often cannot then be applied to IC filter design. Consequently, other methods must be developed to allow active circuits to be implemented on an IC chip and lead to a broad range of frequency performance.

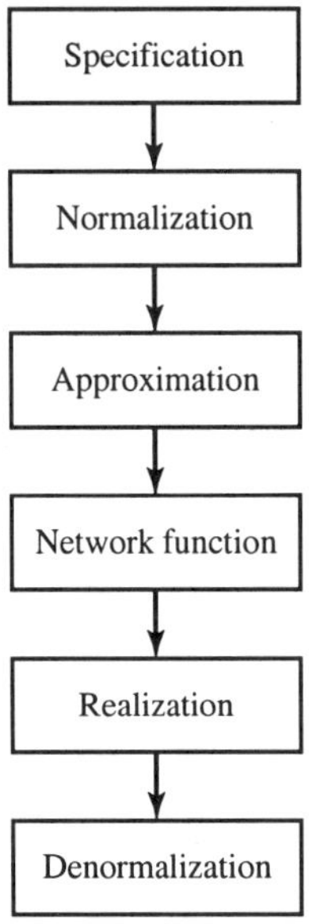


Figure 11.2: General procedure for filter synthesis.

## 11.2 OVERVIEW OF FILTER SYNTHESIS PROCEDURES

**Important Concepts**

**1.** The starting point in filter design is the specification of the filter. This is often done in terms of attenuation of the filter as a function of frequency.

**2.** A transfer function with a response similar to the desired response is chosen. This step is referred to as *approximation.*

**3.** A network is then developed that has the chosen transfer function. This step is known as the *realization* of the filter.

In many university engineering classes, it is easy to get the impression that analysis of circuits is more important than design of circuits. Although analysis is necessary for good design, only rarely is engineer assigned to analyze some circuit. Normally, the engineer is assigned a new circuit to design. After all, the company derives income from the sales of circuits or systems required by the customer.

This chapter will consider an overview of the filter design process and discuss some steps of this process in detail. A reasonable procedure for building most filters is given by Fig. 11.2 [6].

### 11.2.1 Specifications

The specifications for the filter may come from someone who needs a filter and knows the characteristics of the desired magnitude as a function of frequency. Alternately, we may analyze a noise problem and decide how the magnitude response must vary with frequency to attenuate the noise while passing the desired frequency components to the output.

Many methods are available for determining appropriate specifications for the filter response. In some cases, the band of frequency that contains troublesome noise components may be known, especially if the spectral contents of the noise source or sources have been predetermined. An example of known spectral content would be the AM, FM, or television frequency bands. In AM, the major noise generally comes from AM channels

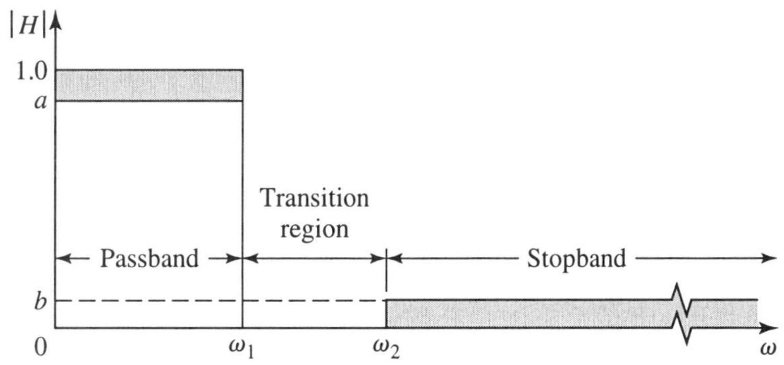


Figure 11.3: A frequency specification for a lowpass filter.

adjacent to the desired channel. The FCC assigns carrier frequencies with either 20-kHz separation or, in very densely populated areas, 10-kHz separation. Consequently, filters must be constructed that can pass the desired frequency with little attenuation while greatly attenuating the signals that result from adjacent channel frequencies. If the noise is generated by unknown sources, a spectral analysis of the incoming signal can provide information on which to base the filter specifications. A typical frequency specification may appear as shown in Fig. 11.3.

In the passband, the magnitude of the transfer function must fall within the shaded region between values of $a$ and 1. This guarantees that its value is near unity, passing input signals within this frequency range with little attenuation to the filter output.

The stopband response is also required to fall within a shaded region between the values of 0 and $b$. This guarantees that input signals within the stopband will be decreased by a factor of at least $b$ before reaching the output. The value of $b$ might be 0.01, 0.001, or lower.

The transition region consists of the frequency range required to drop $|H|$ from its minimum passband value to the maximum stopband value. The edge of the passband, $\omega_1$, is generally called the cutoff frequency and is often denoted $\omega_c$. An example of a lowpass filter transfer function magnitude is

$$|H(\omega)| = \left[\frac{1}{1 + \left(\frac{\omega}{6280}\right)^8}\right]^{1/2} \tag{11.1}$$

If $a$ is defined to be 0.707 (3 dB down from $|H(0)|$), the cutoff frequency is seen to occur when $\omega = \omega_1 = 6280$ rad/s. If $b$ is defined to be 0.04, Eq. (11.1) can be used to write

$$|H(\omega_2)| = 0.04 = \left[\frac{1}{1 + \left(\frac{\omega_2}{6280}\right)^8}\right]^{1/2} \tag{11.2}$$

Solving for $\omega_2$ gives

$$\omega_2 = 1.404 \times 10^4 \text{ rad/s}$$

This frequency is about 1.16 octaves above the cutoff frequency $\omega_1$. A plot of this transfer function magnitude is shown in Fig. 11.4.

All signals at frequencies below $\omega_1 = 6280$ rad/s result in an attenuation by a factor falling between 0.707 and 1. Signals with frequencies above 14,040 rad/s will have attenuation factors falling between 0 and 0.04. Other possible lowpass filter responses are indicated in Fig. 11.5. The slope of the response with frequency in the passband or the stopband may not be monotonic.

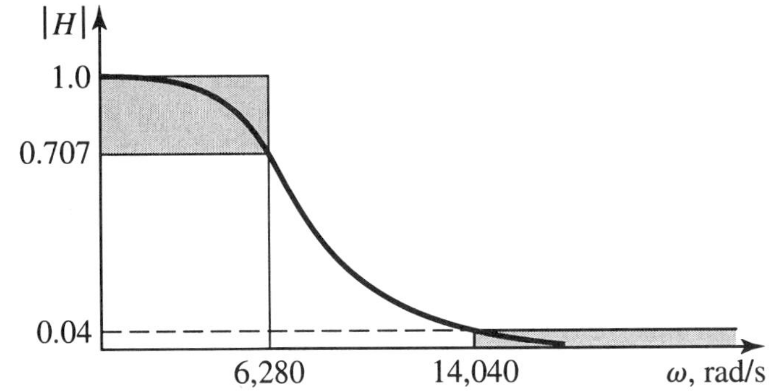


Figure 11.4: Magnitude response of a lowpass filter.

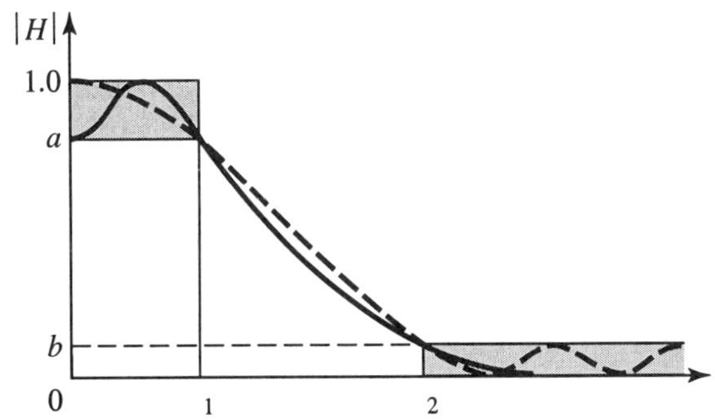


Figure 11.5: Other acceptable lowpass filter responses.

# Example 11.1

A filter has a transfer function given by

$$|H(\omega)| = \left[\frac{1}{1 + \left(\frac{\omega}{1000}\right)^{2n}}\right]^{1/2}$$

If $a$ and $b$ as defined in Fig. 11.3 are 0.707 and 0.002 while $\omega_1 = 1,000$ rad/s and $\omega_2 = 3000$ rad/s, find the minimum integer value of $n$ that satisfies the specifications.

**Solution** By substituting the value of $\omega_1$ into the preceding equation, we note that the specification for $a$ is satisfied regardless of the value of $n$. We now substitute $\omega_2$ into the equation to satisfy the specification for $b$. This results in the inequality

$$\frac{1}{\sqrt{1 + \left(\frac{3000}{1000}\right)^{2n}}} \leq 0.002$$

After some manipulation, we get

$$1 + 3^{2n} \geq 2.5 \times 10^5$$

The minimum integer value of $n$ is then found to be 6.

**Practical Considerations**

In Example 11.1, the falloff between 1000 rad/s and 3000 rad/s is about 51 dB. This frequency interval is 1.59 octaves resulting in an average rate of falloff in the transition region of 32 dB/octave. Except for resonant circuits, each reactive component (inductor or capacitor) can have a maximum or asymptotic falloff of 6 dB/octave. Thus, in this example, at least six reactive elements would be required to synthesize the actual filter. The order of the filter is given by the value of $n$ and this often equals the number of reactive elements needed.

**Practice Problems**

**11.1** In Fig. 11.3, $a = 0.9$, $b = 0.05$, $\omega_1 = 1$, and $\omega_2 = 3$. What is the falloff in dB between $\omega_1$ and $\omega_2$? *Ans:* 25.1 dB.

**11.2** In Practice Problem 11.1, what is the average falloff in dB/octave between $\omega_1$ and $\omega_2$? *Ans:* 15.8 dB/octave.

In the succeeding discussion, we will assume that a lowpass filter is to be constructed. This is a reasonable assumption in light of the fact that

1. Lowpass functions are rather popular in filter design
2. A lowpass prototype filter can be transformed to highpass or other types of responses using simple frequency transformations

### 11.2.2 The Advantage of Normalization

When the filter response has been specified, certain key frequencies can be translated to other frequencies that lead to simpler calculations. For example, a lowpass filter may have a desired cutoff frequency at 10.8 kHz or 67,858 krad/s. The frequency axis can then be translated to that of a similar response having a cutoff frequency of 1 rad/s. This process is referred to as frequency normalization. The filter is first realized to result in a cutoff frequency of 1 rad/s and will generally use impractical element values. The element values will later be denormalized to lead to the actual desired cutoff frequency. This frequency denormalization process will result in more practical element values.

In some filters, an impedance normalization is also effected. A load resistance may be normalized to a 1-Ω value for the normalized filter. An impedance denormalization is then used to change filter element values to those that will result in the wanted response when the load resistance has its actual value.

One advantage in normalizing the frequency and impedance of the prototype filter to fixed values is that standard equations for transfer functions have been developed for this normalized situation. These standard equations have been published and allow a designer to write normalized transfer functions for various types of popular filter responses. Much time can be saved by avoiding the necessity to derive these equations each time a filter design is encountered. These standard functions are then used to aid the filter designer in obtaining the normalized prototype. The prototype can be denormalized to meet the specific cutoff frequency and impedance requirements.

**Normalization/Denormalization**

A particular filter frequency response may have an infinite number of transfer functions, depending on the cutoff frequency. In order to consolidate this high number of functions into a single function, frequency normalization is used. In this method, it is assumed that a particular response is normalized to have a cutoff frequency of $\omega_c = 1$ rad/s. Furthermore, for similar reasons, in filters that are terminated in a resistance, it is assumed that this termination has been normalized to a value of 1 Ω.

If a lowpass filter is to be realized with a cutoff frequency of 12 kHz and a 610-Ω termination, the filter is first designed with a cutoff frequency of 1 rad/s and a termination of 1 Ω. The element values calculated are normalized values. These values are then denormalized to produce a cutoff frequency of 12 kHz, with a 610-Ω terminating resistance.

The key to impedance normalization or denormalization is that transfer functions are based on ratios of impedances. The transfer function of a network remains constant as impedances change, if all impedances change by the same ratio. In order to denormalize a resistance $R_n$ (usually 1 Ω) to a new value $R$, all impedances are scaled up by the factor $k_i$ where $k_i = R/R_n$. This means that inductors are increased by a factor of $k_i$ since inductive impedance varies directly with $L$. Capacitor values are decreased by the factor $k_i$ since capacitive impedance varies inversely with $C$. Thus, the impedance scaling equations are

$$R = k_i R_n \tag{11.3}$$

$$L = k_i L_n \tag{11.4}$$

$$C = \frac{C_n}{k_i} \tag{11.5}$$

These equations allow element values to be changed while keeping the transfer function constant.

Frequency normalization or denormalization is based on the fact that corner frequencies depend on the relative values of reactance and resistance. If a cutoff frequency is related to a capacitor and resistance by

$$\omega_{cn} C_n = R_n$$

then a new cutoff frequency that is $k_\omega$ times higher than $\omega_{cn}$ can be produced by using a capacitance that is a factor of $k_\omega$ times smaller than $C_n$. The capacitive reactance of the denormalized circuit at the new cutoff frequency of $k_\omega \omega_{cn}$ is made to equal the capacitive reactance of the normalized circuit at $\omega_{cn}$. The frequency denormalizing equations are

$$C = \frac{C_n}{k_\omega} \tag{11.6}$$

$$L = \frac{L_n}{k_\omega} \tag{11.7}$$

Both impedance denormalization and frequency denormalization can be accomplished in one step by using the following equations:

$$R = k_i R_n \tag{11.8}$$

$$L = \frac{k_i}{k_\omega} L_n \tag{11.9}$$

$$C = \frac{C_n}{k_i k_\omega} \tag{11.10}$$

## Example 11.2

The circuit of Figure 11.6 has a corner frequency of 1 rad/s. Find the element values required to denormalize this frequency to 12 kHz with a terminating resistance of 610 Ω.

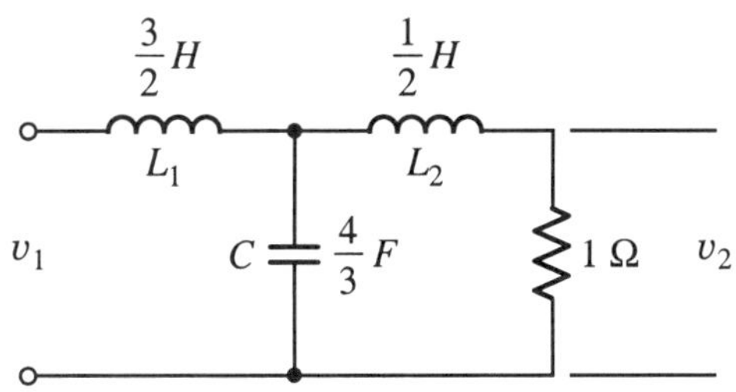


Figure 11.6: A lowpass, normalized filter prototype.

**Solution** The two normalizing factors in this problem are

$$k_i = \frac{610}{1} = 610$$

and

$$k_\omega = \frac{12,000 \times 2\pi}{1} = 7.54 \times 10^4$$

The terminating resistance becomes 610 Ω and the denormalized values of $L_1$, $L_2$, and $C$ are found from Eqs. (11.9) and (11.10) to be

$$L_1 = \frac{610}{7.54 \times 10^4} 1.5 = 12.1 \text{ mH}$$

$$L_2 = \frac{610}{7.54 \times 10^4} 0.5 = 4.05 \text{ mH}$$

$$C = \frac{1.333}{610 \times 7.54 \times 10^4} = 0.0290\ \mu\text{F}$$

These values result in a filter with a cutoff frequency of 12 kHz.

---

We will again emphasize that filter design is based on transfer functions that result in normalized filters. Handbooks publish transfer functions that are normalized for the types of responses of interest. When the appropriate response has been selected, the element values are denormalized to achieve the desired cutoff frequency and impedance level.

**Practice Problem**

**11.3** A filter has values of $L = 120\ \mu$H and $C = 100$ nF. The filter has a 2-kΩ terminating resistance. The corner frequency is 50 kHz. What are the normalized elements for a corner frequency of 1 rad/s and a terminating resistance of 1 Ω? *Ans:* $L = 0.0188$ H, $C = 6.28$ F.

### 11.2.3 The Approximation Problem

The specification of a filter is often given in terms of the magnitude response versus frequency as mentioned earlier. In the ideal case, we draw abrupt line segments to represent the magnitude of the transfer function of the filter as a function of frequency. For example, Fig. 11.7 shows the "brick-wall" or ideal magnitude response of a lowpass filter.

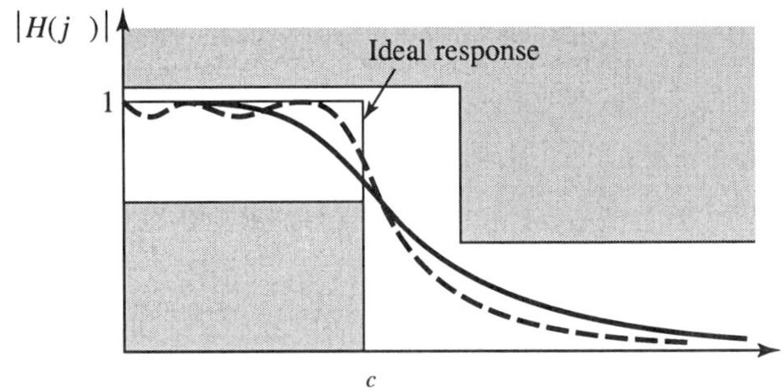


Figure 11.7: An ideal magnitude response of a lowpass filter.

The response of Fig. 11.7 is impossible to achieve with any finite circuit of which we can conceive. Theoretically, an infinite number of ideal elements could result in this response, but we cannot synthesize a practical filter with the magnitude response of Fig. 11.7. On the other hand, the art of approximation is a valuable component of an engineer's repertoire. An engineer should rarely design a system to a higher precision than needed, unless the extra precision comes at no cost. Normally, extra precision increases the cost of the circuit or system. Although the ideal transfer function would result in an excellent filter, the cost of realization is prohibitive. Some rather good approximations to this function are possible and can be realized with a relatively small number of elements. The shaded areas of Fig. 11.7 show the regions into which the approximate response should not fall.

The approximation step consists of selecting a mathematical equation for the magnitude of the filter transfer function that approximates the ideal response. However, there are further constraints in the choice of this approximate function that make the problem slightly more difficult. The response of circuits consisting of resistors, inductors, capacitors and active elements obey certain characteristics. For example, a circuit that has a magnitude response falling at an asymptotic rate of 3.5 dB/octave is essentially impossible to achieve with an electronic circuit. Consequently, the approximation should not only be rather close to the ideal transfer function magnitude at all frequencies, but it should also observe characteristics that are achievable with physical circuit elements. In other words, the approximation problem consists of

1. Approximating some ideal function, often a frequency response, with a nonideal, but realistic equation

2. Choosing an approximation that can be realized or implemented with an actual electronic circuit.

The second part of the approximation problem could be rather difficult or require a lot of trial-and-error effort. Fortunately, this problem was solved years ago in a manner that continues to be acceptable for the majority of filter applications. We will review several of these approaches to the approximation problem.

**Obtaining an Equation for the Transfer Function Magnitude**

The ideal lowpass response has been approximated in at least three ways. The first results in the Butterworth response, but is more properly called the maximally-flat magnitude response. This response is monotonic in the passband and in the stopband. The second is the Chebyshev response that "ripples" in the passband. The ripple width can be controlled to be quite small or can become as large as a 3-dB ripple. Although this response is not monotonic in the passband, it has a sharper falloff in the transition region. The third type of response is the maximally flat phase response. We will discuss each of these important responses in more detail at a later time.

Typically, these responses will be specified in terms of certain key parameters to compare to the ideal response. The cutoff frequency is a key parameter, as also is the asymptotic slope of the response.

In order to demonstrate some of these concepts, let us consider a function given by

$$|H(j\omega)| = \frac{1}{\sqrt{1+\omega^4}} \tag{11.11}$$

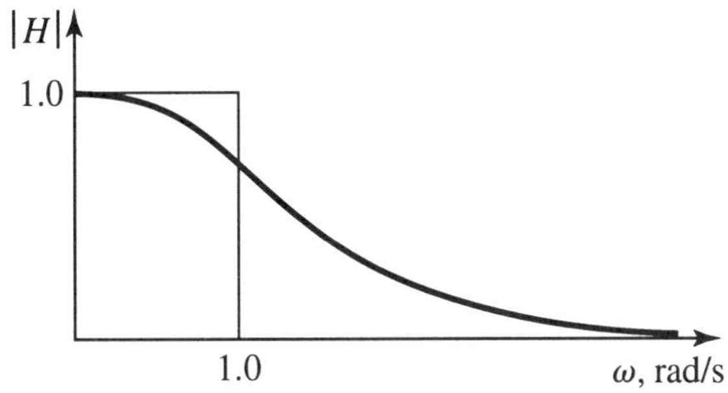


Figure 11.8: A practical frequency response.

This happens to be a second order Butterworth magnitude response, although that is unimportant to the present discussion. A rough plot of this magnitude is shown in Fig. 11.8 along with the ideal response.

The actual magnitude equals unity at zero frequency and remains close to unity for frequencies much smaller than 1 rad/s. In Butterworth filters, the cutoff frequency is defined as the frequency at which the response has dropped 3 dB from its maximum value. The cutoff frequency for this transfer function is

$$\omega_c = 1 \text{ rad/s}$$

The asymptotic falloff is calculated by allowing the frequency to move far above the cutoff frequency. The asymptotic falloff of the response of Eq. (11.11) is

$$20 \log \frac{1}{\sqrt{\omega^4}} = -10 \log \omega^4$$

This falloff is - 40 dB/decade.

If the asymptotic falloff is not sufficient for our purposes, we must modify the magnitude function. Suppose we wanted the response at or above $\omega = 10$ to be equal to or greater than 46 dB below the unity gain value. This function results in a 40-dB attenuation at $\omega = 10$ rather than the desired 46 dB. We can select a new magnitude function to give a greater attenuation. However, the new function must also correspond to a real function that can be implemented by some filter. One possible new function has a magnitude of

$$|H(j\omega)| = \frac{1}{\sqrt{1+\omega^6}} \tag{11.12}$$

This magnitude is 0 dB at zero frequency, - 3 dB at $\omega = 1$, and is $-60$ dB at $\omega = 10$. We will deal with the problem of physical realization of this function in a later section.

**Practice Problem**

**11.4** What is the magnitude of the function in Eq. (11.11) for $\omega = 6.3$ rad/s. Express the answer in dB. *Ans:* -48 dB.

### 11.2.4 Finding the Transfer Function from the Magnitude Function

As implied earlier, the most popular filter specification is given in terms of the frequency response of the magnitude of a transfer function. Another occasionally used specification is the frequency response of the phase with a secondary specification of frequency response of the magnitude. We will look at determining the transfer function of a filter based on the expression for the magnitude response.

Before we discuss this method, a short review of circuit analysis is necessary. When we analyze a network and write the transfer function as a function of frequency, we can calculate the magnitude from

$$|H(s)| = |H(j\omega)|_{s=j\omega} = \sqrt{Re^2(\omega) + Im^2(\omega)} \tag{11.13}$$

where $Re(\omega)$ is the real part and $Im(\omega)$ is the imaginary part of the transfer function $H(j\omega)$.

In order to deal with rational functions, we can conveniently work with the square of the magnitude rather than the magnitude of $H(j\omega)$. We then plot the frequency response of the square of the magnitude, understanding that the desired magnitude is the square root of this value at any frequency. The square of the magnitude is

$$|H(j\omega)|^2 = Re^2(\omega) + Im^2(\omega) \tag{11.14}$$

We also note that this function can be written as

$$|H(j\omega)|^2 = H(j\omega)H(-j\omega) = H(s)H(-s) \tag{11.15}$$

where $s = j\omega$.

To show how these equations are used, suppose that $H(j\omega) = 10 + j6$. Then, using the sum of the squares of the real and imaginary parts, we calculate

$$|H(j\omega)|^2 = 100 + 36 = 136$$

This is also equal to

$$H(j\omega)H(-j\omega) = (10 + j6)(10 - j6) = 136$$

Knowing this key relationship implies that, if we are given some value of $|H(j\omega)|^2$, we can find a corresponding $H(s)$. Then we can design or synthesize a circuit to create this value of $H(s)$ and the desired value of $|H(j\omega)|^2$.

# Example 11.3

The square of a transfer function magnitude is

$$|H(j\omega)|^2 = \frac{1 + \omega^2}{1 - \omega^2 + \omega^4}$$

Find $H(s)$.

**Solution** Substituting for $\omega$ with $\omega = s/j$, gives

$$H(s)H(-s) = \frac{1 - s^2}{1 + s^2 + s^4}$$

Using the quadratic equation to find the denominator factors leads to

$$H(s)H(-s) = \frac{(1-s)(1+s)}{(1+s+s^2)(1-s+s^2)}$$

The transfer function $H(s)$ is now taken as

$$H(s) = \frac{1+s}{1+s+s^2}$$

The remaining numerator and denominator factors are associated with $H(-s)$.

### 11.2.5 Synthesizing the Network

When the transfer function of the filter is known, various procedures can be used to construct the filter. This is a design problem that may have many different solutions. Perhaps a lossless $LC$ network or an active circuit using op amps and $RC$ networks can be used to obtain the desired transfer function. Because this is a complex topic, further discussion of this step of the procedure will be treated in succeeding chapters.

Using an equation for the transfer function to create a physical circuit that possesses this same transfer function is called *network synthesis.* The terms *implementing the network, realizing the network,* or *synthesizing the network,* given the transfer function equation, are equivalent.

**Practice Problem**

**11.5** The square of a transfer function magnitude is

$$|H(j\omega)|^2 = \frac{\omega^2}{1+\omega^2}$$

Find $H(s)$. *Ans:* $s/(1+s)$.

## 11.3 POLES AND ZEROS OF NETWORK FUNCTIONS

**Important Concepts**

1. A network function can be expressed in terms of the Laplace variable, $s$, and the *poles* and *zeros.*

2. The pole and zero locations in the Laplace plane relate to the frequency response of the transfer function.

The poles and zeros of a transfer function determine the frequency, phase, and transient response of the network. These poles and zeros are sometimes called critical frequencies. It is important to understand the relationship between these critical frequencies and the response of the network. This section will review some of the points that are essential to the synthesis of a network.

### 11.3.1 Definition of Poles and Zeros

We will express network functions in terms of the Laplace variable, $s$. A network will always have at least one pair of terminals across which a voltage can be measured, as well as a current into or out of one of these terminals. Each pair of terminals of interest is called a *port.*

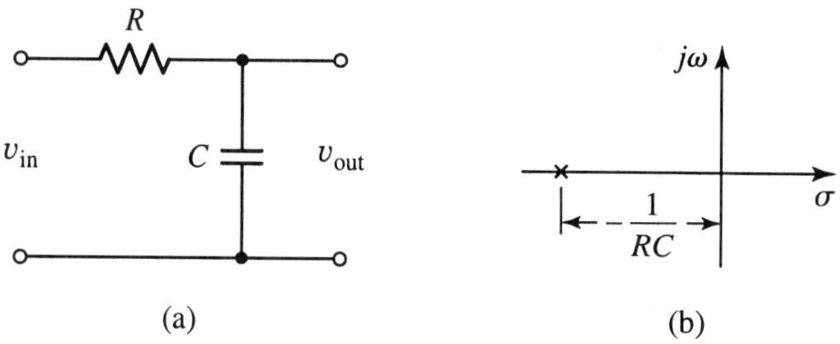


Figure 11.9: (a) A single-pole circuit. (b) The pole position.

A network function is a ratio of the voltage at one port to the voltage at another port, a ratio of current into one port to the current into another port, a ratio of voltage at a port to current into a port, or a ratio of current into a port to voltage at a port.

If the two variables involved are associated with different ports, the function is a transfer function. Examples would include the voltage transfer function from port 1 to port 2. This would be expressed as $V_2/V_1$. Other transfer functions could be

$$\frac{I_2}{V_1} \quad \frac{I_1}{I_3} \quad \frac{V_3}{I_1}$$

If the two variables, current and voltage, are associated with the same port, the function is called a driving-point function. These functions would always be ratios of voltage to current (impedance) or current to voltage (admittance).

A network function can always be expressed in terms of the ratio of two polynomials in $s$, such as

$$\frac{V_2}{V_1} = \frac{a_n s^n + a_{n-1}s^{n-1} + \cdots + a_1 s + a_0}{b_m s^m + b_{m-1}s^{m-1} + \cdots + b_1 s + b_0} \tag{11.16}$$

Both the numerator polynomial, $n(s)$, and the denominator polynomial, $d(s)$, can be factored, at least theoretically, to give

$$\frac{V_2}{V_1} = K\frac{(s-z_1)(s-z_2)\cdots(s-z_n)}{(s-p_1)(s-p_2)\cdots(s-p_m)} \tag{11.17}$$

where $K = a_n/b_m$. The $z_i$ and $p_i$ are complex constants called zeros and poles, respectively. When the variable $s$ takes on one of the $z_i$ values, the function goes to zero. If the variable $s$ approaches one of the $p_i$ values, the function approaches infinity. This, of course, assumes that no $z_i$ equals any $p_i$; that is, common factors in the numerator and denominator have been canceled.

### 11.3.2 Examples of Simple Network Functions

The first example is for the circuit shown in Fig. 11.9, which is a two-port network. Simple circuit analysis results in

$$\frac{v_{\text{out}}}{v_{\text{in}}} = \frac{\frac{1}{RC}}{s + \frac{1}{RC}} \tag{11.18}$$

This term has a constant for the numerator and a single factor in the denominator. There is no finite zero for this function. There is one pole at $s = -1/RC$. This pole position is plotted in Fig. 11.9(b) on the complex Laplace plane. In general, this plot is referred to as a pole-zero diagram.

The second example is the tank circuit or parallel $RLC$ circuit of Fig. 11.10.

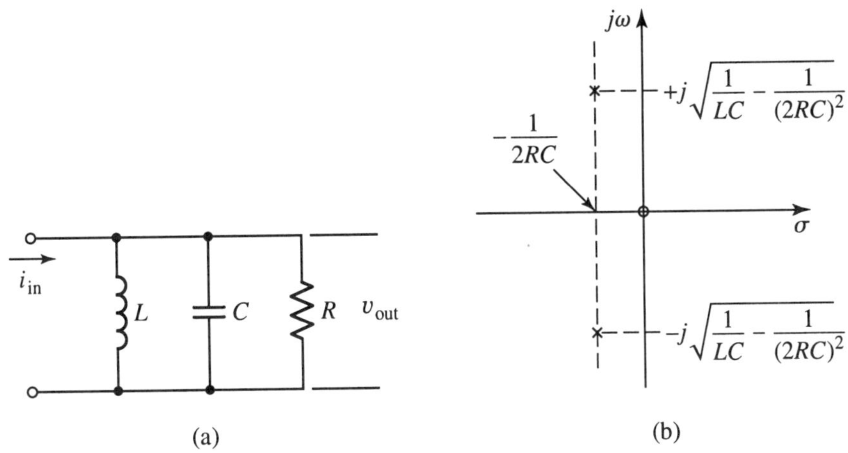


Figure 11.10: (a) A tank circuit. (b) The pole-zero diagram.

The output voltage as a function of input current is found to be

$$v_{out} = i_{in} \frac{\frac{1}{C}s}{s^2 + \frac{1}{RC}s + \frac{1}{LC}} \tag{11.19}$$

Using the quadratic formula, we can factor the denominator to yield

$$v_{out} = \frac{i_{in}\frac{1}{C}s}{\left[s + \frac{1}{2RC} - j\sqrt{\frac{1}{LC} - \frac{1}{(2RC)^2}}\right]\left[s + \frac{1}{2RC} + j\sqrt{\frac{1}{LC} - \frac{1}{(2RC)^2}}\right]} \tag{11.20}$$

The pole-zero diagram shows one zero at $s = 0$, along with a complex conjugate pair of poles.

### 11.3.3 Relationship of Pole and Zero Locations to Frequency Response

The zero and pole factors of a network function can be used to generate a plot of magnitude or phase as a function of frequency. Although other methods might be more convenient, this method helps visualize certain useful results.

The pole-zero diagram of Fig. 11.11 plots a single pole corresponding to the diagram of Fig. 11.9(b). In addition, this diagram includes a component along the imaginary axis corresponding to the excitation frequency of a sinusoidal input signal.

For convenience, we will assume that the network function resulting in the diagram of Fig. 11.11 is

$$N(s) = \frac{1}{s + \sigma_1} \tag{11.21}$$

The pole is located at $s = -\sigma_1$. A vector drawn from the origin to this pole is included in the diagram. For a sinusoidal excitation, the Laplace variable $s$ becomes equal to $j\omega$. The vectors indicated by arrows from the origin along the imaginary axis to $j\omega_1$, $j\omega_2$, and $j\omega_3$ represent three different excitation frequencies.

The magnitude response of Eq. (11.21) is found by evaluating the ratio of the magnitude of the numerator, which is unity in this case, to that of the denominator. The denominator of Eq. (11.21) can be expressed as

$$s - (-\sigma_1) = s + \sigma_1 \tag{11.22}$$

The $s$ term in Eq. (11.22) can be represented by a vector along the imaginary axis. The term $(-\sigma_1)$ is represented by the vector along the real axis to the value $-\sigma_1$. The difference of these two vectors is the pole

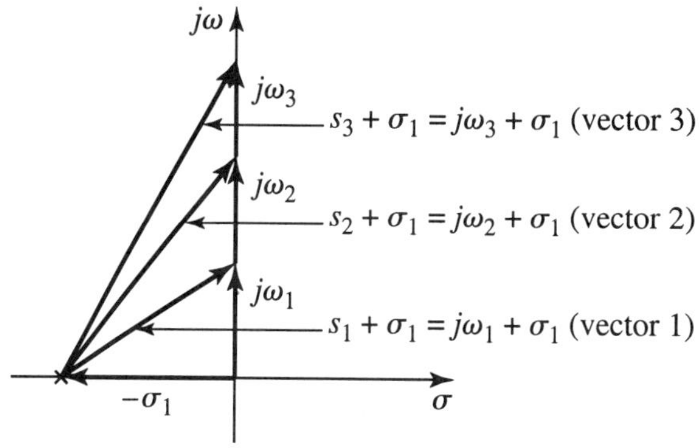


Figure 11.11: Pole-zero diagram with vectors.

factor given by Eq. (11.22). Vector 1 in Fig. 11.11 is the factor corresponding to an excitation frequency of $\omega_1$. Vectors 2 and 3 correspond to frequencies of $\omega_2$ and $\omega_3$, respectively. The magnitudes of these vectors can be used to find the magnitudes of the transfer function at these three frequencies. Simply dividing the vector magnitudes into unity results in the transfer function magnitude at the corresponding frequencies.

The phase response is found in a similar manner. The phase of the transfer function of Eq. (11.21) is given by the negative phase of the denominator factor. The phases of this factor equal the phases of vectors 1, 2, and 3 at the three excitation frequencies shown.

The magnitude response of Eq. (11.21) will equal

$$\frac{1}{\sigma_1} \quad \text{for } \omega_1 \ll \sigma_1$$

$$\frac{1}{\sqrt{2}\sigma_1} \quad \text{for } \omega_2 = \sigma_1$$

$$\frac{1}{\omega_3} \quad \text{for } \omega_3 \gg \sigma_1$$

This behavior describes a lowpass response.

The phase can be seen to shift from 0° at low frequencies to −90° at high frequencies. It will equal −45° when $\omega = \sigma_1$. The phase of the transfer function equals the phase of the numerator minus the phase of the denominator.

A more complex example is represented by the transfer function of Eq. (11.20). A pole-zero diagram is shown in Fig. 11.12. In this diagram, it is assumed that the real part of the pole pair is much smaller than the imaginary part.

With a zero at the origin, the zero factor is represented by vector 1. The two pole factors are represented by vectors 2 and 3. As the excitation frequency increases from zero, the zero factor magnitude increases proportionally. The magnitude of the pole factor given by vector 2 decreases while that given by vector 3 increases. The magnitude of the network function increases approximately as $\omega$ increases.

As the excitation frequency approaches the imaginary part of the pole, the magnitude of this pole factor becomes very small, decreasing to a minimum value equal to the real part of this pole. At frequencies far above the imaginary part of the pole, the magnitudes of both pole factors and the zero factor increase. The network function magnitude falls with $\omega$ in this region. Figure 11.13 shows the magnitude and phase responses of this network function. This filter is a narrowband, bandpass filter.

Although we will not discuss Figs. 11.14 to 11.18 in detail, some comments will be made to relate frequency responses to pole-zero diagrams.

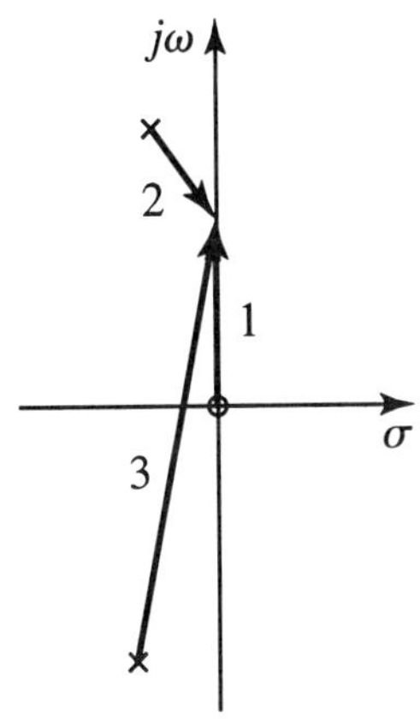


Figure 11.12: Pole-zero diagram for Eq. (11.20).

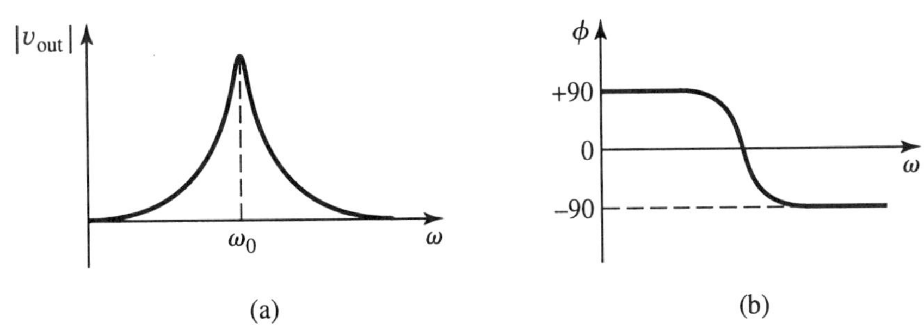


Figure 11.13: Frequency response of the function given by Eq. (11.20). (a) Magnitude response. (b) Phase response.

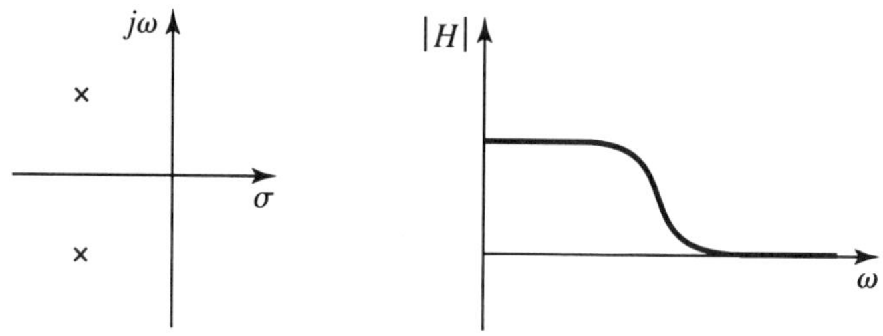


Figure 11.14: Lowpass filter diagrams.

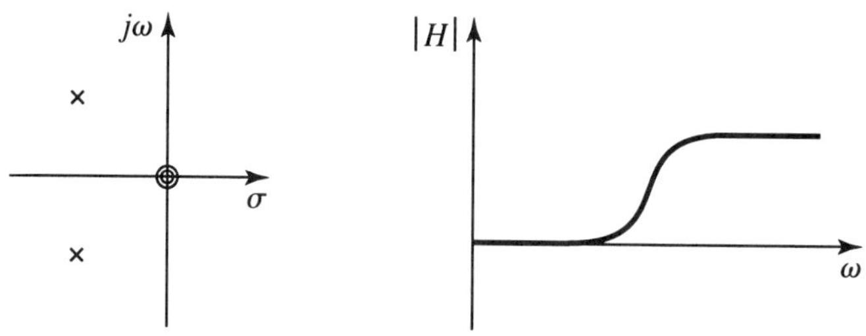


Figure 11.15: Highpass filter diagrams.

### Lowpass Filter

A lowpass filter results from an all-pole circuit. Zero factors cause an asymptotic increase in network or transfer function magnitude, whereas pole factors cause an asymptotic falloff in magnitude with frequency. Figure 11.14 indicates a two-pole lowpass filter along with the magnitude response of the transfer function with frequency. The filter transfer function is given by

$$H(s) = \frac{K}{s^2 + \sqrt{2}s + 1} \tag{11.23}$$

### Highpass Filter

The highpass transfer function includes both zero and pole factors. The numerator and denominator are of the same order. The pole positions may be similar to those of the lowpass transfer function, but multiple zeros may occur at the origin for the highpass circuit. The frequency response of the magnitude of this transfer function shows a zero magnitude at dc and increasing values with $\omega$ up to some maximum constant value as $\omega$ moves to large values. Figure 11.15 shows this behavior for a two-pole, two-zero transfer function. This highpass transfer function is given by

$$H(s) = \frac{Ks^2}{s^2 + \sqrt{2}s + 1} \tag{11.24}$$

### Bandpass Filter

We have seen that the bandpass filter is characterized by a single zero at the origin and two complex poles. The transfer function leads to a narrow passband as the pole locations approach the imaginary axis, that is, as the magnitude of the real part of the poles becomes smaller. As the pole locations move further from the imaginary axis, the passband becomes wider. The diagrams in Fig. 11.16 correspond to a narrowband bandpass filter with a transfer function of

$$H(s) = \frac{Ks}{s^2 + 0.05s + 1} \tag{11.25}$$

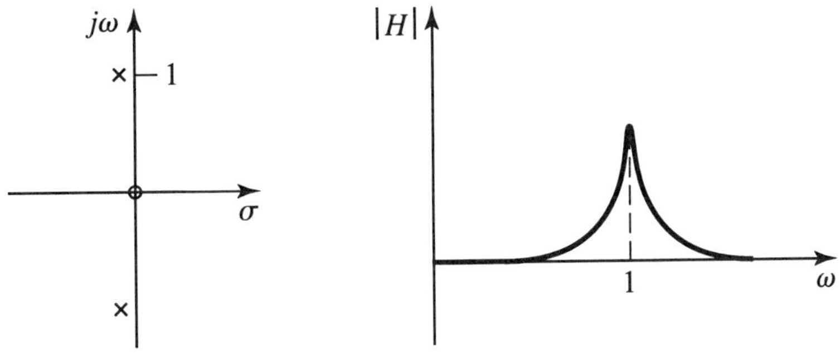


Figure 11.16: Bandpass filter diagrams.

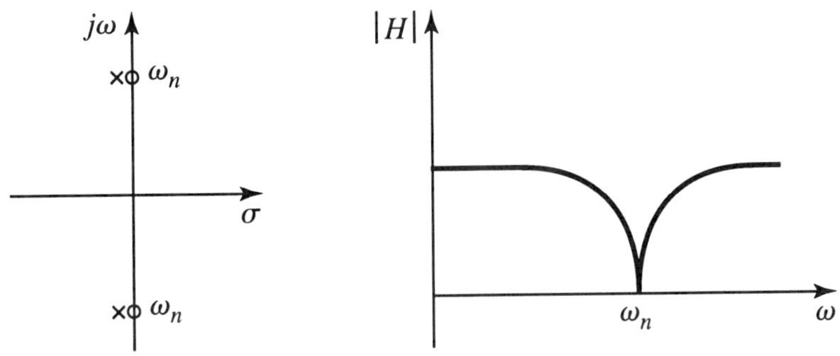


Figure 11.17: Band-reject filter diagrams.

### Band-Reject Filter

The transfer function of a band-reject filter has two imaginary zeros at $\pm j\omega_n$ and two complex poles. When the excitation frequency reaches $+j\omega_n$, the transfer function magnitude becomes equal to zero. Below or above this notch frequency, the transfer function approaches a constant value. The pole-zero diagram and the frequency response are shown in Fig. 11.17 for a transfer function of

$$H(s) = \frac{K(s^2+1)}{s^2+0.05s+1} \tag{11.26}$$

### Allpass Filter

The transfer functions of allpass filters have zeros that are mirror images of the poles across the imaginary axis. As the excitation frequency changes, the magnitudes of the pole factors equal the magnitudes of the zero factors. The overall magnitude of the transfer function is constant with frequency. The phase changes as a function of frequency while the magnitude remains constant. The pole-zero diagram, the magnitude response, and the phase response are shown in Fig. 11.18 for a transfer function of

$$H(s) = \frac{-s+1}{s+1} \tag{11.27}$$

The significant result of this section is that the pole-zero configuration determines the frequency response of a network function. When a certain frequency response is desired, poles and zeros must be selected properly to result in an appropriate network function.

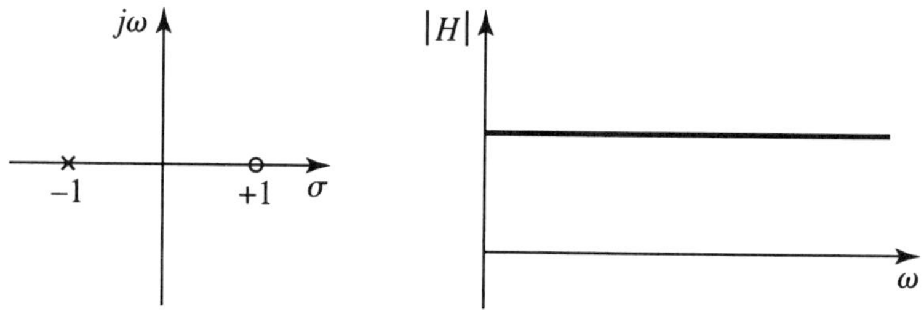


Figure 11.18: Allpass filter diagrams.

**Practice Problem**

**11.6** Calculate the 3-dB frequencies for the transfer function of Eq. (11.25). (bandpass) *Ans:* $\omega_2 = 1.025$ rad/s, $\omega_1 = 0.975$ rad/s.

### 11.3.4 Restrictions on Pole Locations

A circuit with a transfer function having a simple real pole such as

$$\frac{K}{s+a} \tag{11.28}$$

results in an impulse response of

$$Ke^{-at} \tag{11.29}$$

This is a decaying exponential which is a practical result for the case where the pole is at $s_1 = -a$. However, if the pole is at $s_1 = +a$, the transfer function has a factor of

$$\frac{K}{s-a} \tag{11.30}$$

The impulse response of this factor is

$$Ke^{at} \tag{11.31}$$

which is an increasing exponential with time. For a passive network, this result is an impossibility. A passive network cannot respond to an impulse input with an output that continues to increase indefinitely with time. Although an active circuit could approximate this response for a finite time period, this is generally not a desired result.

For transfer functions with complex poles the impulse response leads to exponentially decaying sinusoids if the real part of the poles is negative. If the real part of the poles is positive, an exponentially growing impulse response results.

From a practical standpoint, a passive circuit cannot generate an impulse response that is unbounded and thus cannot have a pole or poles with positive real parts. Practical passive circuits always result in transfer function poles in the left half of the Laplace plane. This same conclusion also applies to practical active circuits with stable, bounded outputs.

What about poles with zero real parts? Poles of some ideal passive circuits can fall on the $j\omega$ axis with no real part. For example, a lossless LC tank circuit would have imaginary poles. An impulse of input current would produce a voltage across the circuit that is a constant-amplitude sinusoid, lasting indefinitely. In a real circuit, there are losses in the inductor and capacitor that can be represented by a resistance. This leads to a small negative real part in the complex poles and produces a slowly-decaying sinusoid.

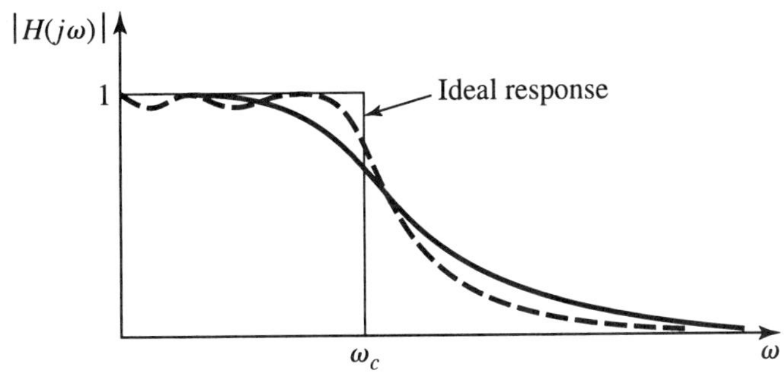


Figure 11.19: Ideal and actual lowpass transfer function responses.

An active circuit, notably an oscillator, can use gain to eliminate the negative real part of the pole and place the poles on the imaginary axis. This leads to an output that continues indefinitely.

We conclude that poles of transfer functions for passive or active filters must fall in the left-half plane (LHP) of the Laplace plane.

## 11.4 APPROXIMATING THE IDEAL FILTER RESPONSE

**Important Concepts**

**1.** Three significant lowpass functions used to approximate the ideal *brick wall* response are the Butterworth, Chebyshev, and Bessel functions.

**2.** The magnitude of the Butterworth function is maximally flat near zero frequency. It falls from its zero-frequency value by 3 dB at the cutoff frequency and continues to fall toward a zero value as frequency increases.

**3.** The magnitude of the Chebyshev function has small ripples in the passband, but falls sharply in the transition region.

**4.** The magnitude of the Bessel function does not fall as sharply as that of the Butterworth or Chebyshev functions. However, the transient response of this function closely reproduces the input signal.

The process of transfer function selection begins with the attempt to duplicate the ideal frequency response characteristics of Fig. 11.19 with a practical transfer function. One of the basic requirements a function must satisfy is that it be expressible as the ratio of two polynomials in the Laplace variable $s$. All actual circuits possess a transfer function that can be expressed in this manner; thus, the function that approximates the ideal characteristics must also be a ratio of polynomials in $s$.

It is well known that actual circuit transfer functions cannot yield discontinuities in the frequency response [5]. Consequently, the actual transfer function can only approximate the ideal function. The sense in which the actual function best matches the ideal function allows latitude for several choices.

One popular choice attempts to approximate the ideal function with one that is as flat as possible in the passband. Even though this function results in a flat function near $\omega = 0$, it does not have a sharp transition to the stopband and is not very flat as the cutoff frequency is approached. This approximation is called the maximally-flat magnitude response or Butterworth response.

A second approximation allows the response to ripple in the passband in order to achieve a sharper (steeper) transition region. This is called the equiripple response or Chebyshev response and maximizes the rate of falloff in the transition region.

The third type of response does not approximate the ideal magnitude response as well as the other two, but exhibits a linear phase characteristic in the passband. This is called a Bessel filter and is selected when phase information rather than amplitude is the prime consideration.

All of these approximations become more accurate as higher order polynomials are used. However, more circuit elements are required to realize these higher order functions. Thus, some compromise between accuracy and filter size is made.

### 11.4.1 The Butterworth (Maximally Flat Magnitude) Response

The Butterworth transfer function is a well-known filter function that has a maximally flat magnitude response.

#### The Definition of Maximal Flatness

The concept of maximal flatness of a magnitude response relates to setting as many derivatives of the magnitude with respect to $\omega$ equal to zero as possible at some frequency. For a lowpass filter, this frequency is $\omega = 0$. Setting these derivatives to zero ensures that the slope of the magnitude will come as close to that of the ideal filter (zero slope) as possible. As $\omega$ increases, the approximation becomes poorer, but near $\omega = 0$, the approximation can be very good.

A general transfer function magnitude can be expressed as

$$|H(j\omega)| = \sqrt{\frac{1 + c_2\omega^2 + c_4\omega^4 + \cdots + c_{2m}\omega^{2m}}{1 + d_2\omega^2 + d_4\omega^4 + \cdots + d_{2n}\omega^{2n}}} \tag{11.32}$$

Since it is easier to work with the square of the magnitude, we write

$$|H(j\omega)|^2 = \frac{1 + c_2\omega^2 + c_4\omega^4 + \cdots + c_{2m}\omega^{2m}}{1 + d_2\omega^2 + d_4\omega^4 + \cdots + d_{2n}\omega^{2n}} \tag{11.33}$$

Dividing the denominator into the numerator, term by term, gives

$$|H(j\omega)|^2 = 1 + (c_2 - d_2)\omega^2 + (c_4 - d_4 - c_2d_2 + d_2^2)\omega^4 + \cdots \tag{11.34}$$

This function can now be differentiated with respect to $\omega$ to give

$$\frac{d|H(j\omega)|^2}{d\omega} = 2(c_2 - d_2)\omega + 4(c_4 - d_4 - c_2d_2 + d_2^2)\omega^3 + \cdots \tag{11.35}$$

$$\frac{d^2|H(j\omega)|^2}{d\omega^2} = 2(c_2 - d_2) + 12(c_4 - d_4 - c_2d_2 + d_2^2)\omega^2 + \cdots \tag{11.36}$$

$$\frac{d^3|H(j\omega)|^2}{d\omega^3} = 24(c_4 - d_4 - c_2d_2 + d_2^2)\omega + \cdots \tag{11.37}$$

$$\frac{d^4|H(j\omega)|^2}{d\omega^4} = 24(c_4 - d_4 - c_2d_2 + d_2^2) + \cdots \tag{11.38}$$

All odd-order derivatives have coefficients that are multiplied by some finite power of $\omega$. All of these derivatives will be zero at $\omega = 0$. To make the even-order derivatives equal zero, the coefficients of the numerator and denominator polynomials of $|H(j\omega|^2$ can be properly chosen. The second derivative can be made to equal zero at $\omega = 0$ if $d_2 = c_2$. The fourth derivative can be zeroed if $d_4 = c_4$. If this process is continued, each denominator coefficient should be equated to the corresponding numerator coefficient. However, for a lowpass transfer function that approaches 0 as $\omega$ approaches $\infty$, the order of the denominator must be higher than that of the numerator. Thus, at least the last coefficient of the denominator has no constraint placed on it. We will choose a transfer function with a constant numerator for our lowpass function.

### The Butterworth Response

There are three constraints applied to the lowpass approximation that creates the Butterworth lowpass transfer function. The major condition applied is the maximally flat condition requiring that as many of the derivatives of the function as possible are set to zero at the $\omega = 0$ point. This condition makes the function as flat as possible near the origin.

The Butterworth approximation also applies the two constraints

$$|H(j0)| = 1 \tag{11.39}$$

and

$$|H(j1)| = 1/\sqrt{2} \tag{11.40}$$

The procedure to create a maximally flat magnitude response is similar to the procedure described in the preceding paragraphs. A lowpass response with zeros only at $\omega = \infty$ can be represented by

$$H(s) = \frac{1}{1 + a_1 s + a_2 s^2 + ... + a_n s^n} \tag{11.41}$$

where $n$ is the order of the denominator polynomial. The square of the magnitude of this transfer function is $H(s)H(-s)|_{s=j\omega}$ and is expressible as

$$|H(j\omega)|^2 = \frac{1}{1 + d_2\omega^2 + d_4\omega^4 + ... + d_{n-2}\omega^{2n-2} + d_{2n}\omega^{2n}} \tag{11.42}$$

We note from this equation that at $\omega = 0$ both the square of the magnitude and the magnitude are unity. As $\omega \to \infty$, the magnitude rapidly approaches zero, behaving asymptotically as $1/d_{2n}\omega^{2n}$ if $d_{2n}$ is nonzero.

We cannot make the slope of the response equal zero throughout the passband, but we can set the slope equal to zero at $\omega = 0$. We can also make the slope depart from zero more slowly as $\omega$ increases by setting higher order derivatives of $|H(j\omega)|^2$ with respect to $\omega$ equal to zero. Again, this can be done most easily by dividing the denominator of Eq. (11.42) into unity. The result of this operation is

$$|H(j\omega)|^2 = 1 - d_2\omega^2 + (d_2^2 - d_4)\omega^4 - (d_2^3 - 2d_2d_4 + d_6)\omega^6 + ...$$

The first derivative of $|H(j\omega)|^2$ is

$$-2d_2\omega + 4(d_2^2 - d_4)\omega^3 - 6(d_2^3 - 2d_2d_4 + d_6)\omega^5 + ...$$

Since each coefficient has some power of $\omega$ as a multiplier, the slope at $\omega = 0$ is always zero.

The second derivative is

$$-2d_2 + 12(d_2^2 - d_4)\omega^2 - 30(d_2^3 - 2d_2d_4 + d_6)\omega^4 + ...$$

This derivative can be set equal to zero if $d_2$ is set equal to zero.

The third and all odd derivatives are always zero at $\omega = 0$ for any values of coefficients. The higher order even derivatives can be zeroed by setting

$$d_4 = d_6 = ... = 0 \tag{11.43}$$

These results correspond to the earlier derivation that sets $d_{2i} = c_{2i}$ since, for this function, $c_{2i} = 0$.

If all coefficients in the denominator were set equal to zero, however, the magnitude could not approach zero at $\omega = \infty$. In fact, it would always be unity. Consequently, $d_{2n}$ must be nonzero, but all others are chosen to be zero; that is

$$d_2 = d_4 = ... = d_{2n-2} = 0 \tag{11.44}$$

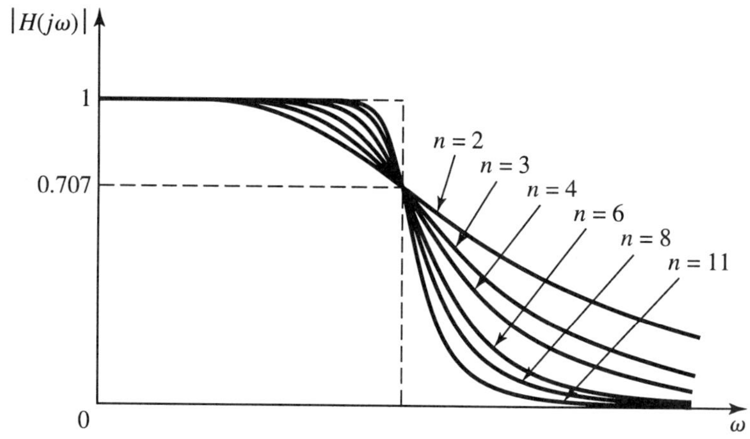


Figure 11.20: Butterworth responses for various values of $n$.

The choice of $d_{2n}$ is somewhat arbitrary, but if chosen to be unity, the response becomes a Butterworth response. This simplifies the final expression for the square of the magnitude to

$$|H(j\omega)|^2 = \frac{1}{1+\omega^{2n}} \tag{11.45}$$

The 3-dB point of the Butterworth transfer function occurs at $\omega_c = 1$.

The concept of analytic continuation can now be applied to the Butterworth response to find a physically realizable transfer function with LHP poles. This procedure has been applied to the various orders up to at least tenth order with the results tabulated [9]. This eliminates the need to recalculate the transfer function each time a filter is desired. These values up to a fifth order response are tabulated in Table 11.3.

**Table 11.3 Butterworth (Maximally Flat) Functions**

| n | $\|H(j\omega)\|$ | $H(s)$ |
|---|---|---|
| 1 | $\frac{1}{\sqrt{1+\omega^2}}$ | $\frac{1}{s+1}$ |
| 2 | $\frac{1}{\sqrt{1+\omega^4}}$ | $\frac{1}{s^2+1.41421s+1.00000}$ |
| 3 | $\frac{1}{\sqrt{1+\omega^6}}$ | $\frac{1}{s^3+2.00000s^2+2.00000s+1.00000}$ |
| 4 | $\frac{1}{\sqrt{1+\omega^8}}$ | $\frac{1}{s^4+2.61313s^3+3.14142s^2+2.61313s+1.00000}$ |
| 5 | $\frac{1}{\sqrt{1+\omega^{10}}}$ | $\frac{1}{s^5+3.23607s^4+5.23607s^3+5.23607s^2+3.23607s+1.00000}$ |

The coefficients of $H(s)$ are tabulated to five decimals because of high accuracy requirements in the synthesis process. Figure 11.20 shows the plots of Butterworth frequency responses of $|H(j\omega)|$ for different values of $n$.

### Pole Positions for Butterworth Transfer Functions

It is certainly not obvious how the functions $H(s)$ in Table 11.3 are obtained from the magnitudes of the functions. We shall now consider this matter in more detail. These functions are found by first calculating the pole positions of the square of the magnitude function [3].

The pole positions for the Butterworth filter are found by the methods considered in the previous chapter. We will demonstrate the method by assuming that we are given a second-order Butterworth filter with a denormalized transfer function magnitude of

$$|H(j\omega)| = \frac{1}{\sqrt{1+\omega^4/\omega_2^4}} \tag{11.46}$$

where $\omega_2$ is a known constant. This function might be derived from knowledge of the necessary frequency response of the desired filter. The problem is to find a transfer function $H(s)$ that can then be used to design or synthesize a physically realizable circuit.

We will use the expression $H(s)$ or $H(j\omega)$ to represent the transfer function of a filter. For most of the filters to be considered in this text, this transfer function will represent the ratio of the output voltage to the input voltage or

$$H(s) = \frac{V_{\text{out}}(s)}{V_{\text{in}}(s)} \tag{11.47}$$

From complex variable theory and from Eq. (11.15) we recognize that

$$|H(j\omega)|^2 = H(j\omega)H(-j\omega) \tag{11.48}$$

For sinusoidal inputs of $s = j\omega$ this equation can be written as

$$|H(j\omega)|^2 = H(s)H(-s)\,|_{s=j\omega} \tag{11.49}$$

The square of the transfer function magnitude for this example is

$$|H(j\omega)|^2 = \frac{1}{1+\frac{\omega^4}{\omega_2^4}}$$

$$= \left.\frac{1}{1+\frac{(s/j)^4}{\omega_2^4}}\right|_{s=j\omega} = \left.\frac{1}{1+\frac{s^4}{\omega_2^4}}\right|_{s=j\omega} = H(s)H(-s)|_{s=j\omega}$$

We now want to find the transfer function $H(s)$ that results in this magnitude function. We can construct $H(s)$ by locating the poles of $H(s)$. In order to do this we first find the poles of $H(s)H(-s)$ by finding the roots of the denominator. Equating the denominator to zero gives

$$\frac{s^4}{\omega_2^4} = -1 = 1\angle 180^\circ = \left(\frac{s}{\omega_2}\right)^4 \tag{11.50}$$

and allows us to find four roots. We find these roots by recalling from complex number theory that the $n$th root of a complex number expressed in polar form is

$$(A\angle\Theta)^{1/n} = A^{1/n}\angle\Phi \tag{11.51}$$

The angle $\Phi$ is given by

$$\Phi = \left[\frac{\Theta}{n} + \frac{360k}{n}\right] \text{ (in degrees)} \tag{11.52}$$

where $k = 0, 1, 2, ..., n-1$. For this case $A = 1$, $n = 4$, and $\Theta = 180^\circ$. The four roots for $s/\omega_2$ are then

$$1\angle 45^\circ,\ 1\angle 135^\circ,\ 1\angle 225^\circ,\ 1\angle 315^\circ$$

or

$$0.707 + j0.707,\ -0.707 + j0.707,\ -0.707 - j0.707,\ 0.707 - j0.707$$

The poles of $H(s)H(-s)$ are then equal to these values multiplied by $\omega_2$. We will assume that $\omega_2$ is normalized to a value of 1. The locations of these normalized poles are shown in Fig. 11.21.

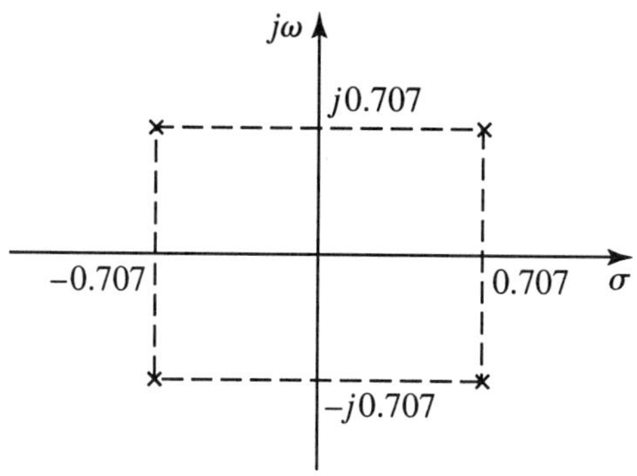


Figure 11.21: Pole locations of $|H(s)H(-s)|$.

The function can be written in terms of factors as

$$H(s)H(-s)$$

$$= \frac{1}{(s - 0.707 - j0.707)(s + 0.707 - j0.707)(s + 0.707 + j0.707)(s - 0.707 + j707)}$$

The four factors in this equation must be associated with $H(s)$ and $H(-s)$.

In order that $H(s)$ be a physically realizable transfer function, its poles must be in the left-hand plane (LHP) of the Laplace plane. Poles in the RHP lead to an impractical rising exponential of the transfer function when excited by a finite impulse. Thus, the LHP poles are associated with $H(s)$ and the RHP poles are associated with $H(-s)$. This gives a transfer function of

$$H(s) = \frac{1}{(s + 0.707 - j0.707)(s + 0.707 + j0.707)}$$

$$= \frac{1}{s^2 + \sqrt{2}s + 1}$$

This function can now be synthesized with passive elements such as resistors, inductors, and capacitors or active elements such as amplifiers along with resistors and capacitors. The topic of synthesis is discussed in Chapter 12. An example will demonstrate the use of the outlined procedure.

# Example 11.4

Derive the expression for $H(s)$ for a third-order Butterworth response.

**Solution** The square of the magnitude for a third-order function is, from Eq. (11.45),

$$|H(j\omega)|^2 = \frac{1}{1 + \omega^6}$$

Substituting $s/j$ for $\omega$ gives

$$H(s)H(-s) = \frac{1}{1 - s^6}$$

The poles for this function are shown in Fig. 11.22.

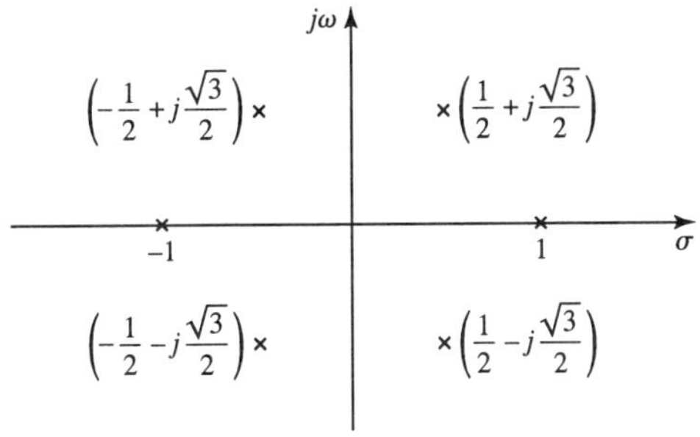


Figure 11.22: Poles for the function of Example 11.4.

We associate the LHP poles with $H(s)$ and write

$$H(s) = \frac{1}{\left(s+\frac{1}{2}-j\frac{\sqrt{3}}{2}\right)(s+1)\left(s+\frac{1}{2}+j\frac{\sqrt{3}}{2}\right)}$$

Multiplying the factors together gives

$$H(s) = \frac{1}{s^3+2s^2+2s+1}$$

which agrees with the information in Table 11.3.

**Practice Problem**

**11.7** Factor the denominator of $H(s)$ for the second-order Butterworth function to obtain the pole locations. *Ans:* $-0.707+j0.707$, $-0.707-j0.707$.

### 11.4.2 The Chebyshev (Equiripple) Response

A response that has equal-amplitude ripples in the passband, but falls sharply above $\omega = 1$, is given by

$$|H(j\omega)| = \frac{1}{\sqrt{1+\epsilon^2 C_n^2(\omega)}} \tag{11.53}$$

To cause equal ripples $\epsilon$ is chosen to be a constant and $C_n(\omega)$ is a polynomial that varies between 0 and 1 as $\omega$ increases from 0 to 1 and becomes much larger than 1 as $\omega$ exceeds 1. Depending on the order of the polynomial, $C_n(\omega)$ can vary between 0 and 1 several times in the passband.

Although the Butterworth filter achieves maximal flatness about the $\omega = 0$ point, the Chebyshev response that we now consider approximates the ideal filter in a different way. This response allows some prescribed deviation from the ideal curve in the passband and the fastest possible rate of cutoff in the transition region. The Chebyshev polynomials have been known for years and were originally named after an individual named Tschebycheff or Chebyshev [8]. The subscript $n$ indicates the order of the polynomial. Table 11.4 shows the Chebyshev polynomials from first to fifth order and indicates the values of $\omega$ for which $C_n$ equals 0 or $\pm 1$.

**Table 11.4 Chebyshev Polynomials**

| n | $C_n(\omega)$ | $\omega$ for $C_n = 0$ | $\omega$ for $C_n = \pm 1$ |
|---|---|---|---|
| 1 | $\omega$ | 0 | 1 |
| 2 | $2\omega^2 - 1$ | 0.707 | 0, 1 |
| 3 | $4\omega^3 - 3\omega$ | 0, 0.866 | 0.500, 1 |
| 4 | $8\omega^4 - 8\omega^2 + 1$ | 0.383, 0.924 | 0, 0.707, 1 |
| 5 | $16\omega^5 - 20\omega^3 + 5\omega$ | 0, 0.588, 0.951 | 0.314, 0.810, 1 |

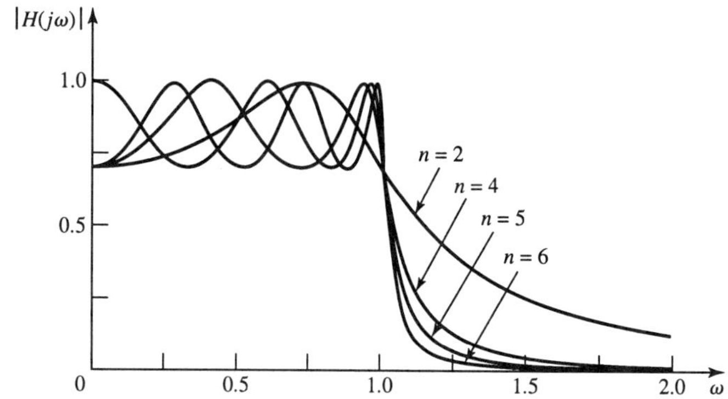


Figure 11.23: Chebyshev responses with $\epsilon = 1$.

As $C_n$ varies between 0 and $\pm 1$ in the passband, the magnitude of the transfer function varies between 1 when $C_n = 0$ and

$$\frac{1}{\sqrt{1+\epsilon^2}}$$

when $C_n = \pm 1$. The ripple width (maximum value minus minimum value) is given by

$$RW = 1 - \frac{1}{\sqrt{1+\epsilon^2}} \tag{11.54}$$

It is common to express this in dB by

$$RW(dB) = 20 \log_{10} \sqrt{1+\epsilon^2} \tag{11.55}$$

The value of $RW$ depends on $\epsilon$ and a few typical values are given in Table 11.5 along with the minimum values of $|H(j\omega)|$ in the passband.

**Table 11.5 Ripple Width as a Function of $\epsilon$**

| $\epsilon$ | **0.34931** | **0.50885** | **0.76478** | **0.99763** |
|---|---|---|---|---|
| RW(dB) | 0.5 | 1.0 | 2.0 | 3.0 |
| Passband min. | 0.944 | 0.891 | 0.794 | 0.708 |

Figure 11.23 shows Chebyshev responses for a ripple width of 3 dB and various values of $n$. The ripple width is typically smaller than the 3 dB shown in Fig. 11.23. We note that the number of peaks in the response equals $n/2$.

If the frequencies at which the polynomials go to 0 or a magnitude of 1 are called critical frequencies, we see that there are $n$ finite critical frequencies with the possibility of one more at $\omega = 0$ for odd-order polynomials. Note also that $C_n = 1$ at $\omega = 1$ for all values of $n$.

An alternative expression for $C_n(\omega)$ is

$$C_n(\omega) = \cos\,(n \cos^{-1}\,\omega) \tag{11.56}$$

for $0 \le \omega \le 1$ and

$$C_n(\omega) = \cosh\,(n \cosh^{-1}\,\omega) \tag{11.57}$$

for $\omega \ge 1$ [6].

The square of the Chebyshev transfer function is given by

$$|H(j\omega)|^2 = \frac{1}{1+\epsilon^2 C_n^2(\omega)} \tag{11.58}$$

As $C_n^2$ varies between 0 and 1 in the passband, the square of the magnitude of $H(j\omega)$ varies between 1 and $1/(1+\epsilon^2)$. This leads to $n/2$ maxima (humps) in the passband. We will again emphasize that at $\omega = 1$, the Chebyshev transfer function is down by $1/\sqrt{1+\epsilon^2}$, compared to 3 dB for the Butterworth function.

The transfer functions for the Chebyshev response depend on the value of $\epsilon$ or ripple width as well as on $n$. Many filter handbooks contain this information, however, the transfer functions for 1-dB ripple width are shown in Table 11.6 for first- through fifth-order filters.

**Table 11.6 Chebyshev Functions for 1-dB Ripple Width**

$\epsilon = 0.50885$, $\epsilon^2 = 0.25893$

| n | $\lvert H(j\omega)\rvert$ | **Denominator of H(s)** |
|---|---|---|
| 1 | $\frac{1}{\sqrt{1+\epsilon^2\omega^2}}$ | $s + 1.9652$ |
| 2 | $\frac{1}{\sqrt{1+\epsilon^2(2\omega^2-1)^2}}$ | $s^2 + 1.0977s + 1.1025$ |
| 3 | $\frac{1}{\sqrt{1+\epsilon^2(4\omega^3-3\omega)^2}}$ | $s^3 + 0.9883s^2 + 1.2384s + 0.4913$ |
| 4 | $\frac{1}{\sqrt{1+\epsilon^2(8\omega^4-8\omega^2+1)^2}}$ | $s^4 + 0.9528s^3 + 1.4539s^2 + 0.7426s + 0.2756$ |
| 5 | $\frac{1}{\sqrt{1+\epsilon^2(16\omega^5-20\omega^3+5\omega)^2}}$ | $s^5 + 0.9368s^4 + 1.6888s^3 + 0.9744s^2 + 0.5805s + 0.1228$ |

### 11.4.3 Comparison of Chebyshev to Butterworth

At frequencies above 1, the term $\epsilon^2 C_n^2$ becomes large compared to 1 and hence is the dominant term. For example, at $\omega = 1.5$, the fourth-order polynomial is $C_4(1.5) = 23.5$. The leading term of $C_n$ dominates the polynomial value as $\omega$ increases. For higher values of $\omega$, the magnitude of the transfer function can be approximated by

$$|H(j\omega)| \approx \frac{1}{\epsilon 2^{n-1}\omega^n} \tag{11.59}$$

We will designate the decibel magnitude of the Chebyshev transfer function by $MH_C$ and the decibel magnitude of the Butterworth function by $MH_B$. Thus,

$$MH_C = 20 \log \frac{1}{\epsilon 2^{n-1}\omega^n} = -20n \log \omega - 6(n-1) - 20 \log \epsilon \tag{11.60}$$

This can be compared to a Butterworth magnitude of

$$MH_B = -20n \log \omega \tag{11.61}$$

The second two terms in $MH_C$ indicate a higher falloff for the Chebyshev response than exhibited by the Butterworth response. The third term in $MH_C$ will always be negative when $\epsilon$ is smaller than unity; thus, the term $-20 \log \epsilon$ will be positive. This term detracts from a sharp falloff, but the second term typically contributes more to the falloff than the third term loses. As an example, for a third-order Chebyshev response with $\epsilon = 0.5$, the magnitude of the transfer function at higher frequencies is

$$MH_C = MH_B - 12 + 6 = MH_B - 6 \tag{11.62}$$

The third-order Chebyshev response is 6 dB lower than the third-order Butterworth response for large values of $\omega$. This additional attenuation takes place primarily in the transition region.

The following table compares the two responses for various orders, with $\epsilon = 0.5$. When $n$ is three or more, the Chebyshev results in more attenuation in the transition region before approaching its asymptotic falloff. The additional falloff can be expressed as $6(n-2)$.

| $n$ | 20 log $(n-1)$ |
|---|---|
| 2 | 0 dB |
| 3 | 6 dB |
| 4 | 12 dB |
| 5 | 18 dB |

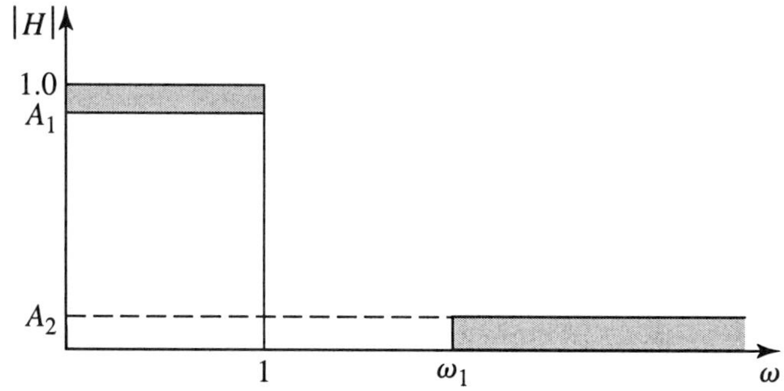


Figure 11.24: Attenuation requirements.

From Fig. 11.23, we see that the cutoff frequency for the Chebyshev filter is the point where the function drops below a value of

$$\frac{1}{\sqrt{1+\epsilon^2}}$$

For values of $\epsilon$ less than 1, the magnitude at $\omega = 1$ is less than 3 dB down from the maximum magnitude in the passband of unity. The Butterworth response is always 3 dB down at $\omega = 1$. Thus, when the Butterworth and the Chebyshev responses are compared, one must not compare on the basis of the magnitude at the normalized cutoff frequency of $\omega = 1$.

As an example of the comparison between the two responses, we will consider the magnitude response of Fig. 11.24 [1]. Let us assume that the response can have a maximum variation of 10% in the passband ($A_1 = 0.9$) and also in the region above $\omega = 2$ ($A_2 = 0.1$).

For a normalized Butterworth response, the curve is 3 dB down at $\omega = 1$. Obviously, this curve is not 3 dB down at this point. If we use the Butterworth response, we must then assume a higher cutoff frequency, $\omega_{cB}$, which will be greater than unity. The Butterworth response can then be expressed as

$$|H(j\omega)|_B = \frac{1}{\sqrt{1+(\omega/\omega_{cB})^{2n}}}$$

At $\omega = 1$ this function must be greater than or equal to 0.9. We can express this as

$$\frac{1}{\sqrt{1+(1/\omega_{cB})^{2n}}} \geq 0.9$$

which implies that

$$(1/\omega_{cB})^{2n} \leq \frac{19}{81}$$

At $\omega = 2$, the function must be less than 0.1 or

$$\frac{1}{\sqrt{1+(2/\omega_{cB})^{2n}}} \leq 0.1$$

This leads to

$$(1/\omega_{cB})^{2n} \geq \frac{99}{2^{2n}}$$

Since

$$\frac{99}{2^{2n}} \leq (1/\omega_{cB})^{2n} \leq \frac{19}{81}$$

then

$$\frac{99}{2^{2n}} \leq \frac{19}{81}$$

The minimum integer value of $n$ that will satisfy this last condition is 5. For the case of $n = 5$, the two inequalities tell us that

$$1.15604 \leq \omega_{cB} \leq 1.26318$$

We will choose $\omega_{cB} = 1.2$ to complete the Butterworth problem.

Now we will consider the realization of the same specifications with a Chebyshev response. The value of the response at $\omega_c = 1$ must exceed 0.9, giving

$$\frac{1}{\sqrt{1+\epsilon^2}} \geq 0.9$$

At $\omega = 2$ the condition is

$$\frac{1}{\sqrt{1+\epsilon^2 C_n^2(2)}} \leq 0.1$$

We solve for $\epsilon$ from the first equality, resulting in

$$\epsilon^2 \leq \frac{19}{81}$$

or

$$\epsilon \leq 0.48432$$

Using the largest allowed value of $\epsilon$ in the second inequality gives

$$|C_n(2)| \geq 20.544$$

From Table 11.4 we find that the third-order polynomial will satisfy this requirement. Thus, a third-order Chebyshev response or a fifth-order Butterworth will satisfy the specifications.

**Practical Considerations**

In this example, the average rate of falloff in the transition region is found to be about 23 dB/octave. Since the maximum asymptotic falloff of the Butterworth filter is $6n$ dB/octave, it is not unreasonable that $n = 5$ for the Butterworth function. On the other hand, the calculation of $n = 3$ for the Chebyshev function might be surprising because the asymptotic falloff of this filter is also $6n$ or 18 dB/octave in this case.

The Butterworth function never falls more steeply than its asymptotic rate of 30 dB/octave, whereas the Chebyshev can fall more steeply than its asymptotic rate of 18 dB/octave in the transition region. This results from the slightly resonant nature of the Chebyshev that leads to ripples in the passband and a steep falloff in the transition region.

**Practice Problem**

**11.8** Calculate the minimum value of a Chebyshev filter response in the passband for a 1.5 dB ripple width. *Ans:* 0.8414.

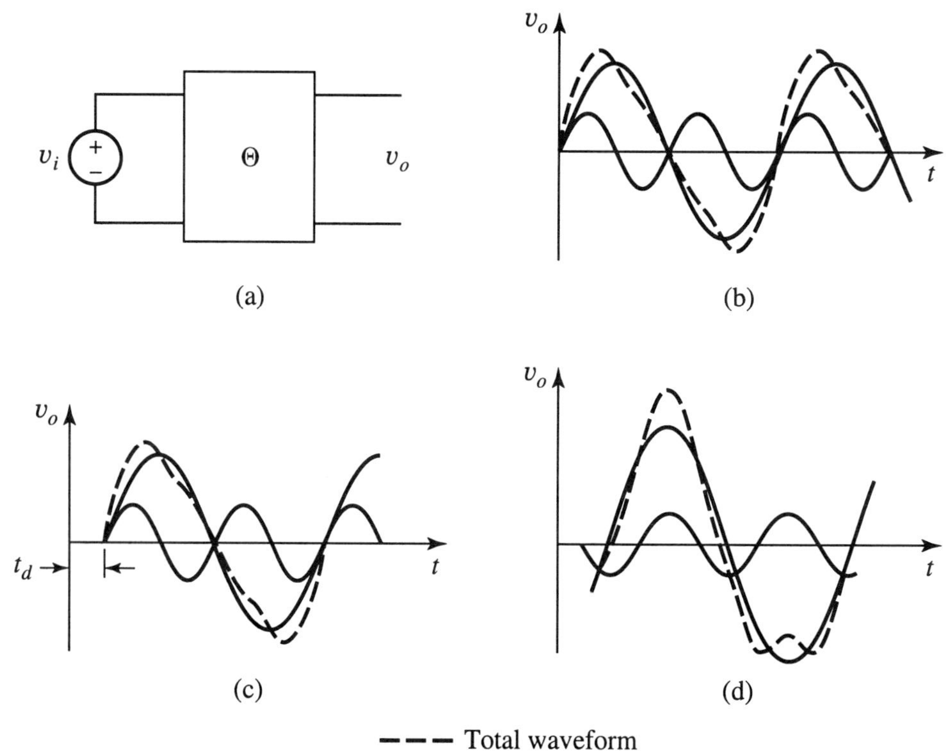


Figure 11.25: (a) A phase-shift network. (b) A waveform with fundamental plus second harmonic. (c) A delayed waveform with no phase distortion. (d) A delayed waveform with phase distortion.

### 11.4.4 Linear Phase (Bessel) Filters

The linear phase filter is also called the Bessel filter. The response of this filter is important in a limited number of applications.

#### The Need for Linear Phase Filters

A waveform consisting of several frequency components can become distorted, or experience a change of shape, if the relative phases of the components are modified. As a simple example, let us assume that a signal consisting of two components is applied to a phase-shifting network as shown in Fig. 11.25.

The input waveform is given by $v = 10 \sin \omega_0 t + 4 \sin 2\omega_0 t$ and is shown in Fig. 11.25(b). If the phase shifter provides a phase shift of

$$\Theta = -\frac{0.045}{2\pi}\omega$$

in degrees, and we set the frequency of the fundamental component to $f = 1$ kHz, the output signal will be

$$v_o = 10 \sin(\omega_o t - 45) + 4 \sin(2\omega_o t - 90)$$

Figure 11.25(c) shows this waveform that is delayed from the original, but is not distorted in shape. The phase shift of the fundamental component is $-45°$ referenced to the fundamental component. The phase shift of the second harmonic is $-90°$ referenced to the second harmonic component. The time delay of both components is the same, although the phase shifts are different. This results from the fact that each phase shift is referenced to a different period.

For a waveform to be delayed, but experience no waveshape distortion, the phase shift must be a linear function of frequency, or

$$\Theta = -k\omega \tag{11.63}$$

Now we will modify the phase-shift network to introduce a phase shift of

$$\Theta = -1.14 \times 10^{-7}\omega^2$$

In this instance, $\Theta$ is a nonlinear function of $\omega$. The output signal is now

$$v_o = 10 \sin(\omega_o t - 45) + 4 \sin(2\omega_o t - 180)$$

This waveform is shown in Fig. 11.25(d), and we see that the peak symmetry about the zero voltage line is destroyed. We emphasize that no change of amplitude of either component takes place in the phase-shift network. The distortion is due only to the change of relative phase between the components.

We can prove the general result that a pure time delay requires a linear phase shift by the use of Laplace transforms. Given a function of time expressed by $f(t)$, the Laplace transform is

$$L[f(t)] = F(s) \tag{11.64}$$

A pure time delay of $t_d$ in $f(t)$ can be expressed by $f(t - t_d)$. The Laplace transform of this delayed function for a sinusoid ($s = j\omega$) is

$$L[f(t - t_d)] = e^{-j\omega t_d} F(j\omega) \tag{11.65}$$

The additional phase shift due to the delay is given by

$$\Theta = -\omega t_d \tag{11.66}$$

In a filter that is designed to attenuate higher frequency components, waveshape change may result from both amplitude and phase distortion. However, the attenuated components are often noise or undesired components that were inadvertently introduced into the system. Therefore, the desired signal is generally not distorted due to amplitude attenuation which eliminates these undesired signals. On the other hand, phase distortion in the passband may result in waveshape change. This phase distortion can be demonstrated from the step responses of Fig. 11.26. The overshoot of the Butterworth filter is considerably greater than that of the linear phase filter. The Chebyshev response results in an even greater overshoot than does the Butterworth filter and is not used in applications requiring low overshoot.

The linear phase filter is also called a maximally-flat time delay filter since it approximates a constant time delay for all components falling in the passband. This filter is designed to attenuate high frequencies while providing for a phase shift that is linear with frequency within the passband. The magnitude response in the transition region is not as sharp as the Chebyshev or the Butterworth filter, but the waveshape change due to phase distortion is minimal. Table 11.7 compares the overshoot for the Butterworth and linear phase filters for filters up to the fifth order.

**Table 11.7 Overshoot of Butterworth and Linear Phase Filters**

| $n$ | Butterworth overshoot-percent | Linear phase overshoot-percent |
|---|---|---|
| 1 | 0.0 | 0.0 |
| 2 | 4.3 | 0.43 |
| 3 | 8.2 | 0.75 |
| 4 | 10.9 | 0.83 |
| 5 | 12.8 | 0.76 |

Figure 11.27 shows a comparison of frequency responses of the Butterworth and linear phase filter for $n = 4$. The linear phase filter obviously is not as sharp in the transition region as the Butterworth filter, but it is sharper than a four-pole filter with equal real poles. It also has a better transient response.

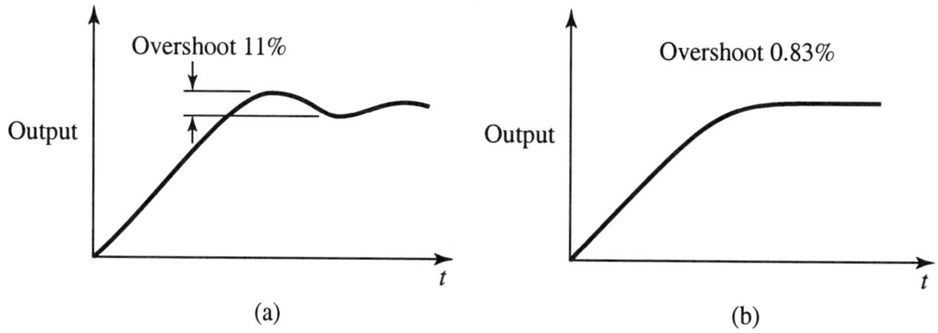


Figure 11.26: Step response of two four-pole filters. (a) Butterworth. (b) Linear phase.

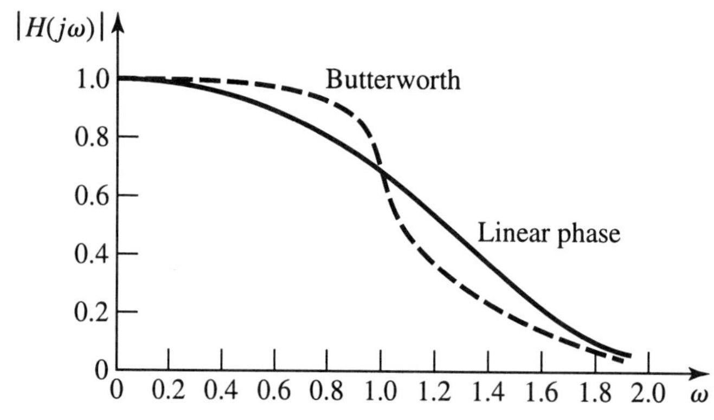


Figure 11.27: Magnitude responses of two four-pole filters.

### 11.4.5 The Transfer Function for the Bessel Filter

Rather than derive the general expression for the linear phase transfer function, we will specifically consider the second-order case to demonstrate the method of obtaining linear phase. A general, second-order, lowpass transfer function can be expressed as

$$H(s) = \frac{1}{s^2 + a_1 s + a_0} \tag{11.67}$$

The phase of this function for sinusoidal excitation is

$$\Theta = -\tan^{-1}\frac{a_1\omega}{a_0 - \omega^2} \tag{11.68}$$

Differentiating $\Theta$ with respect to $\omega$ in Eq. (11.66) gives

$$-\frac{d\Theta}{d\omega} = t_d \tag{11.69}$$

Taking the derivative of $\Theta$ from Eq. (11.68) gives

$$-\frac{d\Theta}{d\omega} = \frac{a_1 a_0 + a_1\omega^2}{a_0^2 + (a_1^2 - 2a_0)\omega^2 + \omega^4} \tag{11.70}$$

In the ideal case, the time delay $t_d$ is constant, but for the actual filter we see that the time delay is a function of $\omega$. Dividing the denominator into the numerator of Eq. (11.70) gives a time delay of

$$t_d = \frac{a_1}{a_0} + \omega^2\frac{(3a_1 - a_1^3/a_0)}{a_0^2} + \cdots \tag{11.71}$$

For small values of $\omega$, the lower order terms are dominant; thus, Eq. (11.71) can approximate a constant if the coefficient of $\omega^2$ is set to zero. Actually, two conditions can be applied to Eq. (11.71). The first condition applied is

$$a_1 = a_0$$

to normalize the approximate time delay to 1 second. The second condition to approximate a constant delay is

$$a_1 = a_0 = 3$$

The second-order transfer function is then

$$H(s) = \frac{1}{s^2 + 3s + 3} \tag{11.72}$$

The denominator in Eq. (11.72) is a Bessel function of the first kind. If this procedure is followed for other orders of filter, the resulting Bessel functions are given in Table 11.8.

**Table 11.8 Denominators of Linear Phase Filters**

| n | Denominator |
|---|---|
| 1 | $s+1$ |
| 2 | $s^2+3s+3$ |
| 3 | $s^3+6s^2+15s+15$ |
| 4 | $s^4+10s^3+45s^2+105s+105$ |
| 5 | $s^5+15s^4+105s^3+420s^2+945s+945$ |

A recursion formula [8] that allows higher order Bessel functions to be calculated from lower order functions is

$$D_{n+1}(s) = (2n+1)D_n(s) + s^2 D_{n-1}(s) \tag{11.73}$$

where $D_n(s)$ is the denominator of the $n$th order transfer function.

The normalization of the time delay to a delay of unity at $\omega = 0$ is essentially equal to a frequency normalization. However, the normalized frequency does not relate to an amplitude condition as $\omega_c$ does in the Butterworth and Chebyshev responses.

We must consider the effect of denormalization on the time delay. The normalized frequency will be designated $u$ so that a third-order transfer function can be expressed as

$$H(ju) = \frac{1}{(ju)^3 + 6(ju)^2 + 15ju + 15}$$

The time delay is expressed in terms of the normalized frequency $u$ as

$$t_d = -\frac{d\Theta}{du} \tag{11.74}$$

If the frequency is denormalized to a value of $\omega = u\omega_o$, we can write the actual time delay as

$$t_o = -\frac{d\Theta}{d\omega} \tag{11.75}$$

where $\Theta$ is the actual phase shift.

It can be shown that the phase shift of the normalized function is the same as that of the denormalized function; that is, $\Theta(u_1) = \Theta(\omega_o u_1)$. It follows from $d\omega = d(u\omega_o) = \omega_o du$, that the actual delay can be written as

$$t_o = -\frac{1}{\omega_o}\frac{d\Theta}{du} = \frac{t_d}{\omega_o} \tag{11.76}$$

Since the normalized delay, $t_d$, is equal to unity, we can write that the actual delay is

$$t_o = \frac{1}{\omega_o} \tag{11.77}$$

Thus, the time delay of the denormalized function at $\omega = 0$ is a factor of $\omega_o$ less than the normalized value at $u = 0$.

As $u$ is increased, the time delay varies from the zero frequency value as indicated in Table 11.9 [8].

**Table 11.9 Deviation in Time Delay from Zero-Frequency Value**

| $n$ | $u$ for 1% deviation | $u$ for 10% deviation | $u$ for 20% deviation | $u$ for 50% deviation |
|---|---|---|---|---|
| 1 | 0.10 | 0.34 | 0.50 | 1.00 |
| 2 | 0.56 | 1.09 | 1.39 | 2.20 |
| 3 | 1.21 | 1.94 | 2.29 | 3.40 |
| 4 | 1.93 | 2.84 | 3.31 | 4.60 |
| 5 | 2.71 | 3.76 | 4.20 | 5.78 |
| 6 | 3.52 | 4.69 | 5.95 | 6.97 |

We note from Table 11.9 that the time delay deviates from its unity zero-frequency value by less than 1% up to $u = 1$ for $n$ greater than 2.

A second table that is important in selecting the order of the linear phase filter is listed as Table 11.10 [8].

**Table 11.10 Attenuation as a Function of $u$**

| $n$ | $u$ for $\frac{1}{50}$ dB | $u$ for $\frac{1}{10}$ dB | $u$ for $\frac{1}{2}$ dB | $u$ for 1 dB | $u$ for 3 dB |
|---|---|---|---|---|---|
| 1 | 0.08 | 0.14 | 0.35 | 0.51 | 1.00 |
| 2 | 0.11 | 0.26 | 0.57 | 0.80 | 1.36 |
| 3 | 0.14 | 0.34 | 0.75 | 1.05 | 1.75 |
| 4 | 0.17 | 0.40 | 0.89 | 1.25 | 2.13 |
| 5 | 0.20 | 0.45 | 1.01 | 1.43 | 2.42 |
| 6 | 0.22 | 0.50 | 1.12 | 1.58 | 2.70 |

The use of these tables is demonstrated in the following two examples.

# Example 11.5

Given a three-pole linear phase filter with a 3-dB frequency of 1 kHz, find the normalizing factor $\omega_o$.

**Solution** For a three-pole filter, Table 11.10 shows that the 3-dB point occurs at $u = 1.75$. Since $u = \frac{\omega}{\omega_o}$ and the desired 3-dB point is $\omega = 2\pi \times 10^3$, the value of $\omega_o$ is

$$\omega_o = \frac{2\pi \times 10^3}{1.75} = 3.59 \times 10^3$$

Once the normalized filter is synthesized, the factor $\omega_o$ can be used to denormalize the element values.

# Example 11.6

A delay of 20 $\mu$s is required along with 3-dB bandwidth of $1.06 \times 10^5$ rad/s ($\pm 5\%$). Find the value of $n$ required.

**Solution** The normalization factor is

$$\omega_o = \frac{1}{t_o} = \frac{1}{20 \times 10^{-6}} = 5 \times 10^4 \text{ rad/s}$$

Since

$$u = \frac{\omega}{\omega_o}$$

the 3-dB value of $u$ must be

$$u_{3db} = \frac{\omega_{3db}}{\omega_o} = \frac{1.06 \times 10^5}{5 \times 10^4} = 2.125$$

From Table 11.10 we see that a fourth-order filter has a 3-dB point of $u = 2.13$; thus, this value of $n$ will satisfy the specifications within the 5% tolerance.

The time delay should be checked from Table 11.9. For $n = 4$, it is seen that the error in time delay is less than 1% up to a value of $u = 1.93$ which is near the 3-dB point. Therefore, this value of $n$ will simultaneously satisfy both specifications.

**Practice Problem**

**11.9** If the actual delay of a third-order Bessel filter is 120 ns, what is the 3-dB frequency of the filter? *Ans:* 14.58 Mrad/s.

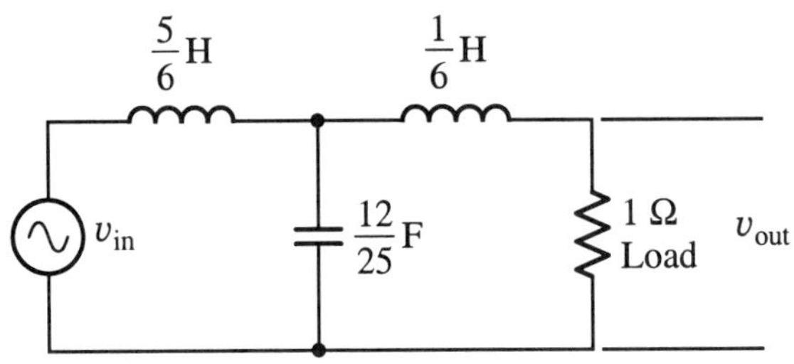


**Figure P11.5**

---

## SUMMARY


**1.** Most electronic filters are designed to pass the desired signals with little attenuation while attenuating the unwanted signals (noise) severely.

**2.** A filter synthesis procedure begins with the specifications of a filter and proceeds to develop the proper transfer function for this response. A circuit configuration that results in this transfer function is then synthesized.

**3.** The poles and zeros of a Laplace transfer function are important in determining the characteristics of a filter.

**4.** The Butterworth, Chebyshev, and Bessel functions are often used to create approximate filter transfer functions. These functions are tabulated in many filter handbooks to aid in filter design.


---

## PROBLEMS

### Section 11.2 Overview of Filter Synthesis Procedures

**11.1** A signal component at 1 kHz is 4.6 V in magnitude. A noise component at 32 kHz has a 0.5 V magnitude. How many dB must this signal be attenuated to make the magnitude be 100 dB below the signal component? What is the minimum number of dB per octave falloff between 1 kHz and 32 kHz required to meet this specification? How many dB per decade is this value?

**11.2** Repeat Problem 11.1 if the noise component has a magnitude of 0.8 V at 28 kHz.

**11.3** Rework Example 11.1 if $b$ is changed to 0.001.

**11.4** A normalized filter with a bandwidth of 1 rad/s and a 1-Ω load resistor has values of $L = 2.62$ H and $C = 1.14$ F. If the actual filter to be used has a bandwidth of $2\pi \times 10^3$ rad/s and a 500-Ω load, find the denormalized values of $L$ and $C$.

**11.5** The filter shown has a 1 rad/s bandwidth. Denormalize the element values to result in a filter with a 2-kHz bandwidth and a 200-Ω load.

**11.6** The second-order normalized Butterworth transfer function, given by Eq. (11.11), has a 3-dB frequency of 1 rad/s. A denormalized second-order Butterworth filter has a 3-dB frequency of 20 kHz.

> **(a)** Calculate the frequency at which the normalized filter is down by 54 dB from its dc value.
> **(b)** Calculate the frequency at which the denormalized filter is down by 54 dB from its dc value.

**(c)** Write the transfer function of the denormalized filter.

**11.7** Repeat Problem 11.6 for the normalized transfer function of Eq. (11.12).

**11.8** Given that

$$|H(j\omega)|^2 = \frac{1}{\omega^4 - 3\omega^2 + 4}$$

Find $H(s)$ where $s = j\omega$.

**11.9** Given that

$$|H(j\omega)|^2 = \frac{1}{\omega^4 + 6\omega^2 + 25}$$

Find $H(s)$ where $s = j\omega$.

**Section 11.3 Poles and Zeros of Network Functions**

**11.10** What kind of filter has a transfer function of

$$H(s) = \frac{1}{s^2 + s + 1}$$

Evaluate the magnitude of the transfer function at $\omega = 0,\ 0.5,\ 1.0$, and 10 rad/s.

**11.11** What kind of filter has a transfer function of

$$H(s) = \frac{s^2}{s^2 + s + 1}$$

Evaluate the magnitude of the transfer function at $\omega = 0,\ 0.5,\ 1.0,\ 2.0$, and 10 rad/s.

**11.12** What kind of filter has a transfer function of

$$H(s) = \frac{s}{s^2 + s + 1}$$

Evaluate the magnitude of the transfer function at $\omega = 0,\ 0.5,\ 1.0,\ 2.0$, and 10 rad/s.

**11.13** What kind of filter has a transfer function of

$$H(s) = \frac{s^2}{s^2 + 0.02s + 1}$$

Evaluate the magnitude of the transfer function at $\omega = 0,\ 0.5,\ 1.0,\ 2.0$, and 10 rad/s.

**Section 11.4 Approximating the Ideal Filter Response**

**11.14** A transfer function is expressed as

$$H(s) = \frac{s^2 + 4s + 3}{s^3 + as^2 + bs + 4}$$

**(a)** Select the coefficients of the denominator, $a$ and $b$, to lead to a maximally flat magnitude response about the $\omega = 0$ point.
**(b)** Calculate the poles of $H(s)$.
**(c)** Calculate the first four derivatives of $|H(j\omega)|^2$ with respect to $\omega$ and evaluate at $\omega = 0$.

**11.15** Repeat Problem 11.14 for a transfer function of

$$H(s) = \frac{s^2 + s + 4}{s^3 + as^2 + bs + 4}$$

**11.16** A filter transfer function is to have a magnitude response that is 0 dB at $\omega = 0$ and falls to $-3$ dB at $\omega = 1$ rad/s. Above $\omega = 3$ rad/s, the magnitude must be down by $-24$ dB or more. Select the minimum order of a Butterworth transfer function that will satisfy these specifications. How many dB/octave will the filter fall as $\omega \to \infty$?

**11.17** Repeat Problem 11.16 if the filter must be down to $-24$ dB at $\omega = 2$ rad/s.

**11.18** Repeat Problem 11.17 if the filter must be down to $-46$ db at $\omega = 3$ rad/s.

**11.19** A Butterworth filter with a 3-dB frequency of 10 kHz is to attenuate a 1-V, 25-kHz signal to a value of less than 0.1 V. Calculate the minimum required value of $n$ (the number of poles).

**11.20** Find the pole positions for a third-order Butterworth lowpass transfer function.

**11.21** If

$$|H(j\omega)|^2 = \frac{1}{9 + 3\omega^2 + \omega^4}$$

find $H(s)$.

**11.22** A filter transfer function is to have a magnitude response with a ripple width of 0.5 dB in the passband. The function is to be down 3 dB at $\omega = 1$ rad/s. Above $\omega = 3$ rad/s, the magnitude must be down by $-24$ dB or more. Select the minimum order of a Chebyshev transfer function that will satisfy these specifications. How many dB/octave will the filter fall as $\omega \to \infty$?

**11.23** A filter is to have a magnitude response that falls between 0.9 and 1.0 in the passband, below $\omega = 1$ rad/s. Above $\omega = 2$ rad/s, the magnitude must be 0.032 or less. Select the minimum order of a Chebyshev filter that will meet these specifications.

**11.24** Will a fifth-order Butterworth filter satisfy the specifications of Problem 11.23? Support your answer with calculations.

**11.25** Find the smallest value of $n$ and the largest value of $\epsilon$ to give a ripple width that is less than or equal to 0.1 and a 3-dB point less than or equal to 1.25 rad/s.

**11.26** A second-order Butterworth filter and a second-order Chebyshev filter are used to satisfy the following specifications. The response should lie between 0.9 and 1 in the passband and should be less than 0.05 above $\omega_1$. Find $\omega_1$ for each case.

**11.27** Three-pole, lowpass Butterworth and Bessel filters are constructed to have a normalized 3-dB bandwidth of 1 rad/s.

**(a)** At what frequencies will each filter pass a signal that equals 90% of the value passed at dc?
**(b)** If the zero frequency time delay of the denormalized Bessel filter is 80 $\mu$s, what is the denormalized 3-dB bandwidth?

**11.28** Repeat Problem 11.16 for a linear phase filter.

## REFERENCES

1. Johnson, D. E., *Introduction to Filter Theory.* Englewood Cliffs, N.J.: Prentice Hall, 1976, Chapter 3.

2. Johnson, D. E., Johnson, J. R., and Moore, H. P., *A Handbook of Active Filters.* Englewood Cliffs, NJ: Prentice-Hall, 1980.

3. Miner, G. F. and Comer, D. J., *Physical Data Acquisition for Digital Processing.* Englewood Cliffs, NJ: Prentice-Hall, 1992.

4. Sallen, R. P and Key, E. L., "A practical method of designing RC active filters," *IRE Transactions on Circuit Theory,* March 1955, pp. 74-84.

5. Schaumann, R. and Van Valkenburg, M. E., *Design of Analog Filters.* New York: Oxford University Press, 2001.

6. Su, K. L., *Analog Filters.* Chapman and Hall, London, UK, 1996.

7. Van Valkenburg, M. E., *Analog Filter Design.* New York: Holt, Rinehart and Winston, 1982.

8. Weinberg, L., *Network Analysis and Synthesis.* New York: McGraw-Hill, 1962, Chapter 11.

9. Zverev, A. I., *Handbook of Filter Synthesis.* New York: Wiley, 1967.

# Chapter 12

# ACTIVE SYNTHESIS

## 12.1 ACTIVE FILTERS

**Important Concepts**

**1.** Active filters are used to create complex poles without using inductors.

**2.** Active filters may also minimize required capacitance values. In some cases, the resulting values can be integrated on an IC chip.

**3.** Positive feedback is often used to create complex poles without inductors.

Active filters are circuits that use active elements such as amplifiers in conjunction with RC circuits to create complex poles in the transfer function. Only real poles can be achieved with passive RC networks, but positive feedback can be used to move poles to a complex location.

Much of the early work in active filters used op amps for the active element with discrete capacitors and resistors. This type of hybrid active filter is still used since it is relatively compact compared to many passive filters. In more recent years, the continuous-time active filter has been implemented on IC chips. We will refer to this implementation as the IC filter and cover this topic in a later chapter.

Although a mild interest in active filters existed prior to the development of the integrated circuit, the major advancements in active filters have been motivated by a desire to construct integrated circuit filters. The results of that work have given the engineer more choices in selecting the type of filter to be used in the system. Completely integrated filters such as the switched-capacitor filter might be used, a hybrid filter combining IC chips with discrete capacitors may be used, or a discrete component, active filter with transistors, resistors, and capacitors may be selected. In more recent years, the continuous-time active filter on an IC chip has become more prominent.

The active filter is used for several reasons. An active filter can eliminate inductors, decrease the size of required capacitors, obtain high values of $Q$ in narrowband applications, and occupy very little physical volume. Of course, these advantages are accompanied by disadvantages. The active filter requires dc power, may experience stability problems, and may be sensitive to temperature change.

Elimination of inductors has been a goal of active synthesis for many years. Of the three passive elements

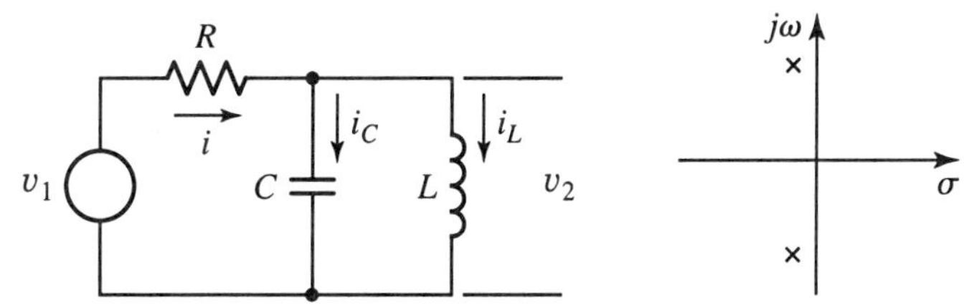


Figure 12.1: A parallel LC circuit and pole locations.

used in filters, the inductor is the most nonideal element. It is used as an energy storage element, but exhibits relatively high resistive energy loss compared to the capacitor. Only small inductors can be integrated on a silicon chip and often have low $Q$s. Even before the field of integrated circuits emerged however, methods of avoiding inductors, especially those of large value, were developed.

One approach to eliminating conventional inductors from circuits is to simulate the inductor by means of integrated circuit components. The gyrator [14] is an example of this type of circuit. A second approach is to simulate the response of circuits that use inductors. Both methods are used at the present time to create filter circuits with complex poles.

Any passive circuit that has complex poles requires both inductors and capacitors. This means that all of the popular response types such as Butterworth, Chebyshev, Bessel, or elliptic can be achieved with passive circuits by using these two elements. In active filter design, a considerably different configuration might be used to create a resonant circuit without using an inductor. Several other configurations can be used to obtain these types of response. The following pages will only summarize a small number of approaches.

The basic idea of an active circuit can be understood by examining the RLC circuit shown in Fig. 12.1.

If $R$ is very large, so that $v_1$ and $R$ simulate a current source of value

$$i = \frac{v_1}{R} \tag{12.1}$$

the output voltage is

$$v_2 = i \times Z_{LC} \tag{12.2}$$

The current $i$ consists of the sum of $i_C$ and $i_L$. If

$$i_L = -i_C \tag{12.3}$$

then $i$ equals zero; that is, no external current enters the LC combination, but the voltage $v_2$ is present. If a voltage appears across an impedance that has no entering current, the impedance is infinite.

At resonance, this ideal situation is only approximated due to losses in the inductor, capacitor, and source resistance, but a very high impedance results. We can write that

$$i_c = \frac{v_2}{1/sC} = sCv_2 \tag{12.4}$$

and

$$i_L = \frac{v_2}{sL} \tag{12.5}$$

Substituting $j\omega$ for $s$ and equating $i_C$ to $-i_L$ allows the resonant frequency to be found. Since

$$j\omega C v_2 = -\frac{v_2}{j\omega L} \tag{12.6}$$

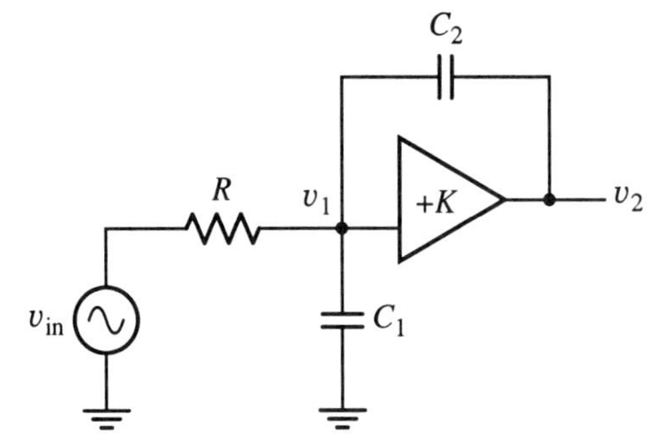


Figure 12.2: A positive feedback circuit.

at resonance, the resonant frequency is

$$\omega_o = \frac{1}{\sqrt{LC}} \tag{12.7}$$

The significant point here is that the current through the inductor is ideally 180° different from the current through the capacitor. It is impossible to get currents in adjacent elements of a passive RC or RL circuit to have a 180° phase difference. However, if positive feedback, implying a gain from an active circuit, is used, this phase shift can be achieved. The circuit of Fig. 12.2 demonstrates this point.

The transfer function of this circuit can be found to be

$$\frac{v_2}{v_{in}} = \frac{K}{s[C_1R + C_2R - KC_2R] + 1} \tag{12.8}$$

The negative sign in the denominator arises since a positive voltage $v_1$ can result in current entering this node through $C_2$ if $K$ is greater than unity. There are three conditions on $K$ that lead to three different results. If $KC_2 < (C_1 + C_2)$, then the transfer function appears to be that of a normal RC circuit. If $KC_2 = (C_1 + C_2)$, the function appears to be a purely resistive function. If $KC_2 > (C_1 + C_2)$, then the circuit can appear to be inductive. This circuit does not simulate an inductor because the impedance value does not vary with frequency in the same way that an inductor does, but the current flow through $C_2$ is in the same direction. Although the circuit shown may have some stability problems, this basic idea can be used to create active circuits.

## 12.2 ACTIVE CIRCUIT PRELIMINARIES

**Important Concepts**

**1.** Complex poles are required for narrowband filters as well as lowpass and highpass filters with Butterworth, Chebyshev, or Bessel responses.

**2.** A lowpass Butterworth filter is more efficient than a filter with real poles only. The response in the transition region is sharper for the Butterworth filter.

**3.** A general biquad function has second-order polynomials in $s$ for both numerator and denominator. Depending on choice of coefficients, this function can be a lowpass, highpass, allpass, bandpass, or band-reject response.

Narrowband bandpass and band-reject filters require complex poles to result in the very steep rolloffs required in some applications. Radio and television circuits have traditionally used resonant circuits that include capacitors and inductors. Whereas narrowband filters require complex poles, lowpass or highpass filters can be constructed with only real poles. However, these filters are not as efficient as Butterworth or Chebyshev filters

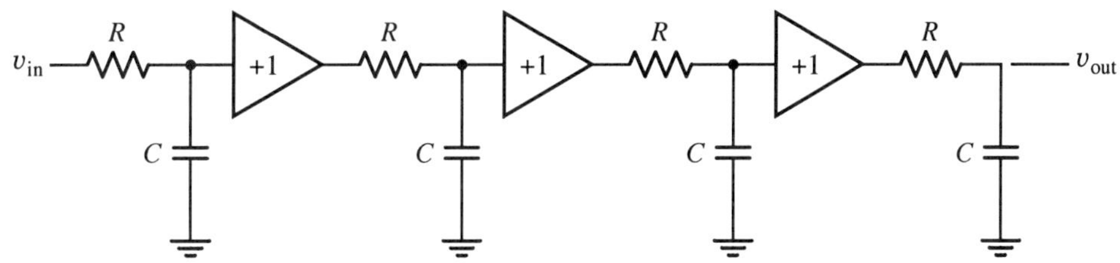


Figure 12.3: A four-pole RC circuit.

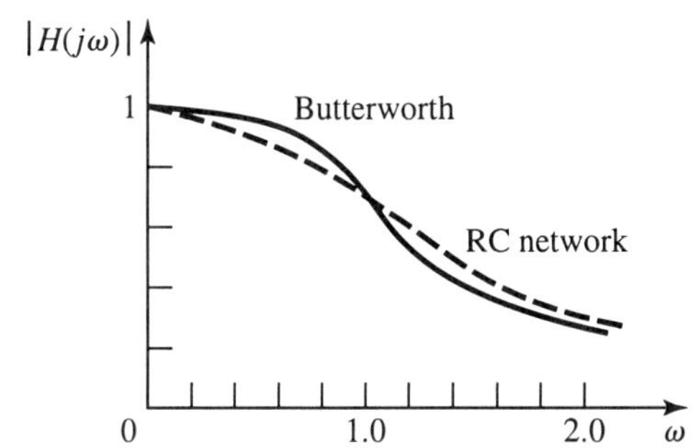


Figure 12.4: Comparison of Butterworth to RC response for four poles.

in passing the desired signals and attenuating the undesired signals. We can compare a lowpass filter realized with four identical, real poles to a Butterworth response using the circuit in Fig. 12.3. The four identical poles are created by RC circuits and isolated by buffer stages.

This circuit can be used in some cases, but is not nearly as sharp as a four-pole Butterworth filter. As seen in Fig. 12.4, the filter with real poles is not very appropriate for applications requiring a sharp transition region. Of course, the four-pole Chebyshev function results in a sharper transition than does the Butterworth function. Complex poles are necessary to implement these higher efficiency filters.

When complex pole filters are to be implemented without inductors, active circuit configurations are necessary. There are basically two architectures applied to the construction of higher order active filters. One is a configuration that implements the overall transfer function desired, with each component affecting the position of more than one pole-pair. We will refer to this type of active synthesis as *multipole realization.*

The second method used realizes each pole-pair by a separate circuit called a *biquad.* The general biquad transfer function can be expressed as

$$H(s) = \frac{a_2 s^2 + a_1 s + a_0}{b_2 s^2 + b_1 s + b_0} \tag{12.9}$$

This function is generally used to create a complex pair of poles. If the transfer function of a filter with complex poles is odd, one pole will be real and the others will be complex. The real pole of the response can be realized by a separate first-order circuit. Each pair of complex conjugate poles can be realized by a biquad circuit. This approach is possible because of the low output impedance of the active circuits used in the individual biquads. Loading effects and interaction between adjacent stages can be completely eliminated with an op amp circuit. The synthesis of an active filter by this method will be referred to as *biquad realization.*

The biquad synthesis method is quite popular and will be considered in a later section. The next section considers a simple method of multipole realization developed many years ago. It will also allow a demonstration of the use of sensitivity analysis in active design.

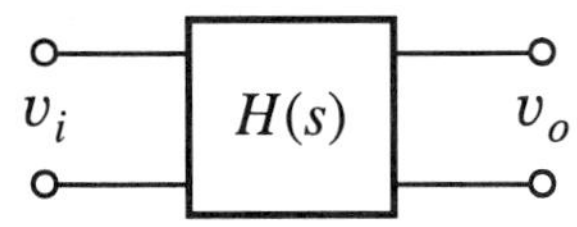


Figure 12.5: Filter network.

**Practice Problems**

**12.1** A four-pole Butterworth filter has a 3-dB frequency of 10 kHz. How many dB is the magnitude of the transfer function down at 8 kHz? *Ans:* 0.67 dB.

**12.2** A filter with four identical real poles has a 3-dB frequency of 10 kHz. How many dB is the magnitude of the transfer function down at 8 kHz? Compare this value to that of Practice Problem 12.1. *Ans:* 1.98 dB.

## 12.3 ACTIVE SYNTHESIS USING BACH FILTERS

**Important Concepts**

1. The Bach filter is an example of an active filter that uses positive feedback to obtain complex poles.
2. The Bach filter requires one op amp stage or buffer stage for each pole implemented.
3. This filter is often used for lowpass or highpass filters.

Though not the most efficient method of creating a Butterworth or Chebyshev response due to the number of op amps required, the Bach filter is a method that allows easy extension to any number of poles desired. This method was originally developed to be used with vacuum tube circuits in the 1940s, but it has been extended to op amp stages. One op amp is required for each pole of the response if more than one pole is to be used.

### 12.3.1 Lowpass Transfer Function

Figure 12.5 indicates a network with a transfer function $H(s) = v_{\text{out}}/v_{\text{in}}$. For sinusoidal excitation, the variable $s$ reduces to $s = j\omega$.

For the Bach filter, it is convenient to use the denormalized transfer functions. The transfer functions for Butterworth filters up to fourth order (four pole) are listed below. These denominators are also called the denormalized Butterworth polynomials.

$$H(s) = \frac{K_1}{s + \omega_c} \tag{12.10}$$

$$H(s) = \frac{K_2}{s^2 + 1.414\omega_c s + \omega_c^2} \tag{12.11}$$

$$H(s) = \frac{K_3}{s^3 + 2\omega_c s^2 + 2\omega_c^2 s + \omega_c^3} \tag{12.12}$$

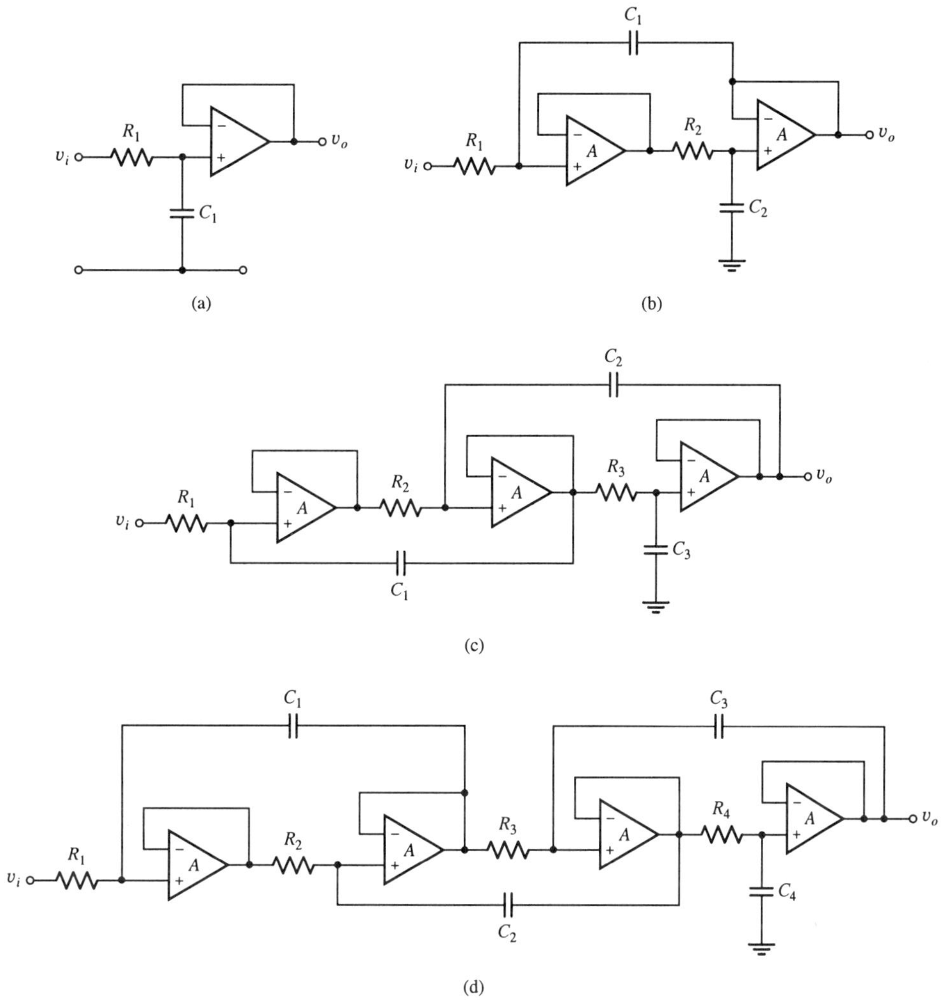


Figure 12.6: Lowpass filters: (a) 1-pole; (b) 2-pole; (c) 3-pole; (d) 4-pole.

$$H(s) = \frac{K_4}{s^4 + 2.613\omega_c s^3 + 3.414\omega_c^2 s^2 + 2.613\omega_c^3 s + \omega_c^4} \tag{12.13}$$

The 3-dB cutoff frequency is designated by $\omega_c$. The first part of the filter synthesis problem, that of selecting an appropriate transfer function, is therefore solved and we can now proceed to the second part. The filter must now be designed to implement the selected function.

There are always several possible solutions to a given synthesis problem. One method of synthesis is based on the circuits of Fig. 12.6 [6]. An $n$-pole lowpass filter can be constructed by simply extending the pattern of Fig. 12.6 to $n$ stages. These filters are referred to as Bach filters.

If we define the parameters $\omega_1 = 1/R_1C_1$, $\omega_2 = 1/R_2C_2$, $\omega_3 = 1/R_3C_3$, and $\omega_4 = 1/R_4C_4$, the transfer functions for the four circuits become

$$H(s) = \frac{\omega_1}{s + \omega_1} \tag{12.14}$$

$$H(s) = \frac{\omega_1\omega_2}{s^2 + \omega_1 s + \omega_1\omega_2} \tag{12.15}$$

$$H(s) = \frac{\omega_1\omega_2\omega_3}{s^3 + \omega_1 s^2 + \omega_1\omega_2 s + \omega_1\omega_2\omega_3} \tag{12.16}$$

$$H(s) = \frac{\omega_1\omega_2\omega_3\omega_4}{s^4 + \omega_1 s^3 + \omega_1\omega_2 s^2 + \omega_1\omega_2\omega_3 s + \omega_1\omega_2\omega_3\omega_4} \tag{12.17}$$

The transfer function for an $n$-pole filter can be found by extending these formulas.

If a maximally flat magnitude response is desired, the transfer function corresponding to the appropriate number of poles is selected from Eqs. (12.10) to (12.13). The corresponding order equation from Eqs. (12.14) to (12.17) is used, and circuit element values are chosen to make the desired transfer function equal to the actual transfer function. An example will demonstrate this method.

# Example 12.1

Design a lowpass, maximally flat magnitude response filter with an asymptotic rolloff of 18 dB/octave. The corner frequency should be 2 kHz. The source resistance is 100 Ω while the load resistance is 10 kΩ.

**Solution** For Butterworth filters the asymptotic rolloff rate is $6n$ dB/octave where $n$ is the number of poles of the transfer function. This design requires three stages to reach an 18 dB/octave rolloff. Equation (12.12) corresponds to the transfer function for this case. For a 2-kHz bandwidth the equation becomes

$$H(s) = \frac{K_3}{s^3 + 8\pi \times 10^3 s^2 + 32\pi^2 \times 10^6 s + 64\pi^3 \times 10^9}$$

Equation (12.16) can be made to equal this equation if $\omega_1 = 8\pi \times 10^3$, $\omega_2 = 4\pi \times 10^3$, and $\omega_3 = 2\pi \times 10^3$ and we assume that $K_3 = \omega_1\omega_2\omega_3 = 64\pi^3 \times 10^9$. The circuit of Figure 12.6(c) has the appropriate transfer function if the element values are properly selected. Let us choose $R_1 = 10$ kΩ to create a relatively high input impedance compared to the 100-Ω source resistance. The required value of $C_1$ is

$$C_1 = \frac{1}{\omega_1 R_1} = 0.0398\ \mu\text{F} \approx 0.04\ \mu\text{F}$$

Selecting $R_2$ and $R_3$ as 10-kΩ resistances leads to values of $C_2 = 0.08\ \mu$F and $C_3 = 0.16\ \mu$F. The transfer function of this circuit is

$$H(s) = \frac{64\pi^3 \times 10^9}{s^3 + 8\pi \times 10^3 s^2 + 32\pi^2 \times 10^6 s + 64\pi^3 \times 10^9}$$

Figure 12.7 shows the frequency response of this filter.

Starting with Bessel or Chebyshev polynomials results in the realization of maximally flat delay or equal-ripple response filters by this same method.

**Practice Problem**

**12.3** Design a three-pole lowpass Chebyshev filter with a 1-dB ripple width in the passband using the Bach configuration and 0.04 $\mu$F capacitors. The cutoff frequency should be 1.5 kHz. Use Table 11.6 for the needed transfer function. *Ans:* $R_1 = 2684\ \Omega$, $R_2 = 2117\ \Omega$, $R_3 = 6686\ \Omega$.

### 12.3.2 Highpass Transfer Function

Given the lowpass transfer function for a desired response, it is a simple matter to obtain the highpass transfer function yielding the same type of response using a frequency transformation. This frequency transformation that

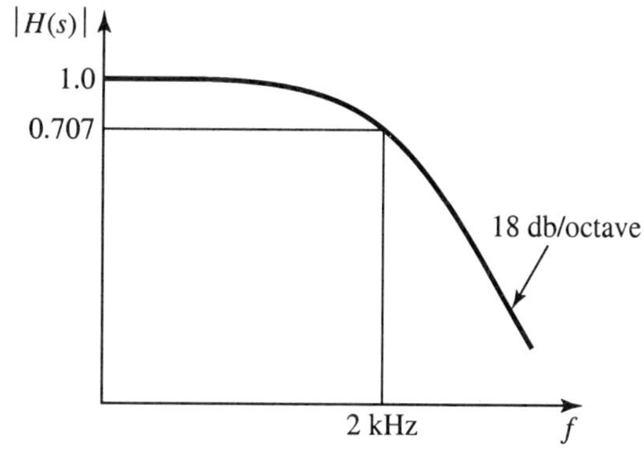


Figure 12.7: Frequency response of a three-pole Butterworth filter.

preserves the same cutoff frequency can be used to convert from the lowpass function to the highpass function. The transformation requires that the variable $s$ be replaced with $\omega_c^2/s$, where $\omega_c$ is the cutoff frequency of both lowpass and highpass functions. For example, the two-pole Butterworth response of Eq. (12.11) will transform as shown below upon applying the preceding frequency transformation.

$$\frac{K_2}{s^2 + 1.414\omega_c s + \omega_c^2} \rightarrow \frac{K_2}{\frac{\omega_c^4}{s^2} + 1.414\frac{\omega_c^3}{s} + \omega_c^2}$$

$$= \frac{(K_2/\omega_c^2)s^2}{s^2 + 1.414\omega_c s + \omega_c^2} = \frac{K_2' s^2}{s^2 + 1.414\omega_c s + \omega_c^2}$$

The Butterworth highpass transfer functions are maximally flat about $s = \infty$ whereas the lowpass functions are maximally flat about $s = 0$. These highpass functions up to the fourth order are

$$H(s) = \frac{sK_1'}{s + \omega_c} \tag{12.18}$$

$$H(s) = \frac{s^2 K_2'}{s^2 + 1.414\omega_c s + \omega_c^2} \tag{12.19}$$

$$H(s) = \frac{s^3 K_3'}{s^3 + 2\omega_c s^2 + 2\omega_c^2 s + \omega_c^3} \tag{12.20}$$

$$H(s) = \frac{s^4 K_4'}{s^4 + 2.613\omega_c s^3 + 3.414\omega_c^2 s^2 + 2.613\omega_c^3 s + \omega_c^4} \tag{12.21}$$

Again the 3-dB frequency is given by $\omega_c$.

The circuits of Fig. 12.6 can be converted to highpass filters simply by interchanging the positions of the resistors and capacitors. The transfer functions then become

$$H(s) = \frac{s}{s + \omega_1} \tag{12.22}$$

$$H(s) = \frac{s^2}{s^2 + \omega_2 s + \omega_2\omega_1} \tag{12.23}$$

$$H(s) = \frac{s^3}{s^3 + \omega_3 s^2 + \omega_3\omega_2 s + \omega_3\omega_2\omega_1} \tag{12.24}$$

$$H(s) = \frac{s^4}{s^4 + \omega_4 s^3 + \omega_4\omega_3 s^2 + \omega_4\omega_3\omega_2 s + \omega_4\omega_3\omega_2\omega_1} \tag{12.25}$$

The highpass filters can be constructed by choosing the parameters $\omega_1$, $\omega_2$, ... , $\omega_n$ such that the applicable equation from Eqs. (12.22) to (12.25) equals the appropriate equation from Eqs. (12.18) to (12.21).

Bessel, Chebyshev, and other responses can be realized for the highpass filter as well, requiring the appropriate starting polynomial. Many other methods can be used to synthesize active filters. Some methods require fewer op amps than the method discussed and these approaches will be discussed in a later section.

## 12.4 DRIFT IN FILTER RESPONSE

**Important Concepts**

**1.** Drift in component values, often resulting from temperature changes, lead to drift in bandwidth or resonant frequency in bandpass networks.

**2.** The sensitivity is a parameter that allows one filter configuration to be compared to another in terms of drift.

**3.** The temperature coefficient can be related to the sensitivity of a circuit.

Some filter designs require very stringent specifications while others may not need precise or stable parameters. A high-$Q$ bandpass filter might need a center frequency stability of 20 parts per million (ppm) and a $Q$ stability of 100 ppm. On the other hand, a lowpass filter could possibly satisfy specifications if the corner frequency drifted by 10%.

Temperature changes often cause variations in parameter values such as $R$, $C$, or $L$ in passive filter circuits. In active filters, the values of $R$ and $C$ continue to change with temperature and the gain of the amplifying elements will also drift with temperature. In addition to temperature drift of element values, there is also a tolerance that applies to each element. In discrete design it is possible to use higher cost elements to limit the absolute accuracy of component values to less than a 1%. In integrated circuits, this is not possible, with typical variations of over 10%.

It is important to understand the parameters that lead to drift in filter performance and to be able to compare, at least in a relative way, the drift of different filters. The sensitivity of some performance figure with respect to the variation in an element value allows this possibility.

### 12.4.1 Passive Circuit Parameter Variation

The simple tank circuit has a resonant frequency determined by

$$\omega_o = \frac{1}{\sqrt{LC}} \tag{12.26}$$

and an effective $Q$ of

$$Q_{\text{eff}} = \frac{R_{sh}}{\omega_o L} \tag{12.27}$$

If the capacitor value changes, the resonant frequency will also change. This variation may be nonlinear making it difficult to characterize in terms of a constant fraction or percent. This is generally the case for active filter circuits, while passive circuits show a simpler variation.

A sensitivity figure has been defined in terms of the fractional variation of a dependent variable divided by the fractional variation of an independent variable. In active filters, the sensitivity value often changes as the nominal value of the independent variable changes.

Assuming $y$ is a function of $x$, we can define the sensitivity of $y$ with respect to $x$ as

$$S_x^y = \frac{\frac{\partial y}{y}}{\frac{\partial x}{x}} = \frac{x}{y}\frac{\partial y}{\partial x} \tag{12.28}$$

We note that $y$ might depend on several independent variables, but the partial derivative of $y$ with respect to $x$ implies that these other variables are constant for the differentiation.

As an example of the use of this definition, we will calculate the sensitivity of $\omega_o$ with respect to capacitance, $C$, for the tank circuit. We find the partial derivative of $\omega_o$ with respect to $C$ as

$$\frac{\partial \omega_o}{\partial C} = \frac{\partial}{\partial C}\left(\frac{1}{\sqrt{LC}}\right) = -\frac{1}{2}\frac{1}{L^{1/2}C^{3/2}} = -\frac{1}{2}\frac{\omega_o}{C}$$

The sensitivity is

$$S_C^{\omega_o} = \frac{C}{\omega_o}\frac{\partial \omega_o}{\partial C} = \frac{C}{\omega_o}\left(-\frac{1}{2}\frac{\omega_o}{C}\right) = -\frac{1}{2}$$

This particular sensitivity is constant and does not depend on the value of $C$. We can write

$$\frac{\partial \omega_o}{\omega_o} = S_C^{\omega_o}\frac{\partial C}{C} = -\frac{1}{2}\frac{\partial C}{C}$$

The fractional change in $\omega_o$ will be one-half the fractional change in $C$ and the change will be in the opposite direction to the change in $C$. If the capacitor value increases by 2%, the resonant frequency decreases by 1%. It can be shown that the sensitivity of $Q_{\text{eff}}$ with respect to $L$ is $-1$ and the sensitivity with respect to $R_{sh}$ is $+1$.

If we know the variation of the capacitance with temperature, often expressed in terms of a temperature coefficient (Tempco), we can calculate the change in $\omega_o$ due to temperature change. The Tempco is defined as

$$TC_x = \frac{\frac{\partial x}{x}}{\partial T} = \frac{1}{x}\frac{\partial x}{\partial T} \tag{12.29}$$

The Tempco is the fractional change in $x$ divided by the temperature change. A capacitor might have a Tempco of $TC_C = 200$ ppm /°C.

If the Tempcos of all the independent variables are known, the Tempco of the dependent variable can be found. For example, if $y$ depends on $w$, $x$, and $z$, and each of these variables changes with temperature, the Tempco of $y$ is

$$TC_y = \frac{1}{y}\frac{dy}{dT} = \frac{w}{y}\frac{\partial y}{\partial w}\cdot\frac{1}{w}\frac{dw}{dT} + \frac{x}{y}\frac{\partial y}{\partial x}\cdot\frac{1}{x}\frac{dx}{dT} + \frac{z}{y}\frac{\partial y}{\partial z}\cdot\frac{1}{z}\frac{dz}{dT}$$

$$= S_w^y TC_w + S_x^y TC_x + S_z^y TC_z \tag{12.30}$$

If the Tempcos of $L$ and $C$ for the tank circuit are $TC_L = 300$ ppm/°C and $TC_C = 200$ ppm/°C, the Tempco of $\omega_o$ will be

$$TC_{\omega_o} = S_L^{\omega_o}TC_L + S_C^{\omega_o}TC_C = -\frac{1}{2}\times 300 - \frac{1}{2}\times 200 = -250 \text{ ppm}/^\circ C$$

# Example 12.2

---

Calculate the maximum change in resonant frequency of the tank circuit over a 20°C temperature change if the maximum Tempcos are $TC_L = 300$ ppm/°C and $TC_C = 200$ ppm/°C.

**Solution** The equation preceding this example indicates a maximum Tempco for $\omega_o$ of

$$TC_{\omega_o} = -250 \text{ ppm}/^\circ C$$

Multiplying this value by a temperature change of $\Delta T = 20$°C gives

$$\frac{d\omega_o}{\omega_o} = -250 \times 20 = -5000 \text{ ppm}$$

The maximum frequency change is then

$$\Delta\omega_{o\text{max}} = \omega_o \frac{d\omega_o}{\omega_o} = -0.005\omega_o$$

---

**Practice Problem**

**12.4** Rework Example 12.2 if $TC_L = 120$ ppm/°$C$ and $TC_C = 180$ ppm/°$C$. *Ans:* $\Delta\omega = -0.003\omega_o$.

### 12.4.2 Active Circuit Parameter Variation

We will begin this subsection by developing the sensitivity figures for the second-order Bach filter shown in Fig. 12.6(b). If the first-stage amplifier gain is called $A_1$ and the second-stage gain is $A_2$, the transfer function can be written as

$$H(s) = \frac{A_1A_2/C_1R_1C_2R_2}{s^2 + \left(\frac{[1-A_1A_2]}{C_2R_2} + \frac{1}{C_1R_1}\right)s + \frac{1}{C_1R_1C_2R_2}} \tag{12.31}$$

When $A_1A_2 = 1$, the transfer function reduces to that of Eq. (12.15). For a second-order transfer function with a high $Q$, the coefficient of the first-order term in the denominator must be small. If we define

$$A = A_1A_2$$

$$\omega_1 = \frac{1}{R_1C_1}$$

and

$$\omega_2 = \frac{1}{R_2C_2}$$

then the transfer function becomes

$$H(s) = \frac{A\omega_1\omega_2}{s^2 + [\omega_2(1-A) + \omega_1]s + \omega_1\omega_2} \tag{12.32}$$

In order to examine the sensitivity of certain important parameters such as $Q$ with respect to $A$, we will consider the standard expression for a two-pole lowpass transfer function. This is,

$$\frac{K}{s^2 + \frac{\omega_n}{Q}s + \omega_n^2} \tag{12.33}$$

Equating coefficients in the denominators of Eqs. (12.32) and (12.33) leads to

$$\omega_n = \sqrt{\omega_1 \omega_2} \tag{12.34}$$

and

$$Q = \frac{\sqrt{\omega_1 \omega_2}}{\omega_2(1-A) + \omega_1} \tag{12.35}$$

The partial derivative of $Q$ with respect to $A$ is

$$\frac{\partial Q}{\partial A} = \frac{\omega_2 \sqrt{\omega_1 \omega_2}}{(\omega_2(1-A) + \omega_1)^2} \tag{12.36}$$

The sensitivity is

$$S_A^Q = \frac{A\omega_2}{\omega_2(1-A) + \omega_1} = A\sqrt{\omega_2/\omega_1}\, Q \tag{12.37}$$

This sensitivity may not be high for a low-$Q$ transfer function, but it becomes larger for high-$Q$ transfer functions. For example, a two-pole normalized Butterworth function is given by

$$\frac{K}{s^2 + \sqrt{2}s + 1} \tag{12.38}$$

which has a value of $Q = 1/\sqrt{2}$. In Eq. (12.32), this function can be obtained if $A = 1$, $\omega_1 = \sqrt{2}$, and $\omega_2 = 1/\sqrt{2}$. From Eq. (12.37), the sensitivity is

$$S_A^Q = 1 \times \sqrt{1/2}\,\sqrt{1/2} = \frac{1}{2}$$

If a high-$Q$ circuit were created with this Bach circuit, say $Q = 100$, using the same values of $\omega_1$ and $\omega_2$, Eq. (12.32) requires that

$$\omega_2(1-A) + \omega_1 = 0.01$$

The value of $A$ required is 2.986 and the sensitivity is found to be 213. This would indicate that a fractional change in $A$ of 1% would lead to a 213% change in $Q$. In many cases, this would be unacceptable; consequently, the Bach circuit is generally limited to low-$Q$ applications. In fact, a fifth-order Butterworth filter requires one pole-pair with $Q = 1.62$, while a fifth-order Chebyshev with $\epsilon = 1$ dB requires a pole-pair with $Q = 5.6$. The sensitivity for the latter case is 10.2.

For lower $Q$ circuits, the Bach method results in a stable circuit for op amp gain elements. When the gain is to be unity, the noninverting configuration with direct feedback from output to inverting input results in a very stable amplifier. In spite of the fact that the op amp gain varies drastically with temperature, the gain of the buffer is very stable. This can be seen by noting that the buffer gain is

$$A = \frac{A_{oa}}{1 + A_{oa}} \tag{12.39}$$

where $A_{oa}$ is the op amp gain. The variation of $A$ with temperature is

$$\frac{dA}{dT} = \frac{1}{1 + A_{oa}}\frac{dA_{oa}}{dT} - \frac{A_{oa}}{(1 + A_{oa})^2}\frac{dA_{oa}}{dT} \tag{12.40}$$

For high op amp gains, this value approaches zero. Practical values of a few parts per million per °C are achieved for the gain of this buffer stage.

## 12.5 SENSITIVITY CALCULATIONS

**Important Concept**

**1.** Sensitivity calculations for certain configurations can be simplified by the equations of this section.

Because sensitivity is an important parameter in active filter design, some relations will be listed at this point. These relations will simplify some difficult calculations. The derivations are found in other texts [10].

### 12.5.1 Useful Sensitivity Relations

1. $S_x^k y = S_x^y$
2. $S_x^y + k = \frac{y}{y+k} S_x^y$
3. $S_x^{y_1 y_2} = S_x^{y_1} + S_x^{y_2}$
4. $S_{x^n}^y = \frac{1}{n} S_x^y$
5. $S_x^{y^n} = n S_x^y$

These relationships simplify the calculations for several commonly occurring sensitivities.

---

**Practice Problem**

**12.5** If $y = (x + a)^{1/2}$, calculate the sensitivity of $y$ with respect to $x$. *Ans:* $\frac{x/2}{x+a}$.

## 12.6 COMPONENTS USED IN BIQUAD SYNTHESIS

**Important Concepts**

**1.** Circuit components used to realize various types of biquad circuits are considered in this section.

**2.** The op amp is used in active filters, primarily in discrete realizations.

**3.** Transconductance amplifiers have higher frequency performance than the op amp and are often used in IC filter design.

**4.** Current conveyors also have good frequency performance and are realizable as IC filters.

Although the field of active filter design continues to evolve, we will consider four methods of biquad synthesis in the following sections that are among the most popular at the present time. The transfer function of a general biquad is defined by Eq. (12.9). A biquad will be defined more fully later, but for now it will suffice to say that these circuits result in second-order transfer functions with a complex pole-pair. Biquad sections can be cascaded to create any number of complex pole-pairs to synthesize a filter. A first-order section is needed to create the one real pole that occurs in odd-order filter responses.

The four types of biquads to be considered are

1. Sallen-Key biquads
2. Two integrator biquads
3. Current conveyors
4. Phase-shift biquads

The Sallen-Key circuits were developed in the 1950s [8] to minimize capacitor size and eliminate inductors from filter circuits. In the 1960s, these methods were applied to op amp active filter circuits.

The two integrator circuits have been applied successfully to op amp active filters as well as integrated circuit filters. The phase-shift biquads and the current conveyor approach are also suitable for op amp active circuits and integrated circuit filters.

A very important component in active filter design is the op amp. A relative of this circuit that is also important is the operational transconductance amplifier (OTA). This circuit is also closely related to the transconductor. Another important circuit used in filter design is the current conveyor. All three of these circuits–the op amp, the OTA and transconductor, and the current conveyor–will be discussed before considering biquad circuits in the succeeding sections.

### 12.6.1 The Op Amp

The op amp has been used for years in active filter synthesis. It is not as useful in high-frequency synthesis, but it has been a workhorse in the low- to intermediate-frequency area since the late 1960s. More recently, high-performance IC op amps have been used to create active filters on a chip.

The history of the op amp predates the IC era. The analog computer, a system that is rarely encountered today, used op amps to simulate the solutions of differential equations very quickly. The op amps used in these systems cost several hundreds of dollars each and occupied 12 to 16 cubic inches of volume. The modern IC op amp costs a few cents and performs at a level that makes it a very important component in filter design.

Although an op amp can be used as an amplifier, very few amplifier circuits can be used as an op amp. The op amp has the unique capability to serve as the basis of electronic circuits that perform mathematical operations. The key to this type of circuit is the virtual short that can exist between inputs when negative feedback is used with the op amp.

Op amp circuits can multiply a signal by a constant value (weight a signal), sum and weight several signals, subtract signals, invert signals, and integrate signals. Summation and integration are made easier with the virtual ground created by the op amp. The name for this amplifier is derived from its ability to create electronic circuits that perform mathematical operations. If these operations cannot be easily done or if the circuit does not create a virtual ground, it is not an op amp.

Because of the great versatility of the op amp, it has become a very popular circuit in many applications, including active circuit design. Unfortunately, the op amp also has some accompanying disadvantages. One major disadvantage that has become somewhat less severe in recent years is the poor frequency response of the amplifier. This component is rarely used in high-frequency filters as a result of the limitations on frequency response.

#### The Virtual Ground

A very important feature of the op amp is the virtual ground resulting from the configuration of Fig. 12.8. A virtual short exists between the two input terminals and, since the noninverting terminal is connected to ground, the inverting terminal is at virtual ground. The term virtual ground simply means that there is a very small impedance between the inverting terminal and ground. The small impedance results from the Miller effect. The feedback resistance bridges the inverting input terminal and the output terminal, resulting in a reflected impedance given by

$$R_{\text{inv}} = \frac{R_F}{1+|A|} \tag{12.41}$$

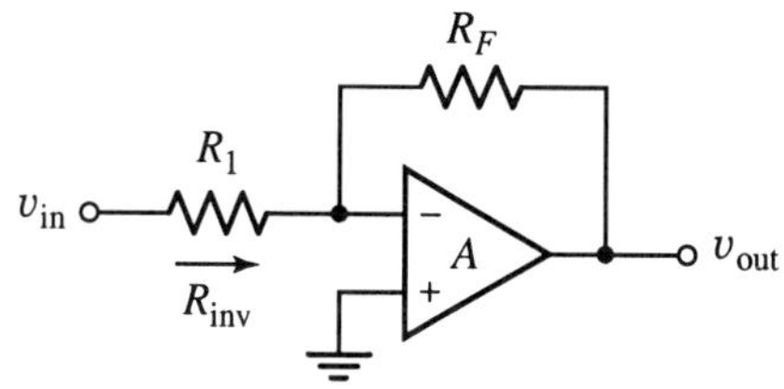


Figure 12.8: An inverting stage.

where $A$ is the op amp gain. If $A_{MB}$ is the midband or low-frequency gain of the op amp, we often approximate this equation as

$$R_{inv} = \frac{R_F}{1 + A_{MB}} \tag{12.42}$$

Of course, the resistance of Eq. (12.41) or (12.42) is actually in parallel with the differential input impedance of the op amp. These equations only apply at low operating frequencies. At higher frequencies, as the op amp gain drops, the input impedance increases and becomes complex rather than real.

At low frequencies, the reflected impedance may be fractions of an Ohm. If the feedback resistance is $R_F = 30\ \text{k}\Omega$ and $A_{MB} = 200,000$ V/V, the reflected value is

$$R_{inv} = \frac{30,000}{200,001} = 0.15\ \Omega$$

This very small resistance leads to the concept of a virtual ground at the inverting terminal. It only exists when negative feedback is applied, as in Fig. 12.8, and it only applies strictly at low frequencies.

### A Closer Look at Virtual Ground

The actual equation for the Miller effect impedance reflected to the inverting input is

$$Z_{inv} = \frac{R_F}{1 + \frac{A_{MB}}{1+jf/f_{oa}}} \tag{12.43}$$

where $A_{MB}$ is the midband gain and $f_{oa}$ is the 3-dB frequency of the op amp gain. The 3-dB frequency is generally a very low value, often less than 100 Hz.

After some manipulation, Eq. (12.43) can be written as

$$Z_{inv} = \frac{R_F}{1 + A_{MB}} \frac{(1 + jf/f_{oa})}{(1 + jf/f_{oa}(1 + A_{MB}))} \tag{12.44}$$

The magnitude of this impedance is plotted in Fig. 12.9. Only at frequencies below $f_{oa}$ does Eq. (12.42) hold. At frequencies above $f_{oa}$, the magnitude of impedance increases directly with frequency. We note that, if the reflected impedance at dc is 0.15 Ω, it has only increased to a magnitude of 15 Ω at a frequency of $100 \times f_{oa}$. It still may appear as a virtual ground if $R_1$ is much larger than this value.

On the other hand, an inverting amplifier has a 3-dB frequency given by

$$f_{inv} = A_{MB} \times f_{oa} \times \frac{R_1}{R_1 + R_F} \tag{12.45}$$

Thus, the available bandwidth may reach values that lead to a relatively high value of reflected impedance, invalidating the concept of a virtual ground at the inverting terminal. An example will demonstrate a typical situation.

## Example 12.3

An op amp has a gain-bandwidth product of $2 \times 10^8$ along with a midband gain of 250,000 V/V. This stage is used in an inverting configuration such as Fig. 12.8 with values of $R_F = 20\ \text{k}\Omega$ and $R_1 = 2\ \text{k}\Omega$.

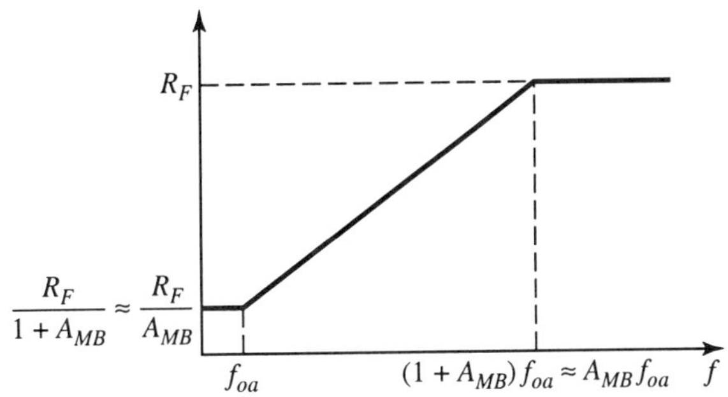


Figure 12.9: Reflected input impedance as a function of frequency.

**(a)** Calculate the dc impedance from the inverting terminal to ground.

**(b)** Calculate the magnitude of the impedance from inverting terminal to ground at the 3-dB frequency of the stage.

**Solution** We will assume that the differential input impedance of the stage is much larger than $R_F$. **(a)** The resistance at dc is given by Eq. (12.42) and is

$$R_{\text{inv}} = \frac{R_F}{1 + A_{MB}} = \frac{20,000}{250,000} = 0.08\ \Omega$$

**(b)** The 3-dB frequency of the inverting stage is found from Eq. (12.45) as

$$f_{\text{inv}} = GBW \times \frac{R_1}{R_1 + R_F} = 2 \times 10^8 \times \frac{2}{22} = 18.2\ \text{MHz}$$

The 3-dB frequency of the op amp is found from

$$f_{oa} = \frac{GBW}{A_{MB}} = \frac{2 \times 10^8}{250,000} = 800\ \text{Hz}$$

Since the magnitude of reflected impedance increases with frequency above 800 Hz, the impedance magnitude at 18.2 MHz is

$$R_{\text{inv}}(18.2) = 0.08 \times \frac{18.2 \times 10^6}{800} = 1.82\ \text{k}\Omega$$

This impedance is comparable to $R_1$ and the circuit can no longer be considered to have a virtual ground at the inverting terminal.

---

A significant question is, "How does this loss of virtual ground affect the performance of the inverting stage?" This question must be answered in terms of the particular application of the op amp. For an amplifier circuit, the gain and bandwidth equations are derived without considering the reflected impedance. Thus, there is little problem in this situation. However, in a summing circuit such as that in Fig. 12.10, the problem is more critical.

This circuit depends on the input currents being proportional to the applied voltages which, in turn, depends on the inverting terminal being a virtual ground. The circuit becomes inaccurate as the frequency is increased to values that make the reflected impedance become significant compared to the summing resistors. Care must be taken to ensure that this situation is avoided by limiting the frequency of operation or using large summing resistors or a combination of both.

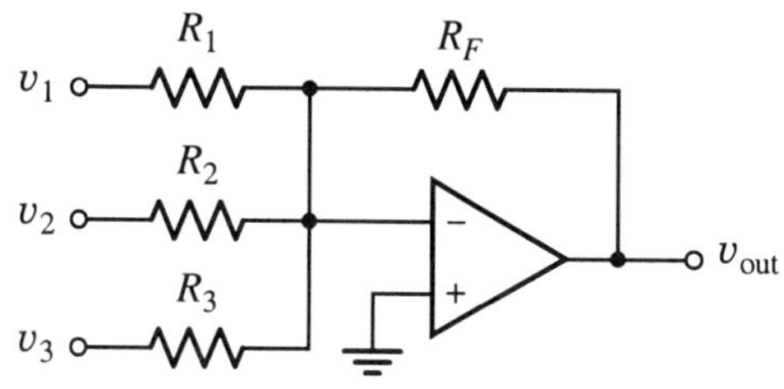


Figure 12.10: A summer.

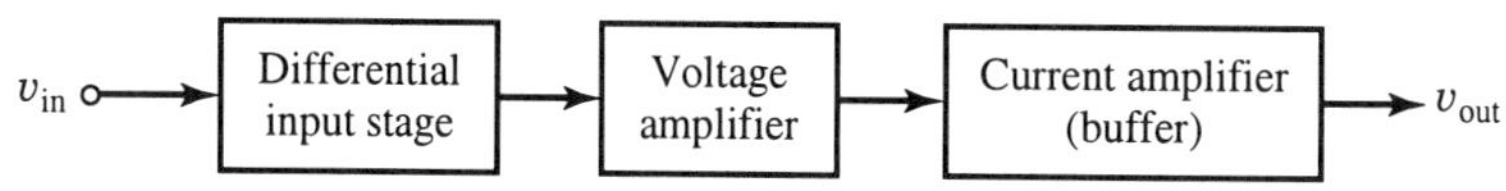


Figure 12.11: Architecture of typical op amp.

The integrator has less of an accuracy problem since the reflected impedance is capacitive. A capacitor exhibits less impedance with frequency which offsets the increase with frequency of the reflected impedance.

### Frequency Limitations of the Op Amp

The op amp does not perform well at high frequencies. In general-purpose op amps, bandwidth is limited internally in order to stabilize the circuit for unity feedback. When an op amp is used in an IC filter, the bandwidth can be considerably greater than that of the general purpose op amp. Although the technology used to implement op amps has changed considerably over the years, the basic architecture has remained remarkably constant.

Many op amp architectures appear as shown in Fig. 12.11. The first section is a differential amplifier required by all op amps to allow a virtual ground and the implementation of mathematical operations. This stage is generally designed to have a very high differential voltage gain, perhaps several thousand. The bandwidth is generally rather low as a result of the high voltage gain. In some cases, this differential amplifier will use an active load that will also convert the double-ended output of the differential stage to a single-ended output that will drive the second voltage amplifier.

The second amplifier stage of Fig. 12.11 is a voltage amplifier with a more modest voltage gain. This gain may range between 10 and 500 V/V. Often this stage is used to compensate the op amp, a topic that will not be discussed further in this section. For now it is sufficient to say that this stage will probably be an inverting stage that can multiply the apparent value of some capacitance placed between the input and output of the amplifier. This is a case wherein the Miller effect is used to advantage. As a result of the Miller effect, the differential amplifier will see a large capacitive load and have a very low upper corner frequency, perhaps 10 to 20 Hz.

The last stage is the output stage. It may be nothing more than an emitter follower or a source follower that has a large current amplification along with a voltage gain that is near unity. This stage will have a very high upper corner frequency.

Most fundamental textbooks on electronic circuit design discuss these topics in detail, thus, further discussion is unnecessary at this point. We simply conclude that the op amp typically is used in filter applications for low and intermediate frequencies up to hundreds of kHz.

### Practice Problem

**12.6** An inverting op amp has a gain of -8 V/V and a $GBW$ of 4 MHz. Calculate the 3-dB frequency of the amplifier. *Ans:* 444 kHz.

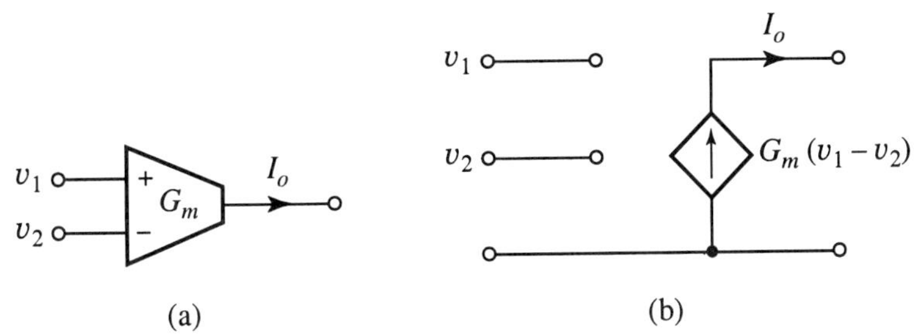


Figure 12.12: (a) Transconductor symbol. (b) Equivalent circuit.

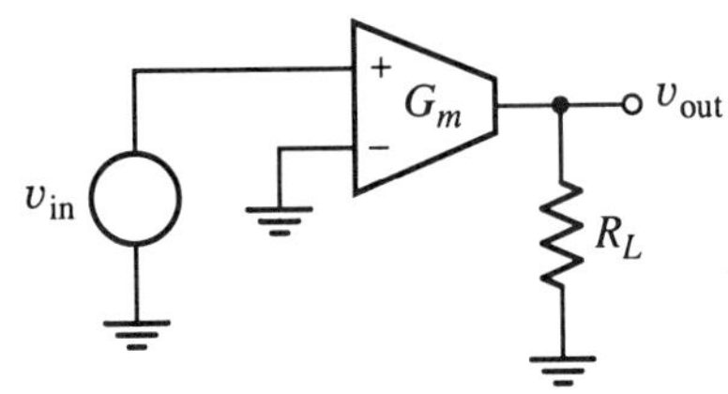


Figure 12.13: An OTA voltage amplifier.

### 12.6.2 Transconductance Amplifiers

The transconductance amplifier or transconductor has recently become more prominent in filter design because of its tunability and its high-frequency response. A simple transconductor has the symbol of Fig. 12.12.

The input impedance and the output impedance of the ideal transconductor are infinite, whereas the transfer function is expressed as

$$i_{\text{out}} = G_m v_{\text{in}} \tag{12.46}$$

An input voltage is transformed into a proportional output current by this circuit. In a practical circuit, the bandwidth of the transfer function will be finite, although it may be relatively large–at least compared to that of the op amp.

One form of transconductor implementation is the operational transconductance amplifier (OTA). Several chip makers have various versions of OTAs available as an IC. These circuits generally have a very high gain and allow only a small voltage to be applied to the input terminals, often limited to a value of approximately 20 mV. A second form of transconductor implementation is based on simple MOSFET circuits. Often scaling is used to achieve a reasonable value of $G_m$ [9], but the values for $G_m$ in these circuits are generally much less than those of the OTA.

There is some controversy over the difference between OTAs and simple transconductors [13]. There are OTA-C filters and $G_m - C$ filters, but the distinction mainly seems to be the range of input signals that can be applied without resulting in significant nonlinear distortion in the circuit. Some applications require a low value of $G_m$, and these circuits use the simple transconductor.

We will now consider some of the functions that can be performed by the OTA or the transconductor.

#### Voltage Amplification with the OTA

Figure 12.13 indicates a voltage amplifier based on an OTA. The simple addition of a resistor at the output generates an output voltage of

$$v_{\text{out}} = i_{\text{out}} R_L = G_m v_{\text{in}} R_L \tag{12.47}$$

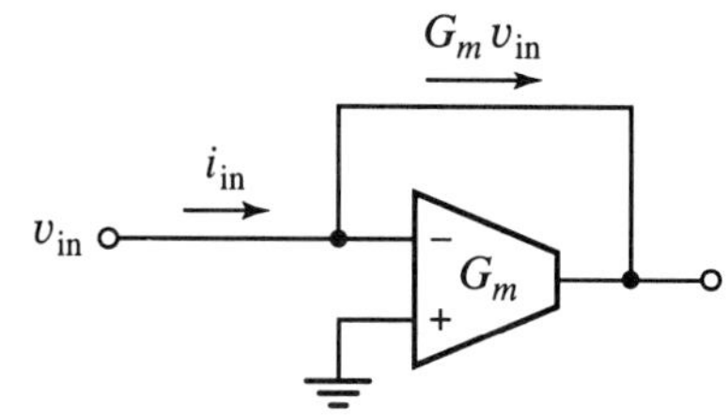


Figure 12.14: A resistance configuration.

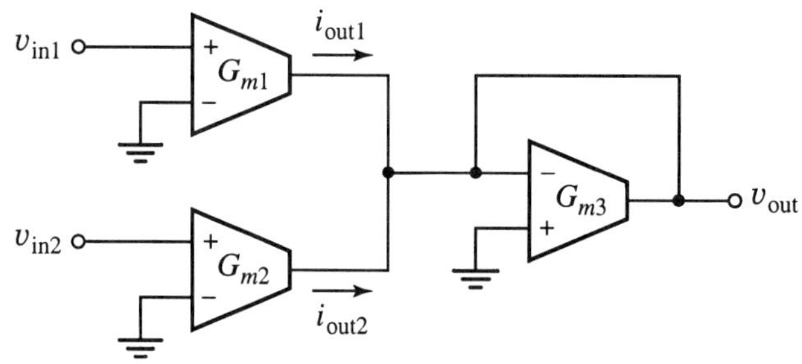


Figure 12.15: A summing circuit.

giving a voltage gain of

$$A_v = G_m R_L \tag{12.48}$$

We note that an inverting amplifier can be created by reversing the connections to the input terminals.

Equation (12.48) has the same form as the voltage gain of a simple MOSFET stage, ignoring second-order effects. If an OTA is used as the transconductor, very high values of voltage gain can be achieved.

### Voltage-Variable Resistor (VVR)

A resistance can be emulated by the OTA using the configuration of Fig. 12.14. An applied voltage of $v_{\text{in}}$ results in an input current of $G_m v_{\text{in}}$ entering the input terminal. This is equivalent to a resistance of

$$\frac{v_{\text{in}}}{i_{\text{in}}} = \frac{v_{\text{in}}}{G_m v_{\text{in}}} = \frac{1}{G_m} \tag{12.49}$$

If the transconductor is tunable, that is, if $G_m$ is a function of some parameter such as a dc bias current, the value of the resistance can be varied.

If the positive and negative input terminals of the circuit in Fig. 12.14 are switched, a negative resistance results. Such a configuration is useful in decreasing the losses in a tank circuit. The parallel resistance of the tank circuit can be decreased by placing a larger, negative resistance in parallel. We will discuss this approach in more detail in the subsection on *current conveyors.*

The resistance created by the OTA can provide the resistive load for a voltage amplifier to create a variable-gain stage. In addition, it can be used to create a summer such as that shown in Fig. 12.15. In this circuit, the output currents of the leftmost OTAs sum together and flow through the resistance created by the last stage. The output voltage is

$$v_{\text{out}} = (i_{\text{out1}} + i_{\text{out2}})\frac{1}{G_{m3}} = \frac{G_{m1} v_{\text{in1}}}{G_{m3}} + \frac{G_{m2} v_{\text{in2}}}{G_{m3}} \tag{12.50}$$

A signal can be subtracted by switching the terminals of one of the input OTAs.

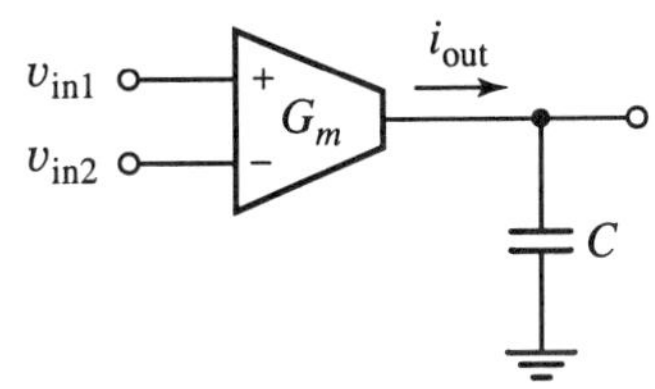


Figure 12.16: An OTA integrator.

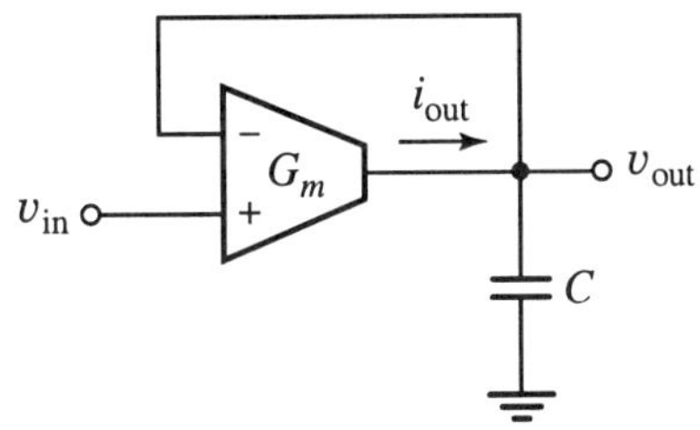


Figure 12.17: A lowpass transfer function.

### An Integrator

The addition of a capacitor at the output of the OTA creates an integrator. This circuit, which is very important in filter design, is shown in Fig. 12.16. The output voltage is

$$v_{out} = \frac{1}{C}\int i_{out}dt = \frac{G_m}{C}\int (v_{in1} - v_{in2})dt \tag{12.51}$$

Often, the inverting terminal is connected to ground to result in an output that is proportional to the integral of an input voltage.

A lossy integrator or lowpass filter can be created with the configuration of Fig. 12.17. The Laplace output voltage is given by

$$v_{out}(s) = \frac{i_{out}(s)}{sC} = \frac{G_m(v_{in} - v_{out})}{sC} \tag{12.52}$$

Solving for the Laplace transfer function gives

$$\frac{v_{out}}{v_{in}} = \frac{1}{s\frac{C}{G_m} + 1} \tag{12.53}$$

This lowpass function has a 3-dB frequency of

$$\omega_{3dB} = \frac{G_m}{C} \tag{12.54}$$

### Nonideal Transconductors

Although the frequency response of the transconductor may be one or two orders of magnitude better than the op amp, the finite corner frequency limits the frequency performance of this device. Generally, the OTA or transconductor frequency behavior can be approximated by a one pole response. This pole must be considered in high-frequency filter design.

Another problem with the OTA and transconductor is the limited dynamic ranges of these devices. Nonlinear effects limit the maximum signal that can be filtered without introducing excess distortion in the output signal.

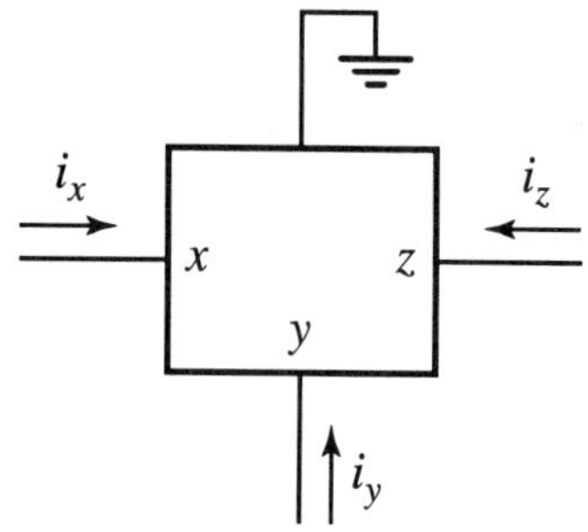


Figure 12.18: A general current conveyor.

More recent transconductors show improved dynamic range [7], but this problem still limits the performance of filters using this component.

**Practice Problem**

**12.7** A transconductor with $G_m = 2$ mA/V is used to construct an integrator with an output of

$$v_{\text{out}} = 10^{-3} \int v_{\text{in}} dt$$

Calculate the value of $C$ used. *Ans:* 2 $\mu$F.

### 12.6.3 Current Conveyors

The current conveyor is a three-port circuit that offers quite different transfer properties between ports. Figure 12.18 shows a block diagram of a current conveyor.

**Mathematical Description of the Current Conveyor**

This device can be described by the matrix equation

$$\begin{bmatrix} i_y \\ v_x \\ i_z \end{bmatrix} = \begin{bmatrix} 0 & a & 0 \\ 1 & 0 & 0 \\ 0 & b & 0 \end{bmatrix} \begin{bmatrix} v_y \\ i_x \\ v_z \end{bmatrix} \tag{12.55}$$

More simply, these equations can be written as

$$i_y = ai_x \quad v_x = v_y \quad i_z = bi_x \tag{12.56}$$

There are several choices for $a$ and $b$. For a current conveyor of the first type (CCI), $a = 1$. For a current conveyor of the second type (CCII), $a = 0$. For either a CCI or CCII, $b$ can be $\pm 1$. If $b = +1$, then we have either a CCI+ or CCII+. If $b = -1$, then we have either a CCI- or CCII-.

For example, a CCI+ gives

$$\begin{bmatrix} i_y \\ v_x \\ i_z \end{bmatrix} = \begin{bmatrix} 0 & 1 & 0 \\ 1 & 0 & 0 \\ 0 & 1 & 0 \end{bmatrix} \begin{bmatrix} v_y \\ i_x \\ v_z \end{bmatrix} \tag{12.57}$$

Before we consider using the current conveyor, we will comment on the implications of some of the equations. If $a = 0$, as in the CCII, the current $i_y = 0$. This implies an infinite impedance at terminal $y$. The relationship

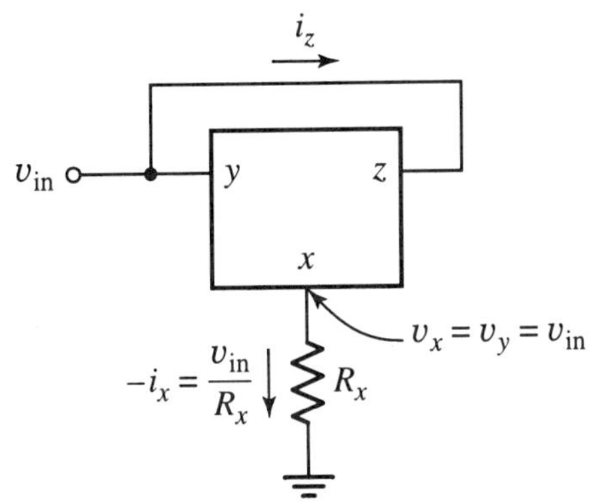


Figure 12.19: An emulated resistance.

$v_x = v_y$ requires a unity voltage gain; however, the impedance at terminal $x$ will not be infinite because a finite current is implied by $i_z = bi_x$.

### Resistance Generation

The current conveyor can be used to emulate a resistance in the configuration of Fig. 12.19. If a CCII- is used, a positive resistance is emulated. If a CCII+ is used, a negative resistance is emulated. The input voltage is applied to the $y$ terminal and causes $v_x = v_y = v_{\text{in}}$. This voltage at the $x$ terminal leads to a current leaving the terminal of value

$$i_x = \frac{v_{\text{in}}}{R_x} \tag{12.58}$$

For the CCII- circuit, the current $i_z$ will enter the $z$ terminal and will also equal the input current. Thus, the input terminal relation is

$$\frac{v_{\text{in}}}{i_{\text{in}}} = \frac{v_{\text{in}}}{\frac{v_{\text{in}}}{R_x}} = R_x \tag{12.59}$$

This appears to be a rather wasteful method of emulating a resistance of value $R_x$ when the direct use of this resistor would accomplish the same result. The method is more interesting when a CCII+ circuit is used. The derivation now results in an emulated resistance at the input terminals of

$$\frac{v_{\text{in}}}{i_{\text{in}}} = -R_x \tag{12.60}$$

This results from the fact that $i_z$ now flows out of the $z$ terminal and $i_{\text{in}} = -i_z$. We will see the benefit of this negative resistance in a later section.

### Integration

The conveyor circuit of Fig. 12.20(a) shows a noninverting integrator whereas the circuit of part (b) shows an inverting integrator. For the noninverting integrator, the output voltage is

$$v_{\text{out}} = \frac{i_z}{sC} = \frac{i_x}{sC} = \frac{v_{\text{in}}}{sRC} \tag{12.61}$$

The constant of integration is $1/RC$.

The inverting integrator has an output voltage of

$$v_{\text{out}} = -\frac{i_z}{sC} = -\frac{v_{\text{in}}}{sRC} \tag{12.62}$$

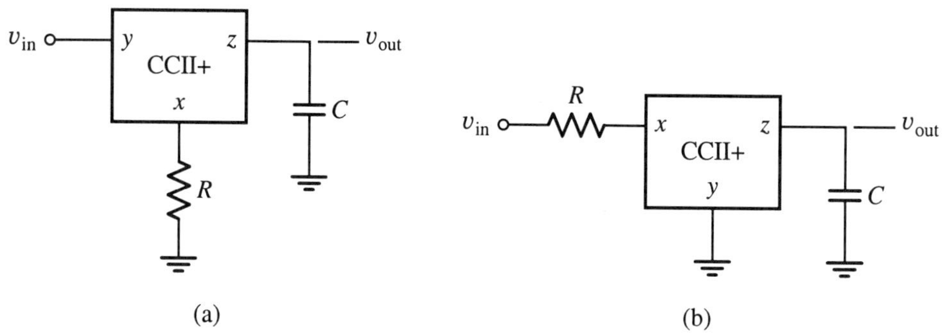


Figure 12.20: (a) An integrator. (b) An inverting integrator.

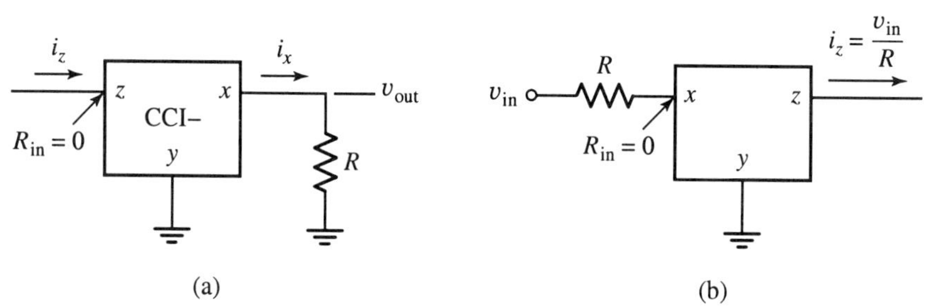


Figure 12.21: CCI converters. (a) Current to voltage. (b) Voltage to current (transconductor).

### Other Uses of the Current Conveyor

The two CCIs of Fig. 12.21 convert between voltage and current. In these configurations, terminal $x$ is a virtual ground, which results in $R_x = 0$. This is because $v_x = v_y$, but since terminal $y$ is grounded, both voltages are zero. In the configuration of part (a), the input current at the $x$ terminal generates an equal output current from the $z$ terminal. This leads to an output voltage of $v_{\text{out}} = i_z R$. The transimpedance of this circuit is $R$. The configuration of Fig. 12.21(b) is a transconductor with a transconductance given by $1/R$.

### The Current-Controlled Current Conveyor

An important variation on the current conveyor is that of current control. In this configuration, some dc bias current controls a resistance that connects to a terminal of the current conveyor. This circuit is designated a CCCII+ or CCCII-. The diagram of Fig. 12.22 represents the CCCII+. If a CCCII+ is used in the circuit of Fig. 12.19, the negative resistance emulated can be controlled by some bias current. This leads to the capability of tuning in some current conveyor filters.

### Realizations of the Current Conveyor

Although current conveyors can be implemented by op amps, the frequency limitations of these devices minimize the use of such realizations. One useful form of a BJT realization is shown in simplified form in Fig. 12.23. This circuit is a CCII+ and was developed by Fabre [3].

The current $i_y$ will not change with $v_y$ since the total current into the $y$ node is determined by the current sources in the collectors of $Q1$ and $Q2$. A positive voltage applied to the $y$ terminal will raise the voltages at the emitters and bases of $Q1$ and $Q2$ by the same amount. This voltage change gets passed on to the emitters of $Q3$ and $Q4$ ensuring that $v_x = v_y$. Current impressed on the $x$ terminal modifies the input currents to the two current mirrors, and this change is reflected to the $z$ terminal.

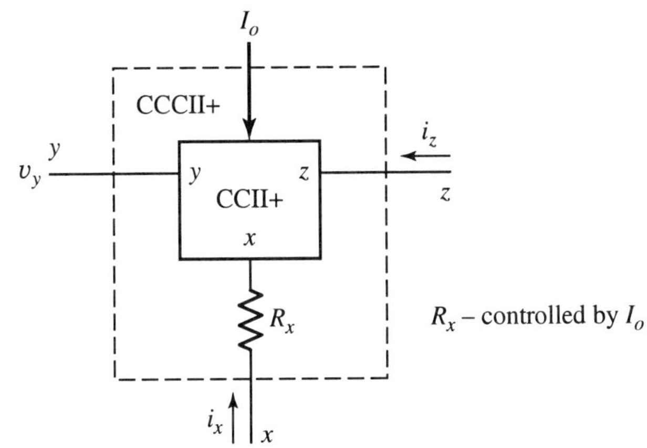


Figure 12.22: The diagram for the CCCII+.

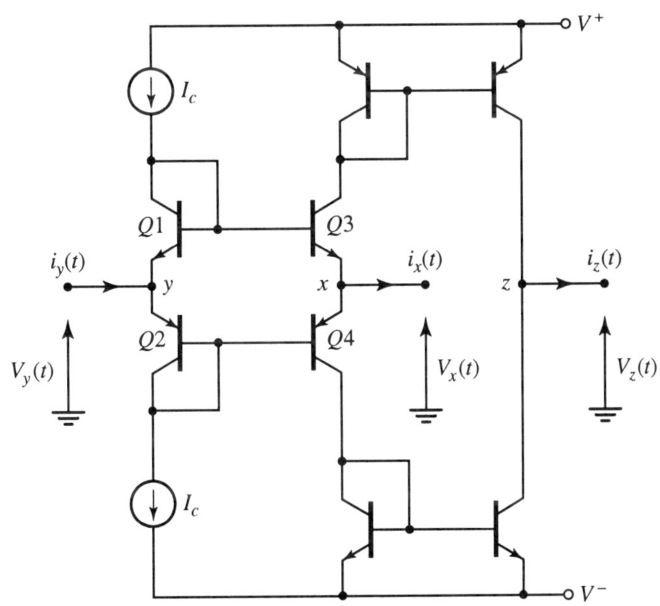


Figure 12.23: A simplified CCII+ circuit.

This version of the CCII+ can be constructed with high-frequency BJT stages and can be used for high-frequency filter design.

## 12.7 THE GENERAL BIQUAD FUNCTION

**Important Concept**

**1.** Circuits that can realize the general biquad function and allow the choice of the coefficients of the numerator or denominator allow the realization of various important filter types.

The general biquad function has the form

$$H(s) = \frac{a_2 s^2 + a_1 s + a_o}{b_2 s^2 + b_1 s + b_o} \tag{12.63}$$

In a given application, some coefficients are chosen as zeros while the finite coefficients are chosen with the necessary values to result in the desired transfer function. For example, if a sixth-order, lowpass, Butterworth filter is to be designed, three biquads can be cascaded. Each biquad must have a normalized transfer function of the form

$$H(s) = \frac{a_o}{s^2 + b_1 s + 1} \tag{12.64}$$

Selecting the $b_1$ coefficient of each biquad function then sets the pole-pairs to the proper values for the Butterworth response.

If a narrowband circuit is desired, the transfer function must take the form

$$H(s) = \frac{a_1 s}{s^2 + b_1 s + 1} \tag{12.65}$$

In this case, the coefficient, $b_1$, must be selected to be small to get a sharp response.

A given configuration may not allow the freedom to select all coefficients or to set the values at the optimum level. When a filter transfer function requires an odd number of poles, a first order section must be cascaded with the biquads. Of course, this section must produce a real pole which can result from a simple RC network or an op amp network with conventional configurations.

## 12.8 SALLEN-KEY BIQUADS

**Important Concepts**

**1.** The Sallen-Key biquad allows the implementation of most important second-order filter responses.

**2.** The Sallen-Key circuits use op amps and consequently have relatively poor frequency responses.

**3.** Large capacitance values are used for the Sallen-Key circuits and are only implemented with discrete circuits.

As mentioned earlier, Sallen and Key proposed several filter configurations involving resistors, capacitors, and vacuum tubes to create complex poles [8]. When integrated circuits became popular in the late 1960s, the Sallen-Key filters were then converted from vacuum tube to op amp circuits.

Biquads have been popular for years, but they have been used primarily in discrete active filter design rather than IC filter design. The limited frequency range in which these filters can operate makes other configurations more attractive to IC design.

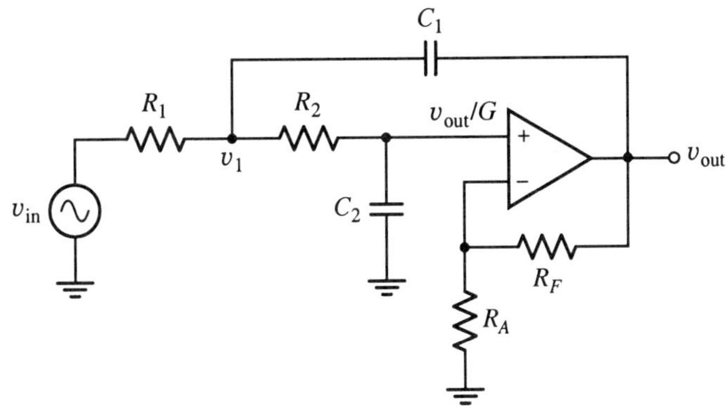


Figure 12.24: A lowpass biquad.

### 12.8.1 A Lowpass Biquad

One of the simplest Sallen-Key lowpass circuits is shown in Fig. 12.24. We assume that the op amp presents infinite input and zero output impedance to the circuit. The gain of the amplifier is

$$G = 1 + \frac{R_F}{R_A} \tag{12.66}$$

We first sum the current into the $v_1$ node to get

$$\frac{v_{in} - v_1}{R_1} + \frac{\frac{v_{out}}{G} - v_1}{R_2} + (v_{out} - v_1)sC_1 = 0 \tag{12.67}$$

We then write the voltage $v_1$ in terms of $v_{out}/G$ as

$$v_1 = v_{out}(sC_2R_2 + 1) \tag{12.68}$$

Combining these two equations to eliminate $v_1$ results in a transfer function of

$$\frac{v_{out}}{v_{in}} = \frac{\frac{G}{C_1C_2R_1R_2}}{s^2 + \left[\frac{1}{C_1R_1} + \frac{1}{C_1R_2} + \frac{1-G}{C_2R_2}\right]s + \frac{1}{C_1C_2R_1R_2}} \tag{12.69}$$

This transfer function has the form of a lowpass biquad. There are several degrees of freedom in selecting components to lead to the proper coefficient values. In order to provide enough constraints to pick reasonable values, several approaches can be taken [10]. Reasonable possibilities are to choose (1) $C_1 = C_2$ and $R_1 = R_2$; (2) $C_1 = C_2$ and $R_F = R_A$; or (3) low-sensitivity designs.

One important point relating to the denormalization of element values is the extra degree of freedom that results from high-input impedance/low-output impedance op amps. Since the input impedance of the following stage will almost always be significantly more than the output impedance of the op amp, very little loading will take place. Furthermore, the filter impedances will see a very high input impedance from the op amp used in the noninverting configuration. The impedance scaling factor is then arbitrary, within reason, and can be used to get desired $R$ or $C$ values.

## Example 12.4

---

Design a fifth-order Chebyshev lowpass filter using Sallen-Key biquads. The cutoff frequency should be 6 kHz and the ripple in the passband should be 1 dB.

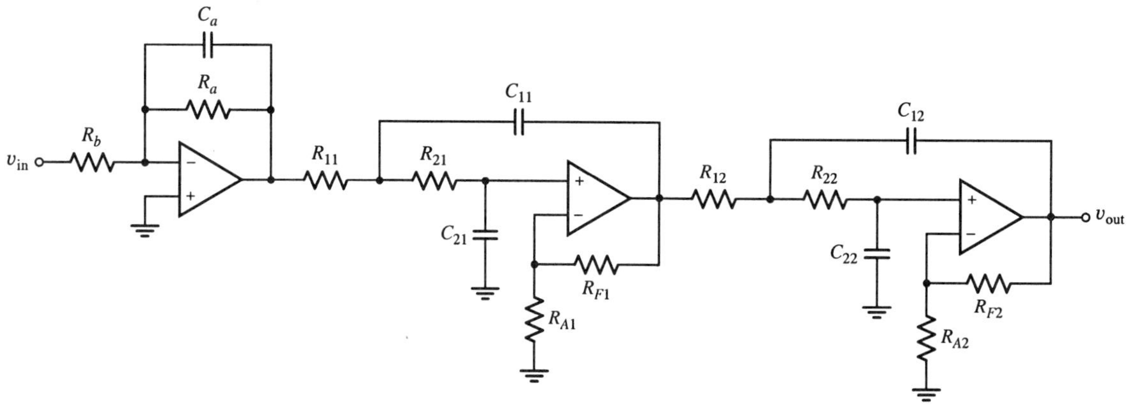


Figure 12.25: A five-pole Chebyshev filter.

**Solution** The normalized transfer function, expressed in terms of a real pole factor and two pole-pairs is found from a table to be

$$\frac{K}{(s+0.2895)(s^2+0.1789s+0.9983)(s^2+0.4684s+0.4293)}$$

This circuit can be realized by the circuit in Fig. 12.25 that includes a first-order section followed by two lowpass biquads. The first section has a transfer function of

$$H_1 = -\frac{R_a}{R_b}\frac{1}{sR_aC+1} = -\frac{1}{CR_b}\frac{1}{s+\frac{1}{R_aC}}$$

The value of $1/R_aC$ is constrained to equal the real pole magnitude from the desired transfer function. Therefore,

$$\frac{1}{R_aC} = 0.2895$$

There is no specific constraint on the value of $R_b$ at this point.

We can now equate the transfer function of each biquad to the corresponding second-order factor of the desired transfer function. Before we do this, let us make the assumptions that $R_{11} = R_{21} = R_1$, $R_{12} = R_{22} = R_2$, $C_{11} = C_{21} = C_1$, and $C_{12} = C_{22} = C_2$. This leads to the constraints that

$$\frac{3-G_1}{R_1C_1} = 0.1789$$

$$\frac{1}{(R_1C_1)^2} = 0.9983$$

$$\frac{3-G_2}{R_2C_2} = 0.4684$$

$$\frac{1}{(R_2C_2)^2} = 0.4293$$

The first two equations have three unknowns as also do the last two equations. We can therefore select one of the unknowns in each pair of equations. Let us select $C_1$ and $C_2$ to be 1 F.

The four equations are now satisfied by the following values:

$$\begin{aligned} R_1 &= 1.001\ \Omega \\ G_1 &= 1.209\ \text{V/V} \\ R_2 &= 1.526\ \Omega \\ G_2 &= 2.285\ \text{V/V} \end{aligned}$$

The amplifier gain values are established in the normal way and will be unaffected by the denormalization. Although we know the frequency denormalization factor is

$$k_\omega = \frac{6000 \times 2\pi}{1} = 37.70\ \text{krad/s}$$

we have not determined the impedance scaling factor. We will assume that we want to use 10-nF capacitances for both $C_1$ and $C_2$. The scaling equation is

$$C = \frac{C_n}{k_i k_\omega}$$

Equating $C$ to 10 nF and solving for $k_i$ results in

$$k_i = \frac{1}{10^{-8} \times 3.770 \times 10^4} = 2653$$

The denormalized values for $R_1$ and $R_2$ are

$$R_{1den} = k_i \times R_1 = 2653 \times 1.001 = 2656\ \Omega$$

and

$$R_{2den} = k_i \times R_2 = 2653 \times 1.526 = 4048\ \Omega$$

---

### 12.8.2 A Highpass Biquad

The Sallen-Key highpass biquad is shown in Fig. 12.26. The positions of the $R$s and $C$s are reversed in the lowpass biquad to create the highpass biquad. The transfer function can be found to be

$$\frac{v_{\text{out}}}{v_{\text{in}}} = \frac{Gs^2}{s^2 + \left[\frac{1}{C_2R_2} + \frac{1}{C_1R_2} + \frac{1-G}{C_1R_1}\right] s + \frac{1}{C_1C_2R_1R_2}} \tag{12.70}$$

Again, there is a lot of flexibility in choosing the element values. The same possibilities mentioned in connection with the lowpass biquad can be applied to the highpass biquad also.

### 12.8.3 The Bandpass Biquad

The Sallen-Key bandpass biquad is shown in Fig. 12.27. This circuit uses three resistors and two capacitors to create the transfer function. The additional two resistors, $R_F$ and $R_A$ are used to set the gain. One version of this circuit uses a unity-gain stage, requiring no resistors. It is also possible to use three capacitors and two resistors in the network to create a bandpass transfer function.

The circuit of Fig. 12.27 has a transfer function of

$$\frac{v_{\text{out}}}{v_{\text{in}}} = \frac{\frac{G}{R_1C_1} s}{s^2 + \left[\frac{1}{C_1R_1} + \frac{1}{C_1R_3} + \frac{1}{C_2R_3} + \frac{1-G}{C_1R_2}\right] s + \frac{R_1+R_2}{C_1C_2R_1R_2R_3}} \tag{12.71}$$

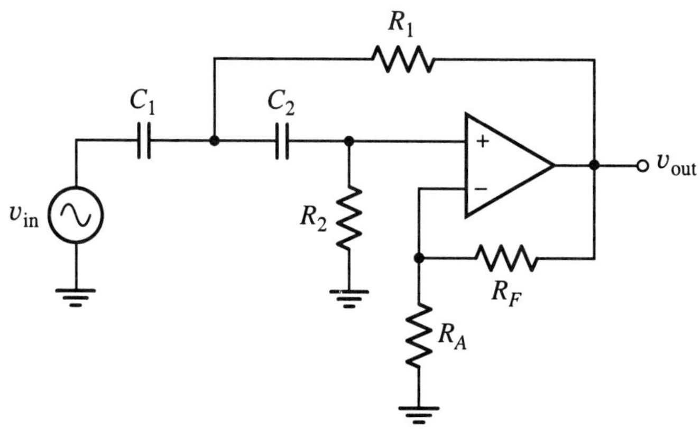


Figure 12.26: The Sallen-Key highpass biquad.

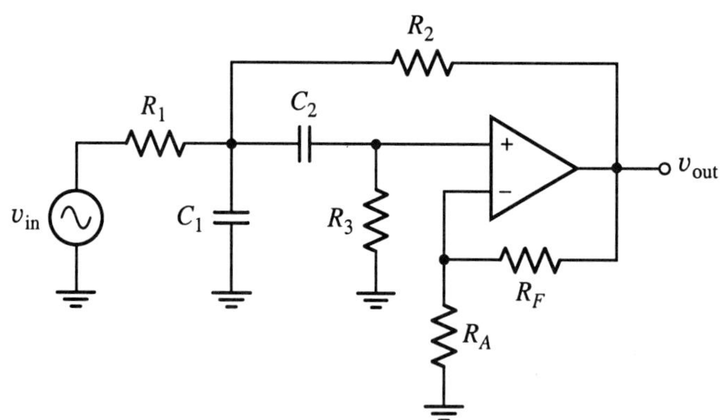


Figure 12.27: The Sallen-Key bandpass biquad.

The denominator of this transfer function can be equated to the standard form of a bandpass function; that is,

$$\text{Denom} = s^2 + \frac{\omega_o}{Q}s + \omega_o^2$$

This allows the following identifications to be made:

$$\omega_o = \sqrt{\frac{R_1 + R_2}{C_1 C_2 R_1 R_2 R_3}} \tag{12.72}$$

and

$$Q = \frac{\sqrt{\frac{R_1+R_2}{C_1 C_2 R_1 R_2 R_3}}}{\frac{1}{C_1 R_1} + \frac{1}{C_1 R_3} + \frac{1}{C_2 R_3} + \frac{1-G}{C_1 R_2}} \tag{12.73}$$

Suppose we let $R_1 = R_2 = R_3 = R$ and $C_1 = C_2 = C$. This allows the resonant frequency and the $Q$ to be written as

$$\omega_o = \frac{\sqrt{2}}{RC} \tag{12.74}$$

and

$$Q = \frac{\sqrt{2}}{4 - G} \tag{12.75}$$

From Eq. (12.75), the value of $G$ for a specified $Q$ can be expressed as

$$G = 4 - \frac{\sqrt{2}}{Q} \tag{12.76}$$

The gain at resonance can be found from Eq. (12.70) by letting the real part of the denominator go to zero to give

$$\left.\frac{v_{\text{out}}}{v_{\text{in}}}\right|_{\omega_o} = \frac{G}{4 - G} = 2\sqrt{2}Q - 1 \tag{12.77}$$

# Example 12.5

Design a Sallen-Key bandpass biquad with a resonant frequency of 10 kHz and a 3-dB bandwidth of 800 Hz. Calculate the gain at resonance for your circuit. Use equal $R$s and equal $C$s.

**Solution** The required $Q$ for this circuit is

$$Q = \frac{f_o}{\Delta f} = \frac{10{,}000}{800} = 12.5$$

We will first find the normalized values of the circuit of Fig. 12.27 and then denormalize these values. We will set a resonant frequency of 1 rad/s from Eq. (12.74). This requires that

$$\omega_o = 1 = \frac{\sqrt{2}}{RC}$$

If $C$ is selected to be 1 F, then $R = \sqrt{2} = 1.414\ \omega$. The required gain of the amplifier is calculated from Eq. (12.76) as

$$G = 4 - \frac{\sqrt{2}}{Q} = 4 - \frac{\sqrt{2}}{12.5} = 3.887 \text{ V/V}$$

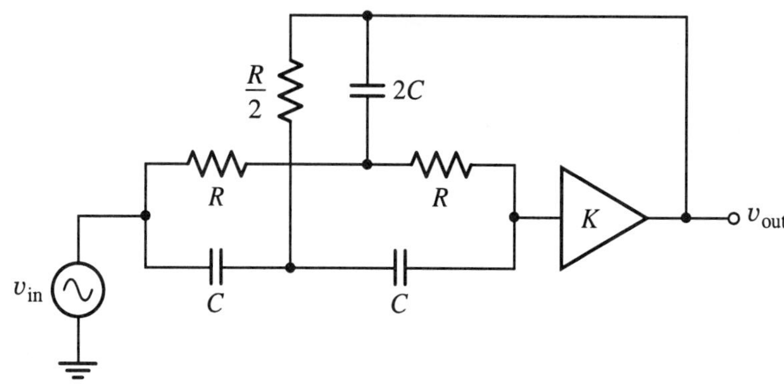


Figure 12.28: A Sallen-Key band-reject biquad.

The values of $R_F$ and $R_A$ can be selected to satisfy this gain requirement. Values of $R_F = 28.87$ kΩ and $R_A = 10$ kΩ will set the proper gain.

The element values must now be denormalized using a value of

$$k_\omega = 10,000 \times 2\pi = 6.283 \times 10^4 \text{ rad/s}$$

The impedance scaling factor is arbitrary. We will set the actual capacitance to 10 nF. The impedance scaling factor is then found from

$$C_{den} = \frac{C}{k_i k_\omega}$$

Solving for $k_i$ gives

$$k_i = 1592$$

The value of resistance to be used is

$$R_{den} = R \times k_i = 1.414 \times 1592 = 2250\ \Omega$$

---

### 12.8.4 A Band-Reject Biquad

The circuit of Fig. 12.28 is a band-reject biquad.

This circuit can be analyzed to find a transfer function of

$$\frac{v_{out}}{v_{in}} = \frac{K\left(s^2 + \frac{1}{R^2C^2}\right)}{s^2 + \frac{4(1-K)}{RC} + \frac{1}{R^2C^2}} \tag{12.78}$$

This function results in a notch frequency of

$$\omega_n = \frac{1}{RC} \tag{12.79}$$

The $Q$ of the notch response is

$$Q = \frac{1}{4(1-K)} \tag{12.80}$$

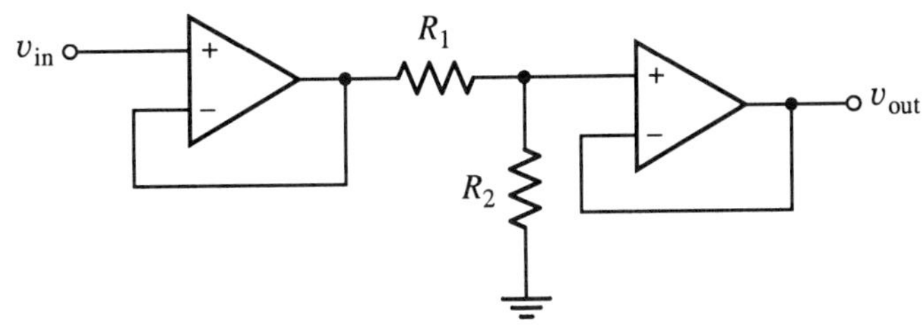


Figure 12.29: An attenuation circuit.

In order that the pole-pair be in the LHP, the value of $K$ must be less than unity. A noninverting op amp has a minimum gain of unity. One simple solution is to use a pair of unity-gain noninverting amplifiers with an attenuator between the stages as shown in Fig. 12.29.

One advantage of the active notch circuit over the twin tee circuit that is often used to obtain a notch response is the adjustability of $Q$. For the twin tee circuit, the $Q$ is fixed at $Q = 1/4$. The $Q$ of the pole factor determines the sharpness of the notch. With the active notch filter, the $Q$ can be adjusted over a practical range that allows a very sharp notch. For example, if $K = 0.95$, the resulting value of $Q$ is

$$Q = \frac{1}{4(1 - 0.95)} = 5$$

This value of $Q$ results in a much sharper notch than that of the twin tee circuit.

There are other versions of active band-reject filters [10] that use a single op amp, but these will not be discussed here.

### 12.8.5 Problems of Sallen-Key Circuits

From the biquad equations for the lowpass, highpass, and bandpass Sallen-Key transfer functions, it appears that these configurations are very useful. Unfortunately, the major limitation on these circuits has been neglected in the derivations of the equations.

The problem with the Sallen-Key biquads is the frequency limitation imposed by the op amp. In Example 12.5, the design resulted in a $Q$ of 12.5 and an op amp gain of $G = 3.89$ V/V. If this design had been done for the IF strip in an AM radio, the center frequency would have had to be 455 kHz. In order that the op amp not affect the performance of this circuit, the bandwidth should be about 10 times the resonant frequency. This would require an op amp bandwidth of 4.55 MHz and a $GBW$ of $3.89 \times 4.55 = 17.7$ MHz. Op amps with this $GBW$ are available, but if the filter is to be used in an FM receiver with an IF of 10.7 MHz, a $GBW$ exceeding 400 MHz will be required, which is out of the question.

If a precision amplifier without feedback could be used for an IC filter, this configuration might be useful. However, it tends to use relatively high values of capacitance that make integration difficult. Thus, the major use of Sallen-Key circuits is in low-frequency discrete active filters.

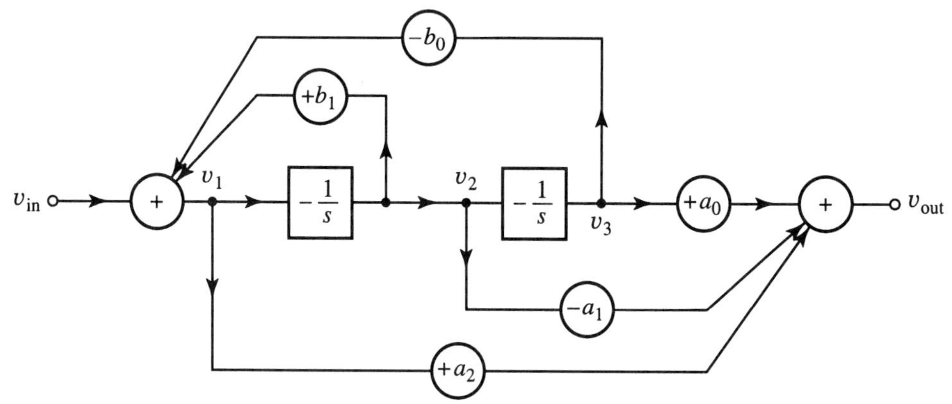


Figure 12.30: General two-integrator system.

## 12.9 TWO-INTEGRATOR ACTIVE BIQUADS

**Important Concepts**

**1.** Two-integrator biquads can realize many important types of second-order responses.

**2.** These biquads can be realized with discrete circuits or on IC chips.

**3.** Several versions of the two-integrator biquad have been developed.

The two-integrator biquads are used in discrete active filters as well as IC active filters. Many variations on this basic method are used to enhance certain advantages. This chapter will consider some of the basic ideas of the two-integrator configuration.

Figure 12.30 shows a general block diagram of the two-integrator system.

The weighting circuits are basically amplifiers with gains equal to $a_i$ or $b_i$. In this block diagram, we assume ideal integrators, summing, and weighting circuits. That is, these functions are performed with no loading of adjacent elements. This general configuration is used in other areas besides filters, for example, in control systems.

If $v_{in}$ is the Laplace of the input voltage and $v_{out}$ is the Laplace output voltage, we can write the voltage at node 1 as

$$v_1 = v_{in} - b_0 v_3 + b_1 v_2 \tag{12.81}$$

The integrators result in the relations

$$v_3 = -\frac{v_2}{s} \tag{12.82}$$

$$v_2 = -\frac{v_1}{s} \tag{12.83}$$

and

$$v_3 = \frac{v_1}{s^2} \tag{12.84}$$

We can now write Eq. (12.81) as

$$v_1 = v_{in} - b_0 \frac{v_1}{s^2} - b_1 \frac{v_1}{s} \tag{12.85}$$

The transfer function relating $v_1$ to $v_{\text{in}}$ can be found as

$$\frac{v_1}{v_{\text{in}}} = \frac{s^2}{s^2 + b_1 s + b_0} \tag{12.86}$$

The output voltage can be expressed as

$$v_{\text{out}} = a_0 v_3 - a_1 v_2 + a_2 v_1 \tag{12.87}$$

Using Eqs. (12.83) and (12.84), we can manipulate Eq. (12.87) to give

$$v_{\text{out}} = a_0 \frac{v_1}{s^2} + a_1 \frac{v_1}{s} + a_2 v_1 = \left( \frac{a_2 s^2 + a_1 s + a_0}{s^2} \right) v_1 \tag{12.88}$$

We can now write the transfer function

$$\frac{v_{\text{out}}}{v_1} = \frac{a_2 s^2 + a_1 s + a_0}{s^2} \tag{12.89}$$

The two transfer functions can be multiplied together to eliminate the variable $v_1$ giving

$$\begin{aligned} \frac{v_{\text{out}}}{v_{\text{in}}} = \frac{v_{\text{out}}}{v_1} \frac{v_1}{v_{\text{in}}} &= \frac{a_2 s^2 + a_1 s + a_0}{s^2} \frac{s^2}{s^2 + b_1 s + b_0} \\ &= \frac{a_2 s^2 + a_1 s + a_0}{s^2 + b_1 s + b_0} \end{aligned} \tag{12.90}$$

Equation (12.90) is recognized as a general biquad expression. Any two of the numerator coefficients can be set to zero to get a lowpass, bandpass, or highpass biquad transfer function.

An interesting result can be achieved with the circuit of Fig. 12.30. A highpass, bandpass, and lowpass function can be realized simultaneously at node 1, node 2, and the output. This results when the weights $a_1$ and $a_2$ are chosen to be zero. With this choice, the output is

$$\frac{v_{\text{out}}}{v_{\text{in}}} = \frac{a_0}{s^2 + b_1 s + b_0} \tag{12.91}$$

We note that $v_2$ is related to $v_{\text{out}}$ by

$$v_2 = -\frac{s}{a_0} v_{\text{out}} \tag{12.92}$$

Combining this with Eq. (12.91) gives

$$\frac{v_2}{v_{\text{in}}} = -\frac{s}{s^2 + b_1 s + b_0} \tag{12.93}$$

Noting that $v_1 = -s v_2$ allows us to write

$$\frac{v_1}{v_{\text{in}}} = \frac{s^2}{s^2 + b_1 s + b_0} \tag{12.94}$$

This result is probably more interesting than useful, especially in the case of the bandpass function. The $Q$ of the poles of all transfer functions is equal, thus, the bandpass cannot be very sharp.

### 12.9.1 Op Amp Two-Integrator Biquads

The op amp can be used to construct two-integrator biquads in low- to intermediate-frequency ranges.

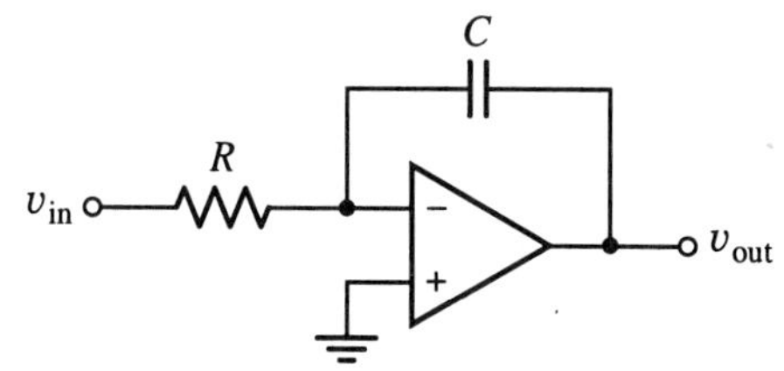


Figure 12.31: An op amp integrator.

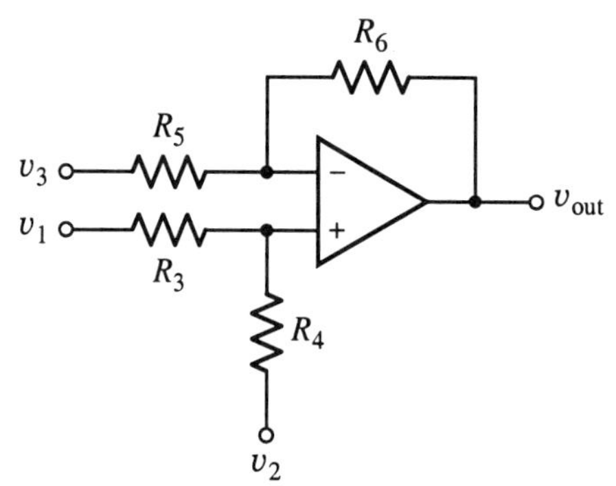


Figure 12.32: A difference/weighting circuit.

**Practical Discrete Two-Integrator Building Blocks**

The most basic building block of the two-integrator circuit is the op amp integrator of Fig. 12.31. The transfer function of this circuit is

$$\frac{v_{out}}{v_{in}} = -\frac{1}{sRC} \tag{12.95}$$

Selection of the $RC$ product allows a very broad range of integration constant. Figure 12.32 indicates a weighting network that can be used in the two-integrator biquad. The output signal can be expressed in terms of the three input signals as

$$v_{out} = v_1\left(\frac{R_4}{R_3+R_4}\right)\left(1+\frac{R_6}{R_5}\right) + v_2\left(\frac{R_3}{R_3+R_4}\right)\left(1+\frac{R_6}{R_5}\right) - v_3\frac{R_6}{R_5} \tag{12.96}$$

With this circuit, two signals can be weighted and added with a third weighted signal subtracted.

**The Kerwin-Huelsman-Newcomb General Biquad**

The KHN [5] general biquad is an implementation of the block diagram of Fig. 12.30. This circuit is shown in Fig. 12.33. Op amp 1 weights and sums the input signal and the voltage $v_2$. The signal $v_3$ is weighted and subtracted by this op amp. Op amps 2 and 3 are integrators. Op amp 5 sums the three required signals while op amp 4 is an inverter with unity gain.

The coefficients of weighting in Fig. 12.30 can be identified by analyzing the circuit of Fig. 12.33. The input signal is weighted by a factor of

$$G = \frac{R_4}{R_3+R_4}\left(1+\frac{R_6}{R_5}\right) \tag{12.97}$$

This factor will multiply the overall transfer function.

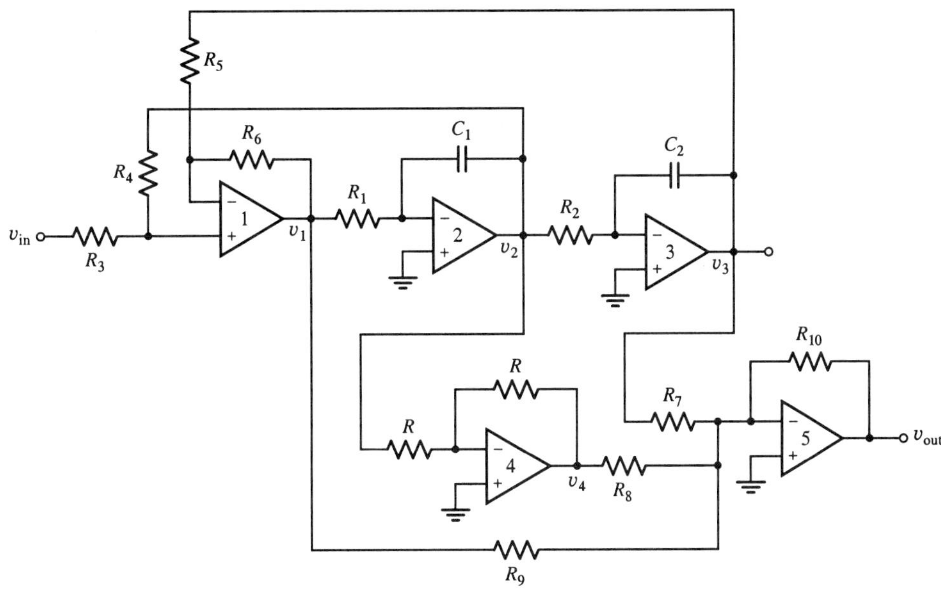


Figure 12.33: A general KHN biquad.

The coefficient $b_0$ is

$$b_0 = \frac{R_6}{R_5} \tag{12.98}$$

The coefficient $b_1$ is

$$b_1 = \frac{R_3}{R_3 + R_4}\left(1 + \frac{R_6}{R_5}\right) \tag{12.99}$$

The coefficient $a_0$ is

$$a_0 = \frac{R_{10}}{R_7} \tag{12.100}$$

The coefficient $a_1$ is

$$a_1 = \frac{R_{10}}{R_8} \tag{12.101}$$

and $a_2$ is

$$a_2 = \frac{R_{10}}{R_9} \tag{12.102}$$

Note that the output of the KHN biquad is an inverted version of the block diagram output. This general biquad realization is used with small modifications to create more specific biquads.

### The Kerwin-Huelsman-Newcomb Lowpass Biquad

Figure 12.34 shows a modification of the general biquad to achieve a lowpass response. This circuit is also called a KHN biquad. By eliminating the circuitry that realizes finite values of $a_1$ and $a_2$, a summer is not required. The value of $a_0$ is unity in this circuit. If the $RC$ product is chosen to be unity also, the transfer function can be

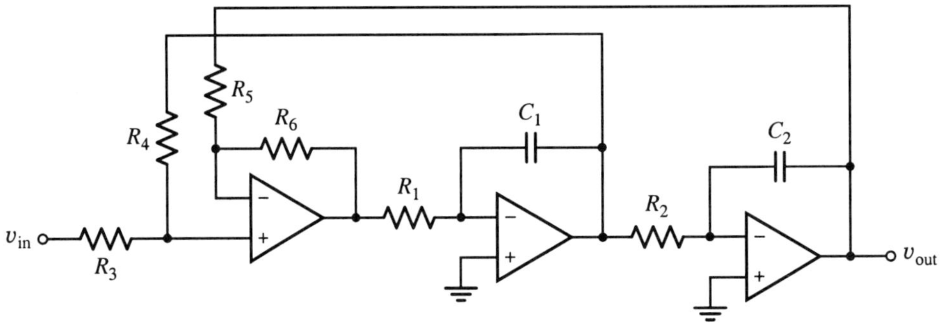


Figure 12.34: A KHN lowpass biquad.

found by substituting the values for $b_0$ and $b_1$ given by Eqs. (12.98) and (12.99) into Eq. (12.90) and multiplying by $G$. This gives

$$\frac{v_{out}}{v_{in}} = \frac{\frac{R_4}{R_5}\frac{R_5+R_6}{R_3+R_4}}{s^2 + \frac{R_3}{R_5}\frac{R_5+R_6}{R_3+R_4}s + \frac{R_6}{R_5}} \tag{12.103}$$

If the $RC$ product is not unity, the transfer function can be found from Eq. (12.103) by replacing $s$ by $sRC$. The result is

$$\frac{v_{out}}{v_{in}} = \frac{\frac{R_4}{R_5}\frac{R_5+R_6}{R_3+R_4}\frac{1}{C^2R^2}}{s^2 + \frac{R_3}{R_5}\frac{R_5+R_6}{R_3+R_4}\frac{1}{RC}s + \frac{R_6}{R_5}\frac{1}{C^2R^2}} \tag{12.104}$$

# Example 12.6

Design a two-pole Butterworth lowpass biquad using the KHN circuit. The 3-dB cutoff frequency is to be 20 kHz.

**Solution** The normalized two-pole Butterworth transfer function is

$$\frac{K}{s^2 + 1.414s + 1}$$

Equating the denominator of this function to that of Eq. (12.103) leads to the conditions

$$\frac{R_6}{R_5} = 1$$

$$\frac{R_3}{R_3 + R_4}\frac{R_5 + R_6}{R_5} = 1.414$$

The first condition can be satisfied by choosing $R_5 = R_6$. Using this relation in the second condition gives $R_4 = 0.414R_3$.

Choosing resistances that are compatible with op amp circuits, we select

$$R_5 = R_6 = 10 \text{ k}\Omega$$

$$R_3 = 10 \text{ k}\Omega \text{ and } R_4 = 4.14\ k\Omega$$

With these values, along with a unity $RC$ product in the integrators, the biquad has a 3-dB frequency of 1 rad/s. This product can be changed to denormalize the filter to the desired corner frequency by making

$$\frac{1}{RC} = \omega_c = 2\pi \times 2 \times 10^4$$

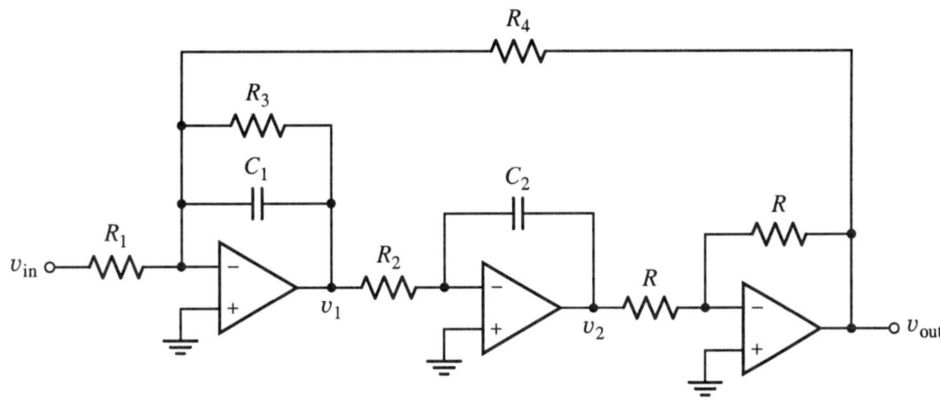


Figure 12.35: The Tow-Thomas biquad.

Selecting $R = 7.96$ kΩ and $C = 1$ nF satisfies this denormalization condition.

---

The bandpass or highpass KHN biquad can be constructed by making either $a_1$ or $a_2$ the only nonzero coefficient in the numerator.

### The Tow-Thomas Biquad

The biquad of Fig. 12.35 [11,12] represents a slight improvement over the KHN biquad in terms of reducing the number of elements by one. However, it affords a simpler determination of gain and pole-$Q$. Each value can be set by the selection of a single resistor value.

The transfer function can be found by writing the voltage $v_1$ in terms of $v_{\text{in}}$. The resulting equation is

$$v_1 = -\frac{R_3/R_1}{sC_1R_3+1}v_{\text{in}} - \frac{R_3/R_4}{sC_1R_3+1}v_{\text{out}} \tag{12.105}$$

Since $v_{\text{out}} = -v_2$ and $v_2 = -v_1/sR_2C_2$, we can eliminate $v_1$ from Eq. (12.105) to get

$$\left[sR_2C_2 + \frac{R_3/R_4}{sC_1R_3+1}\right]v_{\text{out}} = -\frac{R_3/R_1}{sC_1R_3+1}v_{\text{in}} \tag{12.106}$$

Solving for the transfer function gives

$$\frac{v_{\text{out}}}{v_{\text{in}}} = -\frac{\frac{1}{C_1C_2R_1R_2}}{s^2 + \frac{1}{C_1R_3}s + \frac{1}{C_1C_2R_2R_4}} \tag{12.107}$$

This lowpass biquad transfer function can be equated to the standard expression of

$$\frac{v_{\text{out}}}{v_{\text{in}}} = G(0)\frac{\omega_o^2}{s^2 + \frac{\omega_o}{Q}s + \omega_o^2} \tag{12.108}$$

where $G(0)$ is the dc value of the transfer function. Equating Eqs. (12.107) and (12.108) allows the identifications

$$\omega_o = \sqrt{\frac{1}{C_1C_2R_2R_4}} \tag{12.109}$$

$$G(0) = -\frac{R_4}{R_1} \tag{12.110}$$

$$Q = \sqrt{\frac{C_1 R_3^2}{C_2 R_2 R_4}} \tag{12.111}$$

Once $\omega_o$ is selected, the gain can be set at the desired value by the resistance $R_1$. This value does not affect the frequency, $\omega_o$. The pole-$Q$ can be determined by the value of $R_3$ which, again, does not affect the other two parameters.

Returning to the KHN biquad, we can identify these three parameters with the elements in Fig. 12.34 as

$$\omega_o = \sqrt{\frac{R_6}{C_1 C_2 R_1 R_2 R_5}} \tag{12.112}$$

$$G(0) = \frac{R_4}{R_6} \frac{R_5 + R_6}{R_3 + R_4} \tag{12.113}$$

$$Q = \frac{R_4}{R_6} \frac{R_3 + R_4}{R_5 + R_6} \sqrt{\frac{R_6 R_1 C_1}{R_5 R_2 C_2}} \tag{12.114}$$

All elements appear in at least two of the three equations, allowing less freedom of choice in selecting values of $G(0)$, $Q$, and $\omega_o$.

### 12.9.2 Two-Integrator Integrated Circuits

Two types of IC two-integrator circuits that are often used are the MOSFET-C filter and the $G_m - C$ filter. Although the MOSFET-C filter is important, we will cover only the $G_m - C$ filter here.

#### The $G_m - C$ Filter

We have discussed a configuration of the transconductor circuit that leads to the lowpass circuit shown in Fig. 12.17. This first-order circuit can be used with biquads to implement an odd-order transfer function. The transconductor biquads to be used are often based on the inductance simulation circuit of Fig. 12.36.

For the first transconductor, a positive input voltage results in a positive output current or current out of the transconductor output. For the second transconductor, the small circle at the output indicates that a positive voltage at the input of this device causes negative output current or current flowing into the transconductor output. We can write the voltage $v_2$ as

$$v_2 = -\frac{i_{\text{out1}}}{sC} = \frac{G_{m1} v_{\text{in}}}{sC} \tag{12.115}$$

The current $i_{\text{out2}}$ is

$$i_{\text{out2}} = G_{m2} v_2 = \frac{G_{m1} G_{m2} v_{\text{in}}}{sC} \tag{12.116}$$

After noting that the input current, $i_{\text{in}} = i_{\text{out2}}$, we can write the input impedance as the ratio of input voltage to input current, giving

$$Z_{\text{in}} = \frac{v_{\text{in}}}{i_{\text{in}}} = \frac{sC}{G_{m1} G_{m2}} \tag{12.117}$$

This impedance represents an inductor that has a value of

$$L = \frac{C}{G_{m1} G_{m2}} \tag{12.118}$$

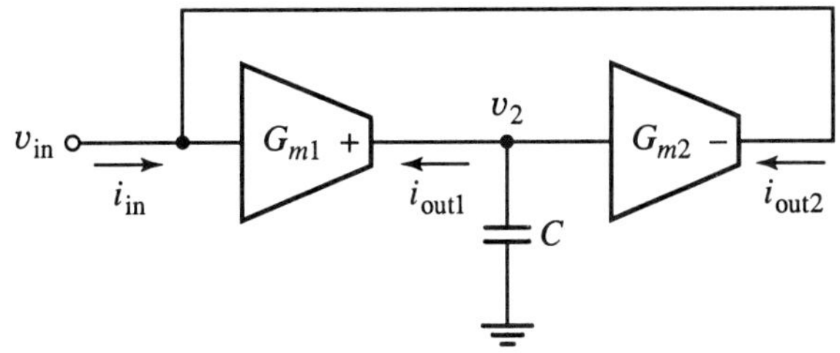


Figure 12.36: An inductance simulation circuit.

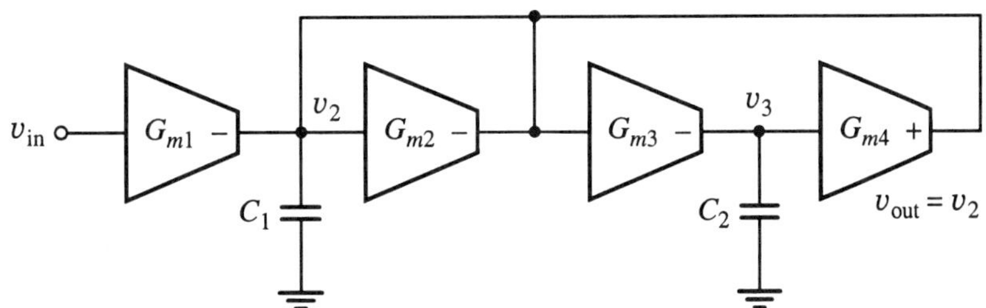


Figure 12.37: A transconductance biquad.

A narrowband biquad can be based on this circuit by adding a parallel capacitance. In order to achieve good buffering and more flexibility in setting circuit parameters, the circuit of Fig. 12.37 may be used. With the output fed back to two earlier nodes, we write

$$v_{out} = v_2 = \frac{G_{m4}v_3 - G_{m2}v_2 - G_{m1}v_{in}}{sC_1} \tag{12.119}$$

The voltage $v_3$ can be expressed as

$$v_3 = -\frac{G_{m3}v_2}{sC_2} \tag{12.120}$$

Substituting this value into the equation for $v_{out}$ leads to a transfer function of

$$\frac{v_{out}}{v_{in}} = \frac{-\frac{G_{m1}}{C_1}s}{s^2 + s\,\frac{G_{m2}}{C_1} + \frac{G_{m4}G_{m3}}{C_1C_2}} \tag{12.121}$$

The normal identifications of coefficients with a narrowband function result in

$$\omega_o = \sqrt{G_{m4}G_{m3}/C_1C_2} \tag{12.122}$$

If $C_1 = C_2 = C$ and $G_{m1} = G_{m3} = G_{m4} = G_m$, the resonant frequency becomes

$$\omega_o = \frac{G_m}{C} \tag{12.123}$$

The equation for $Q$ is

$$Q = \frac{G_m}{G_{m2}} \tag{12.124}$$

Obviously, a high-$Q$ circuit must create a small value of $G_{m2}$ compared to $G_m$. Fortunately, the value of transconductance can be made very low by MOS device scaling techniques.

Another application requiring low values of $G_m$ is the implementation on an IC chip of a lowpass circuit having a low cutoff frequency. In connection with Fig. 12.17, we found a corner frequency of

$$\omega_{3dB} = \frac{G_m}{C} \tag{12.125}$$

For IC circuits, the value of $C$ may be 50 pF or less. In order to create a low frequency in the kHz range, $G_m$ must be less than 1 $\mu$A/V.

## 12.10 NARROWBAND FILTERS BASED ON CURRENT CONVEYORS

**Important Concepts**

1. This method uses conveyors to create an impedance that is inductive.
2. Negative resistance can be generated to cancel resonant circuit losses and achieve higher $Q$s.

The basic concepts of current conveyor circuits have been covered in an earlier section. This section will discuss the application of this element in relatively high frequency, narrowband, integrated circuit synthesis.

### 12.10.1 A Narrowband Current Conveyor

The basic idea behind this method is to create an impedance that looks inductive, using only resistors, capacitors and active devices. In many respects, it is similar to the gyrator circuit reported by Schaumann and Van Valkenburg [15]. The suggested implementation of this method is quite unique [4] and will be treated in this section.

It has been possible to implement a general impedance converter (GIC) for many years, but the actual construction of the circuit limited overall performance to modest levels. The circuit discussed in this chapter operates at 100 MHz with a relatively high $Q$ and is fabricated on an IC chip.

### 12.10.2 An Earlier Proposal

A proposal in the 1990s suggested the configuration shown in Fig. 12.38 to implement an inductor. For the first conveyor, CC1, $i_{y1} = 0$ and $v_{x1} = v_{y1}$. For the second conveyor, CC2, $v_{y2} = 0 = v_{x2}$ and $i_{z2} = +i_{x2}$.

We are interested in finding the input impedance, thus we assume an input current and calculate the resulting input voltage. We can write the input current as

$$i_{\text{in}} = i_{z2} + i = i_{z2} + \frac{v_{\text{in}}}{Z_2 + Z_3} \tag{12.126}$$

The current $i_{z2}$ equals $i_{x2}$ and is given by

$$i_{z2} = i_{x2} = \frac{v_{x1} - v_{x2}}{Z_4} = \frac{v_{x1}}{Z_4} = \frac{v_{y1}}{Z_4} = \frac{v_{\text{in}}}{Z_4}\frac{Z_3}{Z_2 + Z_3} \tag{12.127}$$

Combining these two equations results in

$$i_{\text{in}} = \frac{v_{\text{in}}}{Z_4}\frac{Z_3}{Z_2 + Z_3} + \frac{v_{\text{in}}}{Z_2 + Z_3} = \frac{v_{\text{in}}}{Z_2 + Z_3}\left(1 + \frac{Z_3}{Z_4}\right) \tag{12.128}$$

The impedance looking into this configuration is

$$Z_{\text{in}} = \frac{v_{\text{in}}}{i_{\text{in}}} = \frac{Z_2 + Z_3}{1 + \frac{Z_3}{Z_4}} = \frac{Z_4(Z_2 + Z_3)}{Z_3 + Z_4} \tag{12.129}$$

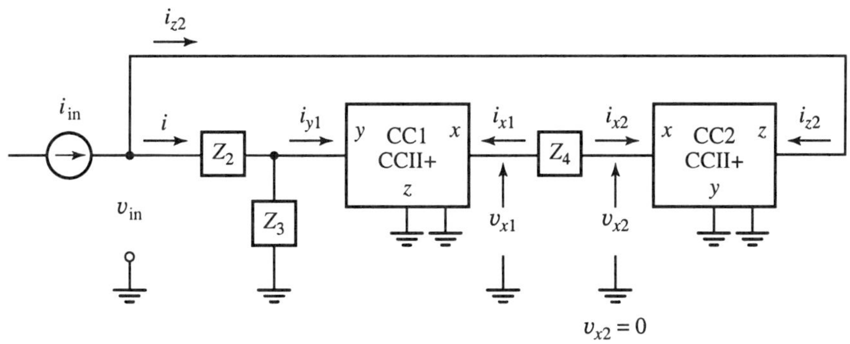


Figure 12.38: An early implementation of an inductor with current conveyors.

We now select impedances that can be realized with resistors or capacitors. For example, we will choose $Z_3 = \frac{1}{sC}$, $Z_2 = R_2$, and $Z_4 = R_4$, with the restriction that $R_4 \ll R_2$. The impedance $Z_{\text{in}}$ is now

$$Z_{\text{in}} = \frac{R_4(sCR_2 + 1)}{sCR_4 + 1} \tag{12.130}$$

This impedance is not exactly that of a lossy inductor, but it matches a lossy inductor over a wide frequency range. The impedance of a lossy inductor is $R(sL/R + 1)$. This impedance and that of the current conveyor circuit are plotted in Fig. 12.39.

If $R_4 = R$ and $CR_2$ is chosen to equal $L/R$, then the current conveyor circuit approximates the inductor in the frequency range

$$\frac{1}{CR_2} \ll \omega \ll \frac{1}{CR_4}$$

We can get a resonance by placing an additional capacitor of value $C$ across the input terminals. When this is done, we get a parallel loss resistance of $R_p = R_4 Q^2$, assuming no capacitive losses. The resonant frequency is

$$\omega_o = \frac{1}{C\sqrt{RR_2}} \tag{12.131}$$

# Example 12.7

For an inductor, assume that $R = 10\ \Omega$ and $L = 100\ \mu$H, resulting in $R/L = 10^5$ rad/s. In the current conveyor circuit, if $R_4 = 10\ \Omega$, select $C$ such that $1/CR_4 \gg 2\pi \times 100$ krad/s. Calculate the resonant frequency.

**Solution** If we select $C = 0.0159\ \mu$F, then $1/CR_4 = 6.289 \times 10^0$ rad/s. Since $CR_2 = L/R - 10^{-5}$, we can pick $R_2 = 628\ \Omega$. We now check to verify that $1/CR_2 = 10^5$ is less than the resonant frequency. The resonant frequency is

$$\omega_o = \frac{1}{C\sqrt{RR_2}} = 794 \text{ krad/s}$$

and is much larger than $1/CR_2$.

Although this circuit has been used in early methods, a better method will be considered in the next subsection.

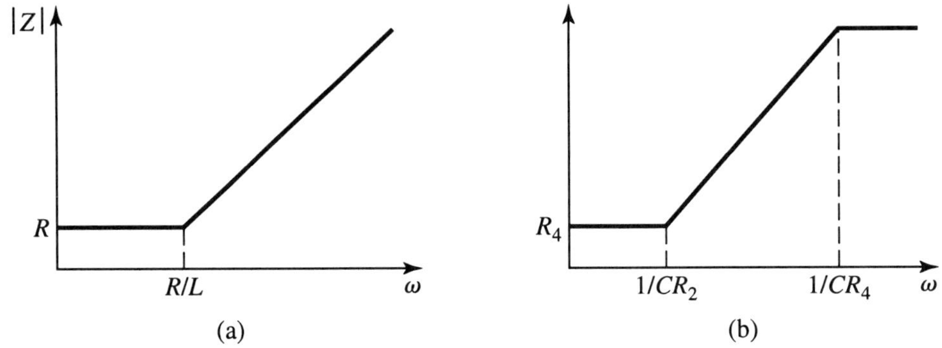


Figure 12.39: Impedance vs. frequency: (a) a lossy inductor; (b) the CC circuit.

### 12.10.3 Synthesis with the Controlled Current Conveyor

We will first consider the circuit of Fig. 12.40 using CCII+ circuits. The input impedance to this circuit can be found by assuming an input voltage, $v_{\text{in}}$, and calculating the resulting current. This impedance is

$$Z_{\text{in}} = \frac{v_{\text{in}}}{i_{\text{in}}} \tag{12.132}$$

The input current can be expressed as

$$i_{\text{in}} = i_{x1} + i_{x2} \tag{12.133}$$

Since the $Y$ terminal of the second current conveyor is tied to ground, the voltage at terminal $X$ is zero since $v_x = v_y$. The current $i_{x2}$ is then

$$i_{x2} = \frac{v_{\text{in}}}{R_{x2}} \tag{12.134}$$

The current $i_{x1}$ can be written as

$$i_{x1} = \frac{v_{\text{in}} - v_{x1}}{R_{x1}} \tag{12.135}$$

This current is more difficult to find since $v_{x1}$ is nonzero in this configuration. This voltage can be written as

$$v_{x1} = v_{y1} = v_C = -i_{z2} \times \frac{1}{sC} \tag{12.136}$$

The current leaving the capacitor is $i_{z2}$, but this current also equals $i_{x2}$. From Eq. (12.134) we can express the voltage $v_{x1}$ as

$$v_{x1} = -\frac{v_{\text{in}}}{sCR_{x2}} \tag{12.137}$$

The total input current is

$$i_{\text{in}} = i_{x1} + i_{x2} = v_{\text{in}} \left( \frac{sC(R_{x1} + R_{x2}) + 1}{sCR_{x1}R_{x2}} \right) \tag{12.138}$$

This allows $Z_{\text{in}}$ to be expressed as

$$Z_{\text{in}} = \frac{v_{\text{in}}}{i_{\text{in}}} = \frac{sCR_{x1}R_{x2}}{sC[R_{x1} + R_{x2}] + 1} \tag{12.139}$$

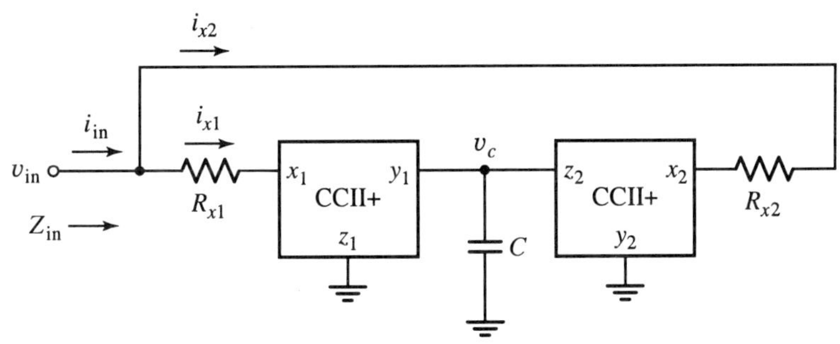


Figure 12.40: Another inductor simulation.

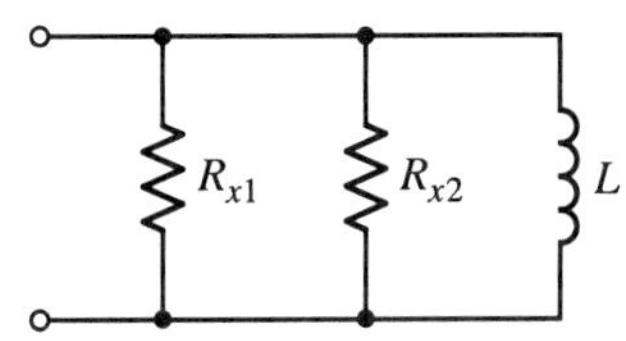


Figure 12.41: A lossy inductor.

Figure 12.41 shows an inductor in parallel with two resistances. The parallel impedance of this circuit can be found to be

$$Z = \frac{sL\frac{R_{x1}R_{x2}}{R_{x1}+R_{x2}}}{sL + \frac{R_{x1}R_{x2}}{R_{x1}+R_{x2}}} \tag{12.140}$$

If $L$ is selected to equal $CR_{x1}R_{x2}$, this impedance $Z$ becomes identical to $Z_{\text{in}}$, the input impedance to the current conveyor circuit. Thus, the current conveyor circuit, using only resistive and capacitive components, generates a lossy inductance.

Often the values of $R_{x1}$ and $R_{x2}$ are small, leading to a low-$Q$ inductor. If $R_p = R_{x1}||R_{x2} = 70\ \Omega$ and the inductor is 70 nH, the $Q$ at 100 MHz is only 1.59.

Although $R_{x1}$ and $R_{x2}$ could be selected to be larger values to improve the inductor $Q$, another method can also be used [4]. The CCII+ circuits of Fig. 12.40 are replaced with CCCII+ circuits. These circuits absorb the resistors into the controlled current conveyors. BJT circuits are used with the emitter resistance of the transistor replacing $R_{x1}$. The second conveyor also replaces $R_{x2}$ with an emitter resistance. These resistances can then be controlled by the dc emitter currents of the respective transistors, since

$$r_e = R_{x1} = R_{x2} = \frac{0.026}{I_E} = \frac{V_T}{I_E} \tag{12.141}$$

where $V_T$ is the thermal voltage of the transistor or $kT/q$. Normally, the dc emitter current is chosen to be relatively large, therefore, $r_e$ is small.

In order to decrease the losses of the circuit and increase the value of $R_{sh}$ in parallel with the inductor, a negative resistance can be placed across the $Z_{\text{in}}$ terminals. The CCCII+ of Fig. 12.42 generates a negative resistance at input terminal $Y$.

The applied voltage, $v_{\text{in}}$, also appears on the interior of the resistance $R_x$ that is equal to $r_e$. The current at the $X$ terminal is

$$i_x = -\frac{v_{\text{in}}}{R_x} \tag{12.142}$$

The current $i_z$ equals this value and is also equal to the input current. The input impedance is then

$$Z_{\text{in}} = \frac{v_{\text{in}}}{i_{\text{in}}} = \frac{v_{\text{in}}}{-v_{\text{in}}/R_x} = -R_x \tag{12.143}$$

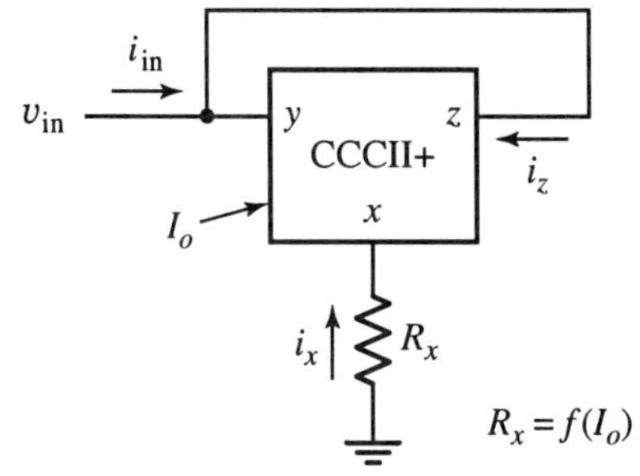


Figure 12.42: A negative resistance generator.

Now let us suppose that the circuit of Fig. 12.42 is placed in parallel with the input terminals of the circuit of Fig. 12.40 with the CCII+ circuits along with $R_{x1}$ and $R_{x2}$ replaced by CCCII+ circuits. If the two conveyors are biased to cause a value of $R_p = 70\ \Omega$ across the inductance, then a negative resistance of $R_x = -72\ \Omega$ will lead to an $R_{sh}$ of

$$R_{sh} = 70\|(-72) = \frac{-72 \times 70}{70 - 72} = 2520\ \Omega$$

This value of resistance leads to a circuit $Q$ at 100 MHz of

$$Q_{\text{eff}} = \frac{R_{sh}}{\omega_o L} = \frac{2520}{2\pi \times 10^8 \times 70 \times 10^{-9}} = 57.3$$

This method is utilized to create an IC with a resonant frequency of 100 MHz and an effective $Q$ of 54 [4].

# 12.11 NARROWBAND FILTERS BASED ON ALLPASS CIRCUITS

**Important Concepts**

**1.** Allpass networks can be used to create the correct phase shift in a feedback loop to lead to lowpass, highpass, or bandpass response.

**2.** Each biquad is based on an approximate bandpass circuit.

## 12.11.1 The Allpass Biquad

For many years the synthesis of biquad functions has been significant in active filter design. Originally, the biquad function was defined in terms of a specific transfer function, but, as mentioned earlier, the meaning has now evolved to apply to a transfer function of the form [10]

$$G(s) = \frac{a_2 s^2 + a_1 s + a_o}{b_2 s^2 + b_1 s + b_o} \tag{12.144}$$

As many as two coefficients in the numerator may be zero. The three types of biquads of interest to this work are of this type; that is, only one numerator coefficient is nonzero. These functions are expressed as

$$\frac{K}{s^2 + b_1 s + b_0} \tag{12.145}$$

$$\frac{Ks}{s^2 + b_1 s + b_0} \tag{12.146}$$

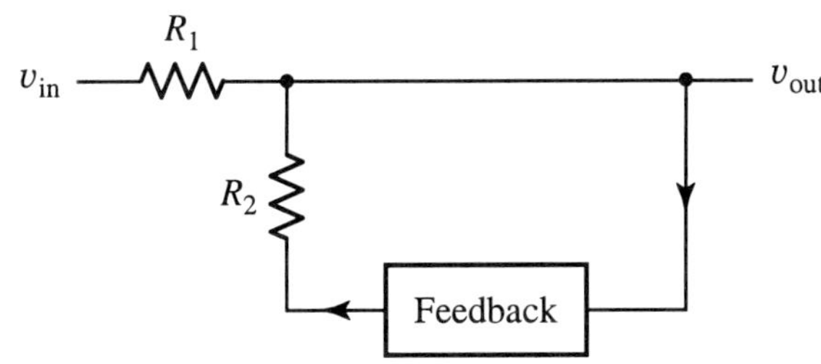


Figure 12.43: Basic configuration of a phase-shift filter.

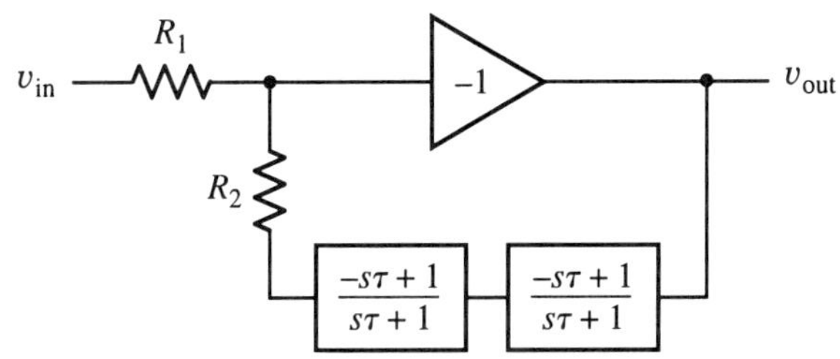


Figure 12.44: An approximate bandpass filter using allpass circuits.

$$\frac{Ks^2}{s^2 + b_1 s + b_0} \tag{12.147}$$

Equation (12.145) is a lowpass, Eq. (12.146) is a bandpass, and Eq. (12.147) is a highpass biquad function. Cascading circuits that implement these biquads allows the synthesis of many important filter functions such as lowpass, highpass, bandpass, and band-reject.

A building block that can be used to create these biquad circuits is the approximate bandpass filter (ABP) based on the allpass network [1,2].

The basic configuration of this phase-shift circuit is shown in Fig. 12.43. A resistive voltage divider forms an output voltage, which is then fed back through some network to the lower side of one of the resistors.

The type of response of the phase-shift filter depends on the characteristics of the feedback network. When the feedback network consists of two allpass circuits and an inverting, unity-gain amplifier as shown in Fig. 12.44, the filter approximates a bandpass filter.

This feedback network has a transfer function of

$$-\frac{-s\tau + 1}{s\tau + 1}\,\frac{-s\tau + 1}{s\tau + 1} \tag{12.148}$$

The resulting filter [1] will then have a transfer function of

$$\frac{R_2}{2R_1 + R_2}\,\frac{(s + \frac{1}{\tau})^2}{s^2 + \frac{2}{\tau}\frac{R_2}{2R_1+R_2}s + \frac{1}{\tau^2}} \tag{12.149}$$

This filter has a resonant frequency of

$$\omega_o = \frac{1}{\tau} \tag{12.150}$$

and a $Q$ of

$$Q = \frac{2R_1 + R_2}{2R_2} \tag{12.151}$$

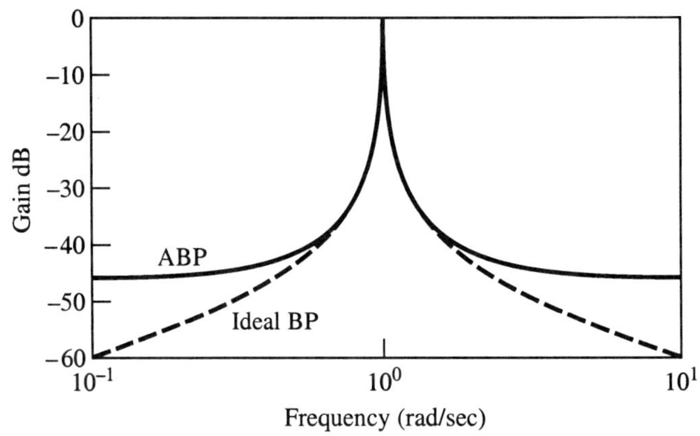


Figure 12.45: Comparison of ABP to bandpass response.

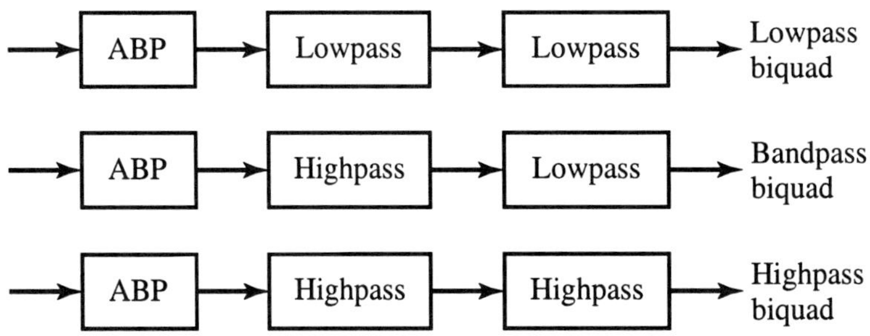


Figure 12.46: Biquad sections using the ABP filter.

The transfer function of Eq. (12.149) shows a double zero at $s = -\tau$ in the numerator rather than a single zero at $s = 0$ as in a true bandpass filter. A comparison of the responses of this filter and a true bandpass filter is shown in Fig. 12.45. Near resonance, the curves are very similar while the ABP function does not show the 6-dB/octave falloff of the true bandpass function at frequencies far away from resonance. This difference is insignificant in many applications, but a modification of the ABP circuit, to be discussed later, converts the circuit to a true bandpass filter.

The basis of operation of the ABP filter is the positive feedback that occurs to create the passband of the response. As the transfer function of the feedback network approaches a phase shift of zero degrees and a magnitude of unity, the effective impedance across $R_2$ approaches an infinite value. The overall transfer function, calculated as

$$\frac{R_{2\text{eff}}}{R_{2\text{eff}} + R_1} \tag{12.152}$$

approaches unity for this condition. When the phase shift moves from the zero value, the overall transfer function decreases from the unity value as $R_{2\text{eff}}$ decreases.

Each of the biquad transfer functions of Eqs. (12.145) - (12.147) can be realized using a modified ABP circuit followed by two one-pole sections as shown in Fig. 12.46.

The lowpass biquad is implemented by cascading two one-pole lowpass sections after the ABP filter. This results in a transfer function of

$$\frac{(s+\omega_o)^2}{s^2 + s\frac{\omega_o}{Q} + \omega_o^2} \times \frac{1}{s+\omega_o} \times \frac{1}{s+\omega_o}$$

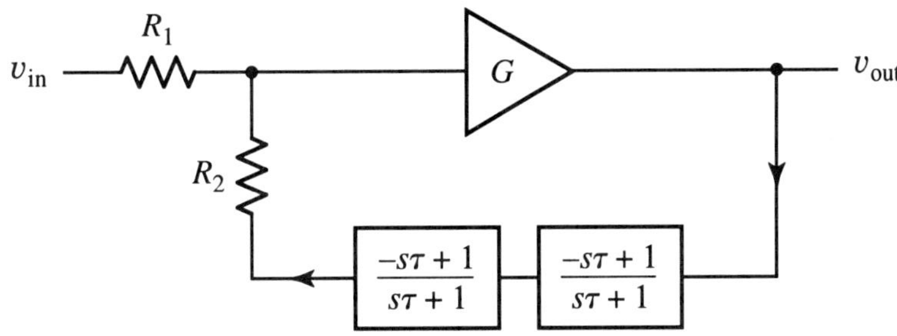


Figure 12.47: A modified ABP filter.

$$= \frac{1}{s^2 + s\frac{\omega_o}{Q} + \omega_o^2} \tag{12.153}$$

The bandpass biquad cascades a one-pole highpass and a one-pole lowpass section after the ABP filter to yield a transfer function of

$$\frac{(s+\omega_o)^2}{s^2 + s\frac{\omega_o}{Q} + \omega_o^2} \times \frac{s}{s+\omega_o} \times \frac{1}{s+\omega_o}$$

$$= \frac{s}{s^2 + s\frac{\omega_o}{Q} + \omega_o^2} \tag{12.154}$$

The highpass biquad is created by cascading two one-pole highpass sections after the ABP to give

$$\frac{(s+\omega_o)^2}{s^2 + s\frac{\omega_o}{Q} + \omega_o^2} \times \frac{s}{s+\omega_o} \times \frac{s}{s+\omega_o}$$

$$= \frac{s^2}{s^2 + s\frac{\omega_o}{Q} + \omega_o^2} \tag{12.155}$$

A modified ABP filter is shown in Fig. 12.47. This circuit replaces the unity-gain inverter of the ABP with an amplifier of gain K.

The resulting transfer function for the modified circuit is

$$\frac{R_2}{R_1 - KR_1 + R_2} \frac{K(s+\frac{1}{\tau})^2}{s^2 + \frac{2}{\tau}\frac{R_1+KR_1+R_2}{R_1-KR_1+R_2}s + \frac{1}{\tau^2}} \tag{12.156}$$

Equation (12.156) allows a greater range of values for the coefficient associated with $s$ in the denominator. For the ABP with an inverter ($K = -1$) in the feedback loop the value of $Q$ from Eq. (12.151) is always greater than 0.5. With the modified circuit of Fig. 12.47, $K$ can be selected to be +1 or -1, resulting in $Q$ values of

$$Q = \frac{1}{2}\frac{R_2}{R_2 + 2R_1} \quad \text{for } K = 1 \tag{12.157}$$

and

$$Q = \frac{2R_1 + R_2}{2R_2} \quad \text{for } K = -1 \tag{12.158}$$

Equation (12.157) results in $Q$ values less than or equal to 1/2, while Eq. (12.158) results in values greater than 1/2.

The value of $Q$ can be set by the values of $R_1$ and $R_2$ while $\tau$ (an $RC$ product of the allpass network) determines the value of $\omega_o$. Thus, the values of $Q$ and $\omega_o$ can be set independently. Setting these values determines the pole locations of the biquad function.

Using cascaded biquads or single-pole filters, one can implement Butterworth, Chebyshev, or other transfer functions. Band-reject or notch filters can be implemented by subtracting the output of the ABP filter from the input signal.

---

## SUMMARY

**1.** Active filters use amplifiers or other active stages, resistors, and capacitors to create filters with complex pole locations in the transfer function.

**2.** Complex poles in the transfer function are required for Butterworth, Chebyshev, Bessel, narrowband, and band-reject filters.

**3.** Circuits that realize a general biquad function can be cascaded to implement pairs of complex poles in a transfer function.

**4.** The Bach filter is a multipole realization that can implement complex poles in its transfer function.

**5.** Parameters such as $Q$ in a narrowband circuit or bandwidth of a lowpass circuit may drift with temperature due to changes in resistances, capacitances, or gains of active elements.

**6.** The sensitivity of one parameter with respect to another allows different filters to be compared.

**7.** The op amp is a component that is often used in low- or intermediate-frequency filter design.

**8.** The operational transconductance amplifier and the transconductor are used in filter design and can extend into the higher frequency range.

**9.** Current conveyors and phase-shift filters can also be used to create biquads that can operate at higher frequencies.

---

## PROBLEMS

### Section 12.3 Active Synthesis Using Bach Filters

**12.1** Design a four-pole, lowpass, Butterworth filter having a 3-dB frequency of 14 kHz using the Bach method. The source resistance is 1 kΩ and the load resistance is 10 kΩ.

**12.2** Design a four-pole, highpass, Butterworth filter with a 3-dB frequency of 14 kHz using the Bach method. The source resistance is 1 kΩ and the load resistance is 10 kΩ.

### Section 12.4 Drift in Filter Response

**12.3** This problem demonstrates that sensitivity figures can be misleading unless related to configuration. An op amp with a voltage gain of 300,000 V/V has a voltage gain change with temperature of

$$\frac{dA}{dT} = 4000 \text{ V/V/°C}$$

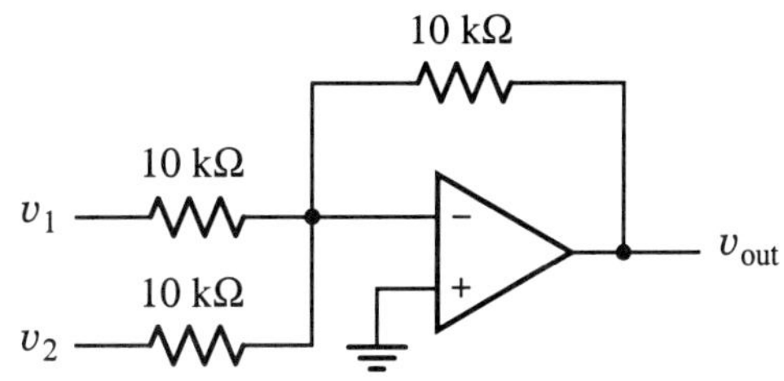


**Figure P12.4**

The sensitivity of $Q$ in a circuit using this op amp is

$$S_A^Q = 1$$

In another configuration, a filter is constructed using feedback that causes $K = A/(A+2)$. In this situation,

$$\frac{dK}{dT} = 6 \times 10^{-8} \text{ V/V/°C}$$

and

$$S_K^Q = 100$$

Assuming these figures are relatively constant over a 10°C change, calculate and compare the fractional changes in $Q$ for both cases for this 10-degree change. What do you conclude from these results?

### Section 12.6 Components Used in Biquad Synthesis

**12.4** The circuit shown is used as a summer. If the op amp has a dc gain of 140,000 V/V and a $GBW = 2$ MHz, at what frequency will the worst case accuracy exceed 3%?

**12.5** The gain of the op amp in the circuit shown is

$$A = \frac{200,000}{1 + j\frac{f}{100}}$$

**(a)** Calculate the low-frequency input impedance of the circuit.

**(b)** Find the input impedance at $f = 100$ kHz.

**12.6** The output resistance of the current mirror stage in the circuit shown is 60 kΩ while the output resistance of the differential stage is 100 kΩ. If $G_m = 1.20$ mA/V for the differential devices, calculate the midband voltage gain at the output.

**12.7** The three upper corner frequencies of an op amp are $P_1 = 10^5$ rad/s, $P_2 = 10^6$ rad/s, and $P_3 = 10^7$ rad/s. The maximum gain of the op amp is 100,000 V/V. Will this circuit be stable in the unity feedback configuration? If not, to what value must $P_1$ be lowered to lead to marginal stability?

**12.8** Show the configuration of an inverting integrator using one CCII+ along with a resistor and a capacitor. The output should be

$$v_{\text{out}} = -\frac{v_{\text{in}}}{s \times 10^{-4}}$$

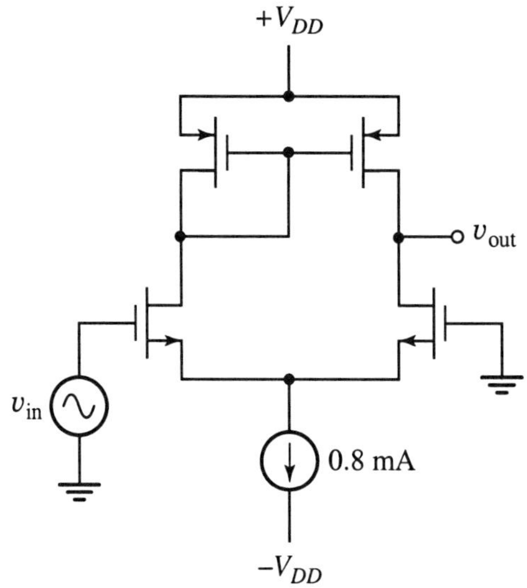


**Figure P12.6**

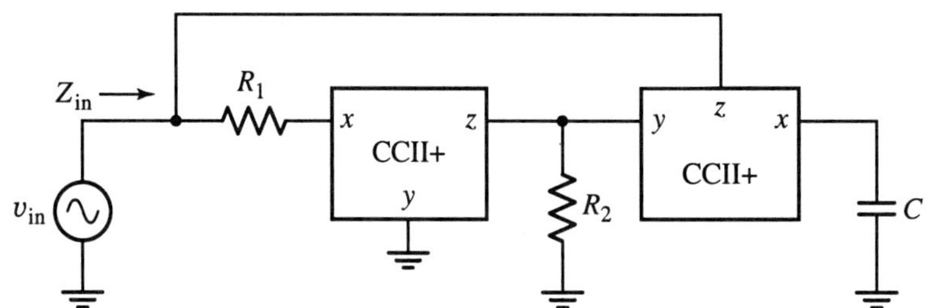


**Figure P12.10**

**12.9** Show the configuration of a noninverting integrator using one CCII+ along with a resistor and a capacitor. The output should be

$$v_{\text{out}} = \frac{v_{\text{in}}}{s \times 10^{-7}}$$

**12.10** Calculate the input impedance of the circuit shown.

### Section 12.7 The General Biquad Function

**12.11** Choose the coefficients in Eq. (12.63) for the two biquads needed for a fourth-order, lowpass, Chebyshev filter with unity gain at $\omega = 0$ and a cutoff frequency of 28 krad/s. Use a 0.5-dB ripple width.

**12.12** Select the coefficients in Eq. (12.63) to create a narrowband circuit that is resonant at $\omega_o = 80$ krad/s with a $Q$ of 24. The resonant gain should be unity.

### Section 12.8 Sallen-Key Biquads

**12.13** Select all element values in the circuit of Fig. 10.1 to produce a two-pole, lowpass, Butterworth filter with a 3-dB frequency of $f_c = 20$ kHz. What is the value of dc gain?

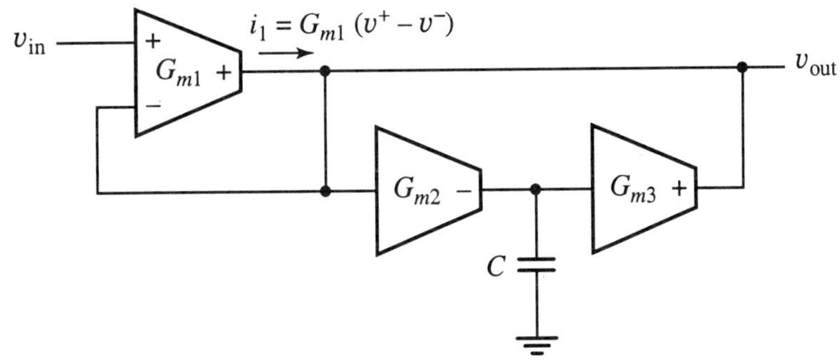


**Figure P12.19**

**12.14** Design a four-pole, lowpass, Chebyshev filter with a 0.5-dB ripple width and a cutoff frequency of $\omega_c = 28$ krad/s. What is the value of dc gain?

**12.15** Rework Problem 12.13 if the biquad is to have a highpass transfer function.

**12.16** Rework Problem 12.14 if the filter is to have a four-pole, highpass, Chebyshev transfer function. What is the value of gain at $\omega = \infty$?

**12.17** Rework Example 12.6 if $Q$ must be 20 with a resonant frequency of $f_o = 10$ kHz.

**12.18** Choose values for the band-reject biquad circuit of Fig. 12.28 for a notch frequency of 60 Hz and a $Q$ of 8.

### Section 12.9 Two-Integrator Active Biquads

**12.19** For the circuit shown

**(a)** derive the transfer function as the ratio of two polynomials in $s$;

**(b)** sketch a Bode diagram of the magnitude of this transfer function.

**12.20** Design a two-pole lowpass Butterworth filter using transconductors in the two-integrator configuration. The corner frequency should be 340 kHz.

### Section 12.10 Narrowband Filters Based on Current Conveyors

**12.21** Find the transfer function of the network shown,

$$H(s) = \frac{v_{out}}{v_{in}}$$

**12.22** For the circuit shown

**(a)** calculate the resonant frequency of the circuit;
**(b)** calculate the circuit $Q$ at resonance;
**(c)** calculate the resonant value of $R_{in}$.

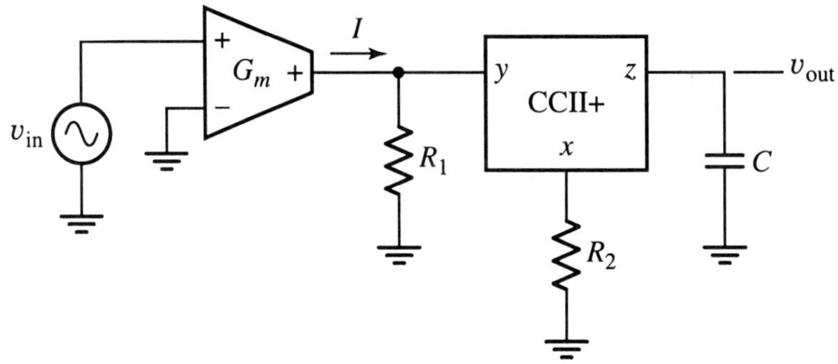


**Figure P12.21**

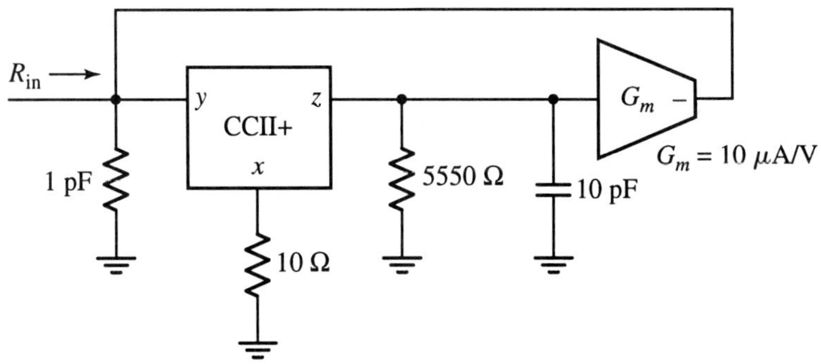


**Figure P12.22**

## Section 12.11 Narrowband Filters Based on Allpass Circuits

**12.23** For the phase-shift circuit shown,

- **(a)** derive the transfer function as the ratio of two polynomials in $s$;
- **(b)** find the pole $Q$.

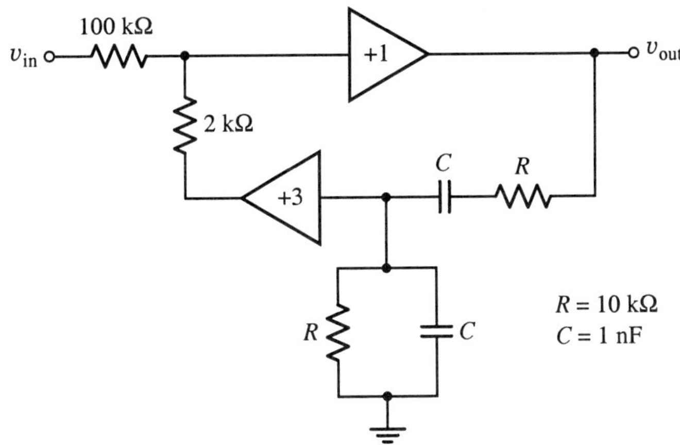


**Figure P12.23**

## REFERENCES


1. Comer, D. J. and McDermid, J. E., "Inductorless bandpass characteristics using allpass networks," *IEEE Transactions on Circuit Theory*, Vol. CT-15 (Dec. 1968), pp. 501-503.

2. Comer, D. T., Comer, D. J., and Gonzalez, J. R., "A high-frequency integrable bandpass filter configuration," *IEEE Transactions on Circuits and Systems, Part II: Analog and Digital Signal Processing,* Vol. 44, No. 10 (October 1997), pp. 856-861.

3. Fabre, A. and Alami, M., "Insensitive current-mode bandpass implementations-based nonideal gyrators," *IEEE Transactions on Circuits and Systems–Part I.* Vol. 39, No. 2 (Feb. 1992), pp. 152-155.

4. Fabre, A., Saaid, O., Wiest, F.,and Boucheron, C., "High-frequency High-Q BiCMOS current-mode bandpass filter and mobile communication application," *IEEE Journal of Solid-State Circuits.* Vol. 33, No. 4 (April 1998), pp. 614-625

5. Kerwin, W. J., Huelsman, L. P., and Newcomb, R. W., "State-variable synthesis for insensitive integrated circuit transfer functions," *IEEE Journal of Solid-State Circuits.* Vol. SC-2 (Sep. 1967), pp. 87-92.

6. Mitra, S. J., *Analysis and Synthesis of Linear Active Networks.* New York: John Wiley and Sons, 1969.

7. Pavan, S. and Tsividis, Y. P., *High Frequency Continuous Time Filters in Digital CMOS Processes.* Boston: Kluwer Press, 2000.

8. Sallen, R. P. and Key, E. L., "A practical method of designing RC active filters," *IRE Transactions on Circuit Theory.* March 1955, pp. 74-84.

9. Steyaert, M., Kinget, P, Sansen, W., Esat-Micas, K. U., and Van Der Spiegel, J., "Full integration of extremely large time constants in CMOS," *Electronics Letters.* Vol. 27 (May 9 1991), pp. 790-791.

10. Su, K. L., *Analog Filters.* London: Chapman and Hall, 1996.

11. Thomas, L. C., "The biquad: part I - Some practical design considerations," *IEEE Trans. on Circuit Theory.* Vol. CT-18 (May 1971), pp. 350-357

12. Tow, J., "Active RC filters - a state-space realization," *Proc. IEEE.* Vol. 56 (June 1968), pp. 1137-1139.

13. Tsividis, Y. P., "Integrated continuous-time filter design–an overview," *IEEE Journal of Solid-State Circuits.* Vol. 29, No. 3 (March 1994), pp. 166-176.

14. Van Valkenburg, M. E., *Analog Filter Design*, New York: Holt, Rinehart and Winston, 1982.

15. Schaumann, R. and Van Valkenburg, M. E., *Design of Analog Filters.* Oxford University Press: New York, 2001.

# Chapter 13

# PRESENT AND FUTURE DIRECTIONS IN ELECTRONICS

## 13.1 PREDICTING THE FUTURE

Only a very wise man or a fool would undertake to predict the future of a field that changes as rapidly as the electronics field. Judge for yourself the US Patent Office worker who in 1905 predicted few further advancements when he said, "Everything that can be invented, has been invented."

Similarly, in the early 1950s, Thomas J. Watson, the cofounder of International Business Machines, is purported to have said, "The world market for computers is about six." In less than a decade, this statement was proven false and today, this prediction is off by a factor of several hundred million.

Perhaps many people familiar with these statements believed they were valid at the time they were made, but after a few years, it became obvious that they were incorrect. On the other hand, Gordon Moore, a cofounder of Intel, noted in 1964 that since integrated circuits had come into existence around 1959, the trend was that the number of components fabricated on a chip doubled every 12 to 18 months. With little history on which to base a prediction, Moore indicated that he thought this trend would continue. In fact, it has continued into the twenty-first century and is referred to as "Moore's law."

Since it is difficult to predict the future of any technological area, the purpose of this chapter is only to indicate possible future directions the electronics field may travel, based on present indications.

## 13.2 ELECTRONIC SYSTEMS

At present, electronic circuits are constructed in several ways as listed.

1. Discrete circuits.
2. Discrete element/IC hybrid circuits.
3. Multiple IC chips.
4. A single IC chip.

### 13.2.1 Discrete Circuits

The traditional means of constructing a circuit is from discrete elements. These circuits consist of individual resistors, capacitors, inductors, and transistors connected by wires or printed conductors on a printed-circuit board (PCB). In spite of the growing popularity and capability of integrated circuits, discrete circuit design continues to be important in specific applications. In analog circuits, initial circuit development may use discrete circuits to demonstrate the viability of the circuit. Small special-purpose circuits, with no IC counterpart, are often implemented with discrete components. Modifications to existing electronic systems are often done with discrete components. High-quality power amplifier design is still done with discrete stages, although lower-quality, lower-power IC amplifiers and hybrid power circuits continue to improve.

### 13.2.2 Discrete Element/IC Hybrid Circuits

This method of circuit construction uses available IC chips along with discrete components for the interfacing and special functions that might be required. Low-quality power amplifiers often use IC preamplifiers and amplifier stages with discrete power output stages to drive the load. For powers in the tens of watts range, an IC chip is physically too small to provide a sufficient thermal path for the power dissipation of the output transistors. Pulse-width modulation schemes minimize the transistor dissipation, but amplifiers delivering 100 W or more to a load require power transistors mounted on a heat sink. The high-frequency, high-power field often uses hybrid circuits. Communication circuits delivering a few watts of power at a frequency of hundreds of MHz generally require special transistor designs for the output stages.

### 13.2.3 Multiple IC Chips

In the early days of the IC era, most digital systems were constructed from small-scale or medium-scale integrated circuits. Several ICs were included on each PCB to create the overall system. As fabrication techniques improved, the computer evolved toward the one-chip system that now exists. Even today, external memory or other peripheral interface chips make most systems a multiple chip system.

The multi-chip approach is popular for smaller companies that derive their business from niche markets. In many cases, the market is not large enough to justify moving to a single ship, even if a single chip is technically feasible. The low cost of individual IC chips, produced for a mass market, often make us forget the high cost of design, layout, fabrication, and testing of a single-chip circuit. Unless the market is large, it is often more cost-efficient to build systems using multi-chip circuits on PCBs. The chips used in these cases are generally small-scale, medium-scale, or large-scale integrated circuit chips.

### 13.2.4 A Single IC Chip

This approach is popularly referred to as a *system-on-a-chip* (SoC). It applies to electronic systems requiring multiple functions that can be fabricated on a single chip. Typically, such systems require both analog and digital functions and are also called *mixed-signal* chips. These chips are aimed at very large markets that allow design and fabrication costs to be spread over a large number of chips. The cell phone that handles both analog and digital signals is a prime candidate for the SoC because of a large cell phone market for small, complex circuitry.

Complex digital systems have been available for many years based on standard cell replication in the layout of the photolithographic masks involved in fabrication. Since digital systems often deal only with two voltage levels requiring minimal voltage accuracy, this approach continues to be feasible.

In analog circuits, an infinite number or, at least, a very large number of voltage levels are significant in normal circuit operation. The standard cell approach for analog design has been slow to evolve owing to the complexity of analog circuits. These circuits must accurately process all voltage levels rather than just the two required for digital circuits. Recently, such cells as amplifiers, digital-to-analog converters, and analog-to-digital converters have become available in the cell libraries of many companies. As digital and analog design techniques become more compatible, the design of SoC electronic systems becomes more viable.

From the beginning of the IC era into the 1980s, the dominant technology was bipolar junction transistor circuits. As MOSFET technology improved, this device began to dominate the computer industry as a result of the capability to pack more devices into a given area. Early mixed-signal circuits utilized both BJT and MOSFET in Bi-MOS op amps. Although this area is still important in high-frequency communication circuits, the mixed-signal area migrated to CMOS in the 1990s because of economics. CMOS processes developed to the point that they can implement almost any analog circuit. Analog BJT designs were converted to equivalent MOSFET designs to allow true mixed-signal development on single chips.

The success of CMOS in analog circuit design in the 1990s led to a prediction that bipolar technology would become obsolete. This prediction overlooked some of the weaknesses of the MOSFET and neglected to anticipate the improvement in performance of the BJT. A significant problem in CMOS technology is the difficulty in driving low-impedance, off-chip loads at high speeds. This problem became more apparent as the wireless/communication market growth accelerated. Again, BiMOS and BiCMOS technologies were utilized to address this problem. Unfortunately, the cost of these processes is relatively high and both CMOS and bipolar device performance often suffer from the engineering tradeoffs involved in merging the two processes.

In the mid-1990s, Motorola and IBM began experimenting with heterojunction bipolar junction devices. Initial devices were constructed using silicon and germanium to create the base-emitter junction and minimize parasitic capacitances and resistances. With conventional silicon devices, the parameter $f_t$ was limited to less than 10 GHz, whereas values for $f_t$ near 100 GHz are practical with heterojunction devices. Because the fabrication process required only a single additional process step to implant the germanium emitter, it was generally compatible with a silicon-based CMOS process. Thus, a series of high-performance, BiCMOS processes have been developed to provide high-frequency bipolar devices on the same chip with deep submicron MOS gate lengths. The Si-Ge BiCMOS process has become very popular in today's communication field.

Additional research in heterojunction technology has shown that both gallium-arsenide and indium-phosphide junctions can produce even higher-speed performance than Si-Ge heterojunctions. These newer heterojunction processes have been commercialized, but production costs are still much higher than the Si-Ge process. In 2001, Motorola reported a process that allows the bonding of gallium-arsenide to a silicon substrate. As this process improves, the mixing of high-frequency gallium-arsenide bipolar circuits with silicon CMOS circuits on a single chip becomes practical.

Another approach that is becoming practical for CMOS is the silicon-on-insulator (SOI) process. Advanced Micro Devices has moved toward a 0.13 $\mu$, SOI chipset in their high-performance workstation and server lines. This approach minimizes device parasitics that limit switching speeds. As processing methods improve in this fabrication procedure, it will continue to rise in popularity. With the many recent advancements in various technologies, the SoC approach to system design is expected to become more prominent.

## 13.3 COMPUTER-AIDED DESIGN AND MODELING

The well-known, rapid development of digital computer-aided design (CAD) tools will not be chronicled here. For many years, digital models required less accuracy than analog models. In recent years, the need for more accurate models in high-performance digital systems has become more apparent. The accurate analog models are often used in digital simulators at the present time.

For many years, the BJT model used for Spice-like programs has been sufficiently accurate for system simulations. Unfortunately, the commercially available MOSFET models have not offered the necessary precision to dependably model complex systems. This problem has forced many companies to develop their own proprietary models. More recently, CAD companies are providing models that can be used for complex submicron devices.

The later Berkeley simulation models such as BSIM3.v3 appear to offer more accurate results even for deep submicron gate lengths. Other parts of the world have had good success with models that are not designed to work with Spice programs. An example of this is the EKV model developed by Enz, Krummenacher, and Vittoz at Ecole Polytechnique Federale de Lausanne in Switzerland. This model has proven to be very useful in the simulation of low-voltage, submicron CMOS mixed-signal circuits.

In order to use the MOSFET models, the proper model parameters must be available for the fabrication process to be used. These parameters are only obtainable from the foundry that will fabricate the circuits. The difficulty in accessing appropriate models and parameters for research purposes is a problem not experienced in the BJT era. It is expected that the trend toward more accurate models and accessibility of model parameters will continue to improve the reliability of MOSFET simulation results.

## 13.4 FUTURE DIRECTIONS

The electronic circuits area presently has a need for good IC designers. This need was generated in the early years of IC development when very few people worked in this area. Many universities continued to teach traditional methods of circuit design, assuming that only a small number of engineers would design IC chips for large companies. This approach was valid for some years, until IC technology improved to the point that even the small companies had access to fabrication facilities. Either through parent companies or private facilities, called foundries, smaller companies can now take advantage of IC capabilities. It is incumbent on the universities that educate engineers to understand the field and understand the trends in order to provide a meaningful education. The following material may help anticipate some important trends.

### 13.4.1 MOSFET Channel Lengths Will Continue to Shrink

The sustained improvement in photolithographic and related masking techniques has allowed a continual shrinkage of MOS device size. When Gordon Moore made his initial statement that is now known as Moore's law, he may not have realized that the MOS device would be responsible for extending the validity of this law for the next 40 years. In 1964, the BJT was the major device used to fabricate IC chips. Although improvement in fabrication methods have led to better BJT performance on chips, the overall improvement has not followed Moore's law. The required resistors and the necessary transistor sizes limit the number of BJTs that can be placed on a chip. A MOSFET chip can contain 100 times the number of devices that can be included on a BJT chip. Thus, to this point in the early twenty-first century, the MOSFET has led to the validity of Moore's law.

Practical systems are now using MOSFET channel lengths of 0.1 $\mu$ or 100 nM. Concern has been voiced over the development of accurate fabrication methods to shrink this channel length further. Limits on power supply voltage and more complex behavior of smaller devices make circuit design more difficult as device size decreases. In addition, the improvement in performance of the MOSFET as channel length shrinks has slowed. However, newer techniques continue to emerge and better models are being developed to utilize the increase in frequency performance and number of devices that can be placed on a chip.

A successful designer of mixed-signal systems needs to know how to design MOSFET circuits for low power supply voltages (less than 1 V), how to minimize the number of resistors on the chip, how to optimize frequency performance, how to match critical components, and how key device parameters change as device size becomes smaller.

### 13.4.2 Higher Performance MOSFET Circuits Will Use Silicon-on-Insulator

In a conventional CMOS process, a lightly doped, p-type, silicon substrate is used to form the nMOS devices. The pMOS devices are formed in n-wells created in the substrate material. This type of construction leads to various problems including parasitic capacitances and conductances that limit circuit operation. A process that has been known for years is that of placing the devices on an insulator rather than a silicon substrate. As the fabrication process improves for SOI, leading to lower costs, this technology will become more important in high-performance MOS circuits. Even now it is being employed in workstations and servers that require high-speed operation.

### 13.4.3 Usage of Heterojunction and Gallium Arsenide BJT Devices Will Increase

Amplifiers and other circuits that operate in the 10-GHz to 50-GHz range are difficult to construct with CMOS technology. The most popular method of implementing these high-frequency circuits uses BJT designs based on either silicon/germanium, gallium/arsenide, or indium phosphide. Analog circuit designers should understand BJT circuits and obtain a background in device physics, modeling, and high-frequency design methods.

### 13.4.4 Mixed-Signal Systems Will Increase in Importance

Just as the personal computer industry made tremendous advancements between 1980 and the present, the mixed-signal area will make great strides over the next several years. Now that some processes are available and others are emerging for combining digital and analog circuits on a single chip, this area promises to be one of rapid development. A prime area that will encourage mixed-signal development is the communications field. The combination of digital and analog signals in these systems makes mixed-signal SoC imperative. A mixed-signal circuit designer should be familiar with digital/analog converters and interface circuits to create efficient designs.

### 13.4.5 Rapid Simulation Tool Development Will Continue

As circuit operation is extended into higher frequency ranges, the layout parasitics become more significant. Simulation programs that model GHz circuits include device models as well as models for interconnects and other parasitics. Refinement of existing programs and development of newer programs that are accurate in the 10-GHz to 50-GHz frequency range must continue.

---

**SUMMARY**

**1.** MOSFET channel lengths will continue to decrease below the 0.1-$\mu$ level. Simulation programs will continue to improve to accurately model these devices.

**2.** As MOSFET devices shrink, silicon-on-insulator processes will become more popular.

**3.** Heterojunction BJT devices and various gallium BJT devices are bringing high-performance mixed-signal systems into prominence.

**4.** A procedure to put gallium arsenide devices on silicon is being developed. This will accelerate the growth of the system-on-a-chip area.

**5.** Simulation tools will continue to improve to account for the effects of layout parasitics.

---

# ANSWERS TO SELECTED PROBLEMS

## CHAPTER 1

**1.2** (a) $V_{GS} = V_T = 0.9$ V (b) $V_{GS} = 3.34$ V

**1.5** $\lambda = 0.0161\ V^{-1}$

**1.8** $r_{ds} \approx 400$ kΩ

**1.11** $g_m = 2.51$ mA/V $\mu C_{ox} W/2L = 1.57\ \text{mA}/V^2$

**1.15** $V_{max} = 8$ V $V_{min} = 1.945$ V

**1.17** $A_{MB} = -22.6$ V/V

**1.20** $R_D = 8$ kΩ $R_1 = 194$ kΩ

**1.24** $A_{MB} = -13.9$ V/V

**1.27** (a) $V_{DSQ} = 7$ V $A_{MB} = -12$ V/V

**1.30** $A_{new} = -45.5$ V/V

**1.34** $V_{GS} = 1.7$ V $A_{MB} = -102.8$ V

**1.38** $f_{pole} = 1.63$ GHz $f_{zero} = 6.72$ GHz $A_{high} = 0.209$ V/V

**1.42** $R_1 = 143$ kΩ $V_{max} = 10.7$ V

**1.45** $A_{MB} = -152$ V/V $r_{in} = 1.3$ kΩ $V_{max} = 8$ V

**1.48** $A_{MB} \approx -77.5$ V/V

**1.52** $f_2 = 80.2$ kHz

**1.55** $A_{MB} = 0.986$ V/V

**1.57** $A_{MB} = -1016$ V/V

## CHAPTER 2

**2.2** Allowable powers: (a) 51.6 W (b) 31.4 W (c) 14.4 W

**2.5** $\Theta_{HS} + \Theta_i = 9.9°\text{C/W}$

**2.8** $P_{ALLOW} = 333$ mW

**2.11** $P_T = 256$ mW $\Theta_{HS} + \Theta_i = 0.173°\text{C/mW}$

**2.13** $P_R = 195$ mW $P_S = 408$ mW

**2.15** $P_R = 93$ mW $P_S = 264$ mW

**2.20** (a) $P_{Tmax} = 2.4$ W (b) $P_{ac} = 1.6$ W (c) $P_{ac} = 0.4$ W

**2.23** $P_{ac} = 389$ mW

**2.26** $\Theta_{HS} + \Theta_i = 16.4°\text{C/W}$

**2.29** $n \approx 3$ $P_{ac} = 6.3$ W

**2.32** (a) Only $n = 2$ delivers more than 4 W.

(b) $n = 4$, $P_{ac-max} = 2.46$ W (c) $I_{CQ} = 0.196$ A (d) $P_{TQ} = 5.30$ W

**2.35** (a) $R_{dc} = 20\ \Omega$ (b) $R_{ac} = 220\ \Omega$ (c) $I_{CQ} = 50$ mA

(d) $V_{CEQ} = 15$ V (e) $P_{ac} = 250$ mW (f) $I_{CQ} = 66.7$ mA (g) $P_{ac} = 445$ mW

**2.37** $n = 3$ $P_{Tmax} = 0.626$ W

**2.42** (a) $P_{ac} = 2.14$ W (b) $P_{Tmax} = 0.488$ W

## CHAPTER 3

**3.1** $A_v = 6.93$ V/V $G_p = 1501$ W/W

**3.4** $THD = 7.88\%$ $THD = -22.1$ dB

**3.7** $\eta = 65.4\%$ $P_{Tmax} = 4.27$ W

**3.10** Use $V_{C1} = 23$ V $P_{Tmax} = 2.65$ W

**3.14** $d = 0.1375$

**3.16** $P_{ac-max} = 3.85$ W $P_{ac} = 0.74$ W

**3.18** $v_c = -2.2$ V

## CHAPTER 4

**4.1** $A = \frac{400{,}000}{1+jf/10}$ $|A(5000)| = 800$ V/V

**4.5** Let $R_F = 74\ \text{k}\Omega$ $R_2 = 1\ \text{k}\Omega$ $f_{2ni} = 30$ kHz

**4.9** (a) $A_{MB} = -8.42$ V/V (b) $f_2 = 382$ kHz (c) $A(f) = \frac{-8.42}{1+jf/382{,}000}$

**4.11** $GBW_{oa} = 900$ kHz

**4.14** $A_{MB} = 10.67$ V/V $f_2 = 125$ kHz

**4.18** $A_{MB} = -20$ V/V $f_2 = 1.195$ GHz

**4.22** $R_C = 404\ \Omega$ $r_\pi = 1700\ \Omega$

**4.25** $C_1 = 173$ pF $A_{MB} = -5.98$ V/V

**4.28** $f_2 = 1.16$ MHz

**4.31** $A_{MBo} = 20.2$ V/V

**4.33** $A_{max} = 860$ V/V

**4.37** (a) 4 stages (b) $A_{MB} = 4.95$ V/V

**4.40** $GBW_{oa-min} = 3.67$ MHz

**CHAPTER 5**

**5.1** $C = 63.3$ pF $Q_{eff} = 38.6$ $\Delta f = 259$ kHz

**5.4** $Q_o = 54.8$

**5.7** $R_p = 241$ k$\Omega$ $R_s = 6\ \Omega$

**5.11** (a) $R_o = 150$ k$\Omega$ (b) $v_{out} = 1.22 \cos \omega_o t$

**5.14** (a) $C = 127$ pF (b) $R_1 = 67.4$ k$\Omega$ (c) $A_{res} = 102$ V/V

**5.17** $C = 246$ pF $C_n = 20$ pF $n = 4$ $R = 90$ k$\Omega$

**5.20** (a) Third harmonic: $v_{out} = 3.68 \sin 3\omega_c t$ (b) $v_{out} = 11.8 \sin 3\omega_c t$

**CHAPTER 6**

**6.2** Oscillation with distortion at $f = 4.50$ MHz

**6.5** $f_o = 35.6$ kHz $|A| = 1.05$ V/V

**6.8** $|v_{out}| = 3.37$ V

**6.11** (a) $f_o = 729.3$ kHz (b) $K = 24.2$ (c) $v_{out} = 8$ V

**6.14** $f = 10{,}206$ Hz

**6.17** $R = 30$ k$\Omega$ $C = 90$ pF $R_1 = 115$ k$\Omega$ $R_2 = 40.7$ k$\Omega$

**6.20** $R = 4000\ \Omega$   $C = 3.98$ nF

**6.24** $L = 100\ \mu$H   $C = 5.23$ nF   $C_c = 100$ pF   $n = 37.2$   $R_1 = 790$ k$\Omega$   $R_2 = 210$ k$\Omega$

**6.28** $V_{DD} = 10$ V   $R_1 = 790$ k$\Omega$   $R_2 = 210$ k$\Omega$   $R_D = 1$ k$\Omega$   $C_1 = 32.4$ pF   $C_2 = 7.36$ pF

## CHAPTER 7

**7.1** $\lambda/10 = 3 \times 10^5$ m

**7.3** $f = 680.3$ kHz

**7.8** $f_{max} - f_{min} = 10$ kHz

**7.12** $V_c = 12.6$ V   $v_{SB} = 5.65$ V

**7.15**

| Freq., kHz | 596 | 598 | 599 | 600 | 601 | 602 | 604 |
|---|---|---|---|---|---|---|---|
| Amplitude, V | 0.796 | 1.59 | 3.18 | 10 | 3.18 | 1.59 | 0.796 |

**7.18** $P_C = 10.67$ W   $P_{SB} = 0.67$ W

**7.23** $Q_{eff} = 42$   $\Delta f = 28.6$ kHz

**7.26** (a) $k_p = 0.0262$ rad/V   (b) $\Delta f = 15.71$ Hz   (c) $\Delta f = 628$ Hz   (d) $\Delta f = 2356$ Hz

## CHAPTER 8

**8.3** (a) $M = 0.13$ V   (b) $R_B = 156$ k$\Omega$

**8.7** (a) $m = 0.2$   (b) $K = 0.5$   (c) $m = 0.6$   (d) $K = 2/3$   (e) $V_c = 0$   (f) $v_{USB} = 1.2$

**8.10** $C_1 = 1.256$ nF   $R_1 = 18.6$ k$\Omega$   $M = 0.92$ V   $R_E = 1.3$ k$\Omega$

**8.13** (a) $M = 12.8$ V   (b) $R_{max} = 36.5$ k$\Omega$

**8.16** $C_{min} = 8.13$ V   $m = 0.667$

**8.20** 25.6 dB and 29.1 dB

**8.22** Use $f_1 = 1255$ kHz   $Q = 23.3$

**8.24** (a) $\Delta V = -0.083$ V   (b) $V = 14.4$ V

## CHAPTER 9

**9.1** $V_c = 2.81$ V   $M = 2.5$ V

**9.5** (a) $M = 5$ V   (b) $M = 1$ V

**9.8** $k_p = 0.025$ rad/V

**9.10** $M = 2$ V

**9.15** $GK \approx 1.6\ \mu$s

**CHAPTER 10**

**10.1** $t = 97.6\ \mu$s

**10.4** 11 bits

**10.10** $V_{max} = 16.384$ V  $V = 8.192$ V

**10.15** Accuracy $\approx 0.2\%$

**10.17** $V_O = 5.90$ V

**10.22** 1000 0000 $\rightarrow$ 1100 0000 $\rightarrow$ 1010 0000 $\rightarrow$ 1001 0000 $\rightarrow$ 1001 1000 $\rightarrow$ 1001 1100 $\rightarrow$ 1001 1110 $\rightarrow$ 1001 1111

**CHAPTER 11**

**11.1** Attenuation = 80.7 dB  53.8 dB/oct

**11.4** $L = 208$ mH  $C = 0.363\ \mu$F

**11.7** (a) $f = 1.264$ Hz  (b) $f = 158,846$ Hz

**11.9** $H(s) = \frac{1}{s^2+4s+5}$

**11.13** Narrowband filter

**11.17** $n = 4$  $-24$ dB/oct

**11.21** $H(s) = \frac{1}{s^2+3s+3}$

**11.25** Two-pole filter works

**CHAPTER 12**

**12.2** Let $R_1 - R_2 - R_3 - R_4 = 10$ k$\Omega$  $C_1 = 2080$ pF  $C_2 = 1040$ pF  $C_3 = 609$ pF  $C_4 = 305$ pF

**12.8** The input voltage connects through $R$ to the x input. A capacitor connects from the z terminal to ground. The y terminal is grounded. The output voltage is at terminal z. The $RC$ product should be chosen to be $10^{-4}$.

**12.10** $Z_{in} = \frac{R_1}{sCR_2+1}$

**12.18** $K = 31/32$  $C = 100$ nF  $R = 26.5$ k$\Omega$

**12.23** $H(s) = 0.0196\frac{s^2+3\times10^5 s+10^{10}}{s^2+5.88\times10^3 s+10^{10}}$

# INDEX